QUANTIZED DETECTOR NETWORKS

Scientists have been debating the meaning of quantum mechanics for over a century. This book for graduate students and researchers gets to the root of the problem: the contextual nature of empirical truth, the laws of observation, and how these impact on our understanding of quantum physics. Bridging the gap between nonrelativistic quantum mechanics and quantum field theory, this novel approach to quantum mechanics extends the standard formalism to cover the observer and their apparatus. The author demystifies some of the aspects of quantum mechanics that have traditionally been regarded as extraordinary, such as wave–particle duality and quantum superposition, by emphasizing the scientific principles. Including key experiments and worked examples throughout to encourage the reader to focus on empirically sound concepts, this book avoids metaphysical speculation and alerts the reader to the use of computer algebra to explore quantum experiments of virtually limitless complexity. This title, first published in 2017, has been reissued as an Open Access publication on Cambridge Core.

GEORGE JAROSZKIEWICZ is a mathematical physicist recently retired from the School of Mathematical Sciences, University of Nottingham. His research interests are focused on the fundamental differences between quantum and classical mechanics.

CAMBRIDGE MONOGRAPHS ON MATHEMATICAL PHYSICS

General Editors: P. V. Landshoff, D. R. Nelson, S. Weinberg

S. J. Aarseth *Gravitational N-Body Simulations: Tools and Algorithms*†
J. Ambjørn, B. Durhuus and T. Jonsson *Quantum Geometry: A Statistical Field Theory Approach*†
A. M. Anile *Relativistic Fluids and Magneto-fluids: With Applications in Astrophysics and Plasma Physics*
J. A. de Azcárraga and J. M. Izquierdo *Lie Groups, Lie Algebras, Cohomology and Some Applications in Physics*†
O. Babelon, D. Bernard and M. Talon *Introduction to Classical Integrable Systems*†
F. Bastianelli and P. van Nieuwenhuizen *Path Integrals and Anomalies in Curved Space*†
D. Baumann and L. McAllister *Inflation and String Theory*
V. Belinski and M. Henneaux *The Cosmological Singularity*†
V. Belinski and E. Verdaguer *Gravitational Solitons*†
J. Bernstein *Kinetic Theory in the Expanding Universe*†
G. F. Bertsch and R. A. Broglia *Oscillations in Finite Quantum Systems*†
N. D. Birrell and P. C. W. Davies *Quantum Fields in Curved Space*†
K. Bolejko, A. Krasiński, C. Hellaby and M.-N. Célérier *Structures in the Universe by Exact Methods: Formation, Evolution, Interactions*
D. M. Brink *Semi-Classical Methods for Nucleus-Nucleus Scattering*†
M. Burgess *Classical Covariant Fields*†
E. A. Calzetta and B.-L. B. Hu *Nonequilibrium Quantum Field Theory*
S. Carlip *Quantum Gravity in 2+1 Dimensions*†
P. Cartier and C. DeWitt-Morette *Functional Integration: Action and Symmetries*†
J. C. Collins *Renormalization: An Introduction to Renormalization, the Renormalization Group and the Operator-Product Expansion*†
P. D. B. Collins *An Introduction to Regge Theory and High Energy Physics*†
M. Creutz *Quarks, Gluons and Lattices*†
P. D. D'Eath *Supersymmetric Quantum Cosmology*†
J. Dereziński and C. Gérard *Mathematics of Quantization and Quantum Fields*
F. de Felice and D. Bini *Classical Measurements in Curved Space-Times*
F. de Felice and C. J. S Clarke *Relativity on Curved Manifolds*†
B. DeWitt *Supermanifolds, 2nd edition*†
P. G. O. Freund *Introduction to Supersymmetry*†
F. G. Friedlander *The Wave Equation on a Curved Space-Time*†
J. L. Friedman and N. Stergioulas *Rotating Relativistic Stars*
Y. Frishman and J. Sonnenschein *Non-Perturbative Field Theory: From Two-Dimensional Conformal Field Theory to QCD in Four Dimensions*
J. A. Fuchs *Affine Lie Algebras and Quantum Groups: An Introduction, with Applications in Conformal Field Theory*†
J. Fuchs and C. Schweigert *Symmetries, Lie Algebras and Representations: A Graduate Course for Physicists*†
Y. Fujii and K. Maeda *The Scalar-Tensor Theory of Gravitation*†
J. A. H. Futterman, F. A. Handler, R. A. Matzner *Scattering from Black Holes*†
A. S. Galperin, E. A. Ivanov, V. I. Ogievetsky and E. S. Sokatchev *Harmonic Superspace*†
R. Gambini and J. Pullin *Loops, Knots, Gauge Theories and Quantum Gravity*†
T. Gannon *Moonshine beyond the Monster: The Bridge Connecting Algebra, Modular Forms and Physics*†
A. García-Díaz *Exact Solutions in Three-Dimensional Gravity*
M. Göckeler and T. Schücker *Differential Geometry, Gauge Theories, and Gravity*†
C. Gómez, M. Ruiz-Altaba and G. Sierra *Quantum Groups in Two-Dimensional Physics*†
M. B. Green, J. H. Schwarz and E. Witten *Superstring Theory Volume 1: Introduction*
M. B. Green, J. H. Schwarz and E. Witten *Superstring Theory Volume 2: Loop Amplitudes, Anomalies and Phenomenology*
V. N. Gribov *The Theory of Complex Angular Momenta: Gribov Lectures on Theoretical Physics*†
J. B. Griffiths and J. Podolský *Exact Space-Times in Einstein's General Relativity*†
S. W. Hawking and G. F. R. Ellis *The Large Scale Structure of Space-Time*†
F. Iachello and A. Arima *The Interacting Boson Model*†
F. Iachello and P. van Isacker *The Interacting Boson-Fermion Model*†
C. Itzykson and J. M. Drouffe *Statistical Field Theory Volume 1: From Brownian Motion to Renormalization and Lattice Gauge Theory*†
C. Itzykson and J. M. Drouffe *Statistical Field Theory Volume 2: Strong Coupling, Monte Carlo Methods, Conformal Field Theory and Random Systems*†
G. Jaroszkiewicz *Principles of Discrete Time Mechanics*
G. Jaroszkiewicz *Quantized Detector Networks*
C. V. Johnson *D-Branes*†

P. S. Joshi *Gravitational Collapse and Spacetime Singularities*†
J. I. Kapusta and C. Gale *Finite-Temperature Field Theory: Principles and Applications, 2nd edition*†
V. E. Korepin, N. M. Bogoliubov and A. G. Izergin *Quantum Inverse Scattering Method and Correlation Functions*†
J. Kroon *Conformal Methods in General Relativity*
M. Le Bellac *Thermal Field Theory*†
Y. Makeenko *Methods of Contemporary Gauge Theory*†
S. Mallik and S. Sarkar *Hadrons at Finite Temperature*
N. Manton and P. Sutcliffe *Topological Solitons*†
N. H. March *Liquid Metals: Concepts and Theory*†
I. Montvay and G. Münster *Quantum Fields on a Lattice*†
P. Nath *Supersymmetry, Supergravity, and Unification*
L. O'Raifeartaigh *Group Structure of Gauge Theories*†
T. Ortín *Gravity and Strings, 2nd edition*
A. M. Ozorio de Almeida *Hamiltonian Systems: Chaos and Quantization*†
M. Paranjape *The Theory and Applications of Instanton Calculations*
L. Parker and D. Toms *Quantum Field Theory in Curved Spacetime: Quantized Fields and Gravity*
R. Penrose and W. Rindler *Spinors and Space-Time Volume 1: Two-Spinor Calculus and Relativistic Fields*†
R. Penrose and W. Rindler *Spinors and Space-Time Volume 2: Spinor and Twistor Methods in Space-Time Geometry*†
S. Pokorski *Gauge Field Theories, 2nd edition*†
J. Polchinski *String Theory Volume 1: An Introduction to the Bosonic String*†
J. Polchinski *String Theory Volume 2: Superstring Theory and Beyond*†
J. C. Polkinghorne *Models of High Energy Processes*†
V. N. Popov *Functional Integrals and Collective Excitations*†
L. V. Prokhorov and S. V. Shabanov *Hamiltonian Mechanics of Gauge Systems*
S. Raychaudhuri and K. Sridhar *Particle Physics of Brane Worlds and Extra Dimensions*
A. Recknagel and V. Schiomerus *Boundary Conformal Field Theory and the Worldsheet Approach to D-Branes*
R. J. Rivers *Path Integral Methods in Quantum Field Theory*†
R. G. Roberts *The Structure of the Proton: Deep Inelastic Scattering*†
C. Rovelli *Quantum Gravity*†
W. C. Saslaw *Gravitational Physics of Stellar and Galactic Systems*†
R. N. Sen *Causality, Measurement Theory and the Differentiable Structure of Space-Time*
M. Shifman and A. Yung *Supersymmetric Solitons*
H. Stephani, D. Kramer, M. MacCallum, C. Hoenselaers and E. Herlt *Exact Solutions of Einstein's Field Equations, 2nd edition*†
J. Stewart *Advanced General Relativity*†
J. C. Taylor *Gauge Theories of Weak Interactions*†
T. Thiemann *Modern Canonical Quantum General Relativity*†
D. J. Toms *The Schwinger Action Principle and Effective Action*†
A. Vilenkin and E. P. S. Shellard *Cosmic Strings and Other Topological Defects*†
R. S. Ward and R. O. Wells, Jr *Twistor Geometry and Field Theory*†
E. J. Weinberg *Classical Solutions in Quantum Field Theory: Solitons and Instantons in High Energy Physics*
J. R. Wilson and G. J. Mathews *Relativistic Numerical Hydrodynamics*†

† Available in paperback

Quantized Detector Networks

The Theory of Observation

GEORGE JAROSZKIEWICZ

CAMBRIDGE
UNIVERSITY PRESS

CAMBRIDGE
UNIVERSITY PRESS

Shaftesbury Road, Cambridge CB2 8EA, United Kingdom

One Liberty Plaza, 20th Floor, New York, NY 10006, USA

477 Williamstown Road, Port Melbourne, VIC 3207, Australia

314–321, 3rd Floor, Plot 3, Splendor Forum, Jasola District Centre, New Delhi – 110025, India

103 Penang Road, #05-06/07, Visioncrest Commercial, Singapore 238467

Cambridge University Press is part of Cambridge University Press & Assessment, a department of the University of Cambridge.

We share the University's mission to contribute to society through the pursuit of education, learning and research at the highest international levels of excellence.

www.cambridge.org
Information on this title: www.cambridge.org/9781009401456

DOI: 10.1017/9781009401432

When citing this work, please include a reference to the DOI 10.1017/9781009401432

First published 2017
Reissued as OA 2023

A catalogue record for this publication is available from the British Library.

ISBN 978-1-009-40145-6 Hardback
ISBN 978-1-009-40142-5 Paperback

Contents

Preface

In a routine optical telescope scan of a distant galaxy, astronomer Alice saw nothing unusual. Her radio astronomer colleague Bob, however, reported intense radio activity in that galaxy. Who had the true view of the galaxy?

This is the sort of question discussed in this book. If you said that Bob had the "true" view of the galaxy, you would be quite normal. *Normal*, in the sense of *average*, or *typical*, or even *reasonable*. But if you went on to read the rest of this book and understand its main message, you might then give a different answer to that question.

It is not a trick question, however. The "correct" answer is **not** "*Alice has the true view of the galaxy.*" Neither is it "*Neither of them*" nor is it "*Both of them.*"

This preface is not the place to discuss possible alternative answers to the above question; you should be able to work one out based on the principles discussed in the main text of this book. Although the question is easy to state, the answer we give in the last chapter is simple neither to explain nor to justify. It is best discussed using a lot of words and rather sophisticated mathematical models and technologies. These are introduced, developed, and applied after intensive preliminary discussions of the issues concerned.

Our answer is intimately bound up with the *laws of observation* as they pertain to quantum processes, the subject matter of this book. These laws are the rules that underpin modern, empirically based perceptions of physical reality (our term for the world of experience).[1] It has taken over two thousand years of philosophical, natural philosophical, and empirical inquiry into the physical universe for some of these rules to be discovered, particularly the ones involving quantum processes. These latter have been understood for only the last hundred years or so, and what they mean is still an active subject of debate. The old question of how many angels can dance on the head of a pin is nothing compared with the question of what the wave function means in quantum mechanics.

The problem is that most rules of quantum mechanics are counter intuitive and may even appear wrong and unphysical to the person in the street. But then, quantum mechanics has appeared baffling ever since it was stumbled on by Planck in 1900. But as with Alice and Bob above, appearances can be deceptive:

[1] I don't imply there are other forms of reality. I can say nothing about that.

it is quantum mechanics that continues to give verified predictions, while our classical intuition, experience, and expectations all continue to be confounded.

It is my intention that this book be of value to a wide spectrum of readers: refined quantum theorists, philosophers of science, experimentalists, students, and the well-educated person in the street. If nothing else, I want to provoke readers in two ways: first, to question their own ingrained belief structures about the world they live in and second, to be alerted to the fact that there are a lot of speculative concepts and theories that are discussed by some scientists as if they were scientifically meaningful, when in reality they have no empirical basis whatsoever. I call these theories *vacuous* and they are dangerous, because they equate theory with empirical evidence. They lure the unsuspecting into unfounded quasi-religious modes of thinking that may be good science fiction but have no place in science proper. A sound interpretation of quantum mechanics really does matter, because that involves our perceptions of reality. After all is said and done, that is really what distinguishes us from other creatures and affects everything that we do.

This book is timely in two ways. First, after a century of attempts to understand quantum theory from a classical, noncontextual perspective, intense activity into the neurological basis of human psychology has started to reveal a starting truth: classical reality exists only in the mind. Every day, the results of new laboratory experiments demonstrate more and more convincingly just how misleading the classical perception of the world really is. We humans are now being shown up for what we are: we live in illusory worlds of our imaginations, constructed by our brains as they attempt to match the vast flood of sensory data streaming in through our senses with patterns of preconditioned rationalization.

Second, there is intense interest and work worldwide in the development of quantum computers. In this book I discuss the application of computer algebra to quantum registers containing many quantum bits (qubits). Although my focus in this book is on the description of quantum experiments, there will be a great overlap between quantum computation and that focus, so that I expect this book to be of some interest, and I hope value, to quantum computation theorists.

A preface is the place for an author to give their thanks and acknowledgment. Here are mine. First, I have been overawed by the fact that the underlying theme of this book, which is the debate about the nature of perception, has been going on for many centuries. Some of the views that I have read greatly impressed me and I realized that there were individuals who saw through the fog of conditioning that surrounds all of us. In particular, I greatly admire the works of David Hume, whose views about the nature of perception I found remarkably in tune with this book. As an undergraduate, I was greatly influenced by the lectures of Nicholas Kemmer. Later, I encountered the brilliant works of Julian Schwinger, and even had the great pleasure of meeting him and talking with him in 1993 in Nottingham and London, during the bicentenary celebrations of the birth of George Green. These extraordinarily generous individuals in particular and many others influenced me in one way or another in the most significant of ways.

Finally, I express my gratitude to all members of my extended family, past and present. My parents gave me the opportunities to start writing and the arrival of my granddaughter Julia gave me the motivation to continue when that writing became tedious. Without them this book would not exist. In particular, my wife, Małgorzata, has been immensely supportive and helpful in all my efforts; I thank her for that and for her great patience.

Acronyms

CA	computer algebra
CBR	computational basis representation
CM	classical mechanics
CPT	charge, parity, and time
CST	causal set theory
CTC	closed timelike curve
DS	double-slit
EPR	Einstein, Podolsky, and Rosen
eQDN	extended quantized detector networks
FLS	Feynman, Leighton, and Sands
GP	generalized proposition
GPC	general proposition classification
GR	general relativity
HV	hidden variables
LG	Leggett–Garg
LSZ	Lehmann–Symanzik–Zimmermann
MDS	monitored double-slit
MS	Misra and Sudarshan
NEO	null evolution operator
ONB	orthonormal basis
POVM	positive operator-valued measure
PVM	projection-valued measure
QDN	quantized detector network
QED	quantum electrodynamics
QFT	quantum field theory
QM	quantum mechanics
REC	relative external context
RIC	relative internal context
RQFT	relativistic quantum field theory
SBR	signal basis representation
SG	Stern–Gerlach
SR	special relativity
SUO	system under observation
ToE	Theory of Everything

1
Introduction

1.1 Motivation

The aim of this book is to introduce, develop, and apply *quantized detector networks* (QDN), an information-based, observer-centric approach to *quantum mechanics* (QM). Six reasons motivating our development of QDN are the following.

Avoidance of Metaphysical Speculation

There is such a variety of unproven (and unprovable) speculation concerning the interpretation of QM that the subject of this book, QDN, may appear at first sight to be yet another in this growing branch of metaphysics. In fact, our motivation is precisely the opposite. QDN was intended from the outset to reduce the level of metaphysics in the application of QM. To achieve this, our strategy is to move the traditional focus of attention away from *systems under observation* (SUOs) and toward the observers of those SUOs. In QDN, wave functions represent not states of SUOs but states of apparatus. We shall call such states *labstates*, to distinguish them from states of SUOs (which do have a place in QDN). It is only labstates that observers can ever deal with directly.

In this respect, QDN is an attempt at a more laboratory-based description of quantum physics than standard QM, focusing as much on *how* an experiment is done as on the results of that experiment. For example, instead of talking about the wave function of an electron, QDN talks about the labstate of the signal detectors that allow us to say anything about that electron in the first place. That is all, there is nothing deeper. Whether or not electrons actually "exist" is thereby relegated to an inessential metaphysical issue that we can choose to ignore.

It is important to understand what QDN does not say as much as what it does say. QDN does not say that electrons do not "exist". QDN merely asks for an empirical definition of "existence".

QM Is Much More Than a Calculational Device

Many physicists hold the utilitarian view that the success of quantum theory in predicting experimental data is sufficient for their purposes, and that further inquiry is really the business of metaphysics and therefore outside the scope of science proper. We cannot say that this is entirely unreasonable. However, we take the view that QM represents such a radical departure from the principles of classical mechanics (CM) that this pragmatical view cannot be all there is to the subject.

New Interpretations Lead to New Experiments

The history of QM is littered with the debris of various interpretations and paradigms. Planck's quantization of energy (Planck, 1900b), Bohr's radiation damping veto (Bohr, 1913), de Broglie's pilot waves (de Broglie, 1924), matrix mechanics (Heisenberg, 1925), wave mechanics (Schrödinger, 1926), the Copenhagen interpretation (Heisenberg, 1930), Hidden Variables (Bohm, 1952), the Relative State interpretation (Everett, 1957), decoherence (Zurek, 2002), and the Multiverse (Deutsch, 1997) are some of the most discussed views of what quantum theory means. We do not think any of these ideas are right or wrong in an absolute sense, because we do not know what that means. It is clear, however, as we shall argue later, that some interpretations are empirically vacuous, meaning that they can be neither proved nor disproved according to scientific empirical protocols. It is our inclination to avoid vacuous concepts and theories.

When examined in detail, each interpretation of QM has led to new insights and questions that have motivated new experiments, resulting in advances in physics.[1]

Strength in Diversity

There is no evidence for the view that there is, or even should be, a single, best way to do QM. For instance, while we believe that quantum electrodynamics (QED) can explain almost everything about the hydrogen atom, the Schrödinger wave equation does a good enough job most of the time and far more economically. QDN is but another way to describe and formulate quantum physics; it will have advantages in some contexts and disadvantages in others.

Logical Development

QDN is not intended to replace QM. On the contrary, QM works brilliantly and it would be foolish to think we could do better. QDN is intended to enhance QM in areas where little or no attention has been paid hitherto: the observer and their apparatus. QDN seems to us to be a logical application of successful principles that sooner or later would be developed further along the lines taken in this book.

[1] Not in every case, unfortunately.

Human Conditioning
There is now a great deal of empirical evidence for the proposition that humans see the world not as it is but how they have been conditioned to see it (Halligan and Oakley, 2000; Kahneman, 2011). This conditioning occurs because of various forces: evolutionary, genetic, cultural, linguistic, familial, political, and so on. CM is based on ordinary, everyday-life human experience. QM was developed recently, only just over a hundred years ago, and only once sufficient technology had been developed that could reveal the deficiencies in our classical conditioning. We have not caught up with QM in our modes of thinking, and it is not obvious that we ever will. It remains a natural tendency for even the most experienced quantum theorist to explain things in familiar terms, such as space, time, and particles. QDN can be seen as an attempt to steer the theorist away from those conditioned thoughts whenever it is wise to do so.

1.2 Physics, Not Metaphysics

To motivate the QDN approach, we review in this chapter some of the conceptual foundations of QM that are relevant to us here. Before we do that, however, a few words of caution about the relationship between physics and metaphysics seem advisable.

In this book, whenever we refer to *metaphysics*, we shall mean the study of propositions and assertions that cannot be empirically tested, that is, cannot be assigned a truth value relative to any physical context. In our view, it would be a serious error to leave the interpretation of QM to metaphysicists and philosophers. If they had something to contribute that was empirically testable, then they would be physicists, not metaphysicists and philosophers.

However, while it is true that physics is properly concerned only with empirically testable propositions, this does not mean that physicists should never deal with abstract concepts. Physicists after all only do physics because of the way they think and are motivated. Those processes are not currently regarded as legitimate concerns in physics, but that is probably due to the status of technology at this time.[2]

Moreover, physicists use mathematics all the time, because it is the best language in which to formulate theories and propositions about the subject. But all of that takes place in the theorist's mind; theories do not exist by themselves. Mathematics is based on axioms and postulates that are tested not empirically but only for internal logical consistency.

Furthermore, the process of objectivization (the attribution of material existence and properties to recurring and persistent physical patterns and processes)

[2] This may well change with advances in the neurosciences. Observers are fundamental to physics, so how observers operate and fit into the grand scheme of things seems a reasonable subject for investigation.

is the only way that physics is done. Significant examples are the concepts of SUO, observer, particle, atom, molecule, wave function, space, time, apparatus, and so on. These are all mental constructs and essentially metaphysical in nature.

There is a paradox of sorts here involving the use of mathematics in physics. A conventional, indeed perhaps ubiquitous, view about empirical physics is that the empirically discovered laws of physics are independent of observers. Yet observers are indispensable. Observers are needed first to formulate theories (which do not exist except in the minds of theorists) and then to test those theories empirically. It is a metaphysical proposition to assert or to believe the proposition that the properties of objects "exist" independently of observers or observation. How could we prove that proposition? We shall call this idea the *realist* interpretation of physics, otherwise known as realism.

The paradox is that the realist interpretation has been remarkably successful in classical physics, that is, the interpretation of phenomena according to the laws of CM based on the Newtonian space-time paradigm or the Minkowski/general relativistic spacetime paradigm. That success has led many physicists to believe that this interpretation should apply to the whole of physics.

The paradox we refer to above has an additional twist: the classical realist position simply does not work when applied to real physics experiments that go beyond a certain basic level of sophistication. It just does not. The empirical evidence against the classical realist position as the be-all and end-all is now virtually unbreakable, although there will always be theorists who do not accept that statement. We have in mind the Bell inequality-type experiments that are discussed in Chapter 17. These have now established beyond much reasonable doubt that classical intuition is deficient when it comes to quantum physics (by which term we mean those experiments where quantum phases are significant). Therefore, something has to give.

This book is an explanation of our interpretation of what quantum mechanics really means, which is that it is a theory of observation. This involves the observer as well as the observed.

The resolution of the above-mentioned paradox that led us to QDN was the painful acceptance of the fact that physics is done by humans. This fact is painful in that the current tendency in physics is to avoid reference to humans specifically, and to marginalize observers as much as possible. There is a standard reason given for this, often referred to as *the principle of general covariance*. This is the assertion that the laws of physics are intrinsic and independent of observers.

The principle of general covariance is, when examined properly, a vacuous proposition, but the paradox is that it works in CM, particularly in general relativity (GR). Some theorists even employ it in their formulations of quantum theory. Such efforts invariably end up as branches of mathematical metaphysics, because QM is all about real world physics and experimentation. We think that perhaps science has reached the point where observers should be factored much more into the equations somehow. That agenda is not easy, and is what this book is all about. What helps greatly here is the hard fact that, logically, there can be no empirical evidence for the realist position or for the universality of the principle of general covariance, because the very notion of empirical evidence necessarily involves human observers. You cannot have your cake and eat it. You cannot use observation to prove that the laws of physics are independent of observation.

If there is one thing that QM teaches us, then, it is the uncomfortable lesson that not all intuition stands up. Sometimes, particularly when it comes to quantum processes, the natural human interpretation of what is believed to be going on is inconsistent, or just meaningless.

Our considered view, then, is that QM is not a theory describing objects per se but a *theory of entitlement.* QM provides us with a set of rules that tells us what we as observers are entitled to say in any particular context, *and no more.* The good news is that this theory of entitlement is self-consistent and has definite mathematical rules that have never yet been found to fail.

We illustrate the sort of scientific philosophy that guides us in this book with the following, scenario A. Suppose a hundred, a thousand, a million observers had independently measured some quantifiable property about some SUO S and had come up with the same numerical value to within experimental error. Then it would be natural to assert that S "has" that property. That is something we do all the time. We say that our house "has" four bedrooms, that our car "is" white, that John "is" tall, and so on. The logic seems inescapable. The world around us "has" properties that we discover.

But when it comes to quantum physics, there is now enough evidence that tells us that we cannot always rely on the above assertion. We cannot exclude the possibility that the very processes used to "see" those presumed properties of SUOs create, in some way, those very properties. If we take this as a warning, then the only thing that we would be entitled to say about scenario A with real confidence is only what we could be sure of: that a hundred, a thousand, a million observers had independently performed some procedure and obtained the same number to within experimental error.

This may sound limited, cautious, lacking in vision, and so on. But the history of empirical quantum science over the last several decades has shown unambiguously that if we made the realist assertion that SUOs "have" measurable properties, then sooner or later we would come to an empirically observed breakdown of some prediction based on that assertion. The theorist Wheeler went so far as to formulate this principle of entitlement in the form of what he called the *participatory principle*:

> Stronger than the anthropic principle is what I might call the participatory principle. According to it we could not even imagine a universe that did not somewhere and for some stretch of time contain observers because the very building materials of the universe are these acts of observer-participancy.

> You wouldn't have the stuff out of which to build the universe otherwise. This participatory principle takes for its foundation the absolutely central point of the quantum: No elementary phenomenon is a phenomenon until it is an observed (or registered) phenomenon.
>
> (Wheeler, 1979)

So how should we interpret the empirical agreement of the observers' measurement on system S? The answer is to look only at what the facts tells us, and in this case, they tell us no more than that a bunch of observers agree on a measured value. The important point is that there is consistency (or reproducibility) in their observations. It is *consistency* that is the key here and in all empirical science (and indeed mathematics). Consistency requires observers, for if we did not have observers, who is the judge of consistency? Therefore, we should interpret the laws of physics not in terms of absolute properties of systems under observation, but as consistency relationships between what observers do.

This means that we are proposing a psychological shift in emphasis, a change in our perspective of what physics is about. In QDN, physics is not the study of SUOs and the determination of their assumed properties, but the study of the relationship between observers and their apparatus. That is the basic logic on which QDN has been developed. QDN is essentially a theory of *entitlement*, a set of rules for what observers can say with relative confidence about their apparatus under contextually relevant circumstances.

We need to add one critical comment here. The QDN approach emphatically does *not* assert or require us to believe that nothing exists outside of observation. Indeed, were we to take such an attitude, we would have no way of calculating the transition amplitudes needed to make predictions. As theorists, we are entitled to model the *information void*, the unobserved regime between state preparation and outcome detection, in any way that leads to successful predictions. To that extent, QM really is like a black box of tricks, whose workings we can only guess at.

1.3 A Brief History of Quantum Interpretation

Before 1900

To understand the impact QM had on physics, we should first understand the principles of CM that most physicists accepted prior to 1900, the year in which Planck stated his quantum hypothesis.

CM is a view of physical reality that is predicated on the way that humans normally perceive and interact with their environment. Without any additional equipment, a typical human interacts with their surroundings by vision, by touch, by sound, by taste, and by smell. Several critical facts about each of these senses conspire to lead the human brain to subconsciously create mental models of their environment, and it is these models that were used by theorists such as Newton, Lagrange, and Hamilton as the basis on which to construct CM. Four of these critical facts are the following.

Length and Time Scales
Humans are relatively enormous compared with atomic scales, so much so that the existence of atoms was not established conclusively until just over a hundred years ago. Matter therefore gives the impression of being continuous. Moreover, the dynamical processes responsible for the stability of atoms and molecules have time scales that are long enough to create the illusion that we shall refer to as *persistence*, the idea that SUOs and observers exist as meaningful entities over significant intervals of time and therefore can be objectified (that is, can be given individual identities).

Daylight Is Bright
To the human eye, daylight is relatively bright. The human brain interprets objects that it sees visually not in photonic, that is, discrete, terms but rather as a continuous process. The net effect is to create the illusion that well-defined objects or SUOs exist over extended time scales in continuous space and time.

Apparent Observer Independence
Different human observers looking visually at an object will generally agree that they see the same thing, albeit with different relative positions and velocity components. This belief leads to the principle of general covariance mentioned above, the idea that intrinsic properties of SUOs are independent of observers and frames of reference.

Zero Observational Cost
When humans see and hear objects, there is generally no noticeable effect back on those objects due to that observation. In quantum physics this is referred to as *noninvasive measurability*.

These facts and others, such as the way that the human brain processes information, all conspire to give the impression that objects under observation "have" physical properties independent of any observer. Moreover, these properties can be measured nondestructively and understood by the rules of classical mechanics. For example, in Newton's laws of motion, there is no mention of any apparatus or observer, apart from the implication that these laws are stated relative to inertial frames of reference.

After 1900

Up to 1900, virtually all physicists thought about space, time, matter, and the way physics was to be done in the above classical terms. The advent of special relativity (SR) in 1905 in no way changed this perspective: special relativity is all about the differences between observers and how they see things classically, and not about those things themselves. Indeed, theorists such as FitzGerald (FitzGerald, 1889), Larmor (Larmor, 1897), and Lorentz (Lorentz, 1899) had been working out many of the details of SR well before 1900, basing their work on the Newtonian space-time paradigm. Their work and that of Einstein in his landmark 1905 paper on SR (Einstein, 1905b) had nothing to do with what Planck started in 1900. So little, in fact, that bridging the gap between SR and QM remains perhaps the greatest challenge in mathematical physics.

Up to 1900, Planck had been trying to understand the empirical data from experiments on black body cavities, using Newtonian physical principles and Maxwell's theory of electromagnetism. That approach led to the prediction that the intensity of light emitted from a black body would grow with frequency, but the data indicated otherwise. In 1900, in order to understand the data, he approached the problem in a new way, departing from conventional principles.

To understand properly what Planck did *not* say in 1900, we need to understand what Einstein *did* say in 1905. In that year, Einstein explained the photoelectric effect in terms of particles of light (Einstein, 1905a). These particles subsequently became known as photons (Lewis, 1926). What seemed paradoxical, and still does to this day, is that photons are regarded as particles associated with Maxwell's theory of electromagnetism, a classical theory that predicts that light is a wave process and not a particulate process.

Planck did *not* assert that light itself was quantized. What he said was that the black body data could be explained if the assumed atomic oscillators lining the walls of a black body cavity could *absorb and/or emit* electromagnetic radiation only in well-defined amounts, known as quanta:

> Let us consider a large number of monochromatically vibrating resonators – N of frequency ν (per second), N' of frequency ν', N'' of frequency ν'', ..., with all N large number – which are at large distances apart and are enclosed in a diathermic medium with light velocity c and bounded by reflecting walls. Let the system contain a certain amount of energy, the total energy E_t (erg) which is present partly in the medium as travelling radiation and partly in the resonators as vibrational energy. The question is how in a stationary state this energy is distributed over the vibrations of the resonators and over the various of the radiation present in the medium, and what will be the temperature of the total system.
>
> To answer this question we first of all consider the vibrations of the resonators and assign to them arbitrary definite energies, for instance, an energy E to the N resonators ν, E' to the N' resonators ν', ... The sum
>
> $$E + E' + E'' + \cdots = E_0$$
>
> must, of course, be less than E_t. The remainder $E_t - E_0$ pertains then to the radiation present in the medium. We must now give the distribution

> of the energy over the separate resonators of each group, first of all the distribution of the energy E over the N resonators of frequency ν. If E considered to be continuously divisible quantity, this distribution is possible in infinitely many ways. We consider, however – this is the most essential point of the whole calculation – E to be composed of a definite number of equal parts and use thereto the constant of nature $h = 6.55 \times 10^{-27}$ erg·sec. This constant multiplied by the common frequency ν of the resonators gives us the energy element ε in erg, and dividing E by ε we get the number P of energy elements which must be divided over the N resonators. If the ratio is not an integer, we take for P an integer in the neighbourhood.
>
> (Planck, 1900a)

Planck's idea was altogether conceptually different from that of Einstein. We can see the difference directly in the above quote from Planck's paper: Planck is concerned there with E_0, the energy associated with the oscillators, whereas Einstein discussed $E_t - E_0$, the energy of the radiation in the cavity. These are very different things.

Einstein was a confirmed realist who never accepted Bohr's interpretation of QM, that is, the set of ideas generally referred to as the Copenhagen interpretation. We have to side with Bohr here: the assertion that photons "exist" is a vacuous one. How could we confirm that there are "particles of light" *without* using detectors that are themselves subject to Planck's quantum principle, absorbing and/or emitting energy in discrete amounts only?

QDN can be understood as a return to Planck's original position, away from the objectification of quanta as "things" and toward the view that quanta characterize processes of observation as they are in the real world of physics. QDN is fully in accord with the principles laid down by Heisenberg, Born, and Bohr, but not with the diametrically opposing views held by followers of HV or the Many Worlds paradigm. We regard those as vacuous, reflecting the classical conditioning that humans are generally subjected to: no more than contextually incomplete attempts to explain the nonclassical rules of quantum mechanics in realist terms.

1.4 Plan of This Book

The nature of the subject we are discussing makes a division into four themes helpful. Each theme has its own particular characteristics and flavor.

The first theme, *Basics*, runs from Chapters 1 to 9. This theme builds up the formalism of QDN from basics, starting with a review of what QDN is about, then turning to the mathematics of classical bits, then quantum registers, and finally the positive operator-valued measure (POVM) approach to quantum physics.

The second theme, *Applications*, runs from Chapters 10 to 20. In it, QDN is applied to a number of experiments that show how QDN differs from standard quantum formalism. We include recent, exciting experiments, typically in quantum optics, such as quantum eraser, delayed choice, and Bell and Leggett–Garg inequality experiments.

The third theme, *Prospects*, runs from Chapters 21 to 26, and is more speculative than the previous themes. In it, we discuss the prospects for future application of the QDN formalism, such as the possibility of constructing a generalized theory of observation.

The final theme, *Appendices*, consists of material that stands alone but is referred to at various places in the other themes.

1.5 Guidelines for Reading This Book

The subject of quantum mechanics and its interpretation has had a long and tortuous history. There is now good empirical evidence for the statement that quantum processes cannot be explained using classical modes of thinking.

It is inevitable that we should comment in this book on topics that may appear metaphysical; that is unavoidable because there is no point in presenting a detailed mathematical formalism (which we do in due course in this book) if we do not say what we think it means in real-life terms. This book is aimed squarely at representing what is actually done in a laboratory, not what is imagined is done.

Although the substantive mathematical formalism starts in Chapter 3, we would advise the reader to go over the first two chapters, in order to gain a feeling for what we are about in this book. The mathematics relating to those concepts is developed in the main body of the text. Most of that mathematics will be familiar and not reviewed in much detail, but where novelty is encountered, mathematical structures are explained more fully.

An important aspect of our approach in this book is the concept of *architecture*, by which we mean a verbal or diagrammatical description of the processes involved in an experiment. Different experiments may have different architectures. For example, an elementary particle scattering experiment will usually involve an initial *in* state, an intermediate scattering regime, and a final *out* state. That architecture is sufficient to describe electron–proton scattering at relatively low energies, but if unstable particles are emitted in what appears to be the final state, then their decay processes require the architecture to be extended in time to include the detection of their decay products. Another important example of architectural differences concerns the distinction between *pure* states and *mixed* states in QM. Architecture is involved intimately in delicate subjects such as Bell inequality experiments, where failure to appreciate details of architecture, particularly those involving counterfactual arguments, can lead to confusion and misleading conclusions to be arrived at.

Acronyms are used extensively throughout this book. We follow the general convention that the first instance of an acronym in any chapter is defined at that point and the acronym used thereafter in that chapter. For convenience, there is a list of acronyms after the Preface.

Where important concepts, paradigms (modes of thinking), and theories are encountered, we will capitalize their first letters. For instance, we shall write Absolute Time and not absolute time, Universe and not universe, and so on.

1.6 Terminology and Conventions

Basic mathematical concepts such as vector spaces and Hilbert spaces are reviewed in the Appendix.

Maps, Operators, Transformations

If U and V are two vector spaces over the same field, then any process, linear or otherwise, that takes any vector in U into V or back into U will be referred to as a *map*, *operator*, or *transformation*, depending on context.

Retractions

If f is a function that maps a set A into a set B, we denote the set of image points in B by $f(A)$. In this book we encounter situations where $f(A)$ is a proper subset of B, which means that there are elements in B that are not images of any element in A. If f is an injection, which means that there is a one-to-one correspondence between elements in A and elements in $f(A)$, then the retraction $\overline{f}$ of f is the well-defined function with domain $f(A)$ that takes elements from $f(A)$ back into A, given by $\overline{f}(f(a)) = a$ for any element a of A. We shall use retractions a lot in this book.

The Two Worlds

As QDN was being developed, it became clear that the theory was dealing with two separate worlds: the inner world of the system under observation and the outer world of the observer and their apparatus. Linking these two worlds are the quantum processes of interest. Therefore, it became necessary to devise a notation that could readily distinguish between these two worlds, yet retain all the characteristics of quantum mechanics.

Our notational solution to this problem is the following. We shall deal frequently with a *total Hilbert space*, the tensor product $\mathcal{H} \otimes \mathcal{Q}$ of two Hilbert

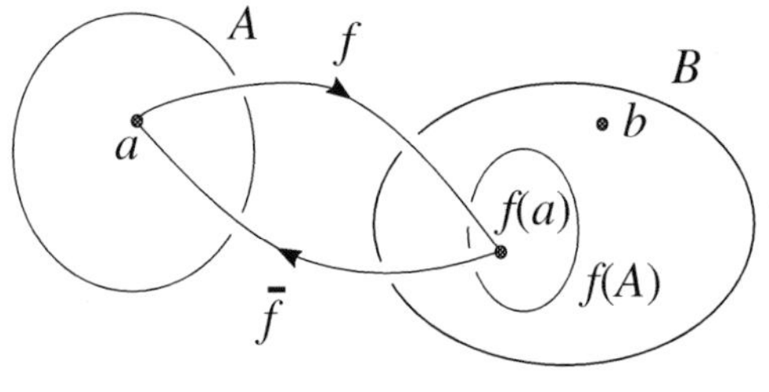

Figure 1.1. If $f : A \to B$ is an injection and $f(A)$ is a proper subset of B, then there are elements such as b in B for which the retraction $\overline{f}$ of f is not defined.

spaces. The first Hilbert space $\mathcal{H}$ will contain the *system states*, the imagined states of the SUO, and for those we shall use the conventional and familiar Dirac bra-ket notation $|\Psi\rangle$, with inner product $\langle\Phi|\Psi\rangle$, and so on. The second Hilbert space, $\mathcal{Q}$, will be a quantum register containing the *labstates*, the quantum states of the apparatus. For these labstates we shall use bold notation $\boldsymbol{j}$ and their duals will be denoted $\overline{\boldsymbol{j}}$. Labstate inner products are denoted $\overline{\boldsymbol{i}}\boldsymbol{j}$, and so on. *Total states* are elements of the total Hilbert space and denoted by $|\Psi, i) \equiv |\Psi\rangle \otimes \boldsymbol{i}$, noting the round bracket on the left-hand side of this definition. Inner products of total states are given by

$$(\Phi, i|\Psi, j) = \langle\Phi|\Psi\rangle\overline{\boldsymbol{i}}\boldsymbol{j}. \tag{1.1}$$

The use of bold font for labstates and overlined bold font for their duals is consistent with the question and answer formalism described in the next chapter.

2
Questions and Answers

2.1 What Is Physics?

Throughout this book we shall be concerned greatly with *questions* and *answers*. The *subject* that an *observer* asks a question of will be denoted by a bold symbol such as $\boldsymbol{\Psi}$. On the other hand, the *question* that is asked of a subject will be denoted by a barred bold symbol such as $\overline{\boldsymbol{Q}}$. Then the *answer* A that is obtained when that question is asked of that subject will be denoted by $A \equiv \overline{\boldsymbol{Q}}\boldsymbol{\Psi}$. This notation will be used to explain what quantized detector networks (QDN) is all about and in its mathematical implementation. Note that the left-right ordering here is significant: $\boldsymbol{\Psi}\overline{\boldsymbol{Q}}$ will mean something very different.

If **I**, the author, were asked the question $\overline{\boldsymbol{Q}} \equiv$ **What is physics?**, my answer $\overline{\boldsymbol{Q}}\mathbf{I}$ would be

$\overline{\boldsymbol{Q}}\mathbf{I}$ = *Physics is the process whereby observers first formulate, then ask, and then answer, empirical questions.*

We need to clarify some points about this answer. By *empirical*, we mean according to the established principles of science, whereby theorists make propositions or assertions and then establish the truth values of those propositions relative to experiments carried out by unbiased experimentalists. This is done to acquire information about systems under observation (SUOs) in order to add to our knowledge about the relative physical Universe (that part of the Universe empirically accessible to us). We are motivated to do all this for numerous reasons, not the least of which is basic curiosity. This is the view of physics taken in this book.

Physics is emphatically *not* a discipline that philosophers and metaphysicists can meaningfully contribute to, because without empirical testing, any proposition has to be regarded as scientifically vacuous (empirically meaningless).

The questions relevant to physicists are posed through experiments and consequently will be referred to as empirical questions. Other sorts of questions will be referred to as theoretical questions.

Matters are never as simple as that. We will clarify presently what we mean by observers, SUOs, and so on. At this point, we observe that the above answer $\overline{\boldsymbol{Q}}\mathbf{I}$ begs a number of questions that should have been asked and answered first. Specifically, we should also ask the questions

$$\begin{aligned}\overline{\boldsymbol{A}} &\equiv \textbf{Who is asking this question?}\\ \overline{\boldsymbol{B}} &\equiv \textbf{Why is this question being asked?}\\ \overline{\boldsymbol{C}} &\equiv \textbf{Who or what is being asked?}\\ \overline{\boldsymbol{D}} &\equiv \textbf{What will be done with the answer?}\end{aligned}$$

and perhaps more. It is the failure to ask questions $\overline{\boldsymbol{A}}$, $\overline{\boldsymbol{B}}$, and so on, that seems to us to be the root cause of many if not all of the conceptual problems people have in the interpretation of quantum mechanics (QM).

We could go further. The *context*, or circumstances, under which we ask our questions is extremely important. For instance, physicists use experiments to answer their empirical questions. The apparatus that they use is part of that context: if the apparatus cannot be constructed, then the questions cannot even be asked. When we look carefully at what we are doing when we do physics, it soon becomes obvious that we have to take great care in how we ask our questions.

We should say at this point that we do have a specific view about how questions should be asked in physics, particularly quantum physics. This view is based on a concept that we call *contextual completeness*, discussed in Section 2.12. Contextual completeness is a measure, based on a simple algorithm, for deciding whether any given assertion or proposition is a reasonable scientific proposition or not. Specifically, we can establish whether it could be investigated according to proper scientific protocols, or whether it is a *vacuous* proposition, meaning that it cannot be empirically tested for a truth value. One of the surprising features of our analysis is that it shows that there are several degrees of contextual completeness, ranging from the completely vacuous (characteristic of metaphysics and philosophy) to the fully contextually complete proposition required in QM.

2.2 Physics and Time

In this book, we take the view that it is necessary and good physics to discuss the meaning of *time*, a central element in all branches of physics and certainly too important to leave to philosophers and metaphysicists.

There are two contrasting natural philosophies of time, called *Manifold Time* and *Process Time* (Encyclopaedia Britannica, 2000). Both of these philosophies are inextricably woven into the fabric of science. Neither is right nor wrong; each has value in the appropriate context. In this book, our focus will be mainly on Process Time, as this is the time most relevant to how humans operate when they do physics experiments.

Manifold Time

Manifold Time is the view that time is a "thing," an objectivized parameter modeled as a real geometrical dimension. An important development in Manifold Time occurred in 1908, when Minkowski proposed that time could be modeled geometrically, as one of the dimensions of *spacetime*, the four-dimensional Lorentz-signature manifold of special relativity (SR), in which all past, present, and future events are contained (Minkowski, 1908; Petkov, 2012). This model of the universe is often referred to as the *Block Universe* (Price, 1997).

There are several significant factors commonly used by physicists to support Manifold Time and the Block Universe concepts. Some of these are the following.

Relativity
Manifestly Block Universe theories such as SR and general relativity (GR) are based on spacetime manifolds with Lorentzian signature metrics. These are discussed in the Appendix, Section A.2.

Reversibility
The laws of Newtonian classical mechanics excluding friction are generally time-reversible. Lines in geometry do not carry any intrinsic direction. Therefore, if time is reversible, it may be possible to model it geometrically.

Unitary Evolution
In QM, states of SUOs evolving without observational intervention are described by Schrödinger unitary evolution, and this is regarded as reversible.

CPT Theorem
In high energy particle physics, the well-known charge, parity, time (CPT) reversal theorem involves symmetry operations in spacetime that are consistent with all experiments to date, particularly particle scattering amplitudes, which have empirically confirmed crossing symmetries (Eden et al., 1966).

Differential Equations
Many differential equations in mathematical physics, such as Maxwell's equations for electrodynamics and Einstein's field equations in GR, contain no dissipative terms, leading to both advanced and retarded solutions to the field equations with no natural (to the theory) reason to eliminate the classically unacceptable advanced solutions.

Process Time

In contrast to Manifold Time, Process Time (Encyclopaedia Britannica, 2000) is based on an acknowledgment that there are irreversible processes going on in the Universe and that humans experience a feeling that there is a "moment of the now." Time is no longer viewed as a *thing* but as a *process*. This implies the existence (in this context) of the systems or objects that are processing in time. According to QDN, the observers of those processes are fundamental to this discussion as well. Factors that we take here as supporting the Process Time paradigm are the following.

Irreversibility
Thermodynamics, one of the fundamental branches of physics, has irreversibility built into it.

The Born Rule
The Born rule in QM deals with probability, which is inherently asymmetric regarding time.

Observers
Observers have memory and intention, which are asymmetric regarding the direction of time. Observers operate contextually in an irreversible way, acquiring information as time progresses.

Hubble Expansion
The Universe appears to be expanding in an irreversible way from a definite point in the past. The Block Universe model has no empirical justification for postulating events earlier than that point. Indeed, the Steady State paradigm (Bondi and Gold, 1948; Hoyle, 1948) that postulated an indefinite past has been generally abandoned by cosmologists.

Contextual Incompleteness
The Block Universe paradigm is contextually incomplete with a generalized propositional classification of zero (contextual completeness and generalized propositional classification are discussed further on in this chapter). It is a vacuous, metaphysical concept.

2.3 Reduction versus Emergence

The Manifold Time and Process Time paradigms can be reconciled if context is taken into account. Manifold Time is based on a reductionist view of the laws of physics, whereas Process Time is associated with emergence (or complexity). Reductionism is the principle that all complex SUOs and indeed the universe can all be explained by a relatively few basic laws. It is a bottom-up view of science. Perhaps the purest form of reductionism is seen in elementary particle physics, which presupposes that there is a single Lagrangian that can be used to describe all phenomena by a Theory of Everything (ToE).

Emergence, on the other hand, is an acceptance that there are situations that transcend reductionism. For example, life is a process, not a collection of atoms; there is no mention of life in the standard model of high energy physics.

Situations in physics that fall out of the scope of reductionism and require emergent thinking generally involve observers and large-scale processes typically involving numbers of particles on the scale of Avogadro's number and more, that is, of the order 10^{26}. Examples in the physical sciences that require emergent concepts are thermodynamics, gravitation, cosmology, and quantum physics. We take the view in this book that it is a fundamental category error to believe that reductionism alone can explain any of those concepts.

The question remains as to the status of reductionism in QDN. Does the relatively successful reductionist Standard Model of particle physics have a place in QDN?

Our answer is emphatically *yes*. Reductionism has a specific role in those aspects of observation that do not involve apparatus explicitly, and that is in the *information void*, the conceptual region between state preparation and outcome detection. We shall be greatly preoccupied with the information void, for it is in that region where the idealized Lagrangians of particle physics can be explored and used to work out the quantum signal transition amplitudes that QDN needs. Indeed, QDN has no way by itself of generating those amplitudes. In this respect, QDN is analogous to scattering matrix (S-Matrix) theory[1] in hadronic physics (Eden et al., 1966), which deals with the architecture of scattering processes but cannot account for the explicit details of amplitudes without introducing Lagrangians.

An important and useful point here is that, by definition, we cannot "know" what the information void is. We can only make models for the transmission of information from preparation device to outcome detector. This is useful because it means we are free in our choice of the models we use to calculate the transmission amplitudes from source to detector across information voids. These models need not be of the standard Lorentzian signature, spacetime manifold type that are used in relativistic quantum field theory. They could be more complicated, such as Snyder's quantized spacetime algebra (Snyder, 1947a,b), or even manifestations of superstrings. At the end of the day, such models simply provide "black box" recipes for the calculation of amplitudes that will have various degrees of empirical correctness but should never be regarded as reflecting absolute truth in any way. Of course, some models will be much better than others.

2.4 Peaceful Coexistence

Peaceful coexistence is a term often used to express the view that the two great scientific paradigms of the twentieth century, quantum theory and relativity, can

[1] *S* stands for *scattering*.

be reconciled without modification to either, and that neither is subordinate to the other. The problems encountered in failed attempts to reconcile these excellent theories arise because of a belief that they can each be explained in reductionist terms. Our view is that each is an aspect of emergent physics, distinguished by their contexts. To date, these contexts appear to be nonintersecting. General relativity (GR) holds as a classical description of processes involving spacetime and matter on large scales, while QM is an observer-centric description of processes where phase degrees of freedom are under the control of observers.

Denote the set of empirical contexts where GR appears necessary by $\mathcal{G}$, and the set of empirical contexts where QM appears necessary by $\mathcal{Q}$. Then the only place where a unified approach might be needed would be the intersection $\mathcal{G} \cap \mathcal{Q}$, assuming it is not empty. Currently, little is known empirically about such an intersection. Specifically, there seems to be no way at this time of empirically testing any of the proposed theories of quantum gravitation.

One area of great current interest is in the question of Beckenstein–Hawking radiation from black holes. If detected, such radiation would "merely" serve to confirm peaceful coexistence and not some incompatibility between GR and QM. Black hole thermodynamics, as generally formulated, involves QM on a classical curved background spacetime, is contextually incomplete (because observers are inadequately treated), and is therefore not the fully quantized theory of gravitation that many theorists would like to have.

We take the view in QDN that GR and QM are emergent descriptions in physics involving different empirical contexts, so there is no clash of principles.

2.5 Questions and Answer Sets

In physics, questions are asked by observers of states of SUOs. These states will typically be represented by elements of spaces such as Hilbert spaces. We shall use the *question and answer* notation introduced at the start of this chapter. We shall represent a given state of some classical or quantum SUO[2] S by a bold symbol such as $\boldsymbol{\Psi}$ and represent any question that we ask of that state by a barred bold symbol such as $\overline{\boldsymbol{Q}}$. Then our convention is that the answer to question $\overline{\boldsymbol{Q}}$ asked of state $\boldsymbol{\Psi}$ will be denoted by $\overline{\boldsymbol{Q}}\boldsymbol{\Psi}$ and referred to as a *contextual answer*.

Now the whole point of asking a question is to reduce an observer's uncertainty. Before a question $\overline{\boldsymbol{Q}}$ is asked of state $\boldsymbol{\Psi}$, the observer (the entity asking the question) will believe that the answer (still to be obtained) is one element of a set of possible answers. We shall call this set an *answer set* and denote it by $\mathcal{A}(Q|\Psi)$.

In principle, answer sets may be quite general. They could be sets of sentences, sets of numbers, or sets of symbols. An answer set may even be empty or consist

[2] By this we mean whether we have decided to use classical mechanics (CM) or QM in our discussion of the experiment.

of an infinite number of real or complex numbers. In this book, we shall focus primarily on answer sets based on real or complex numbers.

A critical, implicit factor in any question and answer process is the role of *contextuality*, that is, the circumstances under which a given proposition may be true or false. Observers do not receive fully formed answers directly from their equipment, such as "*the particle landed here on the screen.*" Rather, an observer will look at a screen, either optically or via electronic equipment, and notice for example that all pixels except one are white. That exceptional pixel requires interpretation by the observer. A typical interpretation would be that it represented the impact site of a particle. That would be valid, however, only if the context of that particular experiment made that a reasonable inference. For instance, the observer might already have determined that the exceptional pixel in question was faulty and could not pick up any signals whatsoever, so did not represent a particle impact.

Context has a critical role, perhaps the most critical role, in quantum physics. In contrast, classical mechanics is based on the realist concept of *noncontextuality*, which asserts that SUOs "have" physical properties regardless of how they are observed or not observed.

2.6 Answer Set Collapse

Time plays a critical role in the concept of question and answer. *Before* we ask a question $\overline{\boldsymbol{Q}}$ of some state $\boldsymbol{\Psi}$, the range of possible answers is generally two or more. Having less than two possible answers in an answer set is of limited practical value in general.

On the other hand, *after* we have obtained the answer value $A \equiv \overline{\boldsymbol{Q}}\boldsymbol{\Psi}$, which is some element of the initial answer set $\mathcal{A}(Q|\Psi)$, our uncertainty has completely disappeared. That disappearance is accompanied by the disappearance, in our minds, of the original answer set. The transition $\mathcal{A}(Q|\Psi)$ to the singleton set $\{\overline{\boldsymbol{Q}}\boldsymbol{\Psi}\}$ is a completely natural result of the questioning and answering process. Unfortunately, it has been objectified by some quantum theorists as the sudden "collapse" of a quantum wave function or state and asserted as something to be concerned about.

The reason for this concern has to do with the interpretation of the wave function/quantum state vector and the role of time in QM. In general, questions cannot always be asked nondestructively. For example, to determine a person's blood type, some blood has to be taken. One of the fundamental differences between CM and QM is that in CM, it is generally asserted that questions can be asked of states of SUOs without affecting those states in any way. That is not always the case in QM.

The problem about *wave function collapse*, or *state reduction*, can be understood as follows. Suppose we ask a question $\overline{\boldsymbol{Q}_1}$ of a state $\boldsymbol{\Psi}$ of an SUO and get the answer $\overline{\boldsymbol{Q}_1}\boldsymbol{\Psi}$. There is no controversy about that, neither classically nor quantum mechanically. Now suppose that we wanted to ask another question, $\overline{Q_2}$ of that state, *after* we have asked $\overline{Q_1}$. There are three possibilities to consider.

Null Tests

The question $\overline{Q_1}$ could be *completely nondestructive*, giving an answer yet leaving the original state unaltered. This scenario is assumed throughout CM, classical questions generally being assumed completely nondestructive. In standard QM this scenario is called a *null test* (Peres, 1995), being described by a state that is an eigenstate of the observable concerned. In his famous book on QM, Dirac discussed this scenario in terms of polarized light passing through two or more polarizing crystals (Dirac, 1958). If an unpolarized monochromatic beam of light is passed through a polarizing crystal, two polarized beams of light are observed to emerge. If either of these beams is then passed through an identical crystal with the same orientation as the first, that beam then goes through the second crystal without further splitting.

A spectacular example of a null test was discussed by Newton in his astonishing book *Opticks* published in 1704 (Newton, 1704). Newton showed that if a beam of sunlight were passed through a prism P^1, then it would be split into a set of sub-beams of different colors, defining a *spectrum*. If that spectrum was then carefully focussed onto a second prism P^2, then P^2 would undo the action of P^1, resulting in a reconstituted beam emerging from P^2 that had the same spectral properties as the initial incident beam. In essence, the combination P^1 followed by P^2 acts as a null test on the original incident beam. This experiment is discussed in detail in Section 11.4.

Partial Destruction

It could be the case that $\overline{Q_1}$ is *partially destructive*, also known as a *measurement of the first kind* (Paris, 2012). After the answer $\overline{Q_1}\boldsymbol{\Psi}$ has been obtained, the SUO is then in some altered state $\boldsymbol{\Psi}'$ different from the original state $\boldsymbol{\Psi}$. Any attempt now by the observer to ask question $\overline{Q_2}$ would then be extracting not the answer $\overline{Q_2}\boldsymbol{\Psi}$ but the answer $\overline{Q_2}\boldsymbol{\Psi}'$.

Total Destruction

It could be the case that $\overline{Q_1}$ is *totally destructive*, also known as a *measurement of the second kind*, or *demolitive measurement* (Paris, 2012). The process of asking $\overline{Q_1}$ totally destroys the original state $\boldsymbol{\Psi}$, so that $\overline{Q_2}$ could not even be asked.

In the real world, human observers get a lot of information about objects around them optically, which is by and large a weakly destructive process that gives the impression of being totally nondestructive. This has led to the classical conditioning that regards the ordering of questions as not significant. In contrast, QM is based on the empirical evidence that all observations that extract genuine information are never nondestructive, Planck's constant being a manifestation of that fact. The temporal ordering of questions is significant in QM and is generally discussed in terms of the commutation properties of the observables associated with the questions involved.

Answer set collapse has everything to do with the observer's state of information and nothing to do with any changes in the state $\boldsymbol{\Psi}$ per se. What happens to a state after a question has been asked is a separate issue and depends on the experiment and on the dynamics assumed.

Standard QM does not in general discuss the third of the above three scenarios; what happens to a quantum state when an experiment has finished and the apparatus has been destroyed or decommissioned is considered outside the remit of standard QM. In contrast, QDN has an agenda to consider such possibilities. Such matters are discussed in Chapter 25.

Whether the processes of observation are destructive or not, answer set collapse will still take place in general, illustrating the QDN view that QM is a theory of observation and not of objects per se.

2.7 Incompatible Questions and Category Errors

Not all questions can be asked sensibly of all states. For instance, suppose $\boldsymbol{\Phi}$ represents the state of an apple resting on my desk. Then if $\overline{\boldsymbol{Q}}$ is the question

$$\overline{\boldsymbol{Q}} \equiv \textbf{How far is the Moon from the Earth?} \tag{2.1}$$

it is obvious that $\overline{\boldsymbol{Q}}\boldsymbol{\Phi}$ is meaningless. This is an example of an *incompatible question*, a mismatch between the concepts used to define the state of the SUO (in this case, the apple on the desk) and the concepts required to formulate the question (in this case, the distance between the Moon and the Earth). Another term for such a conceptual clash is *category error*.

Experimentalists generally have to spend a great deal of time and resources to arrange for their questions and their states of SUOs to be compatible. Compatibility is always contextual: a question $\overline{\boldsymbol{Q}}$ that is compatible with states of one SUO will usually be incompatible with states of some other SUO. For example, a photon detector will click when placed in a laser beam, but will not click when placed in an electron beam.

An important source of incompatibility may be due to the mechanical structure and functionality of apparatus. This can come about in several ways. First, if a detector has not yet been built, or cannot be constructed for some reason, then any attempt to detect a signal is a trivial example of an incompatible question. We may call this *existential incompatibility*. It is the source of a basic error in the interpretation of quantum mechanics. Physics is an empirical subject. If an experiment cannot actually be done in principle or in practice, then theoretical conclusions should not be treated as empirical facts. For example, we have theoretical reasons to believe that no apparatus can be constructed that can detect simultaneously the exact position and exact momentum of an electron. Experimentalists such as Afshar have attempted to circumvent such a conclusion empirically in the case of two-path interferometry (Afshar, 2005), but the general consensus at this time is that they have not succeeded (Jacques et al., 2008).[3]

Suppose a detector does exist in a laboratory. It could be *faulty*, so that any attempt to detect a signal would be doomed to failure. Another possibility is that a detector exists and has registered a signal that the observer knows about (such as in a teleportation experiment), and then that detector has been deliberately *decommissioned* in some way by the observer so as to prevent further signals being generated. Any further attempts to detect a new signal automatically then involve a incompatible question.

Finally, even if a detector exists and is working properly, its compatibility will be contextual on the states being asked. For instance, a photon detector will not detect neutrinos.

There is a class of questions that appear to be nominally physically based, but which have an empty answer set under all known empirical circumstances. This means that there is currently no way known of answering such questions empirically. For example, the question **Is time travel possible?** is one such question. We just don't know the answer. We may refer to such questions as *speculative*. There is a place for them in science, in that they can stimulate reasonable discussion.

A more disturbing class of question is associated with *vacuous theories*, which are theories such that none of their critical propositions can be tested by any known practical means.[4] Such theories are often designated as *not even wrong* (Woit, 2006). Examples currently are the Multiverse paradigm, quantum gravity, and string theory. What is disturbing is not so much that they cannot be tested, but that some of their advocates claim that they represent real physics simply because their apparent mathematical beauty (a subjective opinion) is regarded by their authors as more important than a total lack of critical empirical evidence at this time (2017) (Woit, 2006). Vacuous theories may well be consistent with all currently known empirically established facts, but that is not enough to validate them. A common characteristic is that although they have a generalized proposition classification (GPC, discussed in Section 2.12) of zero or at best one, many of their supporters claim that they represent valid quantum physics (which requires a GPC of three). We disagree. They are speculations and we classify such theories as *mathematical metaphysics*.

2.8 Propositions

In this book, a *proposition* is defined as any string of symbols that may be of interest to an observer. For example, $\boldsymbol{P}_1 \equiv$ **Today is Tuesday**, $\boldsymbol{P}_2 \equiv \boldsymbol{XYZ}$,

[3] We are not at all critical of Afshar's attempts per se. Our view is that such attempts should always be encouraged. That is what science is all about. There *should* always be someone trying to upset the standard paradigms.

[4] A *critical proposition* would be a specific claim made by that theory alone and no other, that would, if confirmed, establish the scientific worth of that theory.

and so on. It is important to note that, by themselves, with no further context, propositions such as P_1 and P_2 are neither true nor false.

Our view of physics will be based on the idea that observers formulate propositions about the Universe and then perform experiments to find out under which contexts those propositions can be deemed to be true or false. We cannot at this time explain *why* observers do such things, and simply take it as a given.[5]

2.9 Negation, Context, Validation

In two-valued classical logic, propositions are assumed to have truth values: given proposition $\boldsymbol{P}$, classical logic examines whether $\boldsymbol{P}$ is *true* or *false.*

To formalize this discussion, we introduce the *validation function*, denoted $\overline{\mathbb{V}}$. Validation is really a question, which is why we put a bar over the symbol $\mathbb{V}$. Validation questions a given proposition and maps it into the binary set $\{0, 1\}$, where the value 0 denotes *false* and the value 1 denotes *true.* If proposition $\boldsymbol{P}$ is false we write $\overline{\mathbb{V}}\boldsymbol{P} = 0$, whereas if $\boldsymbol{P}$ is true we write $\overline{\mathbb{V}}\boldsymbol{P} = 1$.

This leads to the concept of *negation*. Given a proposition $\boldsymbol{P}$, there is often an associated proposition, denoted $\neg\boldsymbol{P}$, called the *negation of* $\boldsymbol{P}$. For example, given $\boldsymbol{P}_1$ is the proposition **Today is Tuesday**, then $\neg\boldsymbol{P}_1$ is the proposition **Today is not Tuesday**.

In classical logic, the relationship between a proposition and its negation is the rule

$$\overline{\mathbb{V}}\boldsymbol{P} + \overline{\mathbb{V}}(\neg\boldsymbol{P}) = 1. \tag{2.2}$$

According to this rule, therefore, we need only establish the validity of a proposition to immediately know the validity of its negation.

There are the following issues to discuss here.

Mathematics Is Not Physics

Eq. (2.2) is a mathematical rule. It is not physics. In physics, we have to be more careful. A proposition such as $\boldsymbol{P}_1 \equiv$ **Today is Tuesday** may have an obvious, physically meaningful negation, but a proposition such as $\boldsymbol{P}_2 \equiv \boldsymbol{XYZ}$ does not have any obvious, physically meaningful negation (but of course, we could always define it symbolically).

Classical Logic Is Insufficient

Classical binary logic deals with mutually exclusive answers, such as *yes* or else *no.* Unfortunately, we cannot be sure that such logic can be applied to every form of proposition in physics. For example, Jaynes put the case that classical probability is an enhanced form of classical binary logic, because in probability theory, answers are not *definitely yes* or else *definitely no*, but *a probability of*

[5] The explanation that *humans do physics in order to understand the universe* does not explain why they find themselves needing to do this. No other species seems to have this ambition.

yes and *a probability of no* (Jaynes, 2003). Another example is QM, where the answers are complex-valued amplitudes.

The Role of Context
Propositions in physics are often by themselves relatively meaningless. For instance, the proposition **Energy is always conserved** is too glib and potentially incorrect. What gives any proposition in physics a meaning and the possibility of a truth value is *context*, by which we mean the circumstances involved in establishing that truth value.

This last point is critical in QM. The rules of logic in QM have to be carefully respected because they are not quite the same as those of CM. The reason has to do with contextuality, which is not factored into classical logic in general. In particular, we should not make the following error of physics. Suppose that $\boldsymbol{P}$ is a physical proposition and we have constructed an apparatus $\mathcal{A}$ that allows us to conclude that $\boldsymbol{P}$ is true. Then we may write

$$\overline{\mathbb{V}}(\boldsymbol{P}, \mathcal{A}) = 1. \tag{2.3}$$

In other words, we have factored into our notation the context (apparatus $\mathcal{A}$) that is needed to empirically confirm the truth of $\boldsymbol{P}$. In words, we would express (2.3) as the statement that *proposition* $\boldsymbol{P}$ *is true relative to context* $\mathcal{A}$.

Example 2.1 In CM, suppose $\boldsymbol{P} \equiv$ **Energy is conserved**, $\mathcal{A}_1$ is a *closed* system, and $\mathcal{A}_2$ is an *isolated* system. Then by definition

$$\overline{\mathbb{V}}(\boldsymbol{P}, \mathcal{A}_1) = 0, \quad \overline{\mathbb{V}}(\boldsymbol{P}, \mathcal{A}_2) = 1. \tag{2.4}$$

2.10 Proof of Negation

Suppose that experiments showed that $\overline{\mathbb{V}}(\boldsymbol{P}, \mathcal{A}) = 0$. Does this imply that $\overline{\mathbb{V}}(\neg \boldsymbol{P}, \mathcal{A}) = 1$?

The answer is that in physics, we cannot assume that

$$\overline{\mathbb{V}}(\boldsymbol{P}, \mathcal{A}) + \overline{\mathbb{V}}(\neg \boldsymbol{P}, \mathcal{A}) = 1, \tag{2.5}$$

because the apparatus $\mathcal{A}$ that allows us to test for the truth of $\boldsymbol{P}$ might be completely useless to test for the truth of $\neg \boldsymbol{P}$. For instance, suppose $\boldsymbol{P}$ is the proposition **Supersymmetric particles exist**. To date (2017), experiments at the Large Hadron Collider have found no sign of such things. That does not imply the truth of the negation of $\boldsymbol{P}$, which is $\neg \boldsymbol{P} \equiv$ **Supersymmetric particles do not exist**. In order to prove empirically that such particles did not exist, we would have to find apparatus $\mathcal{A}'$ such that $\overline{\mathbb{V}}(\neg \boldsymbol{P}, \mathcal{A}') = 1$. It is most unlikely that such apparatus could ever be constructed, because proof of a negation is very often impossible (we cannot prove that fairies do not exist, for instance).

We conclude that the rule (2.2) that holds in classical two-valued logical cannot always be replaced in physics by the rule

$$\overline{\mathbb{V}}(\boldsymbol{P}, \mathcal{A}) + \overline{\mathbb{V}}(\neg \boldsymbol{P}, \mathcal{A}') = 1, \tag{2.6}$$

simply because $\mathcal{A}'$ cannot always be constructed. The rule then in physics is *to say nothing about* $\neg \boldsymbol{P}$. Lack of evidence for a proposition is not evidence in favor of its negation.

2.11 A Heretical View of Reality

Apparatus defines context, and propositions in physics can only be tested by experiments based on apparatus. Therefore, we could logically take the view that scientific propositions have no meaning beyond the experiments that test them. What if the recent "discovery" of the Higgs particle was and will be the only time in the history of the Universe that evidence for the Higgs particle was ever found? We could then make the case for the assertion that the apparatus and its context had created/defined the object of interest, rather than helped in its "discovery."

Of course, such thinking goes against the grain of the physicists' human conditioning, that generally imagines that the Universe is "out there", waiting to be explored. That is the same line of thinking that led to Plato's *theory of forms*, which asserts that mathematics "exists" in its own realm of reality and that mathematicians stumble upon pieces of it (discover preexisting truths) every so often.

2.12 Generalized Propositions and Their Classification

The discussion in the preceding sections suggests that empirical physics involves a great deal of context that is frequently understated in discussions. In fact, the situation is generally worse, as we now suggest, because only the apparatus has been factored into the above discussion, with no mention of any observer. The solution we have developed is to introduce the concept of *generalized proposition* and a classification scheme for generalized propositions, as follows (Jaroszkiewicz, 2016).

Definition 2.2 A *generalized proposition* (GP) $\mathcal{P}$ is of the form

$$\mathcal{P} = (\boldsymbol{P}, C_{\text{int}} | O, C_{\text{ext}}), \tag{2.7}$$

where $\boldsymbol{P}$ is a proposition, C_{int} is the *relative internal context* that an observer O, defined by the *relative external context* C_{ext}, can use to test $\boldsymbol{P}$.

Relative *internal* context includes apparatus and the protocols (including axioms and theory) used to test propositions. Relative *external* context is generally classical information that establishes the relationship between the observer and the wider environment/universe in which they are performing their experiments.

A GP addresses the issues we raised in Section 2.1: ideally, any GP gives information about who is making the proposition, what that proposition is, and how the truth value of that proposition can be determined. This raises the following two points.

Identification
We should always ask the question:

For whom is this GP relevant?

The answer is: for the observer involved and not necessarily for anyone or anything else. It is, after all, an essential requirement of observation that an observer understands their own relationship to their environment. Even in CM, frames of reference have to be identified. No experiment makes sense if any aspect of the GP concerned is unknown to the observer concerned.

Missing Context
What happens if some or all of the contextual information in a GP is missing? We address this fundamental point now.

Definition 2.3 A GP is *contextually complete* if both relative internal context and relative external context have been supplied.

The problem is, there are various degrees of contextual completeness. We propose the following schema to classify any GP, allowing us a mechanism to decide whether a proposition is scientifically useful or not.

Definition 2.4 Given a GP $\mathcal{P} \equiv (P, C_{\text{int}}|O, C_{\text{ext}})$, its *generalized proposition classification* (GPC) $\#\mathcal{P}$ is given by

$$\#\mathcal{P} = \alpha + 2\beta, \tag{2.8}$$

where

$$\alpha = \begin{cases} 0, & C_{\text{int}} = \emptyset \\ 1, & C_{\text{int}} \neq \emptyset, \end{cases} \qquad \beta = \begin{cases} 0, & C_{\text{ext}} = \emptyset \\ 1, & C_{\text{ext}} \neq \emptyset, \end{cases} \tag{2.9}$$

where the empty set symbol $\emptyset$ denotes missing context.

With this schema, we classify several classes of generalized proposition as follows.

Metaphysical Propositions
These are completely contextually incomplete GPs of the form

$$(\boldsymbol{P}, \emptyset|\emptyset, \emptyset), \tag{2.10}$$

where $\emptyset$ denotes a complete absence of any contextual information whatsoever. Metaphysical propositions cannot be validated, so the term *not even wrong* applies to them (Woit, 2006). A metaphysical proposition has a GPC of *zero*.

Mathematical Propositions
These are partially contextually complete GPs of the form

$$(\boldsymbol{P}, A|\emptyset, \emptyset), \tag{2.11}$$

where A is a set of axioms relative to which the truth status of proposition P can be established. A mathematical proposition has a GPC of 1.

Classical Propositions
These are partially contextually complete GPs of the form

$$(\boldsymbol{P}, \emptyset|O, C_{\text{ext}}), \tag{2.12}$$

where O is a primary observer described by relative external context C_{ext}, such as a statement of the rest frame of the observer. However, there is generally no mention of the apparatus used in the experiment, so a classical proposition has a GPC of 2. Such GPs are common throughout CM.

Complete Propositions
These are contextually complete GPs of the form

$$(\boldsymbol{P}, \mathcal{A}|O, C_{\text{ext}}), \tag{2.13}$$

where $\mathcal{A}$ is a specification of the *apparatus* that can be used to establish the relative truth status of the proposition $\mathcal{P}$. It is not enough to construct any apparatus; $\mathcal{A}$ has to be compatible with $\boldsymbol{P}$, allowing the truth value of $\boldsymbol{P}$ to be determined. It is not enough, either, to simply assert that such apparatus exists or could be built.

A complete proposition is a GP that has a GPC of 3.

QDN is designed to deal with complete propositions: relative internal context is explicitly modeled by quantum registers directly related to compatible apparatus, with an endophysical observer assumed to be operating in a well-defined laboratory running through a succession of temporal stages.

3

Classical Bits

3.1 Binary Questions

A *binary question* is a *yes or no* question. A typical binary question $\overline{\boldsymbol{Q}}$ will be of the form $\overline{\boldsymbol{Q}} \equiv$ **Is it true that?**, asked of some proposition $\boldsymbol{P}$ such as

$$\boldsymbol{P} \equiv \textbf{There is a signal in this detector}.$$

Assuming compatibility of question and proposition, the answer $\overline{\boldsymbol{Q}}\boldsymbol{P}$ will be either *one* or *zero*, interpreted as *yes* and *no*, respectively.

The *answer set* associated with any binary question has two elements, denoted 0 (zero) and 1 (one). Usually the element 0 will be contextually interpreted as *no* and element 1 will be interpreted as *yes*, but we could choose to interpret the answer set elements the other way around.

We will base most of this book on the mathematical structures associated with binary questions. Classical binary questions are discussed in this chapter and their quantum analogues discussed in Chapter 4.

Example 3.1 A trial in an English or American court of law can be thought of as a process that answers the binary question $\overline{\boldsymbol{Q}} \equiv$ **Is it true that?** of the proposition $\boldsymbol{P} \equiv$ **This person committed that crime.**

The *negation* $\overline{\neg \boldsymbol{Q}}$ of a binary question $\overline{\boldsymbol{Q}} \equiv$ **Is it true that?** is defined to be $\overline{\neg \boldsymbol{Q}} \equiv$ **Is it false that?**.

3.2 Question Cardinality

The English and American legal systems are based on the ancient Roman principle of *in dubio pro reo* (when in doubt, judge in favor of the accused). This presumption of innocence gives a binary, or dichotomic, flavor to the proceedings: assuming a verdict is reached, then verdicts in such courts can only be either *guilty*, corresponding to *yes*, or else *innocent*, corresponding to *no*.

In Scotland, however, there is a third possible verdict, known as *not proven*. This and other examples leads us to define the *question cardinality* $\#\overline{\boldsymbol{Q}}$ of a question $\overline{\boldsymbol{Q}}$ as the number of elements in its answer set, *before* that question is asked. According to this definition, a *rhetorical question* has cardinality 1, a binary question has cardinality 2, and a *Scottish law trial* answers a question with cardinality 3.

The cardinality of a question is contextual, in that it depends on the compatibility of a given question and a given proposition or state.

3.3 Classical Binary Questions

Binary questions (*yes/no* questions) are the fundamental building blocks of the quantized detector network (QDN) approach to physics discussed in this book. The reason has to do with the way experiments are conducted. Although modern experimentalists acquire vast amounts of data electronically and interpret them in terms of sophisticated theories, what goes on at the most basic level in any experiment is that a number (possibly vast) of binary questions are answered. For instance, the question of whether a photon detector has clicked or not is a binary question.

Another reason for choosing to work with binary questions is the principle that any empirical question, no matter how complicated, can always be expressed in terms of some number (possibly infinite) of binary questions. This will be referred to as the *bitification principle*. We shall return to this topic in Chapter 5 as it leads directly to the concepts of classical and quantum registers. These are the central mathematical constructs in terms of which the ideas of this book are expressed.

For the rest of this chapter we shall review some of the properties of single binary questions in classical mechanics. In that context, they are referred to as *classical bits*, terminology that comes from the theory and application of classical computation. When quantum rules are factored in, binary questions become *quantum bits*, or *qubits*. These are discussed in the next chapter. A fundamental difference (but not the only significant difference) between classical bits and qubits is that the former have cardinality 2, whereas the latter have infinite cardinality in a particular sense to do with the interpretation of quantum mechanics.

3.4 Classical Bits

We turn now to the mathematical representation of binary questions known as bits. A *classical bit*, or *bit* for short, is a set with two elements denoted **0** and **1**.

The word *bit* is short for *binary digit*. The first recorded use of the word in this context has been dated to 1947 in a Bell Labs memo written by John W. Tukey. It first appeared in public in 1948 in Claude E. Shannon's landmark paper on information theory (Shannon, 1948). Bits are used extensively in classical computation and information theory and are central to our implementation of quantized detector networks.

Bits represent the most elementary, useful form of data variable, one with only two possible, mutually exclusive values. But that is a mathematical statement, insufficient for our purposes. In the sciences, including the theory of computing, a bit is generally more than a mathematical set with two elements; bits generally have some associated context that gives some physical meaning or interpretation to each of the two possible values. For example, the two elements of a bit in two-valued logic might be thought of as *true* and *false*. With context, mathematical data becomes physical information.

3.5 Signal Bits

The sort of experiment we are mainly interested in this book typically involves one or more single-click detectors, each detector having two possible outcome states. These states are associated with a classical bit as follows. If a detector is found in its ground state, or no-click state, then that state is represented by the bit state **0**, whereas if that detector is found in its signal state, or click state, then that state is represented by the bit state **1**.

Single-click detectors should not be thought of as simple. Typical detectors such as Geiger counters involve cascade processes that are irreversible and complex. What is significant is that, like avalanches on mountainsides, they are triggered by the smallest of effects, and that is where their value lies.

A *signal bit* $B \equiv \{\mathbf{0}, \mathbf{1}; C_B\}$ is a set with two elements, denoted **0** and **1**, together with a context C_B that gives each element contextual physical significance relative to the observer involved. A *signal bit state*, or *bit state* for short, is any one of the two elements, **0** or **1**, of a signal bit.

Complexity rules the real world and experiments are vastly complex processes involving many degrees of freedom. Signal bits are best thought of as *equivalence classes* defined by a context with two clear alternatives. For example, in the standard quantum mechanics (QM) description of the Stern–Gerlach (SG) experiment, electrons with arbitrary momentum and position but with spin *down* are identified with bit state **0** while electrons with arbitrary momentum and position but with spin *up* are identified with bit state **1**. Typically, these equivalence class will not depend on the color of the observer's shirt or other factors regarded by the observer as inessential to the experiment. Such equivalence classes are defined by the observer's chosen criteria, which may appear subjective.

In the original SG experiment, the detecting screen was a photographic emulsion film, acting as a battery or register of signal detectors (Bernstein, 2010). At the start of the experiment, the film was prepared by the observers by blowing cigar smoke onto it. Then, after many silver atoms had passed through the main magnet of the SG device and impinged onto the screen, the observers noticed that there were two relatively crude but nevertheless separate spots where the bulk of the atoms had landed. It was the observers' decision to interpret these spots as of empirical significance. Stern and Gerlach did not know that their observed spots were a marker of electron spin. In 1922, electron spin was unknown; Stern and Gerlach were trying to validate Bohr's Old Quantum Mechanics theory of the atom. The electron spin interpretation came into focus a few years later.

Classical bits are sets, not vector spaces, because there is no obvious mathematical or physical meaning to the multiplication of a bit state by a real or complex number, or to the addition of two bit states. This is no longer the case when we generalize bits to their probabilistic counterparts, where they are known as *stochastic bits* (or *s-bits*) and to their quantum counterparts, where they are known as *quantum bits* (or *qubits*). Nevertheless, it is useful and convenient to represent bits via two-dimensional complex vector spaces. This allows us to represent bit states as vectors and bit operators as matrices, but it should be kept in mind that classical bit states cannot be added in principle. An exception is in Boolean algebra, where rules such as $1 + 1 = 0$ have a contextual meaning.

3.6 Nodes

QDN analyzes experiments in the simplest form possible, which is in terms of binary questions and answers. We will show in later chapters how any given apparatus is represented in QDN as a collection of binary questions and answers forming a *stage network*, a collection of *nodes* connected by links across which quantum information is transmitted.

Nodes come in two forms, *external* and *internal.*

External Nodes

External nodes correspond to physically existing equipment and come in two varieties: *sources* (preparation devices) and *detectors*. Observers input contextual information into stage networks via sources and extract signal information via detectors.

Internal Nodes

These occur in the *information void*, the region of space and time where no information is extracted. Internal nodes can be thought of as virtual detectors, as they usually correspond to places in a network where a real detector could have been placed, if the observer had so chosen.

Example 3.2 In the double-slit experiment, the two slits are identified with internal nodes, while the source of the incoming beam and the detecting screen are identified with external nodes.

Nodes are not necessarily localized in space. An example involving highly nonlocalized nodes would be an apparatus for the measurement of particle momentum.

While it will always be easy to identify external nodes in any stage diagram, that will not always be the case for the internal nodes. The rule for assigning internal nodes is that any such node represents a potential *opportunity* for information extraction, *if the observer so chooses.* So for instance in the double-slit experiment example considered above, not only could the observer place detectors at any of the two slits, but the observer could conceivably fill the space between source and screen with a vast number of internal nodes, if that was needed. In the limit of extremely large numbers, the QDN description would begin to look more like quantum field theory. The art in QDN is to find the simplest possible description of an experiment in terms of a limited number of nodes.

3.7 Dual Bits

In anticipation of subsequent developments, we introduce here the notion of a *dual bit*. For every bit $B \equiv \{\mathbf{0}, \mathbf{1}\}$ we postulate the existence of another bit, denoted $\overline{B} \equiv \{\overline{\mathbf{0}}, \overline{\mathbf{1}}\}$, referred to as the *dual bit*, or simply the dual, of bit B. The two elements of a dual bit will be referred to as *dual bit states*.

Bits and their duals are related as follows: $\overline{\mathbf{0}}$ is the dual of $\mathbf{0}$ and $\overline{\mathbf{1}}$ is the dual of $\mathbf{1}$. In anticipation of subsequent developments, we define a function $\overline{\boldsymbol{i}}\boldsymbol{j}$ from the Cartesian product $\overline{B} \times B$ into the set $Z \equiv \{0, 1\}$ by the rule

$$\overline{\boldsymbol{i}}\boldsymbol{j} \equiv \delta^{ij}, \qquad i, j = 0, 1, \tag{3.1}$$

where $\overline{\boldsymbol{i}}$ is an element of $\overline{B}$, $\boldsymbol{j}$ is an element of B, and δ^{ij} is the Kronecker delta.

This notation will be extended to *classical registers*, that is, collections of bits. If $B^k \equiv \{\mathbf{0}^k, \mathbf{1}^k\}$ is the kth bit in a register, then its dual $\overline{B^k}$ is given by $\overline{B^k} \equiv \{\overline{\mathbf{0}^k}, \overline{\mathbf{1}^k}\}$. Then rule (3.1) becomes $\overline{\boldsymbol{i}^k}\boldsymbol{j}^k \equiv \delta^{ij}$.

If $k \neq l$, then $\overline{\boldsymbol{i}^k}\boldsymbol{j}^l$ is undefined. The interpretation of this is that an observer cannot expect to extract any information from one detector by looking at any another detector. Exceptions can occur, provided the right context is in place, such as charge conservation.

3.8 The Interpretation of Bits and Their Duals

In the previous chapter we discussed the role of questions and answers in physics, and in this chapter started to link this to the notion of a detector. We now tie in these two ideas with bits and their duals as follows. The two bit states $\mathbf{0}$, $\mathbf{1}$ in a bit do not individually "have" absolute truth values per se: such truth values are only contextual, relative to the questions asked of the associated detector.

The two questions that could be asked of a detector are

$$\begin{aligned} \overline{\mathbf{0}} &\equiv \textbf{Is this the ground state?}, \\ \overline{\mathbf{1}} &\equiv \textbf{Is this the signal state?}. \end{aligned} \tag{3.2}$$

It should be now clear why we have chosen our bit state notation in the given form. In our notation, the question $\overline{\boldsymbol{i}}$ asked of the bit state $\boldsymbol{j}$ is written in the form $\bar{\boldsymbol{i}}\boldsymbol{j}$, and then the answer is given by δ^{ij}. If $\delta^{ij} = 0$ (the number *zero*), then that means that the answer is *no*, whereas if $\delta^{ij} = 1$ (the number *one*), then that means that the answer is *yes*. We shall refer to each side of expressions such as (3.1) as a *classical answer*.

3.9 Matrix Representation

We introduce here a convenient *matrix representation* of bits and their duals. The rule is that bit states are represented by two-component column matrices as follows:

$$\mathbf{0} \underset{R}{=} \begin{bmatrix} 1 \\ 0 \end{bmatrix}, \quad \mathbf{1} \underset{R}{=} \begin{bmatrix} 0 \\ 1 \end{bmatrix}, \tag{3.3}$$

where $\underset{R}{=}$ denotes "is represented by." For the dual bits, we have the row matrix representation

$$\bar{\mathbf{0}} \underset{R}{=} \begin{bmatrix} 1 & 0 \end{bmatrix}, \quad \bar{\mathbf{1}} \underset{R}{=} \begin{bmatrix} 0 & 1 \end{bmatrix}. \tag{3.4}$$

There is a small technical point concerning this representation that we clarify now. The classical answer $\bar{\boldsymbol{i}}\boldsymbol{j}$ is a number, either zero or one. However, according to the rules of matrix multiplication, the action of a two-dimensional row matrix on a two-dimensional column matrix is a 1×1 *matrix*, not a number. For instance,

$$\bar{\mathbf{0}}\mathbf{1} \underset{R}{=} \begin{bmatrix} 1 & 0 \end{bmatrix} \begin{bmatrix} 0 \\ 1 \end{bmatrix} = \begin{bmatrix} 0 \end{bmatrix} \neq 0. \tag{3.5}$$

We resolve this problem by interpreting the left-hand side of (3.5) as the *component* of the 1×1 matrix on the right-hand side. Henceforth we shall ignore this technical point.

3.10 Classical Bit Operators

The process of asking a binary question $\bar{\boldsymbol{i}}$ of a bit state $\boldsymbol{j}$ gives the answer $\bar{\boldsymbol{i}}\boldsymbol{j}$, which is a number (either zero or one). The ordering here is significant: the binary question $\bar{\boldsymbol{i}}$ is to the left and the bit state $\boldsymbol{j}$ is to the right. It turns out to be useful to define objects known as *transition bit operators*, which have the ordering interchanged, forming an object known as a *dyadic*. There are four such operators, defined as $\boldsymbol{T}^{ij} \equiv \boldsymbol{i}\bar{\boldsymbol{j}}$ for $i, j = 0, 1$. The application rules of these operators are as follows:

Action on Bit States

The action of $\boldsymbol{T}^{ij}$ on bit state $\boldsymbol{k}$ is from the left and is given by

$$\boldsymbol{T}^{ij}\boldsymbol{k} \equiv (\boldsymbol{i}\bar{\boldsymbol{j}})\boldsymbol{k} \equiv \boldsymbol{i}(\bar{\boldsymbol{j}}\boldsymbol{k}) = \delta^{jk}\boldsymbol{i}. \tag{3.6}$$

Action on Dual Bit States

The action of bit operator $\boldsymbol{T}^{ij}$ on dual bit $\overline{\boldsymbol{p}}$ is from the right and is given by

$$\overline{\boldsymbol{p}}\boldsymbol{T}^{ij} \equiv \overline{\boldsymbol{p}}(\boldsymbol{i}\overline{\boldsymbol{j}}) \equiv (\overline{\boldsymbol{p}}\boldsymbol{i})\overline{\boldsymbol{j}} = \delta^{pi}\overline{\boldsymbol{j}}. \tag{3.7}$$

The transition operators have the following matrix representations:

$$\boldsymbol{T}^{00} \underset{R}{=} \begin{bmatrix} 1 & 0 \\ 0 & 0 \end{bmatrix}, \quad \boldsymbol{T}^{01} \underset{R}{=} \begin{bmatrix} 0 & 1 \\ 0 & 0 \end{bmatrix}, \quad \boldsymbol{T}^{10} \underset{R}{=} \begin{bmatrix} 0 & 0 \\ 1 & 0 \end{bmatrix}, \quad \boldsymbol{T}^{11} \underset{R}{=} \begin{bmatrix} 0 & 0 \\ 0 & 1 \end{bmatrix}. \tag{3.8}$$

These four matrices form a basis for the four-dimensional vector space of complex 2×2 matrices, so we can use them to construct other useful matrices. There are four bit operators, labelled $\boldsymbol{I}$, $\boldsymbol{F}$, $\boldsymbol{D}$, and $\boldsymbol{U}$ here, that are occasionally useful. They are defined and represented as follows.

The Bit Identity Operator $\boldsymbol{I}$

This operator is defined as $\boldsymbol{I} \equiv \boldsymbol{T}^{00} + \boldsymbol{T}^{11}$. It leaves bit elements unchanged, i.e.,

$$\boldsymbol{I0} = \boldsymbol{0}, \quad \boldsymbol{I1} = \boldsymbol{1}. \tag{3.9}$$

The Bit Flip Operator $\boldsymbol{F}$

This operator is defined as $\boldsymbol{F} \equiv \boldsymbol{T}^{01} + \boldsymbol{T}^{01}$. It switches bit elements, i.e.,

$$\boldsymbol{F0} = \boldsymbol{1}, \quad \boldsymbol{F1} = \boldsymbol{0}. \tag{3.10}$$

In quantum computation, $\boldsymbol{F}$ is known as the *NOT gate* and denoted X (Nielsen and Chuang, 2000).

The Bit Down Operator $\boldsymbol{D}$

This operator is defined as $\boldsymbol{D} \equiv \boldsymbol{T}^{00} + \boldsymbol{T}^{01}$. It forces all bit states into the ground state $\boldsymbol{0}$, i.e.,

$$\boldsymbol{D0} = \boldsymbol{0}, \quad \boldsymbol{D1} = \boldsymbol{0}. \tag{3.11}$$

The Bit Up Operator $\boldsymbol{U}$

This operator is defined as $\boldsymbol{U} \equiv \boldsymbol{T}^{10} + \boldsymbol{T}^{11}$. It forces all bit states into the signal state $\boldsymbol{1}$, i.e.,

$$\boldsymbol{U0} = \boldsymbol{1}, \quad \boldsymbol{U1} = \boldsymbol{1}. \tag{3.12}$$

The four operators $\boldsymbol{I}$, $\boldsymbol{F}$, $\boldsymbol{U}$, and $\boldsymbol{D}$ will be used in bit state dynamics, and then it will be convenient to define $\boldsymbol{O}^1 \equiv \boldsymbol{I}$, $\boldsymbol{O}^2 \equiv \boldsymbol{F}$, $\boldsymbol{O}^3 \equiv \boldsymbol{D}$, and $\boldsymbol{O}^4 \equiv \boldsymbol{U}$.

3.11 Labstates

Our objective in this book is to interpret quantum mechanics via signal states of apparatus instead of states of systems under observation (SUOs). In order to keep this in mind, we shall use the term *labstate* whenever we refer to the former, reserving the term *system state* for the latter.

In this book, we shall deal with three kinds of labstate associated with a single detector: *classical labstates*, *stochastic labstates*, and *quantum labstates.* Classical and stochastic labstates are discussed in this chapter, quantum labstates are discussed in the next chapter. Each form of labstate has its own dynamical evolution rules, which we shall discuss separately in some detail. Because only one detector is involved in these preliminary discussions, such dynamics will be referred to as *rank one*. If two detectors were involved, then we would be discussing rank two dynamics, and so on.

Given a single detector, the observer would find it either in its signal ground labstate **0** or in its signal labstate **1**, assuming the detector existed, was not faulty, and that the observer actually looked.

3.12 Time and the Stages Concept

Before we can discuss the QDN approach to dynamics, we need some more precision in our modeling of the processes of observation, because detectors are distributed not only in space but also in time. Time is a necessary ingredient in our discussion. Once we start to incorporate that element into the discussion, we are led naturally to the *stage* concept that underpins QDN.

In QDN, time is defined relative to an observer and is generally measured in discrete steps called stages. This requires some explanation. It does not mean that we have dispensed with time as conventionally modeled in standard physics, that is, as a continuous real number–valued parameter via which velocities and other temporal derivatives are calculated.

Contrary to what is implied in conventional formulations of quantum mechanics, such as Schrödinger wave mechanics, the time in the laboratory required to complete any observation of a signal state is *always* nonzero. There are in fact no actual continuous time measurements possible in physics. Any references to continuous time observations, such as in *quantum Zeno* (also referred to as *nondemolition*) experiments (Itano et al., 1990), are to contextually incomplete mathematical approximations that often have great validity and usefulness, but only up to a point and under specific assumptions. Continuous time has much the same status in experimental physics as the concept of temperature: a useful and powerful emergent concept representing a great deal of contextuality, but otherwise not an objective thing in its own right.

In QDN, the concept of events in continuous spacetime is replaced by the concept of stage network.

Definition 3.3 A *stage network* is a conceptual collection of external and internal nodes representing apparatus distributed over time and space in a laboratory. Each node is connected by temporal links to other nodes or to *modules*. Modules represent processes between nodes that influence the transmission of classical or quantum information. Nodes are indexed by integers called *labtimes*. Each labtime is associated with a *stage*, the QDN analog of a hypersurface of simultaneity in relativity.

Quantum state preparation occurs at the source nodes and generally takes place over some contextually "small" or negligible interval of labtime (time as measured in the laboratory). States then evolve undisturbed over temporal links and are detected by the observer at the detectors, again over contextually "small" or negligible intervals of labtime.

The power of QM in general is that detailed modeling of what actually goes on at the nodes seems to be less significant than the detailed modeling of the transition amplitudes evolving over temporal links. For example, Feynman diagrams are used in relativistic quantum field theory to calculate those amplitudes, with virtually no modeling of the detection equipment that would be needed in practice.

This view of quantum processes was taken to an extreme with the development of the Multiverse paradigm (Deutsch, 1997). In that paradigm, only evolution of the wave function for the Multiverse in the information void is asserted to be significant. But because real, empirical information is extracted in the laboratory at nodes only, it should not come as a surprise that the Multiverse concept turns out to be empirically vacuous. So where does QDN stand in relation to these nodes and links?

QDN is an attempt to investigate the nodal aspect of quantum physics more than has been hitherto the case. As with scattering matrix (S-Matrix) theory (Eden et al., 1966), QDN provides a framework for discussing the spatiotemporal architecture of observation but does not provide the dynamical details of the amplitudes involved.

Our aim in this book is to discuss a multi-detector approach to quantum physics. By this we mean to discuss real experiments that involve perhaps many preparation channels; large numbers of modules such as beam splitters, mirrors, and suchlike; and batteries of outcome detectors. In such circumstances, we naturally find ourselves encountering the dictates of special and general relativity (GR). To date there has been no empirical evidence that the principles of relativity and of quantum mechanics are incompatible. There is in practice "*peaceful co-existence*" between GR and QM, and that has to be respected in QDN.

The grouping of nodes into sets called *stages* reflects *classical causality*, the notion that an event can influence some events dynamically but not others. In relativity, events outside each other's light cones cannot be causally related. The analogous concept in QDN is that nodes in the same stage cannot transmit or receive quantum information from each other.

There is a subtlety here, however, to do with *shielding*. This needs some explanation. Consider two detectors, A and B, that are, in relativistic parlance, time-like separated. This means that one of the detectors, say B, is inside the forward light cone of A, viewed from the conventional relativistic perspective. Therefore, by standard causal physics, B could in principle be affected dynamically by whatever was done at A. But suppose B was shielded in some way from any dynamical effects from A (such as being placed inside a Faraday cage, in the case of electromagnetic interactions). Then for all practical purposes, we could regard A and B as if they were dynamically independent. It would not then be inconsistent to assign them to the same QDN stage, even though they were not on any hypersurface of simultaneity in physical spacetime. On this basis the QDN definition of simultaneity is contextual.

Example 3.4 Consider an SG experiment where an electron passes through an inhomogeneous magnetic field and is expected to land on one of two possible sites on a screen. Provided there was no tampering with the screen after the electron had passed through the magnetic field, then the observer could take their time in looking at the two sites to see where the electron actually had landed. The observer could in fact look at one site immediately after the electron had passed though (that is, after it had been calculated to have passed through), and then look at the other site 20 years later. The acts of looking at the two sites would take place 20 years of real time apart, but provided the screen had not been tampered with over those 20 years, the two site examinations could legitimately be regarded as having taken place in the same stage.

In fact, all experiments are conducted in this way. Signals are registered irreversibly in detectors, and the observer generally looks at those memories usually much later.

The stage concept is designed to reflect the inherent certainty/uncertainty dichotomy in any experiment: an observer may be quite sure that a signal has been detected in a detector (simply because they looked and found a signal), but the actual exact laboratory time when the signal was triggered could be quite uncertain. Indeed, it is a vacuous concept to imagine that signals trigger instantaneously. How could that be proved? At best, approximate time intervals of triggering could be determined.

The stage concept is naturally tuned in to the notion of wave-function collapse, or state reduction. Nothing physical actually collapses when an observer looks at a detector and finds a positive signal there: it is true that there will be some quantifiable changes, both in the apparatus and in the observer's information store, but these changes are not manifestations of anything that happened in the information void, merely interpreted as evidence that something had happened.

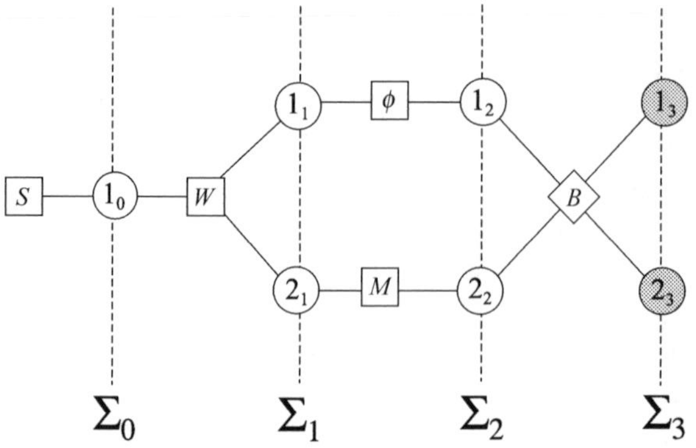

Figure 3.1. A typical stage diagram. Dotted lines represent individual stages, labeled by subscripts, the nth stage being denoted Σ_n. Circles represent nodes, where information either enters or leaves the network, or passes on to other nodes. Shaded circles represent actual outcome detectors. Boxes represent modules such as the source (S), a Wollaston prism (W), a phase-changer (ϕ), a mirror (M), and a beam splitter (B). Solid lines represent transmission in the information void between nodes and modules.

3.13 Stage Diagrams

All experiments have a spacetime architecture that can be represented diagrammatically. In QDN, we use *stage diagrams*. These are simplified diagrams showing the information flow between components of apparatus over the course of an experimental run. Figure 3.1 is a typical stage diagram. Numbered circles represent individual information gates or nodes, either real or virtual (explained later). Shaded circles represent real detectors, that is, nodes that the observer actually extracts signal information from. Boxes represent various modules, such as mirrors and beam splitters. Modules are discussed in Chapter 11. Dotted lines represent the stages, indexed by subscripts.

3.14 Measurements and Observations

To explain more fully the points we are making, we first need to pin down some of the terms we shall use to describe any experiment.

Intervention
An *intervention* is a single act of information extraction from a set of detectors at a single stage.

Run
A *run* is the complete process involved in a given sequence of actions, which starts with initial labstate preparation and ends with final labstate detection.

Experiment
An *experiment* consists of a given number of runs, each following an identical protocol, or experimental procedure.

Runtime
Runtime is time required to perform a given run, as measured by the observer's laboratory clocks.

Measurement
A *measurement* is the statistical result of data accumulated over one or more runs of a given experiment.

3.15 Transtemporal Identity

With the introduction of time, we come to an important question in the development of our formalism: does an detector have a *transtemporal identity* (an identity that persists over some interval of time), or is it something that exists only at a specific time?

Such a question goes to the heart of an ancient debate concerning the nature of reality. In *The Way of Truth*, a surviving fragment from a poem of the ancient Greek philosopher *Parmenides* (ca. 520–450 BCE), it was argued that change is impossible and that existence is timeless, uniform, and unchanging. Parmenides also suggested that the world of appearance (or by our interpretation, *observation*) is false and deceitful. He claimed that truth could not be known through sensory perception and that only pure reason could lead to a proper understanding of reality. In essence, this was an argument for not doing experimental physics.

Parmenides's ideas had an enduring effect on subsequent philosophy, physics, and mathematics. One of his pupils, *Zeno of Elea* (ca. 490–430 BCE) took Parmenides's line of reasoning further and created a number of paradoxes about motion, such as the race between the Tortoise and Achilles. Zeno's paradoxes could not be fully resolved until the modern mathematical understanding of the limit concept was developed. Parmenides also denied the existence of *nothingness*, or the *void*, which stimulated *Leucippus* to propose the existence of atoms.

In contrast to Parmenides, *Heraclitus* suggested that *everything flows, nothing stands still*, and that *change is the only constant.*

We can make sense of some of the ideas of Parmenides, Leucippus, and Heraclitus provided we avoid introducing absolute truths and base our discussion on contextuality, the proper basis for theories of observation of reality.

Any answer to the question of temporality determines the way in which dynamics is thought about and represented. If SUOs and/or detectors have enduring temporal identities, then it is reasonable to imagine that they "evolve" over time in fixed spaces, changing perhaps their states but retaining sufficient attributes to justify giving them their particular identities. This is related to the phenomenon of *persistence*, which depends on the time scales over which objects can be taken reliably to have significance.

Such an approach is taken in conventional classical mechanics (CM), where SUOs are represented by points moving about phase space, and in QM, where state vectors move about in Hilbert space.

The alternative to this is to imagine, like Heraclitus, that everything changes, nothing persists. According to this view, objects such as SUOs and apparatus only appear to persist because sufficient patterns of mass and energy are repeated sufficiently unchanged over certain time scales as judged by some observer, and it is this that gives that observer the impression that "objects" exist in the universe. We shall adopt the Heraclitian point of view, because we need to consider the possibility that observers and apparatus can be created and destroyed. This will mean changing, over time, the dimensions of the mathematical spaces used to model processes of observation.

Therefore, when we discuss a rank-one labstate evolving from stage Σ_M to stage Σ_N, we shall think of it as a succession of detectors, denoted, say, by $\Delta_M, \Delta_{M+1}, \dots, \Delta_N$. Each detector Δ_n is associated with a classical bit B_n and its dual $\overline{B}_n$, and that detector exists only at stage Σ_n. But if persistence is assumed, then the whole set $\{\Delta_i : i = M, M+1, \dots, N\}$ of detectors may be thought of as a single detector with an enduring, transtemporal identity existing over the time interval $[t_M, t_N]$.

It is important not to mix different theoretical spaces associated with different times. While we can ask the question $\overline{\boldsymbol{i}_n}\boldsymbol{j}_n$ for $M \leq n \leq N$, we are not allowed to ask the question $\overline{\boldsymbol{i}_n}\boldsymbol{j}_m$ for $n \neq m$. The reason is obvious when stated in words: *we cannot observe today what does not yet exist or what used to exist.* All actual observations are done in process time. Past and future are inferred from the data so acquired: archaeologists do not dig up the past – they dig up traces of the past embedded in the present. Improper questions such as $\overline{\boldsymbol{i}_n}\boldsymbol{j}_m$, $n \neq m$, are not defined mathematically either, in the same way that dual vectors (one-forms) are defined only by their action on their individual, associated vector space.

3.16 Typical Experiments

A typical rank-one experiment of the classical type starts with a definite initial labstate $\boldsymbol{j}_M$ at initial stage Σ_M. Usually we shall take $M = 0$, but this is not essential. We shall not ask how this initial labstate was created but, in this section, require it to be definite; that is, there is no element of probability here other than certainty. There are therefore only two possible initial labstates in any such experiment: $\boldsymbol{j}_M$ can take the value 0 or the value 1.

Next, we imagine that the observer has no further interaction with "the" detector until stage Σ_{M+1}, but by this time, something will have acted on it to possibly change its signal state. We shall represent this by the action of one of the classical bit operators $\boldsymbol{O}^k$, $k = 1, 2, 3, 4$, defined in Section 3.10, i.e., we suppose the transition

$$\boldsymbol{i}_M \to \boldsymbol{i}_{M+1} \equiv \boldsymbol{O}^{k_M}_{M+1,M}\boldsymbol{i}_M, \tag{3.13}$$

where $\boldsymbol{O}^{k_M}_{M+1,M}$ is given by

$$\boldsymbol{O}^{k_M}_{M+1,M} \equiv \sum_{a=0}^{1}\sum_{b=0}^{1} \boldsymbol{a}_{M+1} O^{k_M}_{ab} \overline{\boldsymbol{b}_M} \tag{3.14}$$

and $O^{k_M}_{ab}$ are the ab components of the corresponding classical bit matrix. This operator takes us from the two-dimensional vector space $\mathcal{Q}_M$ in which we have embedded $\mathbf{0}_M$ and $\mathbf{1}_M$ into the two-dimensional vector space $\mathcal{Q}_{M+1}$ in which we have embedded $\mathbf{0}_{M+1}$ and $\mathbf{1}_{M+1}$.

This process can be continued. Specifically, for any time n such that $M \leq n < N$ we may write

$$\boldsymbol{i}_n \to \boldsymbol{i}_{n+1} \equiv \boldsymbol{O}^{k_n}_{n+1,n}\boldsymbol{i}_n, \tag{3.15}$$

from which we find

$$\boldsymbol{i}_M \to \boldsymbol{i}_N \equiv \boldsymbol{O}_{N,M}\boldsymbol{i}_M, \tag{3.16}$$

where the complete evolution operator $\boldsymbol{O}_{N,M}$ is given by

$$\begin{aligned} \boldsymbol{O}_{N,M} &\equiv \boldsymbol{O}_{N,N-1}\boldsymbol{O}_{N-1,N-2}\ldots\boldsymbol{O}_{M+1,M} \\ &= \sum_{a=0}^{1}\sum_{b=0}^{1} \boldsymbol{a}_N \left[\boldsymbol{O}^{k_{N-1}}\boldsymbol{O}^{k_{N-2}}\ldots\boldsymbol{O}^{k_M}\right]_{ab} \overline{\boldsymbol{b}_M}. \end{aligned} \tag{3.17}$$

If now at time N, the observer stepped in and looked at the detector, they would normally ask the question $\overline{\mathbf{1}_N} \equiv$ **Is there a signal here?** The answer is given by $\overline{\mathbf{1}_N}\boldsymbol{i}_N$, where a value one represents an answer *yes* while a value zero represents an answer *no*.

Each set of integers $\{k_M, k_{M+1}, \ldots, k_{N-1}\}$ in (3.17) represents a specific "operator chain" of labstate dynamical changes. Each of the integers k_n can be 1, 2, 3, or 4, so there is in total 4^{N-M} different operator chains. However, because the multiplication of the operator matrices is closed, the net result is that there are only four possible overall complete evolution operators, given by

$$\boldsymbol{O}^k_{N,M} \equiv \sum_{a=0}^{1}\sum_{b=0}^{1} \boldsymbol{a}_N O^k_{ab} \overline{\boldsymbol{b}_M}, \quad k = 1, 2, 3, 4. \tag{3.18}$$

This form of dynamics is relatively simple but there is a surprising aspect to it: it is entirely deterministic, in that a given present labstate unambiguously determines the labstate at any time in the future. However, if any one or more of the $N - M$ matrices in (3.17) is of $\boldsymbol{D}$ or $\boldsymbol{U}$ type, then the complete dynamics is irreversible. What this means is that even if the observer knew the final labstate and every detail of the evolution operator from initial to final times, but not the initial labstate, they could not say for sure what that initial state was. Retrodiction is therefore impossible if a $\boldsymbol{D}$ or $\boldsymbol{U}$ transition occurs even once.

The mathematical reason why the occurrence of just a single $\boldsymbol{D}$ or $\boldsymbol{U}$ transition generates irreversibility is that each of these maps is two-to-one, and so their matrix representations are singular.

3.17 Rank-One Stochastic Evolution

In the real world, classical certainty is an exception and we have to use concepts of probability to discuss most situations. In this section we extend the ideas of the previous section to incorporate this requirement.

There are two competing philosophies or schools of thought about probability: the *Frequentist school* and the *Bayesian school.* These are in principle quite different in their core philosophies of what probability means, although some convergence of thinking appears to be taking place among the experts, and the differences are at times too subtle to be of much significance to us here.

There is however one clear difference, analogous to the difference between the CM noncontextual view of reality and the contextual QM view. Frequentists talk about probabilities as if they were intrinsic to the events taking place, such as a fair coin "having" a probability of $\frac{1}{2}$ landing on a head. Bayesians require a context to be supplied before they presume to make such an assertion. In an absence of such context, *Bayesians* will make a *prior*, or educated starting guess, such as $\frac{1}{2}$ for the probability p of a head. A Bayesian would then throw the coin a few times and make some outcome observations. With this new information, the Bayesian would make an updated estimate of p based on a well-known formula attributed to Bayes.

For Bayesians, probabilities are contextual. For example, a Bayesian who had thrown a coin 10 times and not observed a single tail would calculate the probability of the eleventh throw landing on a tail to be much less than $\frac{1}{2}$. This is because the 10 observations (i.e., the 10 runs of the basic experiment) had provided new information about the SUO (the coin) that could not be discounted. The specific details of any calculation as to the likely outcome of the eleventh throw would depend on the sort of assumptions made, such as whether successive throws were truly independent. We shall not discuss those details further here. Suffice it to say that *prior information*, or as we would put it, the *context* of the eleventh throw, would have significant bearing on the probability calculation. Also, *how* the coin was thrown would have an important bearing on the probability outcome calculations and this is part of the context as well.

The pragmatic view of a person unfamiliar with the rules of probability would be that the observation of ten successive heads is reasonably convincing evidence for the hypothesis that the coin is in fact double-headed. On that basis, the probability of getting a tail on the eleventh throw would be zero. A more careful analysis based on particular "reasonable" assumptions about the prior gives a probability of about 0.95 of getting a head on the eleventh throw.

We see here a similarity between our approach to observation and the Bayesian approach to probability. In both cases, prior information or context is crucial to the predictions. Our first principle of observation is that all truths in physics are contextual. The equivalent principle in Bayesian statistics is that all probabilities are conditional.

Probability in process physics has a different flavor compared with probability in block world physics. In block world physics, probability has to be discussed in terms of limits of ratios of large numbers of outcomes, presumably counted by some unspecified exophysical observer over some number of runs, or repetitions, that were embedded in the Block Universe. This is the Frequentist approach to probability and is typical of the way probability is discussed in standard QM.

In *Process Time*, however, we may find ourselves in a situation where we have no more than one opportunity to throw a coin. Real life is usually like that and it is that aspect of physics that we are trying to develop. Under such circumstances, the term *propensity* may be used rather than probability. Propensity is a gambler's view of probability as opposed to an accountant's view. We shall use the term "probability" to represent both kinds of concept.

Randomness and uncertainty enter into our discussion in two ways: the observer may be uncertain as to the initial labstate of the detector, and also be uncertain as to which dynamical operators are acting. We need to discuss both aspects. We consider first random labstates.

3.18 Stochastic Bits

Previously, we represented the ground state $\mathbf{0}$ and signal state $\mathbf{1}$ by column vectors, as in (3.3). Now suppose the observer was unsure as to which initial labstate they had started with, to the extent that they could only assign a probability of p for it to be in the ground state $\mathbf{0}$ and a probability $q \equiv 1 - p$ to be in its signal state $\mathbf{1}$. We shall represent such an uncertain labstate by

$$\boldsymbol{\Psi} = p\mathbf{0} + q\mathbf{1} \underset{R}{=} \begin{bmatrix} p \\ q \end{bmatrix}. \tag{3.19}$$

Now suppose that the observer wanted to know if there was a signal in the detector. They need to ask the question

$$\overline{\mathbf{1}} \equiv \textbf{Is the detector in its signal state?} \tag{3.20}$$

In our formalism, the answer is given by

$$\overline{\mathbf{1}}\boldsymbol{\Psi} \underset{R}{=} \begin{bmatrix} 0 & 1 \end{bmatrix} \begin{bmatrix} p \\ q \end{bmatrix} = q, \tag{3.21}$$

where we interpret the right-hand side as a number. In this case, the answer is not zero or one as in the classical case but interpreted as the probability of there being a signal in the detector. Likewise we find $\overline{\mathbf{0}}\boldsymbol{\Psi} = p$, which is the probability that the detector would be found in its ground state.

The sum of the elements of the column vector (3.19) is unity. Such a vector will be called a *stochastic vector* and the associated bit will be called a *stochastic bit*, or s-bit. For such bits, we shall refer to expressions such as $\overline{\boldsymbol{i}}\boldsymbol{\Psi}$ as *stochastic answers*, as they are generally neither zero nor unity, and are interpreted as probabilities.

3.19 Left-Stochastic Matrices

We turn now to the other possibility where randomness may occur: the dynamics affecting an detector may be random. We suppose now that the observer is uncertain as to which of the four possible evolutions $\boldsymbol{I}$, $\boldsymbol{F}$, $\boldsymbol{D}$, or $\boldsymbol{U}$ actually has occurred. Consider the operator $\boldsymbol{S}$ defined by

$$\boldsymbol{S} \equiv p\boldsymbol{I} + q\boldsymbol{F} + r\boldsymbol{D} + s\boldsymbol{U}, \tag{3.22}$$

where p, q, r, and s are probabilities summing to unity. With the matrices as defined previously, we find

$$\boldsymbol{S} \underset{R}{=} \begin{bmatrix} a & b \\ 1-a & 1-b \end{bmatrix}, \tag{3.23}$$

where $a \equiv p + r$ and $b \equiv q + r$ are in the interval $[0, 1]$.

Such a matrix is a *left-stochastic matrix*, i.e., one that has the property that each element lies in the interval $[0, 1]$ and the sum of elements in each column is unity.

3.20 Stochastic Jumps

We now consider stochastic labstates jumping under the influence of left-stochastic operators, from stage to stage. The labstate at stage Σ_n is given by

$$\boldsymbol{\Psi}_n = p_n^0 \mathbf{0}_n + p_n^1 \mathbf{1}_n = \sum_{i=0}^{1} p_n^i \boldsymbol{i}_n, \tag{3.24}$$

where $0 \leqslant p_n^i \leqslant 1$ and $\sum_{i=0}^{1} p_n^i = 1$. This state evolves to $\boldsymbol{\Psi}_{n+1} \equiv \boldsymbol{S}_{n+1,n}\boldsymbol{\Psi}_n$, where $\boldsymbol{S}_{n+1,n}$ is a left-stochastic operator given by

$$\boldsymbol{S}_{n+1,n} \equiv \sum_{i=0}^{1}\sum_{j=0}^{1} \boldsymbol{i}_{n+1} S_n^{ij} \overline{\boldsymbol{j}}_n, \tag{3.25}$$

where S_n^{ij} are the components of the left-stochastic matrix S_n, where

$$S_n \equiv \begin{bmatrix} a_n & b_n \\ 1-a_n & 1-b_n \end{bmatrix}. \tag{3.26}$$

Then $\boldsymbol{\Psi}_{n+1}$ is also a stochastic labstate. The determinant $|S_n|$ of the stochastic matrix S_n in (3.26) is given by $|S_n| = a_n - b_n$, which means that stochastic evolution is irreversible for $a_n = b_n$.

Exercise 3.5 Prove that the product of any two left-stochastic matrices is also a left-stochastic matrix. Prove also that the inverse of a nonsingular left-stochastic matrix is also a left-stochastic matrix.

3.21 Stochastic Questions

In much the same way that we can ask a definite binary question about a stochastic bit state, we may consider the possibility of asking *stochastic binary questions*. By this we mean the following.

Suppose an observer has prepared a stochastic state $\boldsymbol{\Psi}$ of a system under observation and could ask two different binary questions $\overline{\boldsymbol{Q}^1}$ and $\overline{\boldsymbol{Q}^2}$ in principle. Suppose that for some reason outside their control, every time the observer asked a question, there was a probability p that it was actually $\overline{\boldsymbol{Q}^1}$ being asked and not $\overline{\boldsymbol{Q}^2}$, and a probability $q \equiv 1 - p$ that it was actually $\overline{\boldsymbol{Q}^2}$ being asked and not $\overline{\boldsymbol{Q}^1}$. Suppose further that the observer asked a large number of such questions, not knowing precisely which question was being asked each time, but observing the answer each time. Then the average of the observed answers would be given by $p\overline{\boldsymbol{Q}^1}\boldsymbol{\Psi} + q\overline{\boldsymbol{Q}^2}\boldsymbol{\Psi}$, which we could use to define the stochastic binary question $\overline{p\boldsymbol{Q}^1 + q\boldsymbol{Q}^2}$.

4
Quantum Bits

4.1 Quantum Bits

We stressed previously that although classical bits (or equivalently, bits) are not vectors, embedding them into a two-dimensional vector space $\mathcal{Q}$ gives several advantages. The first advantage is that we can represent bit operators by either dyadics or matrices, which allows an efficient encoding of bit dynamics in familiar linear algebraic terms. The second advantage is that it allows us to generalize bits to stochastic bits directly and efficiently.

A third advantage that is of fundamental importance in quantized detector networks (QDN) and which we explore in this chapter is that we can generalize bits to their quantum counterparts, known as *quantum bits* (or equivalently, *qubits*).

A qubit is a complex two-dimensional Hilbert space, denoted $\mathcal{Q}$. That is a mathematical statement, but we need more. A *signal qubit* $(\mathcal{Q}, C)$ is a qubit $\mathcal{Q}$ with an empirical context C that defines a preferred basis $B \equiv \{\mathbf{0}, \mathbf{1}\}$ for that particular Hilbert space $\mathcal{Q}$.

Normally, we shall denote a signal qubit $(\mathcal{Q}, C)$ by $\mathcal{Q}$ whenever the context C is understood and kept in mind. It is implicitly assumed that all of this discussion is relative to some observer conducting experiments in a real laboratory.

4.2 Preferred Bases

Viewed in the right empirical context, a classical bit can be identified with a preferred basis for a given signal qubit, as follows. First, we shall use the same notation $\mathbf{0}$, $\mathbf{1}$ for the two elements of a given classical bit and for their embedding in the associated qubit $\mathcal{Q}$. In that embedding, we shall take the vectors $\mathbf{0}$ and $\mathbf{1}$ in $\mathcal{Q}$ to satisfy the "inner product" rule

$$\bar{\boldsymbol{i}}\boldsymbol{j} = \delta^{ij}, \qquad i, j = 0, 1, \tag{4.1}$$

where $\overline{\mathbf{0}}$ and $\overline{\mathbf{1}}$ are the duals of $\mathbf{0}$ and $\mathbf{1}$, respectively, and δ^{ij} is the Kronecker delta. Then the relationship between the bit $B \equiv \{\mathbf{0}, \mathbf{1}\}$ and its quantum counterpart $\mathcal{Q}$ is this: the elements of B are identified as *the* natural orthonormal basis for $\mathcal{Q}$, known as the *preferred basis*.

The existence of the preferred basis is fundamental to our approach to physics and is in no way controversial. We do not claim that there is a unique or absolute preferred frame in the Universe. Any preferred frame is preferred only by virtue of the relative context associated with a given observer. By definition of what is meant by "*observer*," each observer is assumed always to know the empirical context associated with each signal qubit that they use, being nothing other than a mathematical representation of some detector in the observer's laboratory. We can be confident that there is always going to be such a local preferred basis, because the idea that an observer could extract real information in an experiment with no knowledge about their apparatus makes no sense whatsoever.

By definition, each detector has only two possible outcome states, known as *ground state* and *signal state*. This contextually defines the preferred basis: $\mathbf{0}$ represents the ground (no signal) state of the apparatus and $\mathbf{1}$ represents the signal state of the apparatus.

We emphasize the following point. There runs throughout QM a strand of thinking, conditioned by experience with CM, that states of SUO have some sort of existence of their own. According to this logic, such states do not need any preferred bases for their mathematical representation. This line of thinking then leads to the notion that observers are not needed either.

It is along such realist lines of thinking that Hidden Variables theory (Bohm, 1952), decoherence (in its original form) (Joos, 2012), and the Multiverse (Many Worlds) (Deutsch, 1997) are based. The problem with those interpretations of QM is that they are each contextually incomplete with a generalized proposition classification[1] of zero, meaning that those theories are empirically vacuous.

4.3 Qubit Properties

The power and mystery of QM stems from the possibility of creating states in the laboratory that are represented by linear superpositions of preferred basis states. Given the preferred basis, then any pure signal qubit state $\boldsymbol{\Psi}$ can be written in the form

$$\boldsymbol{\Psi} = \alpha\mathbf{0} + \beta\mathbf{1}, \tag{4.2}$$

where α and β are complex numbers and $\mathbf{0}$, $\mathbf{1}$ are the two elements of the preferred basis.

The fact that α and β are complex and not real is of fundamental significance here. Recall that in our discussion of stochastic bits in Chapter 3, the components

[1] Generalized propositions and their classification is discussed in Section 2.12.

of a stochastic bit state such as (3.19) are nonnegative real numbers representing conditional probabilities. Because the components of a quantum bit state are not even real, let alone nonnegative, we encounter here the first of several issues in the interpretation of QM. This particular issue is generally regarded as being resolved by Born's interpretation of the above complex components as probability amplitudes (Born, 1926), discussed in Section 4.6.

If we choose to use the matrix representation of bits (3.3), then we may write

$$\boldsymbol{\Psi} \underset{R}{=} \begin{bmatrix} \alpha \\ \beta \end{bmatrix}. \tag{4.3}$$

Given qubit state (4.2), we define its dual $\overline{\boldsymbol{\Psi}}$ by

$$\overline{\boldsymbol{\Psi}} \equiv \overline{\alpha\mathbf{0} + \beta\mathbf{1}} = \alpha^*\overline{\mathbf{0}} + \beta^*\overline{\mathbf{1}}, \tag{4.4}$$

where $\alpha^*$and β^* are the complex conjugates of α and β, respectively. Then the "inner product" $\overline{\boldsymbol{\Psi}}\boldsymbol{\Psi}$ is given, using linearity and (4.1), by

$$\begin{aligned} \overline{\boldsymbol{\Psi}}\boldsymbol{\Psi} &= (\overline{\alpha\mathbf{0} + \beta\mathbf{1}})(\alpha\mathbf{0} + \beta\mathbf{1}) = (\alpha^*\overline{\mathbf{0}} + \beta^*\overline{\mathbf{1}})(\alpha\mathbf{0} + \beta\mathbf{1}) \\ &= \alpha^*\alpha \underbrace{\overline{\mathbf{0}}\mathbf{0}}_{1} + \alpha^*\beta \underbrace{\overline{\mathbf{0}}\mathbf{1}}_{0} + \beta^*\alpha \underbrace{\overline{\mathbf{1}}\mathbf{0}}_{0} + \beta^*\beta \underbrace{\overline{\mathbf{1}}\mathbf{1}}_{1} = |\alpha|^2 + |\beta|^2. \end{aligned} \tag{4.5}$$

A *normalized* signal qubit state is one for which $\overline{\boldsymbol{\Psi}}\boldsymbol{\Psi} = 1$, that is, $|\alpha|^2 + |\beta|^2 = 1$. We shall deal extensively with normalized signal qubit states, as these are associated with probability conservation in QM.

Normalized signal qubit states have been defined in terms of their components relative to the preferred basis. However, we can discuss them more generally as qubit states, that is, drop the observational context and think of them as just elements of some qubit. A qubit is a Hilbert space, a concept that is independent of basis and therefore does not require a preferred basis. We can discuss qubits in this context in a more abstract way as follows.

Given any element Φ of a Hilbert space, then by definition it has a *norm* $\|\Phi\|$ or length given by

$$\|\Phi\| \equiv \sqrt{(\Phi, \Phi)}, \ \geq 0, \tag{4.6}$$

where (Φ, Φ) is the inner product of Φ with itself. This length is basis independent.

A normalized qubit state therefore is an element of a complex two-dimensional Hilbert space and has unit norm.

4.4 Qubit Operators

We saw in the section in Chapter 3 that there are only four bit operators, denoted $\boldsymbol{I}$, $\boldsymbol{F}$, $\boldsymbol{D}$, and $\boldsymbol{U}$, that map bit states to bit states. In contrast, there is an infinite number of qubit operators that map qubit states to qubit states. simply because qubits are vector spaces.

We define a *qubit map* to be any map from a qubit into itself. Specifically, given such a map M, then for any element $\boldsymbol{\Psi}$ of qubit $\mathcal{Q}$, the object $M(\boldsymbol{\Psi})$ is some element of $\mathcal{Q}$. For example, the *identity map* I satisfies the rule $I(\boldsymbol{\Psi}) = \boldsymbol{\Psi}$ for any element $\boldsymbol{\Psi}$ of $\mathcal{Q}$.

This definition of qubit map makes no reference to linearity, so a qubit map need not be linear, as in the following example.

Example 4.1 Given a qubit $\mathcal{Q}$ with preferred basis $\{\mathbf{0}, \mathbf{1}\}$, define the qubit map M by $M(\boldsymbol{\Psi}) = \mathbf{0}$ for every element $\boldsymbol{\Psi}$ of $\mathcal{Q}$. Then M is a nonlinear map, as, for example, we have $M(\mathbf{0} + \mathbf{1}) = \mathbf{0}$ but $M(\mathbf{0}) + M(\mathbf{1}) = \mathbf{0} + \mathbf{0} = 2\mathbf{0}$.

The map in the above example is not unphysical. It has the interpretation of a *resetting* or preparation process that prepares a detector to be in its ground state, ready to receive a signal, regardless of the state it is currently in. We shall use such a process in later chapters.

Because we are concerned in this book with quantum processes, almost all of the qubit maps we shall deal with will be linear. Linear qubit maps will be called *qubit operators*. It is conventional in the case of linear operators to drop the round brackets of the argument; that is, we shall write $\boldsymbol{O\Psi}$ to mean $\boldsymbol{O}(\boldsymbol{\Psi})$ whenever $\boldsymbol{O}$ is a linear operator. Then for linear operator $\boldsymbol{O}$, for any elements $\boldsymbol{\Psi}, \boldsymbol{\Phi}$ of $\mathcal{Q}$, and for any complex numbers α, β, we have the rule

$$\boldsymbol{O}\{\alpha\boldsymbol{\Psi} + \beta\boldsymbol{\Phi}\} = \alpha\boldsymbol{O\Psi} + \beta\boldsymbol{O\Phi}. \tag{4.7}$$

In QM, linear operators are often associated with dynamical variables and so additional mathematical structure is introduced. Given two qubit operators $\boldsymbol{O}_1$ and $\boldsymbol{O}_2$ over a qubit $\mathcal{Q}$, we define the linear combination $\alpha\boldsymbol{O}_1 + \beta\boldsymbol{O}_2$ of these two operators to satisfy the rule

$$(\alpha\boldsymbol{O}_1 + \beta\boldsymbol{O}_2)\boldsymbol{\Psi} \equiv (\alpha\boldsymbol{O}_1\boldsymbol{\Psi}) + (\beta\boldsymbol{O}_2\boldsymbol{\Psi}), \tag{4.8}$$

for arbitrary complex numbers α, β, and arbitrary elements $\boldsymbol{\Psi}$ of $\mathcal{Q}$.[2] Then we state without proof that the set $L(\mathcal{Q})$ of all qubit operators over a qubit $\mathcal{Q}$ has all the properties of a four-dimensional complex Hilbert space (Paris, 2012).

The structure of the vector space $L(\mathcal{Q})$ is relatively easily explored. We saw in the previous chapter that given the preferred basis $\{\mathbf{0}, \mathbf{1}\}$ for $\mathcal{Q}$, the four elementary transition operators (ETOs) $\boldsymbol{T}^{ij} \equiv \boldsymbol{i}\overline{\boldsymbol{j}} : i, j = 0, 1$, form a convenient basis for $L(\mathcal{Q})$, that is, any qubit operator $\boldsymbol{O}$ can be written in the form

$$\boldsymbol{O} = \sum_{i=0}^{1}\sum_{j=0}^{1} O^{ij}\boldsymbol{T}^{ij}, \tag{4.9}$$

where the coefficients O^{ij} are complex.

[2] Note that the + symbol on the left-hand side of (4.8) denotes operator addition, while the + symbol on the right-hand side denotes vector addition.

In addition to their vectorial additive properties, it is useful to define multiplication of qubit operators. We define the multiplicative product $\boldsymbol{O}_1\boldsymbol{O}_2$ of two qubit operators $\boldsymbol{O}_1$, $\boldsymbol{O}_2$ by the rule $(\boldsymbol{O}_1\boldsymbol{O}_2)\boldsymbol{\Psi} \equiv \boldsymbol{O}_1\{\boldsymbol{O}_2\boldsymbol{\Psi}\}$ for any element $\boldsymbol{\Psi}$ of $\mathcal{Q}$. The product of any two ETOs is also an ETO, that is, ETO multiplication is closed. Specifically, the multiplication rule is

$$\boldsymbol{T}^{ij}\boldsymbol{T}^{kl} = \delta^{jk}\boldsymbol{T}^{il}. \tag{4.10}$$

This ETO multiplication rule (4.10) is associative, that is to say, that

$$(\boldsymbol{T}^{ab}\boldsymbol{T}^{cd})\boldsymbol{T}^{ef} = \boldsymbol{T}^{ab}(\boldsymbol{T}^{cd}\boldsymbol{T}^{ef}), \tag{4.11}$$

but not commutative, which means $\boldsymbol{T}^{ab}\boldsymbol{T}^{cd} \neq \boldsymbol{T}^{cd}\boldsymbol{T}^{ab}$ in general.

These properties and the existence of the identity operator $\boldsymbol{I}$ mean that qubit operators form a mathematical structure known as a *unital associative algebra*.

4.5 Signal Bit Operators

The space of operators $O(\mathcal{Q})$ over qubit $\mathcal{Q}$ contains infinitely many elements. Fortunately, the necessary existence, in our approach, of the preferred signal basis $\{\boldsymbol{0}, \boldsymbol{1}\}$ singles out a very small number of special qubit operators that we shall use extensively and refer to as *signal bit operators*. In addition to the identity operator $\boldsymbol{I}$ and the zero operator $\boldsymbol{Z}$ (it maps any vector into the zero vector) there are four important signal bit operators, defined as follows.

The Projection Operators

The qubit projection operators $\boldsymbol{P}$, $\widehat{\boldsymbol{P}}$ are defined by

$$\boldsymbol{P} \equiv \boldsymbol{T}^{00} = \boldsymbol{0}\overline{\boldsymbol{0}}, \quad \widehat{\boldsymbol{P}} \equiv \boldsymbol{T}^{11} = \boldsymbol{1}\overline{\boldsymbol{1}}. \tag{4.12}$$

The Signal (Annihilation and Creation) Operators

The signal operators $\boldsymbol{A}$, $\widehat{\boldsymbol{A}}$ are in conventional parlance adjoints of each other and are defined by

$$\boldsymbol{A} \equiv \boldsymbol{T}^{01} = \boldsymbol{0}\overline{\boldsymbol{1}}, \quad \widehat{\boldsymbol{A}} \equiv \boldsymbol{T}^{10} = \boldsymbol{1}\overline{\boldsymbol{0}}. \tag{4.13}$$

The four operators $\boldsymbol{P}, \widehat{\boldsymbol{P}}$, $\boldsymbol{A}$, and $\widehat{\boldsymbol{A}}$ are by inspection just the four ETOs $\boldsymbol{T}^{ij}$ introduced earlier. The advantage in this new designation is mainly psychological: the projection operators play one role in our formalism while the signal operators play another, and it is very helpful to distinguish between them.

In this new notation, the multiplication rule (4.10) is best expressed in the form of a table, Table 4.1, where the entries are the products $\boldsymbol{LR}$, operator $\boldsymbol{L}$ coming from the left-most column and $\boldsymbol{R}$ coming from the top-most row.

4.6 The Standard Born Interpretation

Although superficially qubits look similar to s-bits mathematically, being representable by two component column matrices, they are very different objects as

Table 4.1 *Products of signal bit operators*

L\R	P	$\widehat{P}$	A	$\widehat{A}$
P	P	Z	A	Z
$\widehat{P}$	Z	$\widehat{P}$	Z	$\widehat{A}$
A	Z	A	Z	P
$\widehat{A}$	$\widehat{A}$	Z	$\widehat{P}$	Z

far as physics is concerned. The components of an s-bit are real and interpreted as conditional probabilities, whereas the components of a qubit relative to its preferred basis are complex amplitudes and so cannot be probabilities. It was Max Born who gave the empirically correct interpretation of such complex amplitudes (Born, 1926), as follows.

The Born Interpretation

In QM, suppose $\mathcal{H}$ is a Hilbert space with inner product of elements ψ, ϕ denoted by (ψ, ϕ). Consider two normalized elements Ψ and Φ in $\mathcal{H}$ that represent physical states of some system under observation. Then the conditional probability $Pr(\Psi|\Phi)$ of finding the system to be in state Ψ, given that the system was prepared to be in state Φ, is given by

$$Pr(\Psi|\Phi) = |(\Psi, \Phi)|^2. \tag{4.14}$$

There is a lot of implicit contextual information not given in such a definition, but physicists generally know what is meant and implied. Specifically, they would understand that the formalism refers to a statistical analysis of a sequence of runs. State Φ is prepared through one device at the start of each run, allowed to evolve undisturbed by the observer over some intermediate space-time regime, and then passed through another device that produces an outcome. Each outcome occurs randomly from a range of potential outcomes, with a countable frequency distribution over the ensemble of runs. In general, the ratios of observed outcome frequencies, when a large number of runs is performed, conform excellently to the probabilities predicted by the above Born rule.

Points to note are the following.

Symmetry
There is an inherent symmetry in the relationship between the two states in rule (4.14), referred to as the *microscopic reversibility of quantum processes* (Bohm, 1952). This is because of the mathematical equality $|(\Psi, \Phi)| = |(\Phi, \Psi)|$, which immediately leads to the physical prediction $Pr(\Psi|\Phi) = Pr(\Phi|\Psi)$. This relation has been confirmed empirically countless times.

Complex Amplitudes

The inner product (Ψ, Φ) is complex, which means it does not have the classical interpretation of a probability. Physicists get around this by referring to such expressions as "the amplitude for Φ to go to Ψ," or similar terminology, thereby endowing it with a touch of familiarity. The fact is, however, no one understands precisely why complex amplitudes occur in QM and it remains one of the enduring mysteries of the subject. Schwinger suggested that the appearance of complex numbers in QM is associated with the existence of antiparticles (Schwinger, 1958). Other physicists have investigated the theoretical and empirical possibility of replacing complex amplitudes in QM with hypercomplex (or quaternionic) amplitudes (Adler, 1995; Procopio et al., 2016; Adler, 2016), but there is at this time no empirical evidence that such a step is necessary.

Origin of the Born Rule

There have been attempts to derive the Born interpretation of the wave function from basic principles, but so far none of these attempts has been satisfactory. An interesting variant of such attempts is the idea that the Born rule is but the first step in a possibly infinite hierarchy of terms, a so-called multiorder interference rule that is a generalization of the above Born rule. Sorkin noted that for a two-slit interference experiment, the standard Born interpretation gives for the probability Pr_{AB} of a particle landing at a point on the detecting screen the formula

$$Pr_{AB} = Pr_A + Pr_B + I_{AB}, \tag{4.15}$$

where Pr_A is the probability when slit B is blocked off, Pr_B is the probability when slit A is blocked off, and I_AB is the so-called second-order interference term. Sorkin considered a three-slit experiment with slits A, B, and C and looked at the case for a generalization of the Born rule of the form

$$Pr_{ABC} = Pr_{AB} + Pr_{BC} + Pr_{AC} - Pr_A - Pr_B - Pr_C + I_{ABC}, \tag{4.16}$$

where I_{ABC} represents some novel third-order interference term not predicted by the above Born rule (Sorkin, 1994). A recent experiment has virtually ruled out such a term (Sinha et al., 2010).

Sorkin's motivation in this discussion is directly opposite to ours in this book: he explicitly states that he does not want to base the interpretation of his generalized probabilities with "some undefined concept of 'measurement made by human observers,"' and takes "the attitude that the ontology of QM is identical to that of classical realism" (Sorkin, 1994).

Dynamics

The Born rule is usually applied to states of SUOs evolving in time. In such a case, care has to be taken with the rules for the conservation of probability. These rules are as contextual as anything else in an experiment. For instance, if a state Φ_i is prepared at initial time t_i, allowed to evolve undisturbed until time t_f, then the amplitude to find the system under observation in state Ψ_f at final time is given by $(\Psi_f, U_{fi}\Phi_i)$, where U_{fi} is the unitary evolution operator taking states from initial time to final time. In this scenario, total probability is conserved. On the other hand, particle decay experiments may appear to involve a loss of total probability, if the mathematical modeling is done in too basic a fashion. For example, the Schrödinger wavefunction for a decaying particle SUO is frequently asserted to be given by a function of the form

$$\Psi(t, \boldsymbol{x}) \simeq e^{-i(E-i\Gamma)t/\hbar}\Phi(\boldsymbol{x}), \tag{4.17}$$

where Γ is a real constant related to the so-called half-life of the particle. We can discuss such a scenario in QDN; the QDN approach to particle decay experiments is covered in Chapter 15.

4.7 The Born Interpretation in QDN

We now consider the Born interpretation from the QDN perspective.

Given a normalized qubit state $\boldsymbol{\Psi} = \alpha\mathbf{0} + \beta\mathbf{1}$, where $|\alpha|^2 + |\beta|^2 = 1$, then the conditional probability $Pr(\mathbf{0}|\boldsymbol{\Psi})$ of the observer finding the associated detector in its ground state $\mathbf{0}$ is given by the rule $Pr(\mathbf{0}|\boldsymbol{\Psi}) = |\overline{\mathbf{0}}\boldsymbol{\Psi}|^2 = |\alpha|^2$, while the conditional probability $Pr(\mathbf{1}|\boldsymbol{\Psi})$ of finding the detector in its signal state $\mathbf{1}$ is $Pr(\mathbf{1}|\boldsymbol{\Psi}) = |\overline{\mathbf{1}}\boldsymbol{\Psi}|^2 = |\beta|^2$. We shall discuss the generalization of this rule to collections of detectors in the chapter on quantum register dynamics, Chapter 7.

There are two notational variants that we can use to discuss these probabilities.

Standard Expectation Value Notation
We may write

$$\begin{aligned} Pr(\mathbf{0}|\boldsymbol{\Psi}) &= |\overline{\mathbf{0}}\boldsymbol{\Psi}|^2 = (\overline{\mathbf{0}}\boldsymbol{\Psi})^*(\overline{\mathbf{0}}\boldsymbol{\Psi}) = (\overline{\boldsymbol{\Psi}}\mathbf{0})(\overline{\mathbf{0}}\boldsymbol{\Psi}) \\ &= \overline{\boldsymbol{\Psi}}(\mathbf{0}\overline{\mathbf{0}})\boldsymbol{\Psi} = \overline{\boldsymbol{\Psi}}\boldsymbol{P}\boldsymbol{\Psi}, \end{aligned} \tag{4.18}$$

interpreted in words as "*$Pr(\mathbf{0}|\boldsymbol{\Psi})$ is the expectation value of the ground state projection operator $\boldsymbol{P}$, contextual on the prepared state $\boldsymbol{\Psi}$*." Likewise, we have the rule

$$Pr(\mathbf{1}|\boldsymbol{\Psi}) = \overline{\boldsymbol{\Psi}}\,\widehat{\boldsymbol{P}}\boldsymbol{\Psi}. \tag{4.19}$$

Density Operator Notation
Given a pure bit state $\boldsymbol{\Psi}$, first define the density operator $\boldsymbol{\varrho} \equiv \boldsymbol{\Psi}\overline{\boldsymbol{\Psi}}$. Then the probabilities $Pr(\mathbf{0}|\boldsymbol{\Psi})$, $Pr(\mathbf{1}|\boldsymbol{\Psi})$ are given by the rules

$$Pr(\mathbf{0}|\boldsymbol{\Psi}) = Tr\{\boldsymbol{P}\boldsymbol{\varrho}\}, \qquad Pr(\mathbf{1}|\boldsymbol{\Psi}) = Tr\{\widehat{\boldsymbol{P}}\boldsymbol{\varrho}\}, \tag{4.20}$$

where Tr denotes the *trace* operation, discussed in Chapter 9.

4.8 Classical and Quantum Ensembles

The crucial differences between stochastic bits and qubits are not easy to see at the rank-one level but one of them is this: in CM, a stochastic bit state represents an observer's epistemic uncertainty as to which bit state a detector is actually in *before* they observe it, whereas in QM an observer can be certain of which qubit state a quantized detector is in and it is only the future *outcome* of an observation that is uncertain. Moreover, the uncertainty in the quantum case is generally regarded as intrinsic, or aleatoric, uncertainty.

There are several other ways of saying much the same thing. We can discuss *state preparation*, the processes that lead up to an observer having a contextual-based belief about the signal state of their detector, before observation. A stochastic bit state represents the observer's uncertainty about the preparation processes, whereas in the quantum case, the observer need have no such uncertainty about the preparation of a qubit state. Such a qubit state is called a *pure* state. The Born rule discussed above makes no reference to state preparation and it is assumed that the qubit state is pure.

Another way of showing the difference between stochastic bits and qubits is in terms of questions and answers. Given a stochastic bit of the form $\boldsymbol{S} \equiv a\mathbf{0} + (1-a)\mathbf{0}$, then $\overline{\boldsymbol{i}}\boldsymbol{S}, i = 0, 1$, represents a stochastic answer, which is a probability. On the other hand, given a qubit $\boldsymbol{\Psi}$, then $\overline{\boldsymbol{i}}\boldsymbol{\Psi}$, $i = 0, 1$, represents a *quantum answer*, which is a complex amplitude. Quantum answers have to be processed according to the Born rule given above in order to extract outcome probabilities.

Yet another way of seeing the difference between a stochastic bit and a qubit comes from the notion of *ensemble*. These are discussed in more detail in the Appendix. An ensemble can be either a real collection of near identical systems under observation, such as atoms in a crystal, or a hypothetical collection of imagined alternative futures, only one of which is going to be realized. Stochastic bits are generally associated with the former type of ensemble, while qubits are associated with the latter. We shall refer to the former kind of ensemble as a *classical ensemble* and refer to the latter kind as a *quantum ensemble*.

The differences between stochastic bits and qubits will become more obvious when we deal with *quantum registers*, or collections of qubits, discussed in Chapter 7.

4.9 Basis Transformations

Given a particular detector, then its associated physics provides the associated empirical context: *if* they looked, the observer would recognize when that detector was in its ground state $\mathbf{0}$ and when it was in its signal state $\mathbf{1}$. If this were not the case, we would have to ask what observation meant in this case.

This unambiguity about the context is reflected in the two questions an observer could ask of any detector: **Is this detector in its ground state?** and **Is this detector in its signal state?** A classical bit will always return an unambiguous answer of *yes* or *no*, a stochastic bit returns a probability, and a quantum bit returns a probability amplitude. This is encoded into the vector space formalism by the fact that stochastic bits and qubits are linear combinations of the classical answer states $\mathbf{0}$ and $\mathbf{1}$.

We return now to the fact that the definition of a Hilbert space is basis independent. Let us explore this further. Ignoring physical context and looking at a qubit strictly as a mathematical vector space, we are entitled to change our basis from the preferred basis. Consider therefore replacing each element $\boldsymbol{i}$ in our original preferred basis with some new vector $\boldsymbol{i}'$, as follows. First, we note that our preferred basis B is *orthonormal*, i.e., $\overline{\boldsymbol{i}}\boldsymbol{j} = \delta^{ij},\ i,j = 0,1$. Anticipating physical applications in later chapters, we shall preserve this relationship. Therefore, we shall require $\overline{\boldsymbol{i}'}\boldsymbol{j}' = \delta^{ij},\ i,j = 0,1$. For a complex Hilbert space, such transformations are called *unitary*.

Given our initial basis $B \equiv \{\mathbf{0}, \mathbf{1}\}$, consider a unitary transformation $B \to B' \equiv \{\mathbf{0}', \mathbf{1}'\}$ such that $\overline{\boldsymbol{i}'}\boldsymbol{j}' = \delta_{ij}$, for $0 \leq i,j \leq 1$. For such a transformation we may write

$$\boldsymbol{i} \to \boldsymbol{i}' \equiv \boldsymbol{U}\boldsymbol{i} = \sum_{j=0}^{1} \boldsymbol{j}'\, U^{ji}, \tag{4.21}$$

where the complex coefficients U^{ij} satisfy the unitarity relations

$$\sum_{j=0}^{1} U^{ij*} U^{jk} = \delta^{ik}, \tag{4.22}$$

where U^{ij*} is the complex conjugate of U^{ij}. These unitarity relations guarantee that orthonormality is preserved.

We may always write a unitary matrix in the form

$$U = \begin{bmatrix} \alpha & \beta \\ \gamma & \delta \end{bmatrix}, \tag{4.23}$$

where the coefficients α, β, γ, and δ are complex. From (4.22) we deduce the important relations

$$|\alpha|^2 + |\beta|^2 = |\gamma|^2 + |\delta|^2 = 1, \quad \alpha\gamma^* + \beta\delta^* = 0. \tag{4.24}$$

4.10 The Preferred Basis Problem

At this point we come across another problem related to the fact that a unitary transformation of the elements of a Hilbert space $\mathcal{H}$ has an implied action on the elements of its dual space $\overline{\mathcal{H}}$. This is because although the original definition of the inner product (ψ, ϕ) of a Hilbert space $\mathcal{H}$ is defined as a map from the Cartesian product $\mathcal{H} \times \mathcal{H}$ into the complex field, it can also be interpreted as a mapping of the vector ϕ in $\mathcal{H}$ into the complex numbers by the action of a one-form (a dual vector) $\overline{\psi}$, which is an element of a different vector space, the dual space $\overline{\mathcal{H}}$. The implied action of the above unitary operator $\boldsymbol{U}$ on elements of the dual space is given by

$$\overline{\boldsymbol{i}} \to \overline{\boldsymbol{i}'} \equiv \overline{\boldsymbol{i}}\boldsymbol{U}^\dagger = \sum_{j=0}^{1} \overline{\boldsymbol{j}'} U^{ji*}, \tag{4.25}$$

where $\boldsymbol{U}^\dagger$ is the *Hermitian conjugate* operator. For finite-dimensional vector spaces, the Hermitian conjugate operator is the same as the *adjoint* operator, problems arising only with nonseparable Hilbert spaces (Streater and Wightman, 1964).

The point is that according to our interpretation of the dual vectors $\overline{\mathbf{0}}$ and $\overline{\mathbf{1}}$, they are associated with questions to be asked of answer states. Given the transformation (4.25), we have to ask what "a linear combination of questions" means. While the Born interpretation gives us a meaning for an answer to a classical question asked of a linear combination of vectors, it is not immediately obvious what the interpretation of an object such as $u\overline{\mathbf{0}}+v\overline{\mathbf{1}}$ is. Our current view is that it is a mathematical artefact devoid of physical significance. What underpins this view is that observers are always sure in their minds which questions they are asking in a laboratory.

Another way of saying this is that we have not considered quantizing observers: they are always regarded as classical and this policy will be maintained throughout this book. This does not mean we shall not consider quantizing *apparatus*.

There will be situations where linear combinations of questions makes physical sense. For instance, we saw in the previous chapter that we could interpret linear combinations of questions of stochastic bits in terms of probabilities. On the quantum side, not only will such a possibility be available to us, but there will be a quantum side to this issue. Linear complex combinations of questions will have a role in the *information void*, the regime between state preparation and state outcome detection. In this regime, quantum rules apply and the concept of observer is not meaningful.

We should add at this point that the linear combination of quantum questions we have just referred to is not the same thing as a *mixed* quantum question. A mixed quantum question would be the analogue of a mixed state in QM, where an observer has an epistemic uncertainty as to which quantum state had been prepared. A mixed quantum question would likewise involve an experiment where an observer had an epistemic uncertainty as to which quantum question was being asked. This is not the same thing as a linear combination of quantum questions (which carries the implication that the observer itself is quantized).

One of the problems of interpretation that arises in the Multiverse paradigm (Deutsch, 1999) is that observers and SUOs are described by the vacuous concept of a wave function for the Universe. Since there is by assertion no primary observer, it is not clear what superposition of different observers means, if anything. In this respect, the original ideas of Everett's "Relative State" interpretation of QM seem less unattractive (Everett, 1957), because of the adjective *relative*.[3] It is not unreasonable to imagine that one observer A could describe what other observers B and C are doing by a quantum state vector. On that

[3] Everett does explicitly postulate an absolute wave function for the Universe, something QDN cannot accept, because such a postulate has a GPC of zero.

basis, relative to A, B and C are systems under observation and not observers. QDN is fully compatible with that idea, but regards the Multiverse concept as anathema.

4.11 Rank-One Qubit Evolution

As with bits and s-bits, we can consider the dynamical evolution of a single qubit state. Given a normalized qubit state $\boldsymbol{\Psi}_n \equiv \alpha_n \mathbf{0}_n + \beta_n \mathbf{1}_n$ in $\mathcal{Q}_n$ at stage Σ_n where $|\alpha_n|^2 + |\beta_n|^2 = 1$, we consider a *linear* map $\boldsymbol{U}_{n+1,n}$ from $\mathcal{Q}_n$ to $\mathcal{Q}_{n+1}$ such that normalization is preserved; i.e., we require

$$\boldsymbol{\Psi}_n \to \boldsymbol{\Psi}_{n+1} \equiv \boldsymbol{U}_{n+1,n}\boldsymbol{\Psi}_n, \qquad \overline{\boldsymbol{\Phi}_n} \to \overline{\boldsymbol{\Phi}_{n+1}} \equiv \overline{\boldsymbol{\Phi}_n}\boldsymbol{U}^{\dagger}_{n+1,n}, \tag{4.26}$$

with $\overline{\boldsymbol{\Psi}_{n+1}}\boldsymbol{\Psi}_{n+1} = 1$.

We come now to an important point. In the conventional theory of Hilbert spaces and in its application to standard QM, unitary transformations are maps from a given Hilbert space back into itself. In our situation this is no longer the case. We are dealing with maps from a Hilbert space at time n to another Hilbert space associated with time $n+1$. For example, the transformation (4.26) gives the dyadic representation

$$\boldsymbol{U}_{n+1,n} = \sum_{i,j=0}^{1} \boldsymbol{i}_{n+1} U^{ij}_{n+1,n} \overline{\boldsymbol{j}_n}. \tag{4.27}$$

This operator is one that not only preserves the norm but also preserves inner products under the transformation from one Hilbert space to the next. Such a transformation will be called a *semi-unitary* transformation, rather than a unitary transformation, for the good reason that in a more general context, the dimensions of the Hilbert spaces need not be the same. When the dimensions of the Hilbert spaces are different, then we need to consider the *retraction* of an evolution operator, rather than its inverse. We shall discuss this important aspect of our dynamics in more detail later.

In standard QM, a dynamics that preserves magnitudes of inner products (as opposed to inner products themselves) is usually realized via either unitary transformations or *anti-unitary* transformations. Anti-unitary transformations are subtle, having a great deal to do with the concept of *time reversal* in standard QM. We shall not consider anti-unitary transformations further.

An important issue that impacts on our approach is that one way of discussing time reversal in QM is to switch bra and ket vectors. Conventionally, this does not seem to matter but it amounts to switching questions and answers in our approach, which requires great care in the interpretation.

We may readily generalize our dynamical rule (4.26) to describe evolution from initial stage S_M to final stage S_N, where $N > M$: the product of two semi-unitary transformations is also a semi-unitary transformation. We find the rule

$$\boldsymbol{\Psi}_M \to \boldsymbol{\Psi}_N = \boldsymbol{U}_{N,M}\boldsymbol{\Psi}_M, \tag{4.28}$$

where $\boldsymbol{U}_{N,M} \equiv \boldsymbol{U}_{N,N-1}\boldsymbol{U}_{N-1,N-2}\dots\boldsymbol{U}_{M+1,M}$. The conditional outcome probabilities $Pr(\boldsymbol{i}_N|\boldsymbol{\Psi}_M)$ as measured at stage Σ_N are then given by

$$Pr(\boldsymbol{i}_N|\boldsymbol{\Psi}_M) \equiv |\overline{\boldsymbol{i}_N}\boldsymbol{\Psi}_N|^2, \quad i = 0, 1. \tag{4.29}$$

4.12 Mixed Qubit States

The uncertainty of s-bits is different from that associated with qubits. The former is due to ignorance on the part of the observer and may be called *classical* (or epistemic) uncertainty, while the latter is considered intrinsic and so referred to as *quantum* uncertainty on that account.

It is possible to encounter situations where both types of probability occur naturally. Whenever this happens, the labstates involved are no longer referred to as *pure* but *mixed*. The following example illustrates what we mean.

Example 4.2 An observer is about to ask the question $\overline{\mathbf{1}}$ of a single detector but is not sure how the labstate was prepared. The information that they do have, however, leads them to believe that the probability of the labstate (prior to observation) being $\boldsymbol{\Psi} \equiv \alpha\mathbf{0} + \beta\mathbf{1}$ is p, where $|\alpha|^2 + |\beta|^2 = 1$, while the probability of the labstate being $\boldsymbol{\Phi} \equiv \gamma\mathbf{0} + \delta\mathbf{1}$ is $1 - p$, where $|\gamma|^2 + |\delta|^2 = 1$. What is the overall probability of finding a signal?

Solution

Consider a very large number N of runs. Of these, approximately pN will involve the labstate $\boldsymbol{\Psi}$. The signal outcome probability for each such run is $|\beta|^2$ according to the Born rule. Therefore the total number of runs that involve the labstate $\boldsymbol{\Psi}$ with a signal outcome is approximately $pN|\beta|^2$. A similar calculation for the labstate $\boldsymbol{\Phi}$ gives the number of signal outcomes as $(1-p)N|\delta|^2$. The total number of signal outcomes is therefore approximately $Np|\beta|^2 + N(1-p)|\delta|^2)$. In the limit $N \to \infty$, the signal outcome probability is therefore $p|\beta|^2 + (1-p)|\delta|^2$ (the answer).

4.13 Density Operators

An efficient and standard way of combining classical and quantum probability is through the use of dyadics called *density operators*. Suppose we have a mixed initial state consisting of k different possible states $\boldsymbol{\Psi}^a$, $a = 1, 2, \dots, k$. The density operator $\boldsymbol{\varrho}$ is defined by the dyadic

$$\boldsymbol{\varrho} \equiv \sum_{a=1}^{k} \omega^a \boldsymbol{\Psi}^a \overline{\boldsymbol{\Psi}^a}, \tag{4.30}$$

where ω^a is the probability that the initial state is actually $\boldsymbol{\Psi}^a$. The expectation value $\langle O \rangle_{\boldsymbol{\varrho}}$ of an observable O is then given by the standard rule

$$\langle O \rangle_{\varrho} \equiv Tr\{O\varrho\}, \tag{4.31}$$

where Tr denotes the trace operation, discussed in Chapter 9.

To demonstrate how this works, we apply this formalism to the example considered above.

Example 4.3 With reference to the above example, we see $k = 2$ and

$$\begin{aligned} \omega^1 &= p, & \mathbf{\Psi}^1 &\equiv \mathbf{\Psi} = \alpha\mathbf{0} + \beta\mathbf{1} \\ \omega^2 &= 1 - p, & \mathbf{\Psi}^2 &\equiv \mathbf{\Phi} = \gamma\mathbf{0} + \delta\mathbf{1}. \end{aligned} \tag{4.32}$$

Hence the density matrix is

$$\begin{aligned} \varrho &= \omega^1\mathbf{\Psi}^1\overline{\mathbf{\Psi}^1} + \omega^2\mathbf{\Psi}^2\overline{\mathbf{\Psi}^2} \\ &= p\{\alpha\mathbf{0} + \beta\mathbf{1}\}\{\alpha^*\overline{\mathbf{0}} + \beta^*\overline{\mathbf{1}}\} + (1-p)\{\gamma\mathbf{0} + \delta\mathbf{1}\}\{\gamma^*\overline{\mathbf{0}} + \delta^*\overline{\mathbf{1}}\} \\ &= \{p|\alpha|^2 + (1-p)|\gamma|^2\}\mathbf{0}\overline{\mathbf{0}} + \{p\alpha\beta^* + (1-p)\gamma\delta^*\}\mathbf{0}\overline{\mathbf{1}} \\ &\quad + \{p\beta\alpha^* + (1-p)\delta\gamma^*\}\mathbf{1}\overline{\mathbf{0}} + \{p|\beta|^2 + (1-p)|\delta|^2\}\mathbf{1}\overline{\mathbf{1}}. \end{aligned} \tag{4.33}$$

The observable to use is $\widehat{\boldsymbol{P}} \equiv \mathbf{1}\overline{\mathbf{1}}$, the projection operator associated with the signal state $\mathbf{1}$. Then we find

$$\widehat{\boldsymbol{P}}\varrho = \{p\beta\alpha^* + (1-p)\delta\gamma^*\}\mathbf{1}\overline{\mathbf{0}} + \{p|\beta|^2 + (1-p)|\delta|^2\}\mathbf{1}\overline{\mathbf{1}}. \tag{4.34}$$

Taking the trace then gives

$$Tr\{\widehat{\boldsymbol{P}}\varrho\} = \{p|\beta|^2 + (1-p)|\delta|^2\}, \tag{4.35}$$

which agrees with the probability found above in Example 4.2.

5
Classical and Quantum Registers

5.1 Introduction

In this chapter we extend the discussion of the previous chapter, from one detector to an apparatus with an arbitrary number of nodes at any given stage. Our labeling convention is followed throughout this book: subscripts always label stages whereas superscripts always label nodes and modules (discussed in Chapter 11). The number of nodes at stage Σ_n will be called the *rank* of the apparatus at that stage and denoted r_n. The ith node at stage Σ_n will be denoted by i_n, which should not be confused with labstate vectors such as $\boldsymbol{i}_n$, which are always denoted in bold font.

Whenever we are using classical mechanics (CM), the collection of classical nodes at stage Σ_n will be called a *classical binary register*, denoted $\mathcal{R}_n$. In that case, i_n is represented by classical bit B_n^i, so $\mathcal{R}_n$ is the *Cartesian product* of all the bits at that stage, that is,

$$\mathcal{R}_n \equiv B_n^1 \times B_n^2 \times \cdots \times B_n^{r_n}. \tag{5.1}$$

The cardinality (number of elements) of the rank r classical binary register $\mathcal{R}_n$ is $d_n \equiv 2^{r_n}$. If we wish to indicate the rank of $\mathcal{R}_n$, we write $\mathcal{R}_n^{[r_n]}$.

On the other hand, whenever we are using quantum mechanics (QM), the collection of nodes at stage Σ_n will be called a *quantum register* and denoted $\mathcal{Q}_n$ or $\mathcal{Q}_n^{[r_n]}$. In the quantum case, i_n is represented by qubit Q_n^i. The quantum register $\mathcal{Q}_n$ is the *tensor product* of all the qubits at that stage, that is,

$$\mathcal{Q}_n \equiv Q_n^1 \otimes Q_n^2 \otimes \cdots \otimes Q_n^{r_n}. \tag{5.2}$$

Such a tensor product space is a Hilbert space of dimension $d_n \equiv 2^{r_n}$.

5.2 Labels versus Ordering

Superscripts are used in quantized detector networks (QDN) to identify individual nodes. Therefore, the standard left-right ordering in Cartesian products such as (5.1) or the left-right ordering in tensor products such as (5.2) is redundant. Provided we retain superscripts, we can drop the $\times$ symbol in Cartesian products and the $\otimes$ symbol in tensor products. Moreover, we can dispense with the left-right ordering rule in such products. Henceforth, we adopt the convention that the Cartesian product $B^1 \times B^2$ can be represented unambiguously by either $B^1 B^2$ or $B^2 B^1$. Moreover, the ordered element $(\boldsymbol{i}^1, \boldsymbol{j}^2)$ of $B^1 \times B^2$ can unambiguously be represented by the notation $\boldsymbol{i}^1\boldsymbol{j}^2$ or $\boldsymbol{j}^2\boldsymbol{i}^1$. A similar convention will be applied to tensor products and to dual spaces.

Example 5.1 If $\boldsymbol{\Psi}^1$ is a vector in qubit Q^1 and $\boldsymbol{\Phi}^2$ is a vector in Q^2, then $\boldsymbol{\Psi}^1\boldsymbol{\Phi}^2 = \boldsymbol{\Phi}^2\boldsymbol{\Psi}^1 \equiv \boldsymbol{\Psi}^1 \otimes \boldsymbol{\Phi}^2$ is an element of the tensor product space $Q^1Q^2 = Q^2Q^1 \equiv Q^1 \otimes Q^2$.

Example 5.2 The rank-two classical register $\mathcal{R}^{[2]} \equiv B^1B^2$ contains four classical states:

$$\mathcal{R}^{[2]} \equiv \{\mathbf{0}^1\mathbf{0}^2, \mathbf{1}^1\mathbf{0}^2, \mathbf{0}^1\mathbf{1}^2, \mathbf{1}^1\mathbf{1}^2\}. \tag{5.3}$$

These elements cannot be added together.

On the other hand, the corresponding rank-two quantum register $\mathcal{Q}^{[2]} \equiv Q^1Q^2$ has a preferred basis $B^{[2]} \simeq \mathcal{R}^{[2]}$ corresponding to the above four classical states. Now, however, arbitrary linear combinations of the form $\alpha \boldsymbol{i}^1\boldsymbol{j}^2 + \beta \boldsymbol{k}^1\boldsymbol{l}^2$, for complex α, β and $i, j, k, l = 0, 1$, are allowed, giving new elements in $\mathcal{Q}^{[2]}$. The four elements $\mathbf{0}^1\mathbf{0}^2, \mathbf{1}^1\mathbf{0}^2, \mathbf{0}^1\mathbf{1}^2, \mathbf{1}^1\mathbf{1}^2$ form the preferred basis for $\mathcal{Q}^{[2]}$, while their duals $\overline{\mathbf{0}^1\mathbf{0}^2}, \overline{\mathbf{1}^1\mathbf{0}^2}, \overline{\mathbf{0}^1\mathbf{1}^2}, \overline{\mathbf{1}^1\mathbf{1}^2}$ form the preferred basis for the dual space $\overline{\mathcal{Q}^{[2]}}$.

5.3 The Signal Basis Representation

For any given rank, there are infinitely many more possible quantum states than possible classical states. Representing these quantum states efficiently therefore requires suitable notation. In this section, we introduce a representation that is based on observational context (what we can see). By this we mean it utilizes the specific details of the detectors involved in the experiments being discussed.

Throughout the rest of this chapter we discuss a collection of qubits seen at a single *stage*, or instant of the observer's time, so in this chapter we suppress any reference to time. However, time will not be overlooked in general. In the dynamical theory given in the next chapter, a temporal subscript n will be introduced that is associated with every labstate and with other quantities such as the rank of the quantum register. Our ultimate aim is to construct a general theory of observation in which apparatus itself becomes a dynamical quantity.

Consider a rank-r quantum register $\mathcal{Q}^{[r]} \equiv Q^1 Q^2 \dots Q^r$. Observational context will inform us about the *preferred basis* $B^{[r]}$ for $\mathcal{Q}^{[r]}$. This consists of all possible *signal basis states* of the apparatus. These will represent all possible classical yes-no configurations of the detectors involved. Each element of this basis set is a tensor product of the form $\boldsymbol{i}^1\boldsymbol{i}^2\boldsymbol{i}^3\dots\boldsymbol{i}^r$, where the $\boldsymbol{i}^a$ are all either the $\mathbf{0}^a$ (ground state) or $\mathbf{1}^a$ (signal state) elements of the preferred basis B^a for the ath detector qubit Q^a. When written in this form, $B^{[r]}$ will be referred to as the *signal basis representation* (SBR).

By inspection, we see there are 2^r distinct elements in $B^{[r]}$ and together they constitute an orthonormal basis for the quantum register $\mathcal{Q}^{[r]}$. We can identify the elements in the classical register $\mathcal{R}^{[r]}$ with the elements of the preferred basis $B^{[r]}$ of the quantum register $\mathcal{Q}^{[r]}$.

Example 5.3 The SBR for a rank-three quantum register has $2^3 = 8$ elements and is given by

$$B^{[3]} = \{\mathbf{0}^1\mathbf{0}^2\mathbf{0}^3, \mathbf{1}^1\mathbf{0}^2\mathbf{0}^3, \mathbf{0}^1\mathbf{1}^2\mathbf{0}^3, \mathbf{1}^1\mathbf{1}^2\mathbf{0}^3, \mathbf{0}^1\mathbf{0}^2\mathbf{1}^3, \mathbf{1}^1\mathbf{0}^2\mathbf{1}^3, \mathbf{0}^1\mathbf{1}^2\mathbf{1}^3, \mathbf{1}^1\mathbf{1}^2\mathbf{1}^3\}. \tag{5.4}$$

To define orthonormality, we introduce the dual basis $\overline{B^{[r]}}$, consisting of elements of the form $\overline{\boldsymbol{i}^1\boldsymbol{i}^2\boldsymbol{i}^3\dots\boldsymbol{i}^r}$, where $\overline{\boldsymbol{i}^a} = \overline{\mathbf{0}^a}$ or $\overline{\boldsymbol{i}^a} = \overline{\mathbf{1}^a}$ for $a = 1, 2, \dots, r$. Inner products in $\mathcal{Q}^{[r]}$ are defined by the action of elements of $\overline{B^{[r]}}$ on elements of $B^{[r]}$, plus linearity:

$$\begin{aligned}\overline{\boldsymbol{i}^1\boldsymbol{i}^2\boldsymbol{i}^3\dots\boldsymbol{i}^r}\boldsymbol{j}^1\boldsymbol{j}^2\boldsymbol{j}^3\dots\boldsymbol{j}^r &= (\overline{\boldsymbol{i}^1}\boldsymbol{j}^1)(\overline{\boldsymbol{i}^2}\boldsymbol{j}^2)(\overline{\boldsymbol{i}^3}\boldsymbol{j}^3)\dots(\overline{\boldsymbol{i}^r}\boldsymbol{j}^r)\\ &= \delta^{i^1 j^1}\delta^{i^2 j^2}\delta^{i^3 j^3}\dots\delta^{i^r j^r}.\end{aligned} \tag{5.5}$$

This rule is interpreted as follows. If the right-hand side of (5.5) is zero, then the elements $\boldsymbol{i}^1\boldsymbol{i}^2\boldsymbol{i}^3\dots\boldsymbol{i}^r$ and $\boldsymbol{j}^1\boldsymbol{j}^2\boldsymbol{j}^3\dots\boldsymbol{j}^r$ of the Hilbert space $\mathcal{Q}^{[r]}$ are orthogonal; otherwise, they are the same element and it has length (norm) of one.

Example 5.4 According to our notation, indices keep track of factor qubit spaces in a quantum register. Hence for a rank-four register, for example, we would find

$$\overline{\mathbf{0}^4\mathbf{1}^1\mathbf{1}^3\mathbf{0}^2}\mathbf{0}^3\mathbf{1}^1\mathbf{0}^2\mathbf{0}^4 = \underbrace{(\overline{\mathbf{1}^1}\mathbf{1}^1)}_{1}\underbrace{(\overline{\mathbf{0}^2}\mathbf{0}^2)}_{1}\underbrace{(\overline{\mathbf{1}^3}\mathbf{0}^3)}_{0}\underbrace{(\overline{\mathbf{0}^4}\mathbf{0}^4)}_{1} = 1 \times 1 \times 0 \times 1 = 0. \tag{5.6}$$

5.4 Maximal Questions

Given an apparatus represented by r detectors at a given stage, the observer has the freedom to ask any one of a number of questions. For instance, the observer may look at a particular detector to ascertain its signal state and *not* look at any of the other detectors. Such a question that does not involve all of the detectors will be called a *partial question*. Partial questions are discussed in Chapter 8.

A *maximal question* is one that asks a binary question of every detector in a register. From (5.5), we see that the elements of the dual basis $\overline{B^{[r]}}$ can be interpreted as maximal questions, and there are 2^r of them. It is important to note that in the absence of any theory that quantizes observers and/or their questions, arbitrary vectors in $\overline{Q^{[r]}}$ are not in general physically meaningful maximal questions; only the elements of the preferred dual basis $\overline{B^{[r]}}$ are maximal questions in the formulation of QDN being discussed at this point.

It is not impossible, however, to imagine a scenario where questions being asked correspond to elements of $\overline{Q^{[r]}}$ that are nontrivial linear combinations of preferred basis elements. For instance, a given observer may be conducting an experiment where there are two possible quantum outcomes, such that each outcome would trigger a different maximal question of some state at a later stage. That would correspond to quantization of apparatus and/or observers. This sort of scenario is essentially the focus of Chapter 21.

Example 5.5 Given the rank-two quantum register $\mathcal{Q}^{[2]} \equiv \mathcal{Q}^1\mathcal{Q}^2$, the preferred basis $B^{[2]}$ has four elements:

$$B^{[2]} = \{\mathbf{0}^1\mathbf{0}^2, \mathbf{1}^1\mathbf{0}^2, \mathbf{0}^1\mathbf{1}^2, \mathbf{1}^1\mathbf{1}^2\}. \tag{5.7}$$

The maximal questions are all the elements of the dual basis

$$\overline{B^{[r]}} \equiv \{\overline{\mathbf{0}^1\mathbf{0}^2}, \overline{\mathbf{1}^1\mathbf{0}^2}, \overline{\mathbf{0}^1\mathbf{1}^2}, \overline{\mathbf{1}^1\mathbf{1}^2}\}. \tag{5.8}$$

5.5 Signality

The *signality* of a basis state corresponds to the total "particle number" being detected by the apparatus concerned, *if* any positive signal in any of the detectors is interpreted as registering a particle in the conventional sense.

Definition 5.6 The *signality* of an element $\boldsymbol{i}^1\boldsymbol{i}^2\boldsymbol{i}^3\cdots\boldsymbol{i}^a$ of the signal basis $B^{[r]}$ is the sum $\sum_{a=1}^{r} i^a$.

Example 5.7 The signality of the element $\mathbf{0}^6\mathbf{1}^2\mathbf{1}^1\mathbf{0}^5\mathbf{0}^3\mathbf{1}^4$ in $B^{[6]}$ is $1+1+0+1+0+0=3$.

The particle interpretation of signality has to be viewed cautiously for two reasons. First, detectors register irreversible processes that may have nothing to do with any identifiable, persistent "particle." For example, a *phonon* is a collective phenomenon, a quantum of excitation in a crystal that does not "exist" in a reductionist sense. Second, it is possible in QDN to deal with labstates that are superpositions of preferred basis states of differing signality. Classically, this would have no particle interpretation but is part and parcel of QDN.

Dynamics may in many situations rule out certain superpositions of labstates in a quantum register. For example, charge conservation rules out superpositions of labstates corresponding to different numbers of electrons. Another example is angular momentum: we are not allowed in conventional quantum mechanics to superpose states of particles with different total spin number, conventionally denoted by j. However, the general principle remains: superposition of labstates of different signality is allowed in principle in QDN; only dynamical context rules out certain mathematical possibilities.

These comments apply to the labstates. As far as the dual basis is concerned, we are not at this stage entitled to create linear superpositions of *any* of the basis elements in the dual basis set $\overline{B^{[r]}}$, as these elements represent questions being asked by the observer. We have introduced no concept yet of a "quantized observer," whatever that might mean. Notwithstanding our comments just before Example 5.5, all questions asked at this stage of this book have to be completely classical. The point about QDN is that the labstates do not have this restriction and can be superpositions of the classical-looking elements of the preferred basis. This remark applies *before* the observer looks at a labstate.

5.6 The Economy of Success

A useful fact now emerges, a fact that underpins the success of CM (a moderately simple theory) in describing a hideously complex reality. Of all the 2^r states in a rank-r classical register, only one of them has a truth value of one relative to any given maximal question, and all the other $2^r - 1$ states have truth value of zero, relative to that question. To see this, consider an arbitrary maximal question $\overline{\boldsymbol{i}}^1\overline{\boldsymbol{i}}^2\ldots\overline{\boldsymbol{i}}^r$ asked of an arbitrary classical register state $\boldsymbol{j}^1\boldsymbol{j}^2\ldots\boldsymbol{j}^r$. The answer is

$$\overline{\boldsymbol{i}}^1\overline{\boldsymbol{i}}^2\ldots\overline{\boldsymbol{i}}^r\boldsymbol{j}^1\boldsymbol{j}^2\ldots\boldsymbol{j}^r = \underbrace{(\overline{\boldsymbol{i}}^1\boldsymbol{j}^1)}_{\delta^{i^1j^1}}\underbrace{(\overline{\boldsymbol{i}}^2\boldsymbol{j}^2)}_{\delta^{i^2j^2}}\ldots\underbrace{(\overline{\boldsymbol{i}}^r\boldsymbol{j}^r)}_{\delta^{i^rj^r}} = \delta^{i^1j^1}\delta^{i^2j^2}\ldots\delta^{i^rj^r}, \tag{5.9}$$

which is zero in general except for the particular case when $i^a = j^a$ for each possible value of a, and then the answer value is one. This result greatly simplifies calculations in QDN.

This result contributes to the inherent economy of the particle description in CM, as the following example illustrates. Suppose we wanted to describe a single point particle in three-dimensional space. The conventional approach would be to define a Cartesian coordinate system with coordinates x, y, and z along three mutually orthogonal axes, relative to some chosen origin of coordinates. Then the position of a particle P would be fully specified by giving only three numbers, (x_P, y_P, z_P), referred to as the *position coordinates of the particle.* The economy of this approach is literally infinite: saying where the particle is immediately says where the particle is not, which is every point in space apart from the one with coordinates (x_P, y_P, z_P).

This economy of information comes about because of context: we have been told that there is only one particle.

But suppose now we are presented with a situation where we were not told beforehand how many particles there were in our laboratory. Then we would have to look at every point in space if we wanted to know the total particle number. This is not a trivial point but a hard fact in astrophysics and cosmology: astronomers have to scan space thoroughly in order to make estimates of matter and dark matter densities on cosmological scales.

In the quantum context, we may use signality to tell us how many signals there are in our detector array. A labstate of signality one corresponds closely to what can be thought of as a single particle state. This means that, if we knew beforehand that we were dealing with a signality-one state, then it would suffice to find a positive signal in just one detector in an array to be sure (by context) that there were no positive signals anywhere else in that array.

It is an intriguing thought that when we specify the classical Cartesian coordinates x, y, z of a point particle, we are not only answering the question **where is the particle?** but also saying that it is not at any of the other points in space, of which there is an uncountable number. Clearly, that is an impressive form of economy, but one occurring only because of our contextual information.

5.7 Quantum Registers

We now extend the discussion from quantum labstates of a single detector to quantum labstates of two or more detectors. These higher rank labstates are elements of some quantum register $\mathcal{Q}^{[r]} \equiv Q^1 Q^2 \ldots Q^r$, where $r \geqslant 1$ is the rank of the register and Q^i is the ith qubit in the register. We have suppressed the tensor product symbol $\otimes$ here.

Tensor products of vector spaces are discussed in the Appendix. Unlike Cartesian products, tensor product spaces are genuine vector spaces, the quantum register $\mathcal{Q}^{[r]}$ being a complex vector space of dimension 2^r. This is one of the places where the fundamental differences between CM and QM manifest themselves. In the classical case, there is only a finite number of different possible states in a classical register. For a system of r classical bits, each of which has two possible mutually exclusive answer states, then the total number of integers 0 or 1 we would need in order to parametrize any classical state of the system is just r. Any state in such a classical register can therefore be represented by an r-tuple of the form $\{i^1, i^2, \ldots, i^r\}$, where $i^a = 0$ or 1, for $a = 1, 2, \ldots r$, and there are just 2^r different such r-tuples. In the quantum case, possible states are elements of a finite-dimensional complex vector space of dimension 2^r, which has infinitely many elements.

Consider an arbitrary state $\boldsymbol{\Psi}$ in $\mathcal{Q}^{[2]}$. Given the above preferred basis, we can always write

$$\boldsymbol{\Psi} = z^1\mathbf{0}^1\mathbf{0}^2 + z^2\mathbf{1}^1\mathbf{0}^2 + z^3\mathbf{0}^1\mathbf{1}^2 + z^4\mathbf{1}^1\mathbf{1}^2, \tag{5.10}$$

where the z^a are labeled complex numbers, not powers of z. The dual state $\overline{\boldsymbol{\Psi}}$ is given by

$$\begin{aligned}\overline{\boldsymbol{\Psi}} &\equiv \overline{z^1\mathbf{0}^1\mathbf{0}^2 + z^2\mathbf{1}^1\mathbf{0}^2 + z^3\mathbf{0}^1\mathbf{1}^2 + z^4\mathbf{1}^1\mathbf{1}^2} \\ &= z^{1*}\overline{\mathbf{0}^1\mathbf{0}^2} + z^{2*}\overline{\mathbf{1}^1\mathbf{0}^2} + z^{3*}\overline{\mathbf{0}^1\mathbf{1}^2} + z^{4*}\overline{\mathbf{1}^1\mathbf{1}^2},\end{aligned} \tag{5.11}$$

where z^{a*} is the complex conjugate of z^a. If the state is normalized, we have the condition

$$\overline{\boldsymbol{\Psi}}\boldsymbol{\Psi} = |z^1|^2 + |z^2|^2 + |z^3|^2 + |z^4|^2 = 1. \tag{5.12}$$

If now we write $z^a = x^a + iy^a$, where x^a and y^a are real, then the normalization condition (5.12) is equivalent to saying that possible states rank-two quantum register can be identified one-to-one with points on S^7, the unit sphere in eight-dimensional Euclidean space $\mathbb{E}^8$.

The above analysis is for just *two* detectors. If we had in mind, say, a screen consisting of a million detectors, or a neural network in a brain, the numbers shoot up beyond imagination.[1] Clearly, qubit register states have much more structure than their classical counterparts. The exploration of this structure is in a real sense in its infancy at this time. For example, there is still a great deal to be understood about the physics of quantum entangled states in low rank quantum registers. We discuss some aspects of this in Chapter 22.

5.8 The Computational Basis Representation

The SBR $\{\boldsymbol{i}^1\boldsymbol{i}^2\boldsymbol{i}^3\ldots\boldsymbol{i}^r : i^a = 0, 1 : a = 1, 2, \ldots, r\}$ for elements of the preferred basis $B^{[r]}$ is useful in some respects but less so in others. A frequently more useful but equivalent notation for the elements of the preferred basis will be called the *computational basis representation* (CBR) and is obtained as follows. For each element $\boldsymbol{i}^1\boldsymbol{i}^2\boldsymbol{i}^3\cdots\boldsymbol{i}^r$ of the SBR, there is precisely one element $\boldsymbol{i}$ of the CBR, where i is an integer in the range $[0, 2^r - 1]$, given by the *computational basis map*

$$i = i^1 + 2i^2 + 4i^3 + \cdots + 2^{r-1}i^r, \tag{5.13}$$

where i^3 means the third element in the SBR and not i to the power of three.

The computational basis map can be inverted. Given an element $\boldsymbol{p}$ of the CBR, where p is an integer in the range $[0, 2^r - 1]$, we can write

$$p = p^{[1]} + 2p^{[2]} + 4p^{[3]} + \cdots + 2^{r-1}p^{[p]} = \frac{1}{2}\sum_{a=1}^{r} 2^a p^{[a]}, \tag{5.14}$$

[1] The average human brain has about a hundred billion neurons.

where the symbol $p^{[a]}$ denotes the signal status (zero or one) of the ath detector when the apparatus is in the labstate $\boldsymbol{p}$. The coefficients $p^{[a]}$ will be referred to as the *binary components* of the integer p and are discussed in further detail later on in this chapter. They play an essential role in the formalism of QDN and in many calculations.

Example 5.8 For the rank-three quantum register $Q^1Q^2Q^3$, the SBR of the preferred basis $B^{[3]}$ is

$$B^{[3]} = \{\mathbf{0}^1\mathbf{0}^2\mathbf{0}^3, \mathbf{1}^1\mathbf{0}^2\mathbf{0}^3, \mathbf{0}^1\mathbf{1}^2\mathbf{0}^3, \mathbf{1}^1\mathbf{1}^2\mathbf{0}^3, \mathbf{0}^1\mathbf{0}^2\mathbf{1}^3, \mathbf{1}^1\mathbf{0}^2\mathbf{1}^3, \mathbf{0}^1\mathbf{1}^2\mathbf{1}^3, \mathbf{1}^1\mathbf{1}^2\mathbf{1}^3\}. \tag{5.15}$$

The CBR of $B^{[3]}$ is the set

$$B^{[3]} = \{\mathbf{0}, \mathbf{1}, \mathbf{2}, \mathbf{3}, \mathbf{4}, \mathbf{5}, \mathbf{6}, \mathbf{7}\}, \tag{5.16}$$

where $\mathbf{0} = \mathbf{0}^1\mathbf{0}^2\mathbf{0}^3$, $\mathbf{1} = \mathbf{1}^1\mathbf{0}^2\mathbf{0}^3$, and so on.

The CBR generally has the advantage over the SBR of being more compact and on that account is better suited in many but not all calculations. The dual preferred basis $\overline{B^{[r]}}$ can also be expressed in CBR terms, and then the inner product relations (5.5) take the form

$$\overline{\boldsymbol{i}}\boldsymbol{j} = \delta^{ij}, \quad 0 \leqslant i, j < 2^r, \tag{5.17}$$

which is very useful.

A disadvantage of the CBR is that it masks the signal properties of a given state. For example, the CBR element $\mathbf{3}$ could represent the SBR element $\mathbf{1}^1\mathbf{1}^2$ for a rank-two apparatus or the SBR element $\mathbf{1}^1\mathbf{1}^2\mathbf{0}^3\mathbf{0}^4\mathbf{0}^5$ of a rank-five apparatus and so on. However, context will generally make it clear what is meant by a given CBR expression.

The CBR is useful for representing linear operators over the register. These will generally be denoted in blackboard font in QDN. For example, the classical or quantum register identity operator $\mathbb{I}^{[r]}$ is expressed in the CBR by

$$\mathbb{I}^{[r]} = \sum_{k=0}^{2^r-1} \boldsymbol{k}\overline{\boldsymbol{k}}. \tag{5.18}$$

More generally, if we know the action of a linear register operator on each element of the signal basis, then we can represent that operator as a dyadic in the computation representation, as we show in the next section.

5.9 Register Operators

A *register operator* is a function that maps elements of a classical or quantum register into itself or another classical or quantum register. Linear register operators will be denoted in blackboard bold font, such as $\mathbb{U}$, $\mathbb{A}$, and so on. The same labeling convention used for states and questions will be used for linear operators: upper indices refer to qubits and, while lower labels indicate stages.

Suppose $\mathbb{O}_{n,m}$ is a linear register operator mapping states from an initial stage Σ_m rank-r_m classical or quantum register into a final stage Σ_n rank-r_n classical or quantum register. Suppose further that we are given its action on each element $\boldsymbol{i}_m$ of the CBR, that is,

$$\mathbb{O}_{n,m}\boldsymbol{i_m} = \sum_{j=0}^{2^{r_n}-1} O^{ji}_{n,m}\boldsymbol{j}_n, \quad i = 0, 1, \ldots, 2^{r_m} - 1, \tag{5.19}$$

where the $O^{ji}_{n,m}$ are complex coefficients.

Now "multiply" each side from the right by $\overline{\boldsymbol{i_m}}$ and sum over i:

$$\sum_{i=0}^{2^{r_m}-1} \mathbb{O}_{n,m}\boldsymbol{i}_m\overline{\boldsymbol{i}_m} = \sum_{i=0}^{2^{r_m}-1}\sum_{j=0}^{2^{r_n}-1} O^{ji}_{n,m}\boldsymbol{j}_n\overline{\boldsymbol{i}_m} \tag{5.20}$$

The left-hand side simplifies because of linearity:

$$\sum_{i=0}^{2^{r_m}-1} \mathbb{O}_{n,m}\boldsymbol{i}_m\overline{\boldsymbol{i}_m} = \mathbb{O}_{n,m}\left\{\sum_{i=0}^{2^{r_m}-1} \boldsymbol{i}_m\overline{\boldsymbol{i}_m}\right\} = \mathbb{O}_{n,m}\mathbb{I}_m = \mathbb{O}_{n,m}, \tag{5.21}$$

using (5.18). Hence finally we arrive at the dyadic representation of a typical register operator:

$$\mathbb{O}_{n,m} = \sum_{i=0}^{2^{r_m}-1}\sum_{j=0}^{2^{r_n}-1} \boldsymbol{j}_n O^{ji}_{n,m}\overline{\boldsymbol{i}_m}. \tag{5.22}$$

We define the *retraction* $\overline{\mathbb{O}}_{n,m}$ of a register operator $\mathbb{O}_{n,m}$ by

$$\overline{\mathbb{O}}_{n,m} \equiv \sum_{i=0}^{2^{r_m}-1}\sum_{j=0}^{2^{r_n}-1} \boldsymbol{i}_m O^{ji*}_{n,m}\overline{\boldsymbol{j}_n}, \tag{5.23}$$

where $O^{ij*}_{n,m}$ is the complex conjugate of $O^{ij}_{n,m}$. A retraction as defined here is not necessarily the equivalent of an *inverse* operator.

5.10 Classical Register Operators

We saw in Chapter 3 that there are only four classical bit operators $\boldsymbol{C}^1 \equiv \boldsymbol{I}$, $\boldsymbol{C}^2 \equiv \boldsymbol{F}$, $\boldsymbol{C}^3 \equiv \boldsymbol{D}$, and $\boldsymbol{C}^4 \equiv \boldsymbol{U}$ that map bit states to bit states. Likewise, a classical register operator maps classical register states to classical register states. We shall discuss a classical register operator $\mathbb{C}_{n,m}$ mapping register states from a rank-r_m classical register $\mathcal{R}_m$ to a rank-r_n classical register $\mathcal{R}_n$.

As before, the most economical way to discuss classical register operators is via the CBR, as follows. For any element $\boldsymbol{i}_m$ in $\mathcal{R}_m$, where $i = 0, 1, 2, \ldots, 2^{r_m} - 1$, the operator $\mathbb{C}_{n,m}$ necessarily maps it into precisely *one* element in $\mathcal{R}_n$, and not into a linear combination of two or more, as in the quantum case. This means that we can always write

$$\mathbb{C}_{n,m}\boldsymbol{i}_m = \sum_{j=0}^{2^{r_n}-1} C^{ji}_{n,m}\boldsymbol{j}_n, \quad i = 0, 1, 2, \ldots, 2^{r_m} - 1, \tag{5.24}$$

where for a given i, every element $C^{ji}_{n,m}$ is zero for every integer j in the interval $[0, 2^{r_n} - 1]$ except for one, which has value one. Using the completeness property of the elements $\{\boldsymbol{i}_m\}$ we arrive at the dyadic representation

$$\mathbb{C}_{n,m} = \sum_{j=0}^{2^{r_n}-1}\sum_{i=0}^{2^{r_m}-1} \boldsymbol{j}_n C^{ji}_{n,m}\overline{\boldsymbol{i}_m}. \tag{5.25}$$

This is a special case of (5.22) and there is a total of $(2^{r_n})^{2^{r_m}}$ different possible such operators.

In this set of operators, there are two subsets that are of particular importance in both classical register mechanics and quantum register mechanics. These are the *register projection operators* and the *register signal operators*.

Register Projection Operators

Given a rank-r register, the observer may be interested in asking questions of an individual bit or qubit and leaving all the other bits or qubits alone. The register projection operators $\mathbb{P}^i$ and $\widehat{\mathbb{P}}^i$ will be used in later chapters to construct partial questions. These operators are defined by

$$\begin{aligned}\mathbb{P}^i &= \boldsymbol{I}^1\boldsymbol{I}^2\ldots\boldsymbol{I}^{i-1}\boldsymbol{P}^i\boldsymbol{I}^{i+1}\ldots\boldsymbol{I}^r,\\ \widehat{\mathbb{P}}^i &= \boldsymbol{I}^1\boldsymbol{I}^2\ldots\boldsymbol{I}^{i-1}\widehat{\boldsymbol{P}}^i\boldsymbol{I}^{i+1}\ldots\boldsymbol{I}^r,\end{aligned}$$

where $\boldsymbol{I}^a$ is the bit identity operator for the ath bit or qubit and $\boldsymbol{P}^i$, $\widehat{\boldsymbol{P}}^i$ are the bit projection operators for the ith bit or qubit, as discussed in the previous chapter.

Register Signal Operators

Associated with a rank-r quantum register $\mathcal{Q}^{[r]}$ are some important operators connected to the physics of observation, and these will appear frequently throughout the formalism. The most important of these are the r signal annihilation operators $\mathbb{A}^i$, $i = 1, 2, \ldots, r$ and the related signal creation operators $\widehat{\mathbb{A}}^i$. These operators are defined in terms of the signal bit operators discussed in the previous chapter, as follows:

$$\begin{aligned}\mathbb{A}^i &\equiv \boldsymbol{I}^1\boldsymbol{I}^2\ldots\boldsymbol{I}^{i-1}\boldsymbol{A}^i\boldsymbol{I}^{i+1}\ldots\boldsymbol{I}^r,\\ \widehat{\mathbb{A}}^i &\equiv \boldsymbol{I}^1\boldsymbol{I}^2\ldots\boldsymbol{I}^{i-1}\widehat{\boldsymbol{A}}^i\boldsymbol{I}^{i+1}\ldots\boldsymbol{I}^r, \quad i = 1, 2, \ldots, r,\end{aligned} \tag{5.26}$$

where the superscripts on the right-hand side label the individual qubits in a given rank-r register and we suppress the tensor product symbol $\otimes$.

5.11 The Signal Algebra

We define S^a, the ath *signal set*, to be the set of register operators

$$S^a \equiv \{\mathbb{P}^a, \widehat{\mathbb{P}}^a, \mathbb{A}^a, \widehat{\mathbb{A}}^a\}, \quad a = 1, 2, \ldots, r. \tag{5.27}$$

Table 5.1 *The signal set algebra*

	$\mathbb{P}^a$	$\widehat{\mathbb{P}}^a$	$\mathbb{A}^a$	$\widehat{\mathbb{A}}^a$
$\mathbb{P}^a$	$\mathbb{P}^a$	0	$\mathbb{A}^a$	0
$\widehat{\mathbb{P}}^a$	0	$\widehat{\mathbb{P}}^a$	0	$\widehat{\mathbb{A}}^a$
$\mathbb{A}^a$	0	$\mathbb{A}^a$	0	$\mathbb{P}^a$
$\widehat{\mathbb{A}}^a$	$\widehat{\mathbb{A}}^a$	0	$\widehat{\mathbb{P}}^a$	0

Then an extraordinarily useful fact is that any two elements from different signal sets commute.

For a given signal set S^a, we have the ath *signal set algebra*, as shown in Table 5.1. For $i = 1, 2, \ldots, r$ we have

$$\mathbb{A}^i\mathbb{A}^i = \widehat{\mathbb{A}}^i\widehat{\mathbb{A}}^i = 0, \qquad \left\{\mathbb{A}^i, \widehat{\mathbb{A}}^i\right\} = \mathbb{P}^i + \widehat{\mathbb{P}}^i = \mathbb{I}^{[r]}, \tag{5.28}$$

where curly brackets denote an *anticommutator*, while for $i \neq j$, we have

$$[\mathbb{A}^i, \mathbb{A}^j] = [\widehat{\mathbb{A}}^i, \widehat{\mathbb{A}}^j] = [\mathbb{A}^i, \widehat{\mathbb{A}}^j] = 0, \tag{5.29}$$

where square brackets denote a *commutator*. Note that in the above, the symbol 0 represents the *zero operator* for the register concerned.

The signal set algebra gives QDN a particular flavor; parts of it are reminiscent of a theory with fermions (anticommuting objects), while other parts have a bosonic (commuting) character. At the signal level, however, we are dealing with neither concept specifically; the signal algebra is determined by the physics of observation as it relates to the apparatus and has its own logic that is distinct from that of conventional particle physics.

5.12 Signal Excitations

The signal operators introduced above are used to construct signal states from the signal ground state **0** as follows.

1. The *signality one states* are of the form $\widehat{\mathbb{A}}^i\mathbf{0} \equiv \mathbf{2}^{i-1}, 1 \leq i \leq r$.
2. The *signality two states* are of the form $\widehat{\mathbb{A}}^i\widehat{\mathbb{A}}^j\mathbf{0} \equiv \underline{\mathbf{2}^{i-1} + \mathbf{2}^{j-1}}, 1 \leq i < j \leq r$.
3. The *signality k states* are of the form

$$\widehat{\mathbb{A}}^{i_1}\widehat{\mathbb{A}}^{i_2}\ldots\widehat{\mathbb{A}}^{i_k}\mathbf{0} \equiv \underline{\sum_{j=1}^{k} \mathbf{2}^{i_j-1}}, \tag{5.30}$$

for $1 \leq i_1 < i_2 < \cdots < i_k \leq r$.

Remark 5.9 In the above, we underline expressions such as $\underline{\mathbf{2}^{i-1} + \mathbf{2}^{j-1}}$ to indicate that this is *not* the vector sum of $\mathbf{2}^{i-1}$ and $\mathbf{2}^{j-1}$ but the vector in the CBR corresponding to the sum of the integers 2^{i-1} and 2^{j-1}.

The same approach can be used to discuss maximal questions. Recall that these are represented by the dual preferred basis elements of the form $\overline{\boldsymbol{i}}$, $0 \le i < 2^r$. Then we can write

$$\overline{\mathbf{0}}\mathbb{A}^{i_1}\mathbb{A}^{i_2}\dots\mathbb{A}^{i_k} \equiv \overline{\sum_{j=1}^{k} \mathbf{2}^{i_j-1}}. \tag{5.31}$$

Exercise 5.10 Prove that

$$\mathbb{A}^i\mathbf{0} = 0, \quad \overline{\mathbf{0}}\widehat{\mathbb{A}}^i = 0, \quad 1 \le i \le r. \tag{5.32}$$

Exercise 5.11 Use the signal algebra to show that

$$\overline{\boldsymbol{i}}\boldsymbol{j} = \delta^{ij}, \quad 0 \le i, j, < 2^r. \tag{5.33}$$

5.13 Signality Classes

For each rank r quantum register $\mathcal{Q}^{[r]}$, we may write its preferred basis $B^{[r]}$ in the form

$$B^{[r]} = \{\mathbf{0}, \widehat{\mathbb{A}}^1\mathbf{0}, \widehat{\mathbb{A}}^2\mathbf{0}, \dots, \widehat{\mathbb{A}}^r\mathbf{0}, \widehat{\mathbb{A}}^1\widehat{\mathbb{A}}^2\mathbf{0}, \dots, \widehat{\mathbb{A}}^1\widehat{\mathbb{A}}^2\dots\widehat{\mathbb{A}}^r\mathbf{0}\}. \tag{5.34}$$

For each element, the *signality* of that element is the number of signal creation operators used to create it from the signal ground state.

Example 5.12 For a rank-11 quantum register, the preferred basis state given by $\underline{\mathbf{851}}$ in the CBR is given by $\mathbf{1}^1\mathbf{1}^2\mathbf{0}^3\mathbf{0}^4\mathbf{1}^5\mathbf{0}^6\mathbf{1}^7\mathbf{0}^8\mathbf{1}^9\mathbf{1}^{10}\mathbf{0}^{11}$ in the SBR, from which we can read off its signality as 6. Equivalently, we can see that this state is given by $\widehat{\mathbb{A}}^1\widehat{\mathbb{A}}^2\widehat{\mathbb{A}}^5\widehat{\mathbb{A}}^7\widehat{\mathbb{A}}^9\widehat{\mathbb{A}}^{10}\mathbf{0}$, which clearly has signality 6.

Such examples soon show the enormous advantage of using the CBR for large rank registers. The CBR is particularly suited for computer algebra calculations.

By inspection, it is easy to see that the preferred basis $B^{[r]}$ separates into $r+1$ disjoint *signality classes*:

$$\begin{aligned}
\text{signality zero } & B^{[r,0]} \equiv \{\mathbf{0}\}, \\
\text{signality one } & B^{[r,1]} \equiv \{\widehat{\mathbb{A}}^a\mathbf{0} : a = 1, 2, \dots, r\}, \\
\text{signality two } & B^{[r,2]} \equiv \{\widehat{\mathbb{A}}^a\widehat{\mathbb{A}}^b\mathbf{0} : 1 \leqslant a < b \leqslant r\}, \\
& \vdots \qquad \vdots \\
\text{signality } r \quad & B^{[r,r]} \equiv \{\widehat{\mathbb{A}}^1\widehat{\mathbb{A}}^2\dots\widehat{\mathbb{A}}^r\mathbf{0}\}.
\end{aligned} \tag{5.35}$$

Then we may write the preferred basis as the union of all of these signality classes, that is,

$$B^{[r]} = \bigcup_{a=0}^{r} B^{[r,a]}. \tag{5.36}$$

We define the *signality count* $\sigma^{[r,a]}$ as the number of elements in $B^{[r,a]}$. Then

$$\sigma^{[r,a]} = \frac{r!}{a!(r-a)!} = \binom{r}{a} \tag{5.37}$$

Hence the *total signality* $\sigma^{[r]}$ is given by

$$\sigma^{[r]} \equiv \sum_{i=0}^{r} \sigma^{[r,a]} = \sum_{a=0}^{r} \binom{r}{a} = 2^a, \tag{5.38}$$

as expected.

A significant feature of QDN is that the addition of register states of different signality is permitted, unless the dynamics specifically rules it out. This will happen, for example, whenever the signals are interpreted as electrically charged particles. Under such circumstances, we expect conservation of total signality.

5.14 Binary Decomposition

Given a nonnegative integer k, the *binary decomposition* of k is the expansion

$$k = \sum_{a=1}^{\infty} k^{[a]} 2^{a-1}, \quad k = 0, 1, 2, \ldots, \tag{5.39}$$

where each of the binary coefficients $k^{[a]}$ is either zero or one. Table 5.2 shows the binary decomposition up to $k = 9$.

From the table, we read off for example that

$$6 = 0 \times 2^{1-1} + 1 \times 2^{2-1} + 1 \times 2^{3-1} + 0 \times 2^{4-1} + 0 \times \cdots = 2 + 4 = 6.$$

For each k, the maximum a for which $k^{[a]} = 1$ is the *minimum rank* of k, denoted $\mu(k)$ and indicated by a box in Table 5.2. The minimum rank of k is the rank of the "smallest" quantum register that contains the basis state $\boldsymbol{k}$.

Table 5.2 *The binary decomposition of the integers up to nine*

$a \rightarrow$	1	2	3	4	...
$2^{a-1} \rightarrow$	1	2	4	8	...
$k \downarrow$ 0	0	0	0	0	...
1	$\boxed{1}$	0	0	0	...
2	0	$\boxed{1}$	0	0	...
3	1	$\boxed{1}$	0	0	...
4	0	0	$\boxed{1}$	0	...
5	1	0	$\boxed{1}$	0	...
6	0	1	$\boxed{1}$	0	...
7	1	1	$\boxed{1}$	0	...
8	0	0	0	$\boxed{1}$	...
9	1	0	0	$\boxed{1}$	...

Example 5.13 From Table 5.2 we read off

$$\begin{aligned}&\mu(1) = 1,\\ &\mu(2) = \mu(3) = 2,\\ &\mu(4) = \mu(5) = \mu(6) = \mu(7) = 3.\end{aligned} \tag{5.40}$$

We see $\mu(9) = 4$ and $9^{[1]} = 1$, $9^{[2]} = 0$, $9^{[3]} = 0$, $9^{[4]} = 1$, so

$$9 = 1 \times 2^{1-1} + 0 \times 2^{1} + 0 \times 2^{3-1} + 1 \times 2^{4-1} = 1 + 8 = 9. \tag{5.41}$$

5.15 Computational Basis Representation for Signal Operators

For a given rank-r register $\mathcal{Q}^{[r]}$, the preferred basis elements $\{\boldsymbol{k} : k = 0, 1, 2, \dots, 2^r - 1\}$ are given in the CBR by

$$\boldsymbol{k} = \underline{\sum_{a=1}^{\mu(k)} k^{[a]}\mathbf{2}^{a-1}}, \quad k = 0, 1, 2, \dots, 2^r - 1, \tag{5.42}$$

where $k^{[a]} = 0$ or 1 is the ath *binary component* of the integer k; i.e., we have

$$k = k^{[1]}2^{1-1} + k^{[2]}2^{2-1} + k^{[3]}2^{3-1} + \dots + 2^{\mu(k)-1}. \tag{5.43}$$

Hence we may equate the CBR and SBR elements in the basis $B^{[\mu_k]}$

$$\boldsymbol{k} = \boldsymbol{k}^{[1]}\boldsymbol{k}^{[2]}\dots\mathbf{1}^{\mu_k} = \left(\widehat{\mathbb{A}}^1\right)^{k_{[1]}}\left(\widehat{\mathbb{A}}^2\right)^{k_{[2]}}\dots\left(\widehat{\mathbb{A}}^{\mu_k}\right)\mathbf{0}. \tag{5.44}$$

Now use the results

$$\begin{aligned}\widehat{\mathbb{A}}^a\boldsymbol{k} &= 0 && \text{if } k^{[a]} = 1,\\ &= \underline{\boldsymbol{k} + \mathbf{2}^{a-1}} && \text{if } k^{[a]} = 0.\end{aligned} \tag{5.45}$$

Defining

$$\widehat{k}^{[a]} \equiv 1 - k^{[a]}, \tag{5.46}$$

then we can readily show

$$\begin{aligned}\widehat{\mathbb{A}}^a\boldsymbol{k} &= \widehat{k}^{[a]}\underline{\boldsymbol{k} + \mathbf{2}^{a-1}}, & \overline{\boldsymbol{k}}\widehat{\mathbb{A}}^a &= k^{[a]}\overline{\boldsymbol{k} - \mathbf{2}^{a-1}}\\ \mathbb{A}^a\boldsymbol{k} &= k^{[a]}\underline{\boldsymbol{k} - \mathbf{2}^{a-1}}, & \overline{\boldsymbol{k}}\mathbb{A}^a &= \widehat{k}^{[a]}\overline{\boldsymbol{k} + \mathbf{2}^{a-1}}.\end{aligned} \tag{5.47}$$

Hence we must have

$$\widehat{\mathbb{A}}^a = \sum_{k=0}^{2^r-1} \widehat{k}^{[a]}\underline{\boldsymbol{k} + \mathbf{2}^{a-1}}\overline{\boldsymbol{k}} = \sum_{k=0}^{2^r-1} k^{[a]}\underline{\boldsymbol{k}}\overline{\boldsymbol{k} - \mathbf{2}^{a-1}}, \tag{5.48}$$

$$\mathbb{A}^a = \sum_{k=0}^{2^r-1} k^{[a]}\underline{\boldsymbol{k} - \mathbf{2}^{a-1}}\overline{\boldsymbol{k}} = \sum_{k=0}^{2^r-1} \widehat{k}^{[a]}\underline{\boldsymbol{k}}\overline{\boldsymbol{k} + \mathbf{2}^{a-1}}, \tag{5.49}$$

which leads to

$$\mathbb{P}^a = \sum_{k=0}^{2^r-1} \widehat{k}^{[a]}\underline{\boldsymbol{k}}\overline{\boldsymbol{k}}, \qquad \widehat{\mathbb{P}}^a = \sum_{k=0}^{2^r-1} k^{[a]}\underline{\boldsymbol{k}}\overline{\boldsymbol{k}}. \tag{5.50}$$

These results are consistent with

$$\mathbb{P}^a + \widehat{\mathbb{P}}^a = \sum_{k=0}^{2^r-1} (k_{[a]} + \widehat{k}_{[a]})\boldsymbol{k}\overline{\boldsymbol{k}} = \sum_{k=0}^{2^r-1} \boldsymbol{k}\overline{\boldsymbol{k}} = \mathbb{I}^{[r]}, \qquad a = 1, 2, \ldots, r. \tag{5.51}$$

Example 5.14 Using the above expansions, we find

Rank one

$$\begin{aligned} \mathbb{P} &= \mathbf{0}\overline{\mathbf{0}}, & \widehat{\mathbb{P}} &= \mathbf{1}\overline{\mathbf{1}} \\ \mathbb{A} &= \mathbf{0}\overline{\mathbf{1}}, & \widehat{\mathbb{A}} &= \mathbf{1}\overline{\mathbf{0}}. \end{aligned} \tag{5.52}$$

Rank two

$$\begin{aligned} \mathbb{P}^1 &= \mathbf{0}\overline{\mathbf{0}} + \mathbf{2}\overline{\mathbf{2}}, & \widehat{\mathbb{P}}^1 &= \mathbf{1}\overline{\mathbf{1}} + \mathbf{3}\overline{\mathbf{3}}, \\ \mathbb{P}^2 &= \mathbf{0}\overline{\mathbf{0}} + \mathbf{1}\overline{\mathbf{1}}, & \widehat{\mathbb{P}}^2 &= \mathbf{2}\overline{\mathbf{2}} + \mathbf{3}\overline{\mathbf{3}}, \\ \mathbb{A}^1 &= \mathbf{0}\overline{\mathbf{1}} + \mathbf{2}\overline{\mathbf{3}}, & \widehat{\mathbb{A}}^1 &= \mathbf{1}\overline{\mathbf{0}} + \mathbf{3}\overline{\mathbf{2}}, \\ \mathbb{A}^2 &= \mathbf{0}\overline{\mathbf{2}} + \mathbf{1}\overline{\mathbf{3}}, & \widehat{\mathbb{A}}^2 &= \mathbf{2}\overline{\mathbf{0}} + \mathbf{3}\overline{\mathbf{1}}. \end{aligned} \tag{5.53}$$

Rank three

$$\begin{aligned} \mathbb{P}^1 &= \mathbf{0}\overline{\mathbf{0}} + \mathbf{2}\overline{\mathbf{2}} + \mathbf{4}\overline{\mathbf{4}} + \mathbf{6}\overline{\mathbf{6}}, & \widehat{\mathbb{P}}^1 &= \mathbf{1}\overline{\mathbf{1}} + \mathbf{3}\overline{\mathbf{3}} + \mathbf{5}\overline{\mathbf{5}} + \mathbf{7}\overline{\mathbf{7}}, \\ \mathbb{P}^2 &= \mathbf{0}\overline{\mathbf{0}} + \mathbf{1}\overline{\mathbf{1}} + \mathbf{4}\overline{\mathbf{4}} + \mathbf{5}\overline{\mathbf{5}}, & \widehat{\mathbb{P}}^2 &= \mathbf{2}\overline{\mathbf{2}} + \mathbf{3}\overline{\mathbf{3}} + \mathbf{6}\overline{\mathbf{6}} + \mathbf{7}\overline{\mathbf{7}}, \\ \mathbb{P}^3 &= \mathbf{0}\overline{\mathbf{0}} + \mathbf{1}\overline{\mathbf{1}} + \mathbf{2}\overline{\mathbf{2}} + \mathbf{3}\overline{\mathbf{3}}, & \widehat{\mathbb{P}}^3 &= \mathbf{4}\overline{\mathbf{4}} + \mathbf{5}\overline{\mathbf{5}} + \mathbf{6}\overline{\mathbf{6}} + \mathbf{7}\overline{\mathbf{7}}, \\ \mathbb{A}^1 &= \mathbf{0}\overline{\mathbf{1}} + \mathbf{2}\overline{\mathbf{3}} + \mathbf{4}\overline{\mathbf{5}} + \mathbf{6}\overline{\mathbf{7}}, & \widehat{\mathbb{A}}^1 &= \mathbf{1}\overline{\mathbf{0}} + \mathbf{3}\overline{\mathbf{2}} + \mathbf{5}\overline{\mathbf{4}} + \mathbf{7}\overline{\mathbf{6}}, \\ \mathbb{A}^2 &= \mathbf{0}\overline{\mathbf{2}} + \mathbf{1}\overline{\mathbf{3}} + \mathbf{4}\overline{\mathbf{6}} + \mathbf{5}\overline{\mathbf{7}}, & \widehat{\mathbb{A}}^2 &= \mathbf{2}\overline{\mathbf{0}} + \mathbf{3}\overline{\mathbf{1}} + \mathbf{6}\overline{\mathbf{4}} + \mathbf{7}\overline{\mathbf{5}}, \\ \mathbb{A}^3 &= \mathbf{0}\overline{\mathbf{4}} + \mathbf{1}\overline{\mathbf{5}} + \mathbf{2}\overline{\mathbf{6}} + \mathbf{3}\overline{\mathbf{7}}, & \widehat{\mathbb{A}}^3 &= \mathbf{4}\overline{\mathbf{0}} + \mathbf{5}\overline{\mathbf{1}} + \mathbf{6}\overline{\mathbf{2}} + \mathbf{7}\overline{\mathbf{3}}. \end{aligned} \tag{5.54}$$

In general, the CBR for any signal operator in a rank-r register consists of a sum of 2^{r-1} transition operators, all of which annihilate each other, including themselves. Likewise, a product $\widehat{\mathbb{A}}^i\widehat{\mathbb{A}}^j$ of two different signal operators can be expressed as a sum of 2^{r-2} transition operators that mutually annihilate, and so on. This process of representation can be continued until we arrive at the *saturation operator* $\widehat{\mathbb{A}}^1\widehat{\mathbb{A}}^2\ldots\widehat{\mathbb{A}}^r$, which when applied to the signal ground state creates the antithesis of the ground state, the *fully saturated signal state* $\underline{\mathbf{2^r-1}} \equiv \mathbf{1}^1\mathbf{1}^2\ldots\mathbf{1}^r$.

A particularly useful expression for the signal creation operators is obtained by writing any of them in the form

$$\widehat{\mathbb{A}}^i = \mathbf{2}^{i-1}\overline{\mathbf{0}} + \mathbb{X}^i, \tag{5.55}$$

where the operator $\mathbb{X}^i \equiv \sum_{k=1}^{2^r-1} \widehat{k}^{[i]}\underline{\boldsymbol{k} + \mathbf{2}^{i-1}}\overline{\boldsymbol{k}}$ annihilates the signal ground state. This expression can be used to greatly simplify calculations for those experiments involving signality-one labstates, as in single-photon quantum optics experiments.

6

Classical Register Mechanics

6.1 Introduction

Classical mechanics (CM) is conventionally formulated according to partially contextually complete Block Universe principles: an exophysical observer sitting in some frame of reference looks in on a system under observation (SUO) and assigns truth values to propositions about states of that SUO, but the details of how these truth values are obtained are not given.

There is a curious aspect to this scenario: the contextual incompleteness of CM is itself contextual. By this we mean that there are two scenarios, with different general propositional classifications (GPCs).

The Experimentalist's Perspective

From a classical experimentalist's perspective, generalized propositions (GPs) in CM take the form

$$\mathcal{P}(\boldsymbol{P}, \emptyset|\emptyset, F). \tag{6.1}$$

Here $\boldsymbol{P}$ is some proposition of scientific interest, such as "**The orbit of Jupiter is an ellipse**," and F is relative external context describing the frame of reference used to describe states of the SUO relative to the wider Universe external to that SUO. This form of GP has a GPC of 2 according to the algorithm discussed in Chapter 2.

Remark 6.1 Suppose an astronomer used a telescope T to test proposition $\boldsymbol{P}$. That would not upgrade the GP to one of the form $\mathcal{P}(\boldsymbol{P}, T|\emptyset, F)$ with a GPC of 3. The reason is that the use of a telescope is not essential in this context. Historically, Kepler did not use a telescope to state his laws of planetary motion: he used the data obtained by Tycho using naked eye observations. On the other hand, the observation of the Hubble red shift using telescopes fitted with spectrum analyzers would qualify for such an upgrade.

The Theorist's Perspective
Newtonian mechanics applied to the description of classical dynamically evolving SUOs has the strength of mathematics: the classical laws of motion can be considered as a set of mathematical principles, axioms, and theorems, relative to which propositions about classical states of SUOs can be tested. From this perspective, the contextual information F about frames of reference becomes part of the mathematical framework, and therefore should be included in *relative internal context*. In this scenario, GPs take the form $\mathcal{P}(\boldsymbol{P}, \text{Laws of motion}, F|\emptyset, \emptyset)$, with a GPC of 1, as befitting mathematical propositions.[1]

6.2 Classical Registers

We shall focus our attention on the second scenario discussed above, that of the theorist's perspective. Therefore, we shall not discuss apparatus or the observer. Instead, we focus our attention on the mathematical structures that we employ to model particles moving about in continuous space and time.

The first thing is to construct a model for classical space, objectifying it as a vast collection of containers into which, and out of which, particles can move.

Now conventionally and until relatively recently, space has always been considered to be continuous. But that is an empirically vacuous assertion. Therefore, we take the liberty of modeling space in discrete and finite terms. This leads us to define a classical register $\mathcal{R}^{[r]}$ of rank r, the Cartesian product of r classical bits. We write

$$\mathcal{R}^{[r]} \equiv B^1 B^2 \dots B^r, \tag{6.2}$$

where the B^i, $i = 1, 2, \dots, r$, are the individual labeled bits and we use bit labeling to bypass the need for the Cartesian product symbol $\times$. We will consider classical registers of sufficiently large rank so that they can model regions of classical physical space over which particles can move. In this approach, particle motion is discussed in terms of the tracking of signals from a vast collection of detectors over time. A particularly useful feature of this approach is that signality need not be conserved, which means that classical particle creation and annihilation is readily incorporated into the formalism.

6.3 Architecture

In any such discussion it is important to establish the relevant spatiotemporal architecture. There are two aspects of architecture relevant to the present discussion. One has to do with the unspecified exophysical observer and their

[1] It seems invalid to argue that in Newtonian mechanics, we have the laws of motion for relative internal context and the observer's frame for relative external context, so that would appear to give a GPC of 3. That would be mixing theory and experiment to an unacceptable degree. We have made the point before that mathematics is not physics.

apparatus, and the other is to do with the dynamics of the system under observation. We shall refer to these aspects as the *external* and *internal* architectures, respectively.

External architecture involves time and persistence.

Time

In classical register mechanics, time is modeled as a succession of discrete stages, $\Sigma_M, \Sigma_{M+1}, \ldots, \Sigma_N$, representing the actions or lack of actions by an observer in a physical laboratory. At stage Σ_n, the state of the SUO will be denoted $\mathbf{\Psi}_n$, an element of a classical register $\mathcal{R}_n$ of rank r_n.

Persistence

This refers to the question of identity of a given detector over a succession of stages. In quantized detector networks (QDN), each detector is identified with a single stage only. Therefore, there is no concept of "existential persistence" in QDN. This applies as much to the observer and their apparatus as it does to the SUOs that they observe. The *first law of time* (the dictum of Heraclitus), that *everything changes*, applies here.

That does not rule out *effective persistence*, meaning that as the observer's time runs from stage to stage, their description of the apparatus seems to be constant, in terms of the rank of the registers concerned and in terms of the labeling of the detectors in those registers.

Effective persistence is the rule in ordinary experience, to the extent that humans are strongly conditioned to believe that they are moving around in time over a fixed spatial background. This belief carries over into conventional descriptions of experiments, in which apparatus is assumed to persist while states of SUOs evolve. What is really going on is that the first law of time always applies, but the time scales for significant change in apparatus are generally so great relative to those associated with the SUOs that the former time scales may be taken to be infinite compared with the latter. This is a common assumption made in high-energy particle physics scattering experiments, for instance.

Internal architecture depends on the assumed laws of mechanics, there being three distinct types.

Time-Dependent Laws of Motion

The laws of mechanics for a given SUO could be contextual, changing in some way from stage to stage, or they could be constant. An example of the former scenario is demonstrated by *phase diagrams* in chemistry. These show under which conditions of pressure and temperature a collection of molecules behaves as a gas, a liquid, or a solid. Such behavior is a manifestation of emergent processes driven by reductionist laws of physics, and the challenge is to explain the emergent behavior in each phase using those reductionist laws.

Autonomy or Not

An SUO could be *autonomous*, meaning that it has been effectively isolated from its environment. Alternatively, the exophysical observer could arrange for external agencies, such as electric and magnetic forces, to influence the dynamics of an SUO.

Signality Conservation

Signality is taken in this chapter as a marker of particle number: a signality-p state of a classical register at a given stage will be interpreted as a state with p particles. Newtonian mechanics does not readily countenance changes of particle number. If that has to be considered, then that can be readily modeled by signality nonconserving register dynamics. The reason for this is that register mechanics is more like a field theory than a particle theory.

For the rest of this chapter we shall restrict our attention to autonomous SUOs moving over a succession of classical registers, each of which has the same rank r, with stage-independent laws of dynamics, as these are generally of most interest. In principle, there would be no problem in dealing with other forms of dynamics, including those where the rank of the physical register changes with time. Indeed, that is a common scenario in the quantum register dynamics we shall discuss in the next chapter.

We could also deal with classical stochastic mechanics, which would incorporate Bayesian principles in a natural way, but for the rest of the chapter we shall deal with deterministic laws of mechanics. In the following, the set of integers $\{0, 1, 2, 4, \ldots, 2^r - 1\}$ is denoted $Z^{[r]}$.

We shall use the computational basis representation (CBR) $B_n \equiv \{\boldsymbol{k}_n : k \in Z^{[r]}\}$ to represent the 2^r labstates in a rank r classical register $\mathcal{R}_n$ at stage Σ_n. Consider the temporal evolution of a system from labstate $\boldsymbol{\Psi}_n$ in register $\mathcal{R}_n$ at stage Σ_n to labstate $\boldsymbol{\Psi}_{n+1}$ in register $\mathcal{R}_{n+1}$at stage Σ_{n+1}. The dynamical rules will be encoded into the expression

$$\boldsymbol{\Psi}_n \to \boldsymbol{\Psi}_{n+1} \equiv \mathbb{C}_{n+1,n} \boldsymbol{\Psi}_n, \tag{6.3}$$

where $\mathbb{C}_{n+1,n}$ is a classical register operator. Such an operator maps any one of the 2^r labstates in B_n into precisely *one* of the 2^r labstates in B_{n+1}.

The reason for this constraint on any classical register operator comes from the basic principles of CM: states in CM are well-defined, single-valued elements of phase space at any given time. This translates in the present context to the requirement that at any stage Σ_n, the classical state of an SUO is precisely one of the 2^r possible states of the register $\mathcal{R}_n$. In consequence, any admissible classical register operator acts on single elements of $\mathcal{R}_n$ and maps them into single elements of $\mathcal{R}_{n+1}$. Note that this does not rule out nonsurjective or noninjective operators; that is, there may be elements in B_{n+1} that are not mapped into (nonsurjective), and it could be the case that more than one element in B_n may be mapped into the same element in B_{n+1} (noninjective).

Taking any initial state, there are 2^r possible final states. Since there is a total of 2^r possible initial states, we immediately deduce that there is a total of $(2^r)^{2^r}$ different possible classical register operators mapping from $\mathcal{R}_n$ into $\mathcal{R}_{n+1}$.

Even for low-rank registers, the number of possible operators can be impressive. For a rank-one system, we deduce that there should be $2^2 = 4$ different possible such operators. Indeed, that is exactly what we saw in Chapter 3, where we identified the four classical bit operators $\boldsymbol{I}, \boldsymbol{F}, \boldsymbol{U}$, and $\boldsymbol{D}$. The number of possibilities grows rapidly with r. For example, there are 256 different operators that can map a rank-two register into another rank-two register, and a total of $8^8 = 16{,}777{,}216$ different operators mapping a rank-three register into another rank-three register.

Given a potentially vast number of possible evolution operators, some criteria need to be imposed in order to reduce the discussion to manageable and realistic proportions. What comes to our assistance here is that most of the possible evolution operators over a classical register will not be useful. Many of them will correspond to irreversible and/or unphysical dynamical evolution and only a small subset will be of interest. We need to find some principles to guide us in our choice of evolution operator.

We turn to standard classical mechanics as discussed in Hamiltonian mechanics. The first thing to note is that classical phase spaces are generally constant in time. This corresponds to taking the rank of successive classical registers to be constant, i.e., $r_n = r_{n+1} \equiv r$ for some integer r and for all n in the temporal interval under discussion. We shall make this assumption from this point on.

Next, we recall that in standard CM of the Hamiltonian variety, Hamilton's equations of motion lead to Liouville's theorem. This tells us that as we track a small volume element along a classical trajectory,[2] this volume remains constant in magnitude though not necessarily constant in shape or orientation. This leads to the idea that a system of many noninteracting particles moving along classical trajectories in phase space behaves as an incompressible fluid, such a phenomenon being referred to as a Hamiltonian flow.

An important characteristic of Hamiltonian flows is that flow lines never cross. We shall encode this idea into our development of classical register mechanics. There are two versions of this mechanics, one of which does not necessarily conserve signality while the other does. We consider the first one now.

6.4 Permutation Flows

A classical rank-r register $\mathcal{R}_n$ contains 2^r labstates denoted by $\boldsymbol{k}_n$, $k \in Z^{[r]}$, in the CBR. Consider a permutation P_n of the integers k, such that under P_n, $k_n \to k_{n+1} \equiv P_n(k_n) \in Z^{[r]}$. Define the evolution of the labstate $\boldsymbol{k}_n$ over one

[2] A classical trajectory is a phase space trajectory that satisfies Hamilton's equations of motion.

stage by $\boldsymbol{k}_n \to \boldsymbol{k}_{n+1} \equiv P_n(\boldsymbol{k}_n)$. Such a process is reversible and will be referred to as a *permutation flow*.

Example 6.2 Mathematicians often represent a permutation of, say, five objects in the form (135)(24), which means the permutation $1 \to 3 \to 5 \to 1, 2 \to 4 \to 2$, and so on, where the notation $a \to b \to c$ means that original element a is replaced by, or goes to, element b, original element b is replaced by element c, and so on. A group of elements within a given pair of brackets is called a *cycle*.

Consider a rank-three classical register $\mathcal{R}^{[3]}$. This has a total of eight states $\{\boldsymbol{k} : k = 0, 1, 2, \ldots, 7\}$. Under the permutation (0346)(1)(25)(7), the evolution is given by

$$\mathbf{0} \to \mathbf{3} \to \mathbf{4} \to \mathbf{6} \to \mathbf{0}, \mathbf{1} \to \mathbf{1}, \mathbf{2} \to \mathbf{5} \to \mathbf{2}, \mathbf{7} \to \mathbf{7}. \tag{6.4}$$

This permutation flow does not conserve signality. The simplest proof of this is to note that the signality zero state $\mathbf{0}$ evolves into a state with nonzero signality.

There is a total of $n!$ distinct permutations of n objects, so there are $(2^r)!$ possible distinct permutation flow processes. For large r, the number of permutation flows is a rapidly decreasing fraction of the number $(2^r)^{2^r}$ of all possible forms of rank-preserving classical register processes.

Permutation flows are restricted to constant rank registers, and even then, are of debatable value in the following sense. Given two successive classical registers $\mathcal{R}_n$ and $\mathcal{R}_{n+1}$ of the same rank, the relationship between the labeling of states in the two registers is contextual. By this we mean that the identification of element $\mathbf{1}_n$ with element $\mathbf{1}_{n+1}$, $\mathbf{2}_n$ with $\mathbf{2}_{n+1}$, and so on, is up to the observer. This is analogous to the parallel transport problem in general relativity, where contextual information relating initial and final tangent spaces along a path is required before a notion of parallelism can be established.

Example 6.3 Consider a permutation written as a transformation from $\mathcal{R}_0$ to $\mathcal{R}_1$ in the form

$$\boldsymbol{i}_0 \to \boldsymbol{p(i)}_1 : i = 0, 1, 2, \ldots, 2^r - 1. \tag{6.5}$$

Now suppose that we passively relabel the 2^r states in $\mathcal{R}_1$ by the rule $\boldsymbol{p(i)}_1 \to \boldsymbol{i}_1$. Then the permutation reduces to the "identity" transformation

$$\boldsymbol{i}_0 \to \boldsymbol{i}_1, \quad i = 0, 1, 2, \ldots, 2^r - 1. \tag{6.6}$$

This relabeling seems to have eliminated the significance of permutation dynamics.

This apparent paradox is resolved by *context*: if we had no knowledge of the original permutation (6.5) or of the physical meaning of the original states and the final states, then (6.6) would indeed be trivial. If we had such contextual information, however, then the relabeling $\boldsymbol{p}(i)_1 \to \boldsymbol{i}_1$ would have no physical significance, and the real dynamics would be understood by the observer.

Permutation flows have the following features with analogues in CM.

Reversibility
Permutations form a group, which means that given a permutation P, then its inverse P^{-1} always exists. Hence permutation flows are reversible.

Orbits
Any SUO evolving under stage-independent permutation dynamics will demonstrate patterns known as *orbits* or *cycles*. Given a permutation of 2^r objects, there will exist subsets known as *cycles* of the objects being permuted such that only elements within a given cycle replace each other under the permutation. This is relevant here because we have chosen to discuss time-independent autonomous systems, the evolution of which is given by repeated applications of the same permutation. Therefore, the structure of the cycles does not change and so each cycle is stable, consisting of the same p elements with a dynamical period p. For example, the identity permutation gives a trivial form of mechanics where nothing changes. It has 2^r cycles each of period 1. At the other end of the spectrum, the permutation denoted by $(0 \to 1 \to 2 \to \cdots \to 2^r - 1 \to 0)$ has no cycles except itself and has period 2^r.

Any physical register evolving under time-independent, autonomous permutation mechanics must return to its initial labstate no later than after 2^r time steps. This is the analogue of the Poincaré recurrence theorem (Poincaré, 1890).

6.5 Signality-Conserving Flows

Most permutation flows will not conserve signality, as Example 6.2 shows. Suppose now we have decided to identify signality with *particle number*. Then experience with Hamiltonian mechanics, where particle conservation is the general rule, leads us to investigate signality-conserving flows.

We can readily identify the subset of the permutation flows that conserve signality by using the signality classes discussed in the previous chapter. Suppose $\mathbb{U}_{n+1,n}$ is an evolution operator with the following characteristics:

1. $\mathbb{U}_{n+1,n}\mathbf{0}_n = \mathbf{0}_{n+1}$: such an evolution is called *calibrated*.
2. $\mathbb{U}_{n+1,n}$ permutes signality-1 states, that is, $\mathbb{U}_{n+1,n}\widehat{\mathbb{A}}_n^i\mathbf{0}_n = \widehat{\mathbb{A}}_{n+1}^{i_P}\mathbf{0}_{n+1}$, for $i = 1, 2, \ldots, r$ and with i_P the number into which i is transformed under the permutation.
3. Likewise for each signality class, until finally, we have for the fully saturated signal state $\mathbb{U}_{n+1,n}\underline{\mathbf{2^r-1}}_n = \underline{\mathbf{2^r-1}}_{n+1}$.

Then clearly, signality is conserved under such a dynamical scheme.

The total number of distinct permutations of r objects is $r!$, so there are that many distinct forms of signal permutation dynamics for a rank-r classical register. Since there are $(2^r)!$ distinct forms of permutation dynamics, the set of signal permutation dynamics forms a rapidly decreasing fraction of the set of all possible permutation dynamics.

6.6 Evolution and Measurement

Any experiment consists of several distinct phases. An important phase is the process of measurement itself, which ends with the extraction of classical information from an SUO. Typically this information will be in the form of real numbers, and these can always be expressed in binary form, justifying our approach.

Context plays a vital role here. When, for example, an observer reports that a particle has been observed at position $x = 1.5$, what they mean is that positive signals have been detected at some normal detector or detectors associated with the number $x = 1.5$. This assignment is based on the context of the experiment: the observer will know on the basis of prior theoretical and empirical knowledge, gained during the process of calibration (preparation of apparatus) what those detectors mean in terms of the physics of the SUO concerned, and therefore, what values of some measurable quantity those signals represent.

There is room here for error, in that the observer could associate the wrong context to the signals being observed. Such a process occurred in the Mars Climate Orbiter disaster in 1999, when there was a "failure to use metric units in the coding of a ground software file" (Mars Climate Orbiter Mishap Investigation Board, 1999). In the following, we assume all contexts have been interpreted correctly by the observer.

So far we have discussed the evolution of labstates. For each run or repetition of the experiment, this is modeled by the action of an evolution operator $\mathbb{U}_{N,0}$ mapping initial labstates at stage Σ_0 into final labstates at stage Σ_N. We need now to discuss how numbers are extracted at the end of an experiment consisting of a number of runs.

With reference to the position measurement discussed above, we model the measurement process in terms of *weighted relevant questions*. What this means is this. Suppose the final physical register $\mathcal{R}_N$ has rank r_N. Assuming the experimentalist has established that each detector is working normally, then there will be a total of 2^{r_N} possible normal labstates in this register. Therefore, the observer could ask a total of 2^{r_N} maximal questions. These questions are represented by the dual labstates $\{\overline{\boldsymbol{k}} : k = 0, 1, \ldots, 2^{r_N} - 1\}$. Given a final labstate $\boldsymbol{\Psi}_N$, the answer *yes* or *no* to each question $\overline{\boldsymbol{k}_N} \equiv \overline{\textbf{is it true that } \boldsymbol{\Psi}_N = \boldsymbol{k}\textbf{?}}$ is represented by the number one or zero, respectively, and given by the answer $\overline{\boldsymbol{k}_N}\boldsymbol{\Psi}_N$.

Now the observer will generally have some theory as to what each answer $\overline{\boldsymbol{k}_N}\boldsymbol{\Psi}_N$ means physically. In many experiments, this will be some real number X^k. Therefore, the actual number $\langle X\rangle_{\Psi_N}$ obtained at time t_N at end of a single run of the experiment can be written in the form

$$\langle X\rangle_{\Psi_N} = \overline{\boldsymbol{\Psi}_N}\mathbb{X}_N\boldsymbol{\Psi}_N, \tag{6.7}$$

where $\mathbb{X}_N \equiv \sum_{k=0}^{2^{r_N}-1} \boldsymbol{k}_N X^k \overline{\boldsymbol{k}_N}$ is an *classical observable*, a sum of dyadics representing a weighted relevant question.

Two comments are relevant. First, despite appearances, this is still a classical theory at this point. The final labstate $\boldsymbol{\Psi}_N$ is a single element in the final physical register, $\mathcal{R}_N$, not a superposition of elements. Second, there is nothing in classical mechanics that rules out weighted sums of dyadics. For any element in $\mathcal{R}_N$, all the possible answers $\overline{\boldsymbol{k}_N}\boldsymbol{\Psi}_N$ are zero except for one of them, so (6.7) returns a physically sensible value for $\langle X\rangle_{\Psi_N}$.

A further refinement, anticipating the possibility of random variations in the initial state and the extension of these ideas to quantum mechanics, is to write

$$\langle X\rangle_{\Psi_N} = Tr\{\mathbb{X}_N \boldsymbol{\varrho}_N\}, \tag{6.8}$$

where Tr represents the familiar trace process and $\boldsymbol{\varrho}_N$ is the dyadic $\boldsymbol{\Psi}_N\overline{\boldsymbol{\Psi}_N}$ analogous to a pure state density operator in quantum mechanics.

We note that $\boldsymbol{\Psi}_N = \mathbb{U}_{N,0}\boldsymbol{\Psi}_0$ and $\overline{\boldsymbol{\Psi}_N} = \overline{\boldsymbol{\Psi}_N}\ \overline{\mathbb{U}}_{N,0}$, where the evolution operator $\mathbb{U}_{N,0}$ maps elements of $\mathcal{R}_0$ into elements of $\mathcal{R}_N$ and similarly for the dual evolution operator $\overline{\mathbb{U}}_{N,0}$. In general, it will be true that

$$\overline{\mathbb{U}}_{N,0}\mathbb{U}_{N,0} = \mathbb{I}_0, \tag{6.9}$$

the identity operator for $\mathcal{R}_0$. However, because there is no requirement formally in this approach for the rank r_N of the final physical register $\mathcal{R}_N$ to equal the rank r_0 of the initial physical register $\mathcal{R}_0$, it is possible that $\mathbb{U}_{N,0}\overline{\mathbb{U}}_{N,0}$ does not equal $\mathbb{I}_N$. This corresponds to irreversible dynamics. In the analogous quantum formalism that we will discuss in the next chapter, such evolution operators are referred to as *semi-unitary*, and $\overline{\mathbb{U}}_{N,0}$ is the *retraction* of $\mathbb{U}_{N,0}$.

Using (6.9) in (6.8), we may write $\langle X\rangle_{\Psi_N} = Tr\{\mathbb{X}_N\mathbb{U}_{N,0}\boldsymbol{\varrho}_0\overline{\mathbb{U}}_{N,0}\}$, where $\boldsymbol{\varrho}_0$ is the initial dyadic $\boldsymbol{\Psi}_0\overline{\boldsymbol{\Psi}_0}$.

6.7 Random Initial States

Real experiments normally consist of a large number of repetitions or runs of a basic process. However, it cannot always be guaranteed that the initial labstate is always the same. In principle, we could start with any one of 2^{r_0} initial labstates. In such a case, a statistical approach can be taken.

Consider an extremely large number R of runs, such that there is a total of R^k runs starting with initial labstate $\boldsymbol{k}_0$, for $k = 0, 1, \ldots, 2^{r_0} - 1$. Clearly, $\sum_{k=0}^{2^{r_0}-1} R^k = R$. Then in the limit of R tending to infinity, we would assign a probability $\omega^k \equiv \lim_{R\to\infty} R^k/R$ for the initial labstate to be in state $\boldsymbol{k}_0$.

In such a scenario we define the initial density matrix

$$\boldsymbol{\varrho}_0 \equiv \sum_{k=0}^{2^{r_0}-1} \omega^k \boldsymbol{k}_0\overline{\boldsymbol{k}_0}, \tag{6.10}$$

where $\boldsymbol{k}_0$ is any one of the 2^{r_0} elements of the initial physical register $\mathcal{R}_0$ and the ω^k are probabilities summing to unity. The formalism outlined above then gives the expectation values of operators.

7
Quantum Register Dynamics

7.1 Introduction

In this chapter we move beyond the classical register scenario discussed in the previous chapter, extending the discussion to experiments described by time-dependent quantum registers of varying rank. We apply our previous discussion of the signal basis representation (SBR), the computational basis representation (CBR), signal operators, signal classes, and the CBR of signal operators to the quantum case. Our discussion of dynamics covers persistence, that is, the stability of apparatus, observers, and laboratory time, and the Born probability rule. We state the principles of quantized detector network (QDN) dynamics and show how they apply to the description of quantum experiments. We discuss the signal theorem and path summations.

7.2 Persistence

Our first step is to clarify what QDN assumes about the evolution of apparatus in time, because this affects the modeling. QDN is designed to reflect the behavior of apparatus in the real world and so it is not assumed in general that a given observer's apparatus is constant in time, even during a given run of an experiment.

Although many experiments appear to be carried out with apparatus that persists over any given run of the experiment, and indeed, perhaps over all the runs of that experiment, that is just an incidental factor that reflects no more than an economy in construction. In practice it is invariably easier and more economical to use the same equipment over and over again rather than use it once, throw it away after each run, and build a new version ready for the next run. Lest this be thought of as a trivial point, it is nevertheless an integral and costly feature of many experiments, involving maintenance and upgrading. Indeed, in laboratories such as the Large Hadron Collider, actual run time is a small fraction of total project time. A related, significant issue has to do with the concept of *ensemble*, discussed in the Appendix.

Persistence has everything to do with *time scale*, specifically, the relative laboratory time over which any piece of equipment can be meaningfully discussed as such. The degree of persistence of apparatus is as contextual as anything else in physics. If a run is relatively brief, say, over in very small fractions of a second, as in high-energy particle scattering experiments, then in such an experiment, the apparatus will behave as if it persists forever. On the other hand, in some experiments, individual runs can involve enormous intervals of time between state preparation and observation, as always happens in the case of astrophysical observations of stars and galaxies. In such cases, light from a distant star may be received by astronomers long after that star had ceased to exist as a star.

A necessary criterion for an experiment to be describable by QDN is that any detector exists at least during a given stage. Indeed, we can take the definition of a stage to be that interval of laboratory time over which a given detector can be assumed to persist, that is, have an identity that has physical relevance in the context of the experiment concerned.

In this respect, a stage is to be identified not with a moment or point in time but with the *interval* of time over which information could be extracted by the observer. Such an interval is of contextual temporal length in the laboratory, being as short or as long as the observer requires to acquire a bit or "quantum" of information. The temporal divide between two stages will be called a *quantum tick*, or *q-tick*. It is necessarily a heuristic concept, because any attempt to empirically measure a q-tick and assign a numerical value to a *chronon*, or assumed fundamental unit of time, will be self-referential in some way or other. We note that the concept of *Planck time*, commonly regarded as the shortest meaningful interval of time in physics, is a theoretical construct that could never be observed directly and has been criticized on such grounds (Meschini, 2007).

7.3 Quantized Detector Networks

In the projection-valued measure (PVM) formulation of standard quantum mechanics (QM) (von Neumann, 1955; Peres, 1995), it is generally asserted that state vectors of systems under observation (SUOs) evolve in Hilbert spaces of fixed dimension. Any time dependence of the apparatus itself, such as externally imposed time-dependent electromagnetic fields, is encoded into an explicit time dependence of the Hamiltonian or whatever observables are involved. This approach encodes the idea that experiments are done in fixed laboratories and that information is extracted from states of SUOs that evolve unitarily in time.

This temporal architecture is known as the Schrödinger picture and is widely used in QM. A variant but nominally equivalent temporal architecture is that of the *Heisenberg picture*, in which the quantum states appear frozen in time but now it is the observables of the theory that have unitary temporal evolution over and above any intrinsic, explicit evolution.

In standard QM, these two related architectures were eventually recognized as too limited. One reason is that in the PVM approach, the number of mutually orthogonal possible outcome states associated with a given observable is the number of its eigenstates, which is fixed and cannot exceed the dimension of the Hilbert space concerned. The positive operator-valued measure (POVM) formalism on the other hand was developed to deal with the possibility that the number of observational outcome channels set up in a laboratory may be different from the dimension of the Hilbert space used (Ludwig, 1983a,b; Kraus, 1983; Peres, 1995) and, indeed, may exceed that number, in the case of finite dimensions.

In contrast, the QDN formalism assumes from the outset that the Hilbert space representing outcome possibilities is always different from one stage to the next, even if the dimensionality remains constant, and even if the detectors involved in the experiment appear to persist over several or all stages of an experiment.

To be specific, let us denote the observer's apparatus at any given stage Σ_n by $\mathcal{A}_n$. Some parts of $\mathcal{A}_n$ will correspond to actual, real detectors, while other parts will be virtual, or potential, detectors. Whatever the case, $\mathcal{A}_n$ will be always be modeled by a countable number r_n of vertices at any stage of the relevant stage diagram for that experiment. In this book, we denote the ith vertex at stage Σ_n by i_n, where the superscript i runs from 1 to r_n and labels all of the distinct vertices in that stage. Note that i_n (a vertex) should not be confused with $\boldsymbol{i}_n$ (bold font), our notation for a CBR element of the preferred basis at stage Σ_n. Note also that detector vertices should not be confused with *modules*, which are nondetecting parts of apparatus that sit in the information void between stages.

In the description of real experiments, r_n will always be finite, in contrast to the situation in QM, where the Hilbert space may be infinite dimensional. The harmonic oscillator in one spatial dimension is an example where QM assumes either that the position observable has a continuous spectrum or, equivalently, that the Hamiltonian has a countable infinity of energy eigenstates. The harmonic oscillator is discussed in Chapter 24. In the world of empirical reality, there are no such harmonic oscillators, just SUOs for which such a description may be a reasonable approximation.

The question as to whether r_n is finite or not is central to many if not all of the technical difficulties encountered in the refinement of standard QM known as quantum field theory. The harmonic oscillator appears to be intimately involved in many of these problems in one way or another. Although the mathematical properties of the quantized harmonic oscillator play an essential role in accounting for the particle concept in free quantum field theory, those same properties generate fundamental problems in interacting field theories. For instance, the ultraviolet divergences encountered in most Feynman loop integrals are linked to the unbounded energy spectrum of the standard QM oscillator, while infrared divergences are linked to the assumed continuity of spacetime and the zero-point energy of the quantized oscillator.

The issue of zero-point energy of the quantized oscillator was explicitly addressed in quantum optics by Glauber in his landmark papers on photon correlations (Glauber, 1963a,b,c). He pointed out that the irreversibility of quantum detection processes involves the non-Hermitian positive energy electric field operator $\boldsymbol{E}^{(+)}(t, \boldsymbol{x})$ rather than the Hermitian electric field operator $\boldsymbol{E}(t, \boldsymbol{x}) \equiv \boldsymbol{E}^{(+)}(t, \boldsymbol{x}) + \boldsymbol{E}^{(-)}(t, \boldsymbol{x})$. It is the vacuum expectation values (VEVs) involving the $\boldsymbol{E}$ field operators that are affected by zero-point energy, whereas VEVs involving the $\boldsymbol{E}^{(+)}$ alone are not. Glauber's conclusion was the following:

> *The electric field in the vacuum undergoes zero-point oscillations which, in the correctly formulated theory, have nothing to do with the detection of photons.* (Glauber, 1963c)

According to Glauber, then, the standard quantum field theory approach to quantum detection, if it involves VEVs of Hermitian operators, will be inadequate to model irreversible, local detection processes as they actually occur in real laboratories. The reason the standard Lehmann–Symanzik–Zimmerman (LSZ) (Lehmann et al., 1955) scattering formalism works is that the *in* and *out* Hilbert spaces representing prepared and outcome states, respectively, are applied at infinitely remote times in the past and future, respectively, in a causally irreversible way. In between those two remote times, a quantum process is technically in what we have called in earlier chapters the information void, a laboratory regime during which no signal detection takes place. During such a regime, unitary evolution can be assumed to take place without violating any of the principles of QM. That is the essential reason why the standard model Lagrangian, which respects CPT[1] inversion symmetry and is used to work out unitary evolution, is a good basis for particle scattering calculations. However, finite time processes, which are generally not considered in high-energy physics, would require the same review that Glauber carried out for photonic processes.

Given the apparatus $\mathcal{A}_n$ at stage Σ_n, its associated detectors, both real and virtual, are represented by a set of signal qubits $\{Q_n^i : i = 1, 2, \ldots, r_n\}$, with qubit Q_n^i being identified with vertex i_n. Here as elsewhere, upper indices label individual qubits and detectors, while lower indices denote stages. The tensor product $Q_n^1 Q_n^2 \ldots Q_n^{r_n}$, plus the information held by the observer about the physical significance of those qubits, constitute a quantum register of rank r_n. This is a Hilbert space of dimension 2^{r_n}.

A fundamental property of any quantum register of rank greater than one is that it contains entangled states as well as separable states. Separability and entanglement are discussed in some detail in Chapter 22.

[1] Charge, space (parity), and time.

Example 7.1 Given a rank-three quantum register $Q^1Q^2Q^3$, the state $\mathbf{\Psi} \equiv \mathbf{1}^1\mathbf{0}^2\mathbf{1}^3 - \mathbf{0}^1\mathbf{0}^2\mathbf{0}^3$ is partially separable relative to the chosen basis, because we can write $\mathbf{\Psi} = \{\mathbf{1}^1\mathbf{1}^3 - \mathbf{0}^1\mathbf{0}^3\}\mathbf{0}^2$. On the other hand, $\mathbf{\Phi} \equiv \mathbf{1}^1\mathbf{0}^2\mathbf{1}^3 - \mathbf{0}^1\mathbf{1}^2\mathbf{0}^3$ cannot be written as a product, so is entangled relative to the chosen basis.

In QDN, entanglement is regarded as contextual on the observer's information about their apparatus, and not as an intrinsic property of SUOs. QDN tries to avoid terms such as "entangled photons," but we reserve the right to use such terminology occasionally, provided it does not mislead. The concept of an entangled labstate is not only perfectly acceptable in QDN but actually essential for the correct calculation of outcome signal detection probabilities in many experiments.

7.4 Persistence and Ensembles

A conventional assumption in QM is that pure states of a system under observation may be represented by time-dependent elements of a fixed Hilbert space. The chosen Hilbert space is usually assumed fixed for two reasons. First, there is the conditioned belief that an SUO "exists" in time as a separate entity long enough for the observer to study it. Another contributory factor is the *persistence of the apparatus*, or the tendency of actual apparatus to exist in its original form and functionality in a laboratory before and after its useful role has ended.

Most physics experiments deal with persistent apparatus. That is generally arranged by the observer as a matter of economy: experimentalists generally do not have the resources to scrap their apparatus at the end of each run and then rebuild it in time for the next run.

There are situations, however, where persistence cannot be assumed. For example, astronomers can catch light from a supernova shock wave only during an extremely limited time, and that particular observation cannot be repeated because the source of the signal has long gone. What helps the observers is the vast numbers of photon signals that they manage to detect during that limited window of opportunity.

A similar issue arises in quantum cosmology. The Universe is believed to be expanding. On that account, any approach to quantum cosmology should take the attendant irreversibility into consideration, and not treat the evolution of the Universe in traditional QM terms as a typical SUO. The Universe is in an ensemble of one, and not only contains the observer and the apparatus, but will outlast both of them.

In QDN, individual detectors are never persistent. Each detector is assigned a particular stage at which it operates as an detector, and outside that time, has no role in the formalism. This is the QDN analogue of the concept of an *event* in relativity. Some applications of QDN will for convenience assume persistence of apparatus (Jaroszkiewicz, 2004), the only effect being to increase the number of qubits used in the formalism.

7.5 Observers and Time

Observers generally come equipped with their own sense of time, and quantum experiments are carried out relative to that time. Relativity teaches that there are two observer-related time concepts with different properties; *coordinate time* and *proper time*. In both special relativity (SR) and general relativity (GR), the former time concept is used to label events in spacetime and is generally locally integrable. This means that spacetime can be discussed in terms of coordinate patches (Schutz, 1980). Within a given coordinate patch, events can be labeled by spacetime coordinates in a path-independent way. When a particular coordinate patch is related to clocks and rods in a specific laboratory, we shall refer to that coordinate time as *labtime*. In GR, the description of labtime requires some assumption about time-like foliations of spacetime and frame fields. On the other hand, proper time is nonintegrable, which is to say that the proper time between two events depends on the particular path taken between those events. In other words, proper time is contextual.

In QDN, the time parameter associated with an experiment can normally be identified with the proper time of an idealized inertial observer moving along a time-like worldline, and for whom their laboratory appears to be at rest at all times. In some situations, we may have to apply QDN to what we call *interframe* experiments. These involve state preparation in one inertial frame and signal detection in another. The Doppler effect in observational cosmology is an example of interframe physics.

What are important in such situations in SR and GR are *space-like hypersurfaces*: these are the analogues of the concept of stages in QDN. On a space-like hypersurface, no two events can be causally connected; that is, no event can be the cause of any other event on the same hypersurface of simultaneity. In QDN, the analogous statement is that no two detectors on a given stage can affect each other in any way. This could be because the detectors are really relatively space-like separated, but the possibility exists, frequently in real experiments, that the detectors are relatively time-like separated but shielded from each other.

In the real world, observers have finite existence: they come and go. Observers and their apparatus are created at certain times and disappear at later times, as seen by other observers in the wider universe. QDN as formulated here allows for a discussion of different observers, each with their individual time parameters and lifetimes. The use of quantum registers also raises the possibility of accounting for the origin of various temporally related concepts such as light cones, time dilation, and other metric-based phenomena in terms of quantum register dynamics. A useful way to discuss what is going on is in terms of *causal sets*, the structures of which arise naturally within quantum register dynamics (Eakins and Jaroszkiewicz, 2005). Causal sets are discussed in Chapter 23.

During their operational lifetimes, observers quantify their time in terms of real numbers, usually read from clocks. Most clocks give only a crude estimate of the passage of time, and as a result, the ordinary human perception of time as a one-dimensional continuum is just a convenient approximation. The classical view of time is that it is a continuum at all scales and for all phenomena. Certainly, things appear consistent with that view in the ordinary world.

In QM, however, the situation is quite different. What matters in a quantum experiment is information acquisition from the observer's apparatus and this can only ever be done in a discrete way, regardless of any theoretical assumption to the contrary (Misra and Sudarshan, 1977). While an observer's effective sense of time can be modeled accurately as continuous, it is certainly the case that an observer can look at a detector and determine its status in a discrete way only. There are no truly continuous-in-time observations. It is important here to distinguish between what experimentalists actually do in experiments and what theorists imagine they do.

The discreteness of the information extraction process forms the basis of the time concept in QDN. In general, a given observer will represent the state of their apparatus (the labstate) at a finite sequence of their own (observer) times, denoted by the integer n. These times will be referred to as *stages.* In QDN, a pure labstate at stage Σ_n will be denoted by $\boldsymbol{\Psi}_n$.

In QDN, stage Σ_{n+1} is always regarded as definitely *later* than stage Σ_n. There is no scope in QDN for the concept of closed time-like curve (CTC) found in some GR spacetimes, such as the Gödel model (Gödel, 1949). Invariably, discussions that do involve CTCs cannot accommodate quantum processes properly, at least not those that involve probabilities.[2]

A final point on this topic: there is no need to assume that the laboratory time interval between stages has a definite value or that a succession of stages involves equal intervals of labtime.

7.6 The Born Probability Rule

One of the most significant attributes of quantum processes is the randomness of quantum outcomes. Given identical state preparation, different runs of a given experiment generally demonstrate *controlled* unpredictability. Controlled, because the observer may know all about the range of possible outcomes and the probabilities of those outcomes before one is actually observed, but unpredictable because the observer cannot in general say beforehand which particular one of those outcomes will occur in any particular run.

[2] This is for the good reason that the probability concept requires an observer accumulating information with an irreversible sense of time, something that a genuine CTC is incompatible with.

Remark 7.2 This subject is a minefield of issues that lurk unseen until we stumble across them. Reference was made to "identical state preparation" in the previous paragraph. That is clearly a vacuous concept. By definition, different runs of a given experiment cannot have absolutely identical state preparation: the Universe will have aged for sure during any two runs, and there will be vast changes in the local environment of any apparatus on atomic scales. Yet clearly, most of those changes will *not* influence the outcome probabilities. As pointed out by Kraus, what appears significant are *equivalence classes* of state preparation processes (Kraus, 1983). Exactly where the line between influence and noninfluence of external factors in state preparation should be drawn is one of those deep questions that no one knows anything much about. The *Heisenberg cut* is a hypothetical line demarcating the classical world of the observer and the quantum world of the SUO. We propose the term *Kraus cut* to denote the hypothetical line between those factors that have no influence on state preparation or outcome and those that do. All we can do is to observe that some things are critical in state preparation, and everything else seems unimportant. Much the same point is emphasized in Peres (1995).

In practice the QM approach to probability works brilliantly and we use it in QDN. The Born probability rule (Born, 1926) in QM states that if a final state $|\Psi\rangle$ is represented by a superposition of the form

$$|\Psi\rangle = \sum_{i=1}^{d} \Psi^i |i\rangle, \tag{7.1}$$

where the possible outcomes are represented by orthonormal vectors $|i\rangle$, $i = 1, 2, \ldots, d$, in some Hilbert space, then the conditional probability P_i of outcome $|i\rangle$ is given by $P_i = |\langle i|\Psi\rangle|^2$, if the final state is normalized to unity.

This rule is used in much the same way in QDN, as follows. Consider a normalized pure labstate $\boldsymbol{\Psi}_n$ at stage Σ_n. This can always be expanded in terms of the computational basis representation (CBR) of the preferred register basis B_n at that stage in the form

$$\boldsymbol{\Psi}_n = \sum_{i=0}^{2^{r_n}-1} \Psi_n^i \boldsymbol{i}_n, \tag{7.2}$$

where the coefficients Ψ_n^i are complex and

$$\sum_{i=0}^{2^{r_n}-1} |\Psi_n^i|^2 = 1. \tag{7.3}$$

In QDN, labstates are usually normalized to unity for convenience. Because the preferred basis states form a complete orthonormal basis set, we may immediately read off from (7.2) the various CBR conditional probabilities $\Pr(\boldsymbol{i}_n|\boldsymbol{\Psi}_n)$, which are given by the rule

$$\Pr(\boldsymbol{i}_n|\boldsymbol{\Psi}_n) \equiv |\overline{\boldsymbol{i}_n}\boldsymbol{\Psi}_n|^2 = |\Psi_n^i|^2, \quad 0 \leqslant i < 2^{r_n}. \tag{7.4}$$

$\Pr(\boldsymbol{i}_n|\boldsymbol{\Psi}_n)$ is the conditional (Bayesian) probability for the observer to find the apparatus in signal state $\boldsymbol{i}_n$ at stage Σ_n, *if the observer looked at their apparatus at that time.* These probabilities are conditional on the observer being sure, just before they looked, that the labstate at stage Σ_n was $\boldsymbol{\Psi}_n$.

There is no natural restriction in QDN to labstates that are eigenstates of signality; i.e., superpositions of basis states from different signality classes are permitted in principle. QDN is analogous in this respect to the extension of Schrödinger wave mechanics to Fock space and to quantum field theory.

7.7 Principles of QDN Dynamics

We are now in a position to discuss the principles of labstate dynamics from the perspective of a single observer. At stage Σ_n, this observer will hold in their memory current information about their apparatus $\mathcal{A}_n$, the associated quantum register $\mathcal{Q}_n$, and the labstate $\boldsymbol{\Psi}_n$ (assumed pure here). An analogous statement will hold for each stage in a finite sequence of stages running from some initial stage Σ_0 to some final stage Σ_N, where $N > 0$. QDN does not assume observers exist over unbounded intervals of time, so the formalism is applied only over a finite number of stages.

We restrict attention in this chapter to pure labstates, that is, labstates fully specified by single elements in a quantum register. A mixed-state, density matrix approach to QDN dynamics is discussed in Chapter 9.

For the most basic sort of experiment, labstate preparation will be assumed to have taken place by initial stage Σ_0 and outcome detection is to take place at final stage Σ_N. For each integer n such that $0 \leqslant n \leqslant N$, the observer associates with their apparatus $\mathcal{A}_n$ at that time a quantum register $\mathcal{Q}_n$. This register is the tensor product $\mathcal{Q}_n \equiv Q_n^1 Q_n^2 \dots Q_n^{r_n}$ of a finite number r_n of qubits, $Q_n^1, Q_n^2, \dots, Q_n^{r_n}$, each qubit Q_n^i representing the ith real or virtual detector i_n in $\mathcal{A}_n$. The quantum register $\mathcal{Q}_n$ is a Hilbert space with preferred basis B_n consisting of the 2^{r_n} CBR signal states.

There is no requirement in QDN or implication in our notation for the detector represented by Q_{n+1}^i to be related in any obvious way to the detector represented by Q_n^i; that is, we do not assume persistence. In other words, successive quantum registers are understood as different Hilbert spaces, even if they have the same rank, that is, if $r_{n+1} = r_n$. This is one of the factors which makes QDN more general in its scope than standard QM, although all the principles of QM are incorporated in QDN.

At stage Σ_n, the observer describes the quantum state of their apparatus at that time by a labstate $\boldsymbol{\Psi}_n$, which is some normalized vector in $\mathcal{Q}_n$. Using the CBR, this state can be written in the form (7.2) and normalized according to (7.3).

A given run of an experiment will be described by the observer in terms of a sequence $\{\boldsymbol{\Psi}_n : 0 \leqslant n \leqslant N\}$ of normalized labstates, each element of which is associated with a particular quantum register $\mathcal{Q}_n$, ending with state outcome observation at the final stage Σ_N.[3] The question now is how successive labstates relate to each other between times M and N.

Provided each run is prepared in the same way, and provided the apparatus during eah run is controlled in the same way, we can discuss a typical labstate $\boldsymbol{\Psi}_n$ as a representative for an ensemble of runs. Indeed, that is the only interpretation that makes sense if an objectivized or hidden-variables interpretation of the wave function is to be avoided. QDN follows the traditional QM view as emphasized by Peres in this respect (Peres, 1995).

The dynamical transition from labstate $\boldsymbol{\Psi}_n$ to labstate $\boldsymbol{\Psi}_{n+1}$ involves a mapping from quantum register $\mathcal{Q}_n$ to quantum register $\mathcal{Q}_{n+1}$ that satisfies two criteria fundamental to quantum mechanics: *linearity* and *norm preservation*. This leads us to give a number of definitions and theorems that have proved central to QDN.

7.8 Born Maps and Semi-unitarity

In the following, we restrict the discussion to Hilbert spaces of finite dimension. This is in line with our general philosophy in QDN that there are no empirically observable infinities in physics.

Definition 7.3 A *Born map* is a norm-preserving map from one Hilbert space $\mathcal{H}$ to some other Hilbert space $\mathcal{H}'$; if Ψ in $\mathcal{H}$ is mapped into $\Psi' \equiv \mathfrak{B}(\Psi)$ in $\mathcal{H}'$ by a Born map $\mathfrak{B}$, then $(\Psi', \Psi')' = (\Psi, \Psi)$.

Born maps are used in QDN in order to preserve total probabilities (hence the terminology), but unfortunately, their properties are insufficient to model all quantum processes. Born maps are not necessarily linear, as can be seen from the elementary example $\mathfrak{B}(\Psi) = |\Psi|\Phi'$ for all Ψ in $\mathcal{H}$, where Φ' is a fixed element of $\mathcal{H}'$ normalized to unity and $|\Psi|$ is the norm of Ψ in $\mathcal{H}$. To go further, it is necessary to impose linearity.

Henceforth, we adopt the rule that if U is a linear map, then we may drop the parentheses and write $U\Psi$ instead of $U(\Psi)$.

Definition 7.4 A *semi-unitary operator* is a linear Born map. If U is such a map, then for any elements ψ, ϕ in $\mathcal{H}$ and complex α, β, we may write $|\alpha\psi + \beta\phi| = |\alpha U\psi + \beta U\phi|'$.

[3] QDN allows partial observations to be made at intermediate stages.

The proofs of the following theorems are left to the reader.

Theorem 7.5 *A semi-unitary operator from Hilbert space $\mathcal{H}$ to Hilbert space $\mathcal{H}'$ exists if and only if* $\dim \mathcal{H} \leqslant \dim \mathcal{H}'$.

Theorem 7.6 *A semi-unitary operator from $\mathcal{H}$ to $\mathcal{H}'$ is an injection; that is, $\{U\psi = U\phi\}$ if and only if $\psi = \phi$.*

Theorem 7.7 *If U is a semi-unitary operator from $\mathcal{H}$ to $\mathcal{H}'$, then its retraction $\overline{U}$ exists; that is, we have $\overline{U}U = I_{\mathcal{H}}$, where $I_{\mathcal{H}}$ is the identity operator over $\mathcal{H}$.*

Corollary 7.8 A semi-unitary operator preserves inner products and not just norms. This means that an orthonormal basis set for Hilbert space $\mathcal{H}$ is mapped by a semi-unitary operator into a mutually orthonormal set of vectors in $\mathcal{H}'$ with the same cardinality.

Theorem 7.9 *If U is a semi-unitary operator from $\mathcal{H}$ to $\mathcal{H}'$ and* $\dim \mathcal{H} = \dim \mathcal{H}'$, *then the retraction $\overline{U}$ of U is also a semi-unitary operator from $\mathcal{H}'$ to $\mathcal{H}$. For such an operator, $\overline{U}U = I_{\mathcal{H}}$ and $U\overline{U} = I_{\mathcal{H}'}$.*

Definition 7.10 An operator U satisfying the conditions of Theorem 7.9 will be called a *unitary*.

Remark 7.11 If U is a semi-unitary operator from $\mathcal{H}$ to $\mathcal{H}'$ and $\dim \mathcal{H} < \dim \mathcal{H}'$, then $U\overline{U} \neq I_{\mathcal{H}'}$, simply because the retraction $\overline{U}$ is defined not over $\mathcal{H}'$ but over the proper subset $U(\mathcal{H})$ of $\mathcal{H}'$.

7.9 Application to Dynamics

It is normally assumed in QDN that a labstate $\boldsymbol{\Psi}_n$ in $\mathcal{Q}_n$ at stage Σ_n is mapped into a labstate $\boldsymbol{\Psi}_{n+1}$ in $\mathcal{Q}_{n+1}$ by some Born map $\mathcal{B}_{n+1,n}$. Because $\overline{\boldsymbol{\Psi}_{n+1}}\boldsymbol{\Psi}_{n+1} = \overline{\Psi_n}\Psi_n$ under such a map, the Born rule used in conjunction with the signal bases B_n and B_{n+1} means that total probability is conserved. This is not the same thing as conservation of signality, charge, particle number, or any other quantum variable.

The following three scenarios are possible.

$\mathfrak{B}_{n+1,n}$ Is Nonlinear

By Theorem 7.5, nonlinearity is unavoidable if the rank r_n of $\mathcal{Q}_n$ is greater than the rank r_{n+1} of $\mathcal{Q}_{n+1}$, but can arise even if this is not the case. Nonlinearity here is interpreted as a marker of classical intervention by the observer. For example, switching off any apparatus at stage Σ_{n+1} would be modeled by the Born map $\mathcal{B}_{n+1,n}(\boldsymbol{\Psi}_n) = \mathbf{0}_{n+1}$ for any state $\boldsymbol{\Psi}_n$ in $\mathcal{Q}_n$, where $\mathbf{0}_{n+1}$ is the signal ground state of the apparatus at stage Σ_{n+1}. Another example is *state reduction* due to observation; that is, if at stage Σ_{n+1} the observer actually looks at the apparatus and determines its signal status, then this would be modeled by the nonlinear Born map $\mathfrak{B}_{n+1,n}(\boldsymbol{\Psi}_n) = \boldsymbol{k}_{n+1}$, where now $\boldsymbol{k}_{n+1}$ is some element of the CBR of the preferred basis B_{n+1}, chosen randomly with a probability weighting given by the Born rule. In this particular case, however, there are actually *two* labstates associated with stage Σ_{n+1}: $\boldsymbol{\Psi}_{n+1}$ representing the state of the apparatus immediately prior to state reduction and $\boldsymbol{k}$ representing the actual observed outcome immediately after. None of this represents anything more than mathematical modeling of the observer's actual or potential knowledge about the signal status of their apparatus. QDN makes no comment on whether anything deeper than changes in apparatus signal status has occurred. In particular, any speculation of superluminal information flow concerns *correlations*, which is a phenomenon that occurs in all forms of mechanics and has everything to do with what the observer did in the past.

$\mathfrak{B}_{n+1,n}$ Is Linear and $r_n = r_{n+1}$

This scenario corresponds to unitary evolution in standard QM, and to reflect this, we use the notation $\mathcal{B}_n(\boldsymbol{\Psi}_n) \equiv \mathbb{U}_{n+1,n}\boldsymbol{\Psi}_n = \boldsymbol{\Psi}_{n+1}$. From Theorem 7.9, $\mathbb{U}_{n+1,n}$ in this case satisfies the rules

$$\overline{\mathbb{U}}_{n+1,n}\mathbb{U}_{n+1,n} = \mathbb{I}_n, \qquad \mathbb{U}_{n+1,n}\overline{\mathbb{U}}_{n+1,n} = \mathbb{I}_{n+1}, \tag{7.5}$$

where $\overline{\mathbb{U}}_{n+1,n}$ is the retraction of $\mathbb{U}_{n+1,n}$. In such a case it is reasonable to call $\mathbb{U}_{n+1,n}$ *unitary*, being the formal analogue of a unitary operator in QM.

$\mathfrak{B}_{n+1,n}$ Is Linear and $r_n < r_{n+1}$

In this case we use the same notation as in the second case above, i.e., $\mathcal{B}_{n+1,n}(\boldsymbol{\Psi}_n) \equiv \mathbb{U}_{n+1,n}\boldsymbol{\Psi}_n = \boldsymbol{\Psi}_{n+1}$, but now $\mathbb{U}_{n+1,n}$ is properly semi-unitary and only the first relation $\overline{\mathbb{U}}_{n+1,n}\mathbb{U}_{n+1,n} = \mathbb{I}_n$ in (7.5) carries over. Such a scenario arises in particle decay experiments, for example. These are discussed in Chapter 15.

We cannot in general expect the rank r_n of the quantum register $\mathcal{Q}_n$ to be constant with n, so if we wish to preserve probability and restrict the dynamical evolution to be linear in the labstate, then we have to assume

$$r_M \leqslant r_{M+1} \leqslant \cdots \leqslant r_n \leqslant \cdots \leqslant r_N, \tag{7.6}$$

where Σ_M is the initial stage and Σ_N is the final stage. From this, we can appreciate that unless experimentalists are careful, their quantum registers will grow irreversibly in rank. On the other hand, the particle decay experiments discussed in Chapter 15 specifically require the rank to increase at each time step.

The use of Born maps means total probability is always conserved, even if linearity is absent. In principle, therefore, QDN allows for a discussion of nonlinear QM, still based on most of the familiar Hilbert space concepts used in QM. As we have mentioned in the case of nonlinear Born maps, necessarily nonlinear processes such as state preparation, state reduction, the switching on and off of apparatus, and so on, which are outside the scope of unitary (Schrödinger) evolution in standard QM, can all be discussed in QDN in terms of nonlinear Born maps.

Our interest will generally be in experiments based on linear quantum processes, so (7.6) will be assumed. For such an experiment running from stage Σ_M to stage Σ_N, for $N > M$, and knowing $r_n \leqslant r_{n+1}$, then the labstate $\boldsymbol{\Psi}_n$ will change according to the rule

$$\boldsymbol{\Psi}_n \to \boldsymbol{\Psi}_{n+1} \equiv \mathbb{U}_{n+1,n}\boldsymbol{\Psi}_n, \quad M \leqslant n < N, \tag{7.7}$$

where $\mathbb{U}_{n+1,n}$ is a semi-unitary operator (this terminology will be used from now on even in the case where $r_n = r_{n+1}$).

The CBR at stages Σ_n and Σ_{n+1} can be used to represent $\mathbb{U}_{n+1,n}$. Specifically, we find

$$\mathbb{U}_{n+1,n} = \sum_{j=0}^{2^{r_{n+1}}-1} \sum_{i=0}^{2^{r_n}-1} \boldsymbol{j}_{n+1} U^{j,i}_{n+1,n} \overline{\boldsymbol{i}_n}. \tag{7.8}$$

This representation can be used to define a retraction operator $\overline{\mathbb{U}}_{n+1,n}$, given by

$$\overline{\mathbb{U}}_{n+1,n} = \sum_{j=0}^{2^{r_{n+1}}-1} \sum_{i=0}^{2^{r_n}-1} \boldsymbol{i}_n U^{j,i*}_{n+1,n} \overline{\boldsymbol{j}_{n+1}}, \tag{7.9}$$

where $U^{j,i*}_{n+1,n}$ is the complex conjugate of $U^{j,i}_{n+1,n}$, if the semi-unitarity condition holds, that is, if

$$\sum_{j=0}^{2^{r_{n+1}}-1} U^{j,i}_{n+1,n} U^{j,k*}_{n+1,n} = \delta^{ik}. \tag{7.10}$$

A useful way of thinking about the semi-unitarity condition (7.10) and the rules of semi-unitary operators is in terms of complex vectors as follows. Let V be a finite-dimensional complex vector space with inner product $\boldsymbol{a}^\dagger \cdot \boldsymbol{b}$. An *orthonormal d-subset* of V is any set of elements $\{\boldsymbol{a}^i : i = 1, 2, \ldots, d\}$ of V that are normalized to unity and mutually orthogonal; that is, we have the rule $\boldsymbol{a}^{i\dagger} \cdot \boldsymbol{a}^j = \delta^{ij}$.

Now let V' be another finite-dimensional complex vector space with inner product $\boldsymbol{a}'^\dagger \cdot \boldsymbol{b}'$. The question of semi-unitarity reduces to the possibility of finding an injection from any given orthonormal d-subset of V into at least one orthonormal d-subset of V'. The above semi-unitarity theorems tell us that this cannot be done if $d > \dim V'$.

7.10 The Signal Theorem

The mathematical properties of semi-unitary operators and their relationship to signal bases have an important bearing on the permitted physics of QDN dynamics. Consider an experiment at stages Σ_n and Σ_{n+1} and assume semi-unitarity. At stage Σ_n the labstate $\boldsymbol{\Psi}_n$ is given by a superposition of signal states from signal basis $B_n \equiv \{\boldsymbol{i}_n : 0 \leqslant i < 2^{r_n}\}$, while the labstate $\boldsymbol{\Psi}_{n+1}$ is given as a superposition of signal states from signal basis $B_{n+1} \equiv \{\boldsymbol{i}_{n+1} : 0 \leqslant i < 2^{r_{n+1}}\}$. Because of linearity, the crucial question as far as the dynamics is concerned is how individual signal states evolve. Semi-unitarity imposes the following constraint, which we call the *signal theorem*:

Theorem 7.12 *Two different preferred basis states $\boldsymbol{i}_n$ and $\boldsymbol{j}_n$ in a signal basis B_n cannot evolve by semi-unitary dynamics into labstates that have only one preferred signal basis state in common.*

Proof Take $0 \leqslant i < j < 2^{r_n}$. Suppose $\boldsymbol{i}_n$ evolves by semi-unitarity dynamics into a labstate according to the rule

$$\boldsymbol{i}_n \to \mathbb{U}_{n+1,n}\boldsymbol{i}_n = \alpha \boldsymbol{k}_{n+1} + \boldsymbol{\phi}_{n+1}, \tag{7.11}$$

while $\boldsymbol{j}_n$ evolves according to the rule

$$\boldsymbol{j}_n \to \mathbb{U}_{n+1,n}\boldsymbol{j}_n = \beta \boldsymbol{k}_{n+1} + \boldsymbol{\psi}_{n+1}. \tag{7.12}$$

Here k is some integer in the semi-open interval $[0, 2^{r_{n+1}})$, α and β are non-zero complex numbers, and $\boldsymbol{\phi}_{n+1}$ and $\boldsymbol{\psi}_{n+1}$ are elements in $\mathcal{Q}_{n+1}$ sharing no signal states in common either with each other or with $\boldsymbol{k}_{n+1}$ in their CBR basis expansions, which means

$$\overline{\boldsymbol{k}_{n+1}}\boldsymbol{\phi}_{n+1} = \overline{\boldsymbol{k}_{n+1}}\boldsymbol{\psi}_{n+1} = \overline{\boldsymbol{\phi}_{n+1}}\boldsymbol{\psi}_{n+1} = 0. \tag{7.13}$$

From Corollary 7.8, semi-unitarity preserves inner products and not just norms, so we must have

$$0 = \overline{\boldsymbol{i}_n}\boldsymbol{j}_n = \overline{\boldsymbol{i}_n \mathbb{U}_{n+1,n}}\mathbb{U}_{n+1,n}\boldsymbol{j}_n = \overline{\alpha \boldsymbol{k}_{n+1} + \boldsymbol{\phi}_{n+1}}(\beta \boldsymbol{k}_{n+1} + \boldsymbol{\psi}_{n+1}) \tag{7.14}$$

$$= (\alpha^* \overline{\boldsymbol{k}_{n+1}} + \overline{\boldsymbol{\phi}_{n+1}})(\beta \boldsymbol{k}_{n+1} + \boldsymbol{\psi}_{n+1}) = \alpha^* \beta, \tag{7.15}$$

using (7.12). This establishes the theorem. □

The signal theorem leads to the following important result for conventional physics. Suppose an observer constructs an apparatus that, if prepared at stage Σ_n to be in its signal ground state $\mathbf{0}_n$, would evolve into $\mathbf{0}_{n+1}$. If the dynamics is semi-unitary, then we may write

$$\mathbf{0}_n \to \mathbb{U}_{n+1,n}\mathbf{0}_n = \mathbf{0}_{n+1}. \tag{7.16}$$

This condition models an important physical property expected of most laboratory apparatus; we would not expect equipment that was in its signal ground state to spontaneously generate outcome signals subsequently, unless it was interfered with by some external agency. Any apparatus that satisfies (7.16) will be called *calibrated* between stages Σ_n and Σ_{n+1} on that account. The analogue of such a situation in Schwinger's source theoretic approach to quantum field theory (Schwinger, 1969) would be one where the external sources were switched off during some interval of time, so that the vacuum (empty space) remained unchanged during that time.

Suppose now that, given such a calibrated apparatus, the observer had instead prepared at stage Σ_n some labstate $\boldsymbol{\Psi}_n$ of the CBR form

$$\boldsymbol{\Psi}_n = \sum_{i=1}^{2^{r_n}-1} \Psi^i \boldsymbol{i}_n, \tag{7.17}$$

that is, a labstate with no signal ground component (note that the summation (7.17) in runs from 1, not 0). Then for calibrated apparatus with semi-unitary labstate evolution, the signal theorem tells us that there can be no signal ground component in the labstate $\boldsymbol{\Psi}_{n+1}$ at time $n+1$, and so we may write

$$\boldsymbol{\Psi}_n \to \mathbb{U}_{n+1,n}\boldsymbol{\Psi}_n = \sum_{j=1}^{2^{r_{n+1}}-1} \Phi^j \boldsymbol{j}_{n+1}, \tag{7.18}$$

where

$$\Phi^j = \sum_{i=1}^{2^{r_n}-1} U^{ji}_{n+1,n}\Psi^i. \tag{7.19}$$

This is an important result, because it tells us that under normal circumstances, calibrated apparatus does not normally fall into its signal ground state during an experiment, unless forced to do so by an external agency, such as the observer switching it off.

Example 7.13 Consider a calibrated rank-one apparatus evolving into a rank-one apparatus between stages Σ_n and Σ_{n+1} under semi-unitary evolution. Then by a suitable choice of phase of basis elements, we may always write

$$\begin{aligned} \mathbf{0}_n &\to \mathbb{U}_{n+1,n}\mathbf{0}_n = \mathbf{0}_{n+1}, \\ \mathbf{1}_n &\to \mathbb{U}_{n+1,n}\mathbf{1}_n = \mathbf{1}_{n+1}, \end{aligned} \tag{7.20}$$

from which we conclude the dynamics is essentially trivial.

The following example is important, as it models what happens in various quantum optics modules such as beam splitters and Wollaston prisms.

Example 7.14 Consider a calibrated rank-two apparatus evolving into a rank-two apparatus between stages Σ_n and Σ_{n+1} under semi-unitary evolution. Then calibration means that we must have $\mathbb{U}_{n+1,n}\mathbf{0}_n = \mathbf{0}_{n+1}$. Suppose further that it is known that any signality-one labstate always evolves into a signality-one labstate. Then we may write

$$\begin{aligned}\widehat{\mathbb{A}}_n^1\mathbf{0}_n &\equiv \mathbf{1}_n \to \mathbb{U}_{n+1,n}\mathbf{1}_n = \alpha\mathbf{1}_{n+1} + \beta\mathbf{2}_{n+1},\\ \widehat{\mathbb{A}}_n^2\mathbf{0}_n &\equiv \mathbf{2}_n \to \mathbb{U}_{n+1,n}\mathbf{2}_n = \gamma\mathbf{1}_{n+1} + \delta\mathbf{2}_{n+1},\end{aligned} \tag{7.21}$$

where the complex coefficients α, β, γ, and δ satisfy the unitarity constraints

$$|\alpha|^2 + |\beta|^2 = |\gamma|^2 + |\delta|^2 = 1, \quad \alpha^*\gamma + \beta^*\delta = 0. \tag{7.22}$$

Now consider the signality-two state $\widehat{\mathbb{A}}_n^1\widehat{\mathbb{A}}_n^2\mathbf{0}_n \equiv \mathbf{3}_n$. Since the apparatus is calibrated, this state must necessarily evolve into a state that has no signal ground component. Therefore, we may write

$$\mathbb{U}_{n+1,n}\mathbf{3}_n = a\mathbf{1}_{n+1} + b\mathbf{2}_{n+1} + c\mathbf{3}_{n+1}, \tag{7.23}$$

where $|a|^2 + |b|^2 + |c|^2 = 1$. Moreover, since the evolution is given as semi-unitary, then inner products are preserved. Hence we deduce

$$\alpha^*a + \beta^*b = \gamma^*a + \delta^*b = 0. \tag{7.24}$$

Writing these relations in matrix form, we find

$$\begin{bmatrix}\alpha^* & \beta^*\\ \gamma^* & \delta^*\end{bmatrix}\begin{bmatrix}a\\ b\end{bmatrix} = \begin{bmatrix}0\\ 0\end{bmatrix}. \tag{7.25}$$

The 2×2 matrix on the left-hand side of this expression is necessarily invertible, leading to the conclusion that $a = b = 0$ and therefore that

$$\mathbf{3}_n \to \mathbb{U}_{n+1,n}\mathbf{3}_n = \mathbf{3}_{n+1}, \tag{7.26}$$

modulo some arbitrary phase.

The following application of the signal theorem is surprising and somewhat counterintuitive, because what appears to be a trivial mathematical result rules out an entire class of physics experiment.

Example 7.15 Suppose an experimentalist prepares a rank-two, signality-one labstate of the form

$$\boldsymbol{\Psi}_n = (\alpha\widehat{\mathbb{A}}_n^1 + \beta\widehat{\mathbb{A}}_n^2)\mathbf{0}_n, \tag{7.27}$$

where $|\alpha|^2 + |\beta|^2 = 1$. Suppose further that the dynamics is semi-unitary and that the apparatus at stage Σ_{n+1} is of rank three. Then by the signal theorem, semi-unitary evolution such that

$$\widehat{\mathbb{A}}_n^1 \mathbf{0}_n \to \mathbb{U}_{n+1,n} \widehat{\mathbb{A}}_n^1 \mathbf{0}_n = (a\widehat{\mathbb{A}}_{n+1}^1 + b\widehat{\mathbb{A}}_{n+1}^2)\mathbf{0}_{n+1}, \quad |a|^2 + |b|^2 = 1,$$
$$\widehat{\mathbb{A}}_n^2 \mathbf{0}_n \to \mathbb{U}_{n+1,n} \widehat{\mathbb{A}}_n^2 \mathbf{0}_n = (c\widehat{\mathbb{A}}_{n+1}^2 + d\widehat{\mathbb{A}}_{n+1}^3)\mathbf{0}_{n+1}, \quad |c|^2 + |d|^2 = 1, \quad (7.28)$$

is not possible.

This result tells us that a double-slit type of experiment where each slit has only one quantum outcome site in common with the other cannot be physically constructed. Experiments where two or more quantum outcome sites are in common are possible, and then inevitable quantum interference terms will occur in final state amplitudes. For example, in a standard double-slit experiment, every site on the detector screen is affected by the presence of each of the two slits.

This result reinforces an important rule in QM: we cannot simply add pieces of apparatus together and expect the result to conform to an addition of classical expectations. A double-slit experiment where both slits are open is not equivalent to two single-slit experiments run coincidentally and simultaneously.

7.11 Null Evolution

There is an interesting class of quantum process described by evolution operators called *null evolution operators*, associated with the concept of a *null test*. Recall that a null test is one that occurs between two or more stages but no information is extracted. The phenomenon of *persistence* is associated with null evolution operators

Persistence

To understand the action of a null evolution operator, consider the idealized scenario of an initial labstate $\mathbf{\Psi}_n$ in a rank r quantum register $\mathcal{Q}_n$ evolving into labstate $\mathbf{\Psi}_{n+1}$ in a quantum register $\mathcal{Q}_{n+1}$ of the same rank r. In the following, we shall use the CBR at all stages.

Suppose we are given that

$$\mathbf{\Psi}_n = \sum_{i=0}^{2^r-1} \Psi_n^i \boldsymbol{i}_n. \quad (7.29)$$

Now consider a particular evolution operator $\mathbb{N}_{n+1,n}$ defined by

$$\mathbb{N}_{n+1,n} \equiv \sum_{j=0}^{2^r-1} \boldsymbol{j}_{n+1} \overline{\boldsymbol{j}_n}. \quad (7.30)$$

Then under this evolution operator,

$$\mathbf{\Psi}_n \to \mathbb{N}_{n+1,n} \mathbf{\Psi}_n = \sum_{j=0}^{2^r-1} \boldsymbol{j}_{n+1} \overline{\boldsymbol{j}_n} \sum_{i=0}^{2^r-1} \Psi_n^i \boldsymbol{i}_n = \sum_{i,j=0}^{2^r-1} \Psi_n^i \boldsymbol{j}_{n+1} \underbrace{\overline{\boldsymbol{j}_n} \boldsymbol{i}_n}_{\delta^{ij}} = \sum_{i=0}^{2^r-1} \Psi_n^i \boldsymbol{i}_{n+1}. \quad (7.31)$$

This is a labstate at stage Σ_{n+1} with exactly the same coefficient profile, that is, set of coefficients $\{\Psi_n^i\}$, as the initial labstate. It is reasonable in this context to refer to this phenomenon as an example of *persistence* and refer to $\boldsymbol{\Psi}_{n+1}$ as a *persistent image* of $\boldsymbol{\Psi}_n$.

Such an evolution operator will be referred to as a *null evolution operator* (NEO). An NEO $\mathbb{N}_{n+1,n}$ makes sense only under particular circumstances: the rank of the quantum register at stage Σ_{n+1} must be the same as that at stage Σ_n *and* there has to be a one-to-one identification of the elements of the preferred bases.

This form of evolution demonstrates the two faces of time: on the one hand, labtime (the time of the observer) goes on as normal, being collated with relative external context such as the expansion of the Universe. The labtime of a given observer cannot be reversed relative to other observers, according to all known current physics, although it can be slowed relative to the labtime of other observers. On the other hand, some objects such as labstate profiles of persistent labstates may appear to be indifferent to labtime.

Dynamical Null Tests

In Chapter 11 we discuss Newton's famous experiment that showed how a light beam incident on one prism would split into a spectrum of subbeams that could subsequently be refocused onto a second prism and recombined once more into a single beam. This is an example of a nontrivial null test. Because the action of the first prism is nontrivial, and therefore requires nontrivial "undoing" by the second prism, we refer to the overall process as an example of a *dynamical null test*. The generic QDN description of such tests is based on the following.

Consider an labstate $\boldsymbol{\Psi}_n$ in initial quantum register $\mathcal{Q}_n$ of rank $r = r_n$, given by (7.29). Now apply semi-unitary evolution from stage Σ_n to stage Σ_{n+1} given by evolution operator

$$\mathbb{U}_{n+1,n} = \sum_{i=0}^{2^{r_{n+1}}-1} \sum_{j=0}^{2^{r_n}-1} \boldsymbol{i}_{n+1} U_{n+1,n}^{i,j} \overline{\boldsymbol{j}_n}, \tag{7.32}$$

where $\widehat{d_n} \equiv \dim \mathcal{Q}_n - 1$, $\widehat{d}_{n+1} \equiv \dim \mathcal{Q}_{n+1} - 1$, and the coefficients $\left\{U_{n+1,n}^{i,j}\right\}$ satisfy the semi-unitary condition (7.10). According to our theorems on semi-unitary operators, we require r_{n+1}, the rank of $\mathcal{Q}_{n+1}$, to satisfy $r_{n+1} \geqslant r_n$.

The labstate $\boldsymbol{\Psi}_{n+1}$ at stage Σ_{n+1} is given by

$$\boldsymbol{\Psi}_{n+1} = \mathbb{U}_{n+1,n} \boldsymbol{\Psi}_n = \sum_{i=0}^{2^{r_{n+1}}-1} \sum_{j=0}^{2^{r_n}-1} \boldsymbol{i}_{n+1} U_{n+1,n}^{i,j} \Psi_n^j. \tag{7.33}$$

Clearly, $\boldsymbol{\Psi}_{n+1}$ will not be a persistent image of $\boldsymbol{\Psi}_n$ in general. Now consider stage Σ_{n+2}, and suppose that the rank r_{n+2} of $\mathcal{Q}_{n+2}$ is given by $r_{n+2} = r_n + p$, where r_n is the rank of $\mathcal{Q}_n$ and $p \geqslant 0$. Then in terms of the qubits making up the quantum register, we can write

$$\mathcal{Q}_{n+2} \equiv \underbrace{Q^1_{n+2} Q^2_{n+2} \dots Q^{r_n}_{n+2}}_{\widehat{\mathcal{Q}}_{n+2}} Q^{r_n+1}_{n+2} \dots Q^{r_n+p}_{n+2}. \tag{7.34}$$

Here $\widehat{\mathcal{Q}}_{n+2}$ is a subspace of $\mathcal{Q}_{n+2}$ of dimension equal to that of $\mathcal{Q}_n$. Note that, by construction, the first 2^{r_n} elements of the CBR for $\mathcal{Q}_{n+1}$ involve signal excitations only of the detectors associated with $\widehat{\mathcal{Q}}_{n+2}$.

Now define the operator

$$\mathbb{V}_{n+2,n+1} \equiv \sum_{i=0}^{2^{r_n}-1} \sum_{j=0}^{2^{r_{n+1}}-1} \boldsymbol{i}_{n+2} U^{j,i*}_{n+1,n} \overline{\boldsymbol{j}_{n+1}}. \tag{7.35}$$

Note that the upper limit on the summation over the index i is $2^{r_n} - 1$, not $2^{r_{n+1}} - 1$. This operator effectively maps states in $\mathbb{U}_{n+1,n}\mathcal{Q}_n$ into $\widehat{\mathcal{Q}}_{n+2}$. Under evolution generated by $\mathbb{V}_{n+2,n+1}$ we find

$$\boldsymbol{\Psi}_{n+2} \equiv \mathbb{V}_{n+2,n+1} \boldsymbol{\Psi}_{n+1} = \sum_{i=0}^{2^{r_n}-1} \Psi^i_n \boldsymbol{i}_{n+2}, \tag{7.36}$$

which is a persistent image, in subspace $\widehat{\mathcal{Q}}_{n+2}$, of the original labstate $\boldsymbol{\Psi}_n$.

This process demonstrates the principle known as *microscopic reversibility*: operator $\mathbb{V}_{n+2,n+1}$ has effectively "undone" the action of $\mathbb{U}_{n+1,n}$ on $\boldsymbol{\Psi}_n$. It is important to understand that as far as QDN is concerned, nothing has remained unchanged: the observer changes from stage to stage and all labstates change with those jumps. Microscopic reversibility is an illusion in a sense, but one with significant empirical content.

7.12 Path Summations

The QDN formulation of dynamics has some of the hallmarks of the Feynman path integral formulation of quantum mechanics (Feynman and Hibbs, 1965), with some significant differences: in QDN, time is not continuous, the Hilbert space changes at each intermediate time step and is assumed finite dimensional, and there is no need to introduce a Lagrangian or Hamiltonian.

A typical run or repetition of a basic experiment will be assumed to start at stage Σ_0 and finish at a later stage Σ_N, for $N > 0$. Given labstate preparation at stage Σ_0, there will be semi-unitary evolution through a sequence of apparatus stages $\{\Sigma_n : 0 < n < N\}$. At these intermediate stages, the observer does not look at their detectors, which are therefore to be regarded as virtual. Outcome detection takes place only at the final stage Σ_N.

At the final stage Σ_N, the observer looks at all of their detectors and works out from rule (5.14) which element of the CBR of the preferred basis B_N corresponds to the observed set of signals, for that given run. The objective in practice is to compare the statistical distribution of observed outcomes with the theoretically derived conditional probability $Pr(\boldsymbol{k}_N|\boldsymbol{\Psi}_0)$ for each of the possible final state signal basis outcomes $\boldsymbol{k}_N$, $0 \leqslant k < 2^{r_N}$.

Semi-unitary evolution will be assumed to hold between stages Σ_0 and Σ_N, i.e., condition (7.6) is valid. Given an initial labstate $\mathbf{\Psi}_0 \equiv \sum_{i=0}^{2^{r_0}-1} \Psi_0^i \boldsymbol{i}_0$, the next labstate is given by $\mathbf{\Psi}_1 = \mathbb{U}_{1,0}\mathbf{\Psi}_0$, where $\mathbb{U}_{1,0}$ is semi-unitary, and so on, until finally we may write

$$\mathbf{\Psi}_N = \mathbb{U}_{N,N-1}\mathbb{U}_{N-1,N-2}\ldots\mathbb{U}_{1,0}\mathbf{\Psi}_0, \qquad N > 0. \tag{7.37}$$

Inserting a resolution of each evolution operator of the form (7.8), the final state can be expressed in the form

$$\mathbf{\Psi}_n = \sum_{j_N=0}^{2^{r_N}-1} \sum_{j_{N-1}=0}^{2^{r_{N-1}}-1} \cdots \sum_{j_0=0}^{2^{r_0}-1} \boldsymbol{j}_N U_{N,N-1}^{j_N,j_{N-1}} U_{N-1,N-2}^{j_{N-1},j_{N-2}} \ldots U_{1,0}^{j_1,j_0} \Psi_0^{j_0}. \tag{7.38}$$

We may immediately read off from this expression the coefficient of the signal basis vector $\boldsymbol{i}_N$. This gives the QDN analogue of the quantum mechanics Feynman amplitude $\langle \Phi_{\text{final}}^i | \Psi_{\text{initial}} \rangle$ for the initial state $|\Psi_{\text{initial}}\rangle$ to go to a particular final outcome state $|\Phi_{\text{final}}^i\rangle$. In our case, what we are actually reading off is $\mathcal{A}(\boldsymbol{i}_N|\mathbf{\Psi}_0)$, the amplitude for the labstate to propagate from its initial state $\mathbf{\Psi}_0$ and then be found in signal basis state $\boldsymbol{i}$ at stage Σ_N. We find

$$\mathcal{A}(\boldsymbol{i}_N|\mathbf{\Psi}_0) = \sum_{j_{N-1}=0}^{2^{r_{N-1}}-1} \sum_{j_{N-1}=0}^{2^{r_{N-2}}-1} \cdots \sum_{j_0=0}^{2^{r_0}-1} U_{N,N-1}^{i,j_{N-1}} U_{N-1,N-2}^{j_{N-1},j_{N-2}} \ldots U_{1,0}^{j_1,j_0} \Psi_0^{j_0}. \tag{7.39}$$

The required conditional probabilities are obtained from the Born rule as discussed above, and so we conclude

$$Pr(\boldsymbol{i}_N|\mathbf{\Psi}_0) = \left| \sum_{j_{N-1}=0}^{2^{r_{N-1}}-1} \sum_{j_{N-2}=0}^{2^{r_{N-2}}-1} \cdots \sum_{j_0=0}^{2^{r_0}-1} U_{N,N-1}^{i,j_{N-1}} U_{N-1,N-2}^{j_{N-1},j_{N-2}} \ldots U_{1,0}^{j_1,j_0} \Psi_0^{j_0} \right|^2. \tag{7.40}$$

By writing the amplitude (7.39) in the form

$$\mathcal{A}(\boldsymbol{i}_N|\mathbf{\Psi}_0) = \sum_{j_{N-1}=0}^{2^{r_{N-1}}-1} U_{N,N-1}^{i,j} \mathcal{A}(\boldsymbol{j}_{N-1}|\mathbf{\Psi}_0), \tag{7.41}$$

it is straightforward to use induction and the semi-unitary matrix conditions (7.10) to prove that

$$\sum_{i=0}^{2^{r_N}-1} Pr(\boldsymbol{i}_N|\mathbf{\Psi}_0) = 1, \tag{7.42}$$

which means total probability is conserved, as expected.

Feynman derived his path integral for continuous time quantum mechanics by discretizing time and then taking the limit of the discrete time interval going to zero. Technical problems occur in the taking of this limit, and because of these, the path integral in its original formulation (Feynman and Hibbs, 1965) is generally regarded as ill-defined. However, it is an invaluable heuristic tool that provides the best way to discuss the quantization of certain classical theories for which other approaches prove inadequate. In QDN, time is discrete and in that sense we follow Feynman's lead while avoiding the pitfalls associated with the continuum limit, which we do not take in QDN.

This completes our introduction to the QDN formalism. An obvious extension is to include *mixed* labstates. These are discussed in Chapter 9.

8

Partial Observations

8.1 Introduction

In this chapter we discuss the quantized detector network (QDN) approach to *partial observations*, or the extraction, during an extended-in-time quantum process, of only some of the quantum information embedded in a detector amplitude. In order to deal with partial questions, we need first to discuss how QDN deals with *maximal questions*, which means looking at all the detectors at a given stage.

8.2 Observables

Our preoccupation with detectors rather than systems under observation (SUOs) is nothing new in quantum mechanics (QM). Indeed, the primary significance of what could be observed was a guiding principle when Heisenberg formulated his matrix mechanics approach to QM (Heisenberg, 1925). The conventional position in QM is that the only important operators in the theory are the *observables*. These correspond closely to the variables used in classical mechanics (CM) to describe measurable quantities, such as energy, momentum, electric charge, and so on.

In standard QM, observables are generally assumed to be self-adjoint operators (Hermitian operators in the case of finite-dimensional Hilbert spaces) so that they have real eigenvalues, and these are the QM analogues of CM variables. Apart from that proviso, operators representing observables have few restrictions.

This generality raises significant questions. A particularly critical question is whether a given "observable" makes empirical sense. There is no theorem that proves that every self-adjoint operator corresponds to something that can be actually observed in the laboratory. For example, in one-dimensional wave mechanics, the operator $\widehat{pppxppp}$ is self-adjoint and its classical analogue, $pppxppp$, is a perfectly regular function over phase space. But we know of no experiment that can measure such a quantity directly, whereas experiments to measure p or x directly could be devised.

The problem as we see it in this book is that the *observable* concept in standard QM puts the cart before the horse: it is first implicitly assumed in QM that states of SUOs can be created, and only then is the question of what can be done to and on those states raised. QDN does things the other way around: the apparatus that creates the states and detects the signals has to come first. Indeed, that is all that is needed. "State preparation" defines the states contextually and the outcome detectors define what information can be obtained. In QDN, "observables" are nothing but the signals in final stage detectors plus the context that informs the observer as to the meaning of those signals.

Transformations and Symmetries

In the twentieth century, great advances in quantum physics were made, driven by relativistic transformation theory, leading to the Lorentz covariant formulation of quantum fields known as relativistic quantum field theory (RQFT). That theory is generally regarded as the best theory to describe the phenomenology of elementary particles. However, there are some deep conceptual issues concerning the interface between the classical world of the observer's knowledge base and the nonclassical behavior of labstates.

An important factor here is the relationship between different observers. In the standard QM approach to observables, the transformation properties of those observables are regarded as crucial. Indeed, Dirac's approach to QM was developed in part as an analogue of CM "transformation theory" (Dirac, 1958; Goldstein, 1964; Leech, 1965).

In QDN the issue of transformation theory is avoided by the assertion that a given laboratory needs no transformation. And if it is required to discuss the relationship between two different laboratories (which we have to remind the reader consist of atoms and molecules that cannot just pass through each other without significant interaction, none of which is taken into account in standard QM), then we simply regard the two laboratories as a single larger laboratory defined by the context of the situation. For example, a Doppler shift experiment is conventionally described in terms of a source of light moving relative to a detector. In QDN terms, all of that can be described as one experiment in a single laboratory, albeit one that changes from stage to stage. Laboratories in QDN are not restricted to single inertial frames.

There is an interesting question that can posed here, and its resolution has everything to do with the dominant principle of contextuality underpinning this book: *Given a distant source of light, such as a remote galaxy on the other side of the Universe, how can astronomers observing light from that source consider themselves to be part of a single laboratory that includes that galaxy (which might have long ago been destroyed)?*

The answer is *context*. A single detected photon, no matter how much "red shifted" it was,[1] could not carry any contextual information. But the actual

[1] Once again, we have to resort to realist language to convey our meaning, though it is an oversimplification to do so.

described scenario gives the game away: how could we even say that there was a galaxy acting as a source of light without having observed sufficient light from it to establish that fact? That degree of observation thereby provides the context that allows us to think of that remote galaxy and our detectors as part of the same experiment.

In fact, such context is generally built up over relatively long periods of time and depends on technology. Thousands of years ago, astronomers could only see via their naked eyes a strange blob of diffuse light where M31, the Andromeda Galaxy, is in the night sky. Then as telescopes were developed, followed by long-exposure photography, sufficient observations were made to establish the context that we know today. Only that relatively recent context allows us to say whether a signal from Andromeda is red-shifted or blue-shifted light.[2]

8.3 Maximal Questions

In this section, dependence on the temporal index n is suppressed, as all questions are asked at some given stage Σ_n.

Given a rank-r quantum register, an arbitrary pure labstate $\boldsymbol{\Psi}$ is of the general form

$$\boldsymbol{\Psi} = \Psi^0\mathbf{0} + \sum_{i=1}^{r} \Psi^i \widehat{\mathbb{A}}^i \mathbf{0} + \sum_{1\leqslant i<j\leqslant r} \Psi^{ij} \widehat{\mathbb{A}}^i \widehat{\mathbb{A}}^j \mathbf{0} + \cdots + \Psi^{12\ldots r} \widehat{\mathbb{A}}^1 \widehat{\mathbb{A}}^2 \ldots \widehat{\mathbb{A}}^r \mathbf{0}, \tag{8.1}$$

where we do not rule out superpositions of elements with different signality. Labstates are generally normalized to unity, so the coefficients in (8.1) satisfy the condition

$$\overline{\boldsymbol{\Psi}}\boldsymbol{\Psi} = |\Psi^0|^2 + \sum_{i=1}^{r} |\Psi^i|^2 + \sum_{1\leqslant i<j\leqslant r} |\Psi^{ij}|^2 + \cdots + |\Psi^{12\ldots r}|^2 = 1. \tag{8.2}$$

Example 8.1 An arbitrary normalized labstate in a rank-two quantum register is of the form

$$\boldsymbol{\Psi} = \{\Psi^0 \mathbb{I}^{[2]} + \Psi^1 \widehat{\mathbb{A}}^1 + \Psi^2 \widehat{\mathbb{A}}^2 + \Psi^3 \widehat{\mathbb{A}}^1 \widehat{\mathbb{A}}^2\}\mathbf{0}, \tag{8.3}$$

with $|\Psi^0|^2 + |\Psi^1|^2 + |\Psi^2|^2 + |\Psi^3|^2 = 1$.

The interpretation of these coefficients is based on the Born rule in standard QM (Born, 1926): if the apparatus is in labstate (8.3) prior to the observer looking at both detectors "simultaneously" (which is possible in QDN by definition), then the probability of each detector being found in its ground state is $|\Psi^0|^2$, the probability of detector 1 being in its signal state *and* detector 2 being in its

[2] In fact, Andromeda is moving *toward* our galaxy, so light from ordinary sources in Andromeda should show a small blue shift.

ground state is $|\Psi^1|^2$, the probability of detector 1 being in its ground state *and* detector 2 being in its signal state is $|\Psi^2|^2$, and the probability of both detectors being in their signal states is $|\Psi^3|^2$. Note that $|\Psi^1|^2$ is *not* the probability that there is a signal in detector 1. That probability is given by $|\Psi^1|^2 + |\Psi^3|^2$.

Example 8.2 An observer prepares a pure labstate $\boldsymbol{\Phi}$ in a rank-four quantum register. Show that the probability $Pr(\mathbf{1}^1\mathbf{0}^2\mathbf{0}^3\mathbf{1}^4|\boldsymbol{\Phi})$ that the observer would find detectors 1 and 4 in their signal states and detectors 2 and 3 in their ground states is given by

$$Pr(\mathbf{1}^1\mathbf{0}^2\mathbf{0}^3\mathbf{1}^4|\boldsymbol{\Phi}) = \overline{\boldsymbol{\Phi}}\widehat{\mathbb{P}}^1\mathbb{P}^2\mathbb{P}^3\widehat{\mathbb{P}}^4\boldsymbol{\Phi}. \tag{8.4}$$

Solution In this case, we need to ask the maximal question $\overline{\mathbf{1}^1\mathbf{0}^2\mathbf{0}^3\mathbf{1}^4}$ of the labstate $\boldsymbol{\Phi}$, giving the amplitude $\mathcal{A}(\mathbf{1}^1\mathbf{0}^2\mathbf{0}^1\mathbf{1}^4|\boldsymbol{\Phi}) \equiv \overline{\mathbf{1}^1\mathbf{0}^2\mathbf{0}^3\mathbf{1}^4}\boldsymbol{\Phi}$. Then the Born rule gives

$$\begin{aligned}
Pr(\mathbf{1}^1\mathbf{0}^2\mathbf{0}^3\mathbf{1}^4|\boldsymbol{\Phi}) &\equiv |\mathcal{A}(\mathbf{1}^1\mathbf{0}^2\mathbf{0}^3\mathbf{1}^4|\boldsymbol{\Phi})|^2 \\
&= (\overline{\mathbf{1}^1\mathbf{0}^2\mathbf{0}^3\mathbf{1}^4}\boldsymbol{\Phi})^*(\overline{\mathbf{1}^1\mathbf{0}^2\mathbf{0}^3\mathbf{1}^4}\boldsymbol{\Phi}) \\
&= (\overline{\boldsymbol{\Phi}}\mathbf{1}^1\mathbf{0}^2\mathbf{0}^3\mathbf{1}^4)(\overline{\mathbf{1}^1\mathbf{0}^2\mathbf{0}^3\mathbf{1}^4}\boldsymbol{\Phi}) \\
&= \overline{\boldsymbol{\Phi}}(\mathbf{1}^1\mathbf{0}^2\mathbf{0}^3\mathbf{1}^4\overline{\mathbf{1}^1\mathbf{0}^2\mathbf{0}^3\mathbf{1}^4})\boldsymbol{\Phi} \\
&= \overline{\boldsymbol{\Phi}}\underbrace{(\mathbf{1}^1\overline{\mathbf{1}^1})}_{\widehat{\boldsymbol{P}}^1}\underbrace{(\mathbf{0}^2\overline{\mathbf{0}^2})}_{\boldsymbol{P}^2}\underbrace{(\mathbf{0}^3\overline{\mathbf{0}^3})}_{\boldsymbol{P}^3}\underbrace{(\mathbf{1}^4\overline{\mathbf{1}^4})}_{\widehat{\boldsymbol{P}}^4}\boldsymbol{\Phi} \\
&= \overline{\boldsymbol{\Phi}}\widehat{\boldsymbol{P}}^1\boldsymbol{P}^2\boldsymbol{P}^3\widehat{\boldsymbol{P}}^4\boldsymbol{\Phi}.
\end{aligned} \tag{8.5}$$

But it is straightforward to show that for a rank-four register,

$$\widehat{\boldsymbol{P}}^1\boldsymbol{P}^2\boldsymbol{P}^3\widehat{\boldsymbol{P}}^4 = \widehat{\mathbb{P}}^1\mathbb{P}^2\mathbb{P}^3\widehat{\mathbb{P}}^4, \tag{8.6}$$

where each of the operators on the right-hand side is a register operator. This then gives the required result (8.4).

For a rank-r register, we shall call a register product of r distinct register projection operators a *maximal question*. The above example illustrates how any maximal question for a rank-r quantum register can be related to the tensor product of r distinct bit projection operators, one for each detector in the register.

There are three points to note here. First, Eq. (8.6) holds only because the left-hand side is a register operator, being the tensor product of r individual bit operators, one for each detector in the register. Second, a maximal question can be identified with a specific element of the preferred basis. Since there are 2^r elements in the latter, we deduce that there is a total of 2^r distinct maximal questions. The third point is a technical one: the product concepts on each side of (8.6) are different. The left-hand side is the tensor product of bit operators, the right-hand side is the product of register operators.

8.4 Partial Questions

The above example shows how to ask a specific question of each and every detector in a quantum register at a given stage. For a rank-r quantum register, any maximal question involves a product of r distinct register projection operators. For each detector i, $1 \leqslant i \leqslant r$, there are two related register operators, $\mathbb{P}^i$ and $\widehat{\mathbb{P}}^i$, which form a *conjugate pair*. Therefore there are exactly 2^r distinct maximal questions, as stated above.

In the real world, however, observers could choose to ask *partial questions*, which involve looking at only some (or even none) of the detectors. The simplest example of a partial question involves the normalization condition

$$\overline{\boldsymbol{\Psi}}\boldsymbol{\Psi} = 1, \tag{8.7}$$

because we can always write $\overline{\boldsymbol{\Psi}}\boldsymbol{\Psi} = \overline{\boldsymbol{\Psi}}\mathbb{I}^{[r]}\boldsymbol{\Psi}$, where $\mathbb{I}^{[r]}$ is the register identity operator. We may identify this operator with the question

What is the probability that every detector is either in its ground state or signal state?

and call this a *rank-zero partial question*, because it involves looking at no (i.e. zero) detectors.

Now suppose we wanted to ask a question involving just one detector, such as the ath, where $1 \leqslant a \leqslant r$. Given (8.7), we insert the register identity operator $\mathbb{I}^{[r]}$ as before and use the property

$$\mathbb{P}^a + \widehat{\mathbb{P}}^a = \mathbb{I}^{[r]}, \qquad a = 1, 2, \ldots, r. \tag{8.8}$$

Using this property and linearity, we find

$$\overline{\boldsymbol{\Psi}}\mathbb{P}^a\boldsymbol{\Psi} + \overline{\boldsymbol{\Psi}}\,\widehat{\mathbb{P}}^a\boldsymbol{\Psi} = 1. \tag{8.9}$$

Each of the terms on the left-hand side is nonnegative. By inspection, $\overline{\boldsymbol{\Psi}}\mathbb{P}^a\boldsymbol{\Psi}$ is the probability that detector a would be found in its ground state, while $\overline{\boldsymbol{\Psi}}\,\widehat{\mathbb{P}}^a\boldsymbol{\Psi}$ is the probability that a would be found in its signal state, regardless of what was going on in any of the other $r-1$ detectors. We will refer to each of these partial questions as a *rank-one partial question.*

This process can be extended naturally to higher rank partial questions, until we reach rank r, which are the maximal questions we discussed in the previous section.

Example 8.3 Given a labstate $\boldsymbol{\Psi}$ prepared in a rank-637 register, what is the probability that if the observer looked only at detectors 99, 323, and 438, they would find 99 and 323 each in its ground state and 438 in its signal state?

Solution

The required probability Pr is given by the expectation value

$$Pr = \overline{\boldsymbol{\Psi}}\mathbb{P}^{99}\mathbb{P}^{323}\widehat{\mathbb{P}}^{438}\boldsymbol{\Psi}. \tag{8.10}$$

It will be clear from the above that the set of all partial questions involves expectation values of all possible products of the register projection operators. This leads to the following theorem.

Theorem 8.4 *For a rank-r classical or quantum register $\mathcal{Q}^{[r]}$, the number of possible partial questions is 3^r.*

Proof To prove the theorem, we determine the number of partial questions of each rank and then add up all those numbers.

There is only one rank zero partial question.

The observer could go to each of the r detectors one by one and ask one of two possible questions of it: the two questions that can be asked at detector i are given by the register projectors $\mathbb{P}^i, \widehat{\mathbb{P}}^i$. These are rank-one partial questions, so we conclude that there is a total of 2^r rank-one partial questions.

Assuming $r > 1$, the observer could now ask rank-two partial questions, involving only two distinct detectors in the register. Given a rank-r register, there is a total of $r(r-1) = \binom{r}{2}$ distinct pairs, and for each pair of choices, 2^2 alternative questions can be asked. For example, for the choice $i < j$, we can ask the four questions $\mathbb{P}^i\mathbb{P}^j, \widehat{\mathbb{P}}^i\mathbb{P}^j, \mathbb{P}^i\widehat{\mathbb{P}}^j$, and $\widehat{\mathbb{P}}^i\widehat{\mathbb{P}}^j$. Therefore, there is a total of $2^2\binom{r}{2}$ such partial questions.

We can continue this argument until we reach the maximal questions, which are rank-r partial questions. There is only one way of choosing r objects from r objects, and 2^r possible maximal questions to be asked of that choice. Hence we find the grand total T_Q of distinct partial question operators to be given by

$$T_Q = 1 + 2\binom{r}{1} + 2^2\binom{r}{2} + \cdots + 2^r\binom{r}{r} = (1+2)^r = 3^r, \tag{8.11}$$

as asserted. □

8.5 Partial Question Eigenvalues

Every partial question has the property that each preferred basis element is an eigenstate of it, with an eigenvalue of either zero or one. Using the signal basis representation (SBR), the eigenvalue in each case can be readily read off. For example, in a rank-five register, the state $\mathbf{0}^1\mathbf{1}^2\mathbf{1}^3\mathbf{0}^4\mathbf{1}^5$ is an eigenstate of the partial question operators $\mathbb{P}^1$ and $\mathbb{P}^1\widehat{\mathbb{P}}^4$, with eigenvalues 1 and 0, respectively. On the other hand, using the computational basis representation (CBR) is not so convenient here. The given state $\mathbf{0}^1\mathbf{1}^2\mathbf{1}^3\mathbf{0}^4\mathbf{1}^5$ has the CBR $\underline{\mathbf{22}}$, so we see

$$\mathbb{P}^1\underline{\mathbf{22}} = \underline{\mathbf{22}}, \quad \mathbb{P}^1\widehat{\mathbb{P}}^4\underline{\mathbf{22}} = 0, \tag{8.12}$$

but the respective eigenvalues cannot now be directly read off. Since the CBR is generally useful, we need to quantify the action of the partial questions on the CBR. We do this as follows. The set $\{\mathbb{I}^{[r]}, \mathbb{P}^1, \ldots, \widehat{\mathbb{P}}^1\widehat{\mathbb{P}}^2\ldots\widehat{\mathbb{P}}^r\}$ of partial questions

Table 8.1 *Question eigenvalues for a rank-two register*

	$\mathbf{0}^1\mathbf{0}^2 = \mathbf{0}$	$\mathbf{1}^1\mathbf{0}^2 = \mathbf{1}$	$\mathbf{0}^1\mathbf{1}^2 = \mathbf{2}$	$\mathbf{1}^1\mathbf{1}^2 = \mathbf{3}$
$\mathbb{Q}^1 \equiv \mathbb{I}^{[2]}$	$Y^{1,0} = 1$	$Y^{1,1} = 1$	$Y^{1,2} = 1$	$Y^{1,3} = 1$
$\mathbb{Q}^2 \equiv \mathbb{P}^1$	$Y^{2,0} = 1$	$Y^{2,1} = 0$	$Y^{2,2} = 1$	$Y^{2,3} = 0$
$\mathbb{Q}^3 \equiv \widehat{\mathbb{P}}^1$	$Y^{3,0} = 0$	$Y^{3,1} = 1$	$Y^{3,2} = 0$	$Y^{3,3} = 1$
$\mathbb{Q}^4 \equiv \mathbb{P}^2$	$Y^{4,0} = 1$	$Y^{4,1} = 1$	$Y^{4,2} = 0$	$Y^{4,3} = 0$
$\mathbb{Q}^5 \equiv \widehat{\mathbb{P}}^2$	$Y^{5,0} = 0$	$Y^{5,1} = 0$	$Y^{5,2} = 1$	$Y^{5,3} = 1$
$\mathbb{Q}^6 \equiv \mathbb{P}^1\mathbb{P}^2$	$Y^{6,0} = 1$	$Y^{6,1} = 0$	$Y^{6,2} = 0$	$Y^{6,3} = 0$
$\mathbb{Q}^7 \equiv \widehat{\mathbb{P}}^1\mathbb{P}^2$	$Y^{7,0} = 0$	$Y^{7,1} = 1$	$Y^{7,2} = 0$	$Y^{7,3} = 0$
$\mathbb{Q}^8 \equiv \mathbb{P}^1\widehat{\mathbb{P}}^2$	$Y^{8,0} = 0$	$Y^{8,1} = 0$	$Y^{8,2} = 1$	$Y^{8,3} = 0$
$\mathbb{Q}^9 \equiv \widehat{\mathbb{P}}^1\widehat{\mathbb{P}}^2$	$Y^{9,0} = 0$	$Y^{9,1} = 0$	$Y^{9,2} = 0$	$Y^{9,3} = 1$

contains 3^r elements. We define $\mathbb{Q}^1 \equiv \mathbb{I}^{[r]}, \mathbb{Q}^2 \equiv \mathbb{P}^1, \ldots, \mathbb{Q}^{3^r} \equiv \widehat{\mathbb{P}}^1\widehat{\mathbb{P}}^2 \ldots \widehat{\mathbb{P}}^r$. This choice of labeling is arbitrary, there being no obvious way to order a complete set of partial questions. Then for any partial question $\mathbb{Q}^P$, $P = 1, 2, \ldots, 3^r$, and any CBR element $\boldsymbol{i}$, $i = 0, 1, 2, \ldots, 2^r - 1$, we write

$$\mathbb{Q}^P \boldsymbol{i} = Y^{P,i} \boldsymbol{i}, \tag{8.13}$$

where the *question eigenvalue* $Y^{P,i}$ is either zero or unity.

Example 8.5 For a rank-two register, there are $3^2 = 9$ distinct partial questions and $2^2 = 4$ distinct CBR elements. Table 8.1 shows the question eigenvalues.

8.6 Identity Classes

A full partial question set can be divided into groups of operators that sum up, in that group, to the register identity. Each such group will be called an *identity class*. A rank-r register has 2^r identity classes.

Example 8.6 A rank-one register has two identity classes, given by $C^{1,1} \equiv \left\{\mathbb{I}^{[1]}\right\}$ and $C^{2,1} \equiv \left\{\mathbb{P}^1, \widehat{\mathbb{P}}^1\right\}$.

Example 8.7 A rank-two register has four identity classes, given by

$$\begin{aligned} C^{1,2} &\equiv \left\{\mathbb{I}^{[2]}\right\}, \\ C^{2,2} &\equiv \left\{\mathbb{P}^1, \widehat{\mathbb{P}}^1\right\}, \\ C^{3,2} &\equiv \left\{\mathbb{P}^2, \widehat{\mathbb{P}}^2\right\}, \\ C^{4,2} &\equiv \left\{\mathbb{P}^1\mathbb{P}^2, \widehat{\mathbb{P}}^1\mathbb{P}^2, \mathbb{P}^1\widehat{\mathbb{P}}^2, \widehat{\mathbb{P}}^1\widehat{\mathbb{P}}^2\right\}. \end{aligned} \tag{8.14}$$

In Table 8.1, the four identity classes are separated by horizontal lines. A given identity class consists of partial questions of the same rank, so we shall refer to each class by its rank. In the above example, $C^{1,1}$ and $C^{1,2}$ are rank-zero identity classes; $C^{2,1}, C^{2,2}$, and $C^{3,2}$ are rank-one identity classes; and $C^{4,2}$ is a rank-two identity class.

Identity classes are related to probability conservation. We shall find that if we want to conserve probability in any calculation, we need to restrict partial questions to the same identity class. It may then be necessary to relabel the partial questions and their question eigenvalues with an extra label identifying individual identity classes.

8.7 Needles in Haystacks

The classification of question rank and identity class sheds some light on how the unimaginable complexity of the real world can be comprehended by intelligent observers. Suppose an observer wanted to model the Universe by an enormously large number N of qubits, giving a quantum register $\mathcal{Q}^{[N]}$ of rank 2^N. Suppose now that that observer was investigating their environment by asking a limited number of partial questions. This is a typical scenario in empirical science: resources are not infinite and experimentalists can only do so much. By inspection of Table 8.1, we see an interesting pattern, one that would be repeated in the general case. If we ask no questions, then that is represented by the rank-zero identity class. We see from the question eigenvalues for such a question, denoted $\mathbb{Q}^1$ in Table 8.1, that the answer for each possible labstate is one, meaning that we can extract no new information about the system under observation (SUO) from such a question. But that question has cost us nothing.

Moving on, we may now decide to ask rank-one questions, represented by $\mathbb{Q}^2$ and $\mathbb{Q}^3$ in Table 8.1. Now we start to get some real information about the state of the SUO, but it is also starting to become expensive.

This process could continue, with greater rank questions being posed, with more information coming out but at ever increasing cost.

There is an interesting trade-off. Looking at Table 8.1, we note that for a given labstate, the maximal rank identity class partial questions all have answer zero except one of them. This enormously simplifies the problem of finding a signal, in that should we find an answer of one halfway through this process, we can immediately stop, because we can be sure all the remaining answers are zero.

We can now appreciate what it means to do experiments and why they are done. Experiments are careful arrangements of partial questions of the same identity class, designed to eliminate as many zero-value answers and home in on unity answers (those for which $Y^{\theta,i} = 1$), guided by theory. Such searches can require planning to choose the right questions and the expenditure of enormous resources as in the case of the search for the Higgs particle at the Large Hadron Collider. That task was far more of a technical problem than searching for a needle in a haystack.[3] At the end of the day, we note that the result of the search was a single number, that is, one, which means *yes, the Higgs particle exists* (according to our contextual interpretation of the data).

This line of thinking also encourages experimentalists not to give up but to continue searches that appear to be unsuccessful. For instance, there is at this time (2017) no direct empirical evidence for supersymmetric partners of various particles such as electrons and photons. Because the number of zero-value question eigenvalues is potentially vast, a lack of confirmation so far does not mean that a *yes* answer does not exist.

[3] Which can be done quickly with the right apparatus, such as a metal detector.

9
Mixed States and POVMs

9.1 Introduction

As the decades of the twentieth century rolled on, quantum mechanics (QM) became more and more sophisticated and mathematical. Understanding what this theory means intuitively was confounded not only by nonclassical concepts such as wave–particle duality and quantum interference, but also by issues to do with observation. The Newtonian classical mechanics (CM) paradigm of reality, wherein reductionist laws of physics describe observer-independent dynamics of systems under observation (SUOs) with observer-independent properties, was found to be inadequate. Quantum theorists were confronted with the *measurement problem*, which attempts to understand, explain, and rationalize the laws of QM that underpin the processes of observation that go on in the laboratory. They are not the same as those of CM in several puzzling respects.

Historically, the first sign of the measurement problem was Planck's quantization of energy (Planck, 1900b) and the second was Bohr's veto on radiation damping in hydrogen (Bohr, 1913). These occurred in the first quarter of the twentieth century, a period in physics often referred to as *Old Quantum Mechanics*. Another indicator that intuition was inadequate was Born's interpretation of the squared modulus of the Schrödinger wave function as a probability density (Born, 1926).[1] That interpretation has everything to do with observers and observation, because probability without an observer is a vacuous concept.

Eventually, the *projection-valued measure* (PVM) formalism emerged, championed by von Neumann in an influential book on the mathematical formulation of QM (von Neumann, 1955). Subsequently, pioneers such as Ludwig (Ludwig, 1983a,b) and Kraus (Kraus, 1974, 1983) refined the theory into the general *positive operator-valued measure* (POVM) formalism that we shall discuss and use.

[1] Born appears to have at first taken the *magnitude* of the wave function as the probability density, but corrected himself in time.

The quantized detector network (QDN) approach to quantum experiments is most naturally expressed in the PVM formalism, as QDN focuses on the individual detectors in the laboratory. However, the POVM formalism is more general than the PVM formalism, giving a description of multiple detector processes similar to QDN. This raises the question of how the two approaches, QDN and POVM, are related. The aim of this chapter is to explore this relationship.

Before we review the PVM and POVM formalisms, we review some essential mathematical concepts.

9.2 Sets and Measures

Most of the mathematics used in this book involves spaces, which are sets with additional structure such as inner products (Howson, 1972).

Definition 9.1 A sigma-algebra on set A is a collection Σ^A of subsets of A such that

1. Σ^A includes the empty subset $\emptyset$.
2. Σ^A is closed under complement: if E is an element of Σ^A, then so is its complement in A, denoted E^c. Since by property 1, the empty set $\emptyset$ belongs to Σ^A and its complement is A, then property 2 means that A itself is a member of Σ^A.
3. Σ^A is closed under countable unions, which means that if we pick any countable number $E^1, E^2, \ldots, E^n$ of elements in Σ^A, then their union $\cup_{i=1}^n E^i$ is also in Σ^A.
4. Σ^A is closed under countable intersections, which means that if we pick any finite number $E^1, E^2, \ldots, E^n$ of elements of Σ^A, then their intersection $\cap_{i=1}^n E^i$ is also in Σ^A. Note that this intersection could be empty, but that possibility is covered by property 1.

Definition 9.2 The *extended reals* $\mathbb{R}^*$ is the set of real numbers $\mathbb{R}$ and two extra elements, $+\infty$ and $-\infty$. These latter two elements are referred to as *plus infinity* and *minus infinity*, respectively, and are interpreted accordingly.

Definition 9.3 Given a sigma-algebra Σ^A over set A, a *measure* on A assigns an extended real number $\mu(E)$ to each element E of Σ^A such that

1. *Nonnegativity*: for any element E in Σ^A, $\mu(E) \geqslant 0$.
2. *Measure of empty set*: $\mu(\emptyset) = 0$.
3. *Countable additivity*: for any countable collection $\{E^1, E^2, \ldots\}$ of pairwise disjoint elements of Σ^A, meaning that $E^i \cap E^j = \emptyset$ for $i \neq j$, then

$$\mu\left(\bigcup_{i=1}^{\infty} E^i\right) = \sum_{i=1}^{\infty} \mu(E^i). \tag{9.1}$$

In applications to quantum physics, the set A referred to in the above definitions will consist of all possible outcomes of an experiment; the sigma-algebra will consist of all possible ways of grouping those outcomes; and the measure μ will be the assignment of probabilities to the elements of Σ^A.

9.3 Hilbert Spaces

Hilbert spaces are complex vector spaces with a suitable inner product concept. In this chapter, we shall use the well-known Dirac bracket notation, following Paris (2012). In most of this book we deal with finite-dimensional Hilbert spaces, as the guiding philosophy of QDN is to model what goes on in the laboratory. No infinities are ever actually encountered in the laboratory; at worst, a readout on a counter goes off-scale.

Definition 9.4 An orthonormal basis (ONB) for a d-dimensional Hilbert space $\mathcal{H}$ is a set of elements $\{|\psi^n\rangle : n = 1, 2, \ldots, d\}$ of $\mathcal{H}$ with the following properties:

1. **Orthonormality**: $\langle\psi^n|\psi^m\rangle = \delta^{nm}$, $1 \leqslant n, m \leqslant d$, where δ^{nm} is the Kronecker delta.
2. **Completeness/resolution of the identity**:

$$\sum_{n=1}^{d} |\psi^n\rangle\langle\psi^n| = I^{\mathcal{H}}, \tag{9.2}$$

where $I^{\mathcal{H}}$ is the identity operator on $\mathcal{H}$.

9.4 Operators and Observables

An *operator* is any rule that assigns an element of one Hilbert space to some element either in the same Hilbert space or in another Hilbert space. There are various kinds of operators that are important to us here. A *linear operator* O from Hilbert space $\mathcal{H}^1$ to Hilbert space $\mathcal{H}^2$ is one such that if $|\Psi\rangle$ and $|\Phi\rangle$ are arbitrary elements of $\mathcal{H}^1$ and α and β are arbitrary complex numbers, then

$$O(\alpha|\Psi\rangle + \beta|\Phi\rangle) = \alpha O|\Psi\rangle + \beta O|\Phi\rangle, \tag{9.3}$$

where we note that addition on the left-hand side of (9.3) is in $\mathcal{H}^1$, while addition on the right-hand side is in $\mathcal{H}^2$.

The linear space of linear operators from $\mathcal{H}$ to $\mathcal{H}$ is denoted $L(\mathcal{H})$ and is itself a Hilbert space (Paris, 2012).

A *positive operator* O over Hilbert space $\mathcal{H}$ is one such that for any element $|\Psi\rangle$ of $\mathcal{H}$, we have

$$\langle\Psi|O|\Psi\rangle \geqslant 0. \tag{9.4}$$

A positive operator is self-adjoint.

An *observable* X is a self-adjoint operator that admits a discrete *spectral decomposition*; i.e., we can write

$$X = \sum_{n=1}^{d} x^n P^n, \tag{9.5}$$

where the x^n are real and the eigenvalues of X, and the *projectors* P^n are given by $P^n \equiv |x^n\rangle\langle x^n|$. Here the normalized eigenvectors $|x^n\rangle$ satisfy the eigenvalue equation

$$X|x^n\rangle = x^n|x^n\rangle \tag{9.6}$$

and form an ONB for $\mathcal{H}$.[2] Orthonormality then leads to the product rule

$$P^n P^m = \delta^{nm} P^n \tag{9.7}$$

and the completeness sum

$$\sum_{n=1}^{d} P^n = I^{\mathcal{H}}. \tag{9.8}$$

9.5 Trace

If $\{|\psi^n\rangle : n = 1, 2, \ldots, d\}$ is an ONB for $\mathcal{H}$, then the *trace* $\mathrm{Tr}\{O\}$ of an operator O is defined by

$$\mathrm{Tr}\{O\} \equiv \sum_{n=1}^{d} \langle\psi^n|O|\psi^n\rangle. \tag{9.9}$$

The trace of an operator is independent of choice of ONB for $\mathcal{H}$.

Given an arbitrary state $|\Psi\rangle$ in $\mathcal{H}$, then we can use any ONB to show that

$$\langle\Psi|O|\Psi\rangle = \mathrm{Tr}\{O|\Psi\rangle\langle\Psi|\}. \tag{9.10}$$

This result is crucial to the density operator and POVM formalisms widely applied in QM.

9.6 Projection-Valued Measure

We are now in a position to discuss PVMs in standard QM.

Given an observable X, the probability $\Pr(x^n|\Psi)$ of final outcome x^n given initial normalized state $|\Psi\rangle$ is according to the Born rule (Born, 1926) given by

$$\begin{aligned}\Pr(x^n|\Psi) &= |\langle x^n|U|\Psi\rangle|^2 = \langle\Psi|U^\dagger|x^n\rangle\langle x^n|U|\Psi\rangle \\ &= \langle\Psi|U^\dagger P^n U|\Psi\rangle = \mathrm{Tr}\{P^n U \varrho U^\dagger\},\end{aligned} \tag{9.11}$$

[2] This can always be arranged, even if some of the eigenvalues are degenerate.

where U is the unitary evolution operator taking the initial state $|\Psi\rangle$ to its final state $U|\Psi\rangle$ at the time of measurement and ϱ is the initial *density operator*, defined by

$$\varrho \equiv |\Psi\rangle\langle\Psi| \tag{9.12}$$

in this instance.

The *expectation value* $\langle X\rangle_\Psi$ of the observable X conditional on Ψ is given by

$$\begin{aligned}\langle X\rangle_\Psi &\equiv \sum_{n=1}^{d} x\Pr(x^n|\Psi) = \sum_{n=1}^{d} x^n \mathrm{Tr}\{P^n U\varrho U^\dagger\} \\ &= \mathrm{Tr}\left\{\left[\sum_{n=1}^{d} x^n P^n\right] U\varrho U^\dagger\right\} = \mathrm{Tr}\{XU\varrho U^\dagger\}.\end{aligned} \tag{9.13}$$

9.7 Mixed States

Suppose an observer carries out an experiment consisting of a large number of runs but is unsure, at the start of each run, of the initial state. Suppose further that the observer's ignorance can be quantified into the statement that for each run, the initial state is taken from a discrete probability space Ω, that is, a finite distribution of possible states, each labeled by superscript κ, running from 1 to some finite integer K, such that the probability of preparing state $|\Psi^\kappa\rangle$ is ω^κ:

$$\Omega \equiv \left\{\omega^\kappa, |\Psi^\kappa\rangle : \sum_{\kappa=1}^{K} \omega^\kappa = 1, \quad \langle\Psi^\kappa|\Psi^\kappa\rangle = 1\right\}. \tag{9.14}$$

Here the possible initial states are normalized but not necessarily mutually orthogonal. The number K of possibilities is arbitrary and can exceed the dimension d of $\mathcal{H}^S$. The probabilities ω^κ are *epistemic* in character, that is, classical probabilities. Such a random initial state is referred to as a *mixed* state when $K >$ 1. When $K = 1$, the observer's ignorance about the initial state is zero and the corresponding unique element of the Hilbert space is referred to as a *pure* state.

The expectation value $\langle X\rangle_\varrho$ of the observable X conditional on a mixed state ϱ is now

$$\begin{aligned}\langle X\rangle_\varrho &= \sum_{\kappa=1}^{K} \omega^\kappa \langle\psi^\kappa|U^\dagger X U|\psi^\kappa\rangle \\ &= \mathrm{Tr}\{XU \underbrace{\sum_{\kappa=1}^{K} \omega^\kappa |\psi^\kappa\rangle\langle\psi^\kappa|}_{\varrho} U^\dagger\} = \mathrm{Tr}\{XU\varrho U^\dagger\},\end{aligned} \tag{9.15}$$

where

$$\varrho \equiv \sum_{\kappa=1}^{K} \omega^\kappa |\psi^\kappa\rangle\langle\psi^\kappa| \tag{9.16}$$

is the appropriate generalization of the density matrix operator (9.12) to the mixed state situation.

It is easy to show that a density operator is a positive operator and has unit trace. This means that the eigenvalues of a density operator are nonnegative and sum to unity.

9.8 Partial Trace

The above formalism is standard quantum theory as applied to a single Hilbert space, typically the space of SUO states. We now extend the discussion to the tensor product of two Hilbert spaces, $\mathcal{H}^S$ and $\mathcal{H}^A$, where in anticipation of our later needs, S will stand for SUO and A will stand for *apparatus* or, more technically, *ancilla*.[3] $\mathcal{H}^S$ will have dimension d and $\mathcal{H}^A$ will have dimension D.

Given two finite-dimensional Hilbert spaces $\mathcal{H}^S$, $\mathcal{H}^A$, not necessarily of the same dimension, consider a state $|\Psi^{SA})$ in the tensor product $\mathcal{H}^S \otimes \mathcal{H}^A$. Note that we shall henceforth use *round* brackets to denote states in such an SUO-apparatus tensor product space, rather than angular brackets, and call them *total states*. Then the associated density operator is defined by

$$\varrho^{SA} \equiv |\Psi^{SA})(\Psi^{SA}|. \tag{9.17}$$

Now suppose $X \in L(\mathcal{H}^S)$ is an observable over $\mathcal{H}^S$. Then we can write

$$X = \sum_{n=1}^{d} x^n P^n. \tag{9.18}$$

Now construct the operator $\widehat{P}^n \equiv P^n \otimes I^A$ in $L(\mathcal{H}^S \otimes \mathcal{H}^A)$, where I^A is the identity operator over $\mathcal{H}^A$. Then

$$\Pr(x^n|\Psi^{SA}) = (\Psi^{SA}|U^\dagger \widehat{P}^n U|\Psi^{SA}) = \mathrm{Tr}_{AB}\{\widehat{P}^n U \varrho^{SA} U^\dagger\}. \tag{9.19}$$

Here the subscript AB reminds us we are taking the full trace, that is, in the tensor product space $\mathcal{H}^S \otimes \mathcal{H}^A$.

The concept of *partial trace* is related to the concept of partial question that we discussed in the previous chapter. Partial traces are defined by constructing ONBs for the component Hilbert spaces and summing only over some of them. Suppose $\{|s^m\rangle : m = 1, 2, \ldots, d\}$ is an ONB for $\mathcal{H}^S$ and $\{|a^n\rangle : n = 1, 2, \ldots, D\}$ is an ONB basis for $\mathcal{H}^A$. Then $\{|s^m, a^n) \equiv |s^m\rangle \otimes |a^n\rangle : m = 1, 2, \ldots, d,\ n = 1, 2, \ldots, D\}$ is an ONB basis for $\mathcal{H}^S \otimes \mathcal{H}^A$.

Suppose V is an operator over $\mathcal{H}^S$ and W is an operator over $\mathcal{H}^A$. Then the *full trace* $\mathrm{Tr}_{SA}\{V \otimes W\}$ of the tensor product operator $V \otimes W$ is given by

$$\begin{aligned} \mathrm{Tr}_{SA}\{V \otimes W\} &\equiv \sum_{m=1}^{d} \sum_{n=1}^{D} (s^m, a^n|V \otimes W|s^m, a^n) \\ &= \left\{ \sum_{m=1}^{d} \langle s^m|V|s^m\rangle \right\} \left\{ \sum_{n=1}^{D} \langle a^n|W|a^n\rangle \right\}. \end{aligned} \tag{9.20}$$

[3] In standard QM, ancillas are often treated as auxiliary, almost incidental aspects. In QDN, they are essential and as important as SUOs.

There are two partial traces available. $\mathrm{Tr}_S\{V \otimes W\}$ is a partial trace over the SUO degrees of freedom, giving

$$\mathrm{Tr}_S\{V \otimes W\} = \sum_{m=1}^{d} \langle s^m|V \otimes W|s^m\rangle = \left\{\sum_{m=1}^{d} \langle s^m|V|s^m\rangle\right\} W, \tag{9.21}$$

and is an operator over $\mathcal{H}^A$. Similarly, $\mathrm{Tr}_A\{V \otimes W\}$ is a partial trace over the apparatus degrees of freedom, giving

$$\mathrm{Tr}_A\{V \otimes W\} = \sum_{n=1}^{D} \langle a^n|V \otimes W|a^n\rangle = V \left\{\sum_{n=1}^{D} \langle a^n|W|a^n\rangle\right\}, \tag{9.22}$$

and is an operator over $\mathcal{H}^S$.

Given a density operator ϱ^{SA} over $\mathcal{H}^S \otimes \mathcal{H}^A$, then we define the partial traces

$$\varrho^S \equiv \mathrm{Tr}_A\{\varrho^{SA}\}, \quad \varrho^A \equiv \mathrm{Tr}_S\{\varrho^{SA}\}. \tag{9.23}$$

Then ϱ^S is a density operator over $\mathcal{H}^S$ and ϱ^A is a density operator over $\mathcal{H}^A$, with

$$\mathrm{Tr}_S\{\varrho^S\} = \mathrm{Tr}_A\{\varrho^A\} = 1. \tag{9.24}$$

Circularity

For a single Hilbert space $\mathcal{H}$, the trace of operators $A, B, \dots, Z$ satisfies the circularity property

$$\mathrm{Tr}_H\{ABC\dots Z\} = \mathrm{Tr}_H\{BC\dots ZA\}. \tag{9.25}$$

This property holds also for tensor products of Hilbert spaces. If $R^1, R^2, \dots, R^N$ are operators over $\mathcal{H}^S \otimes \mathcal{H}^A$, then

$$\mathrm{Tr}_{SA}\{R^1R^2\dots R^N\} = \mathrm{Tr}_{SA}\{R^2R^3\dots R^NR^1\}. \tag{9.26}$$

Circularity does not hold for partial traces.

9.9 Purification

Suppose $\mathcal{H}^S$ is a Hilbert space of dimension d, representing states of some SUO. Consider a density operator ϱ^S over $\mathcal{H}^S$. From previous sections, we can find a set $\{|\psi^m\rangle : m = 1\dots d\}$ of normalized eigenvectors of ϱ^S with nonnegative eigenvalues, that is,

$$\varrho^S|\psi^m\rangle = \lambda^m|\psi^m\rangle, \quad m = 1, 2, \dots, d, \quad \lambda^m \geqslant 0. \tag{9.27}$$

Then we can write

$$\varrho^S = \sum_{m=1}^{d} \lambda^m|\psi^m\rangle\langle\psi^m|. \tag{9.28}$$

We can assume orthonormality, that is, $\langle\psi^m|\psi^n\rangle = \delta^{mn}$.

Now construct another Hilbert space $\mathcal{H}^A$ with $\dim \mathcal{H}^A = D \geqslant d$, with an ONB $\{\theta^n : n = 1, 2, \ldots, D\}$. Next, define a pure state $|\psi^{SA})$ in $\mathcal{H}^S \otimes \mathcal{H}^A$ of the form

$$|\psi^{SA}) \equiv \sum_{n=1}^{d} \sqrt{\lambda^n} |\psi^n\rangle \otimes |\theta^n\rangle. \tag{9.29}$$

The density operator ϱ^{SA} associated with this pure state is then given by

$$\varrho^{SA} \equiv |\psi^{SA})(\psi^{SA}| = \sum_{m,n=1}^{d} \sqrt{\lambda^m \lambda^n} |\psi^m\rangle\langle\psi^n| \otimes |\theta^m\rangle\langle\theta^n|. \tag{9.30}$$

Now partially trace ϱ^{SA} over $\mathcal{H}^A$:

$$\begin{aligned}
\mathrm{Tr}_A\{\varrho^{SA}\} &\equiv \sum_{c=1}^{D} \langle\theta^c|\varrho^{SA}|\theta^c\rangle = \sum_{c=1}^{D} \langle\theta^c| \sum_{m,n=1}^{d} \sqrt{\lambda^m\lambda^n} |\psi^m\rangle\langle\psi^n| \otimes |\theta^n\rangle\langle\theta^n|\theta^c\rangle \\
&= \sum_{c=1}^{D} \sum_{m,n=1}^{d} \sqrt{\lambda^m\lambda^n} |\psi^m\rangle\langle\psi^n| \underbrace{\langle\theta^c|\theta^m\rangle}_{\delta^{cm}} \underbrace{\langle\theta^n|\theta^c\rangle}_{\delta^{nc}} \\
&= \sum_{m=1}^{d} \lambda^m |\psi^m\rangle\langle\psi^m|,
\end{aligned} \tag{9.31}$$

that is,

$$\mathrm{Tr}_A\{\varrho^{SA}\} = \varrho^S. \tag{9.32}$$

In words, we can represent the density operator for a mixed state in one Hilbert space by the density operator for a pure state in a larger Hilbert space. This fundamental result opens the door to Naimark's theorem (below). The state $|\psi^{SA})$ is called a *purification* of ϱ^S. There are infinitely many possible purifications of a given density operator.

9.10 Purity and Entropy

Definition 9.5 Given a density operator ϱ with spectrum of eigenvalues $\{\lambda^k : k = 1, 2, \ldots, d\}$, define the *purity* $\mu[\varrho]$ by

$$\mu[\varrho] \equiv \sum_{k=1}^{d} (\lambda^k)^2. \tag{9.33}$$

It is straightforward to prove that for any density operator, $d^{-1} \leqslant \mu[\varrho] \leqslant 1$.

Mixed states ignore correlation information encoded between an SUO and its environment. Given a density operator ϱ, another measure of information loss associated with ϱ is the *von Neumann entropy*, defined by

$$S[\varrho] \equiv -\mathrm{Tr}\{\varrho \ln \varrho\} = -\sum_k \lambda^k \ln \lambda^k. \tag{9.34}$$

Then it is straightforward to show that $0 \leqslant S[\varrho] \leqslant \ln d$. Von Neumann entropy is a monotonically decreasing function of purity and vice-versa (Paris, 2012). A pure state has purity 1 and von Neumann entropy zero, while a maximally mixed state has entropy $\ln d$ and purity $1/d$.

9.11 POVMs

The Born rule for mixed states can be rewritten in a form that can be generalized, leading to the POVM formalism, a more general approach to quantum measurement than the PVM formalism.

In standard QM, given a density operator ϱ^S on a d-dimensional SUO Hilbert space $\mathcal{H}^S$ and observable $X \equiv \sum_{m=1}^{d} x^m P^m$, then the probability $\Pr(x^m|\varrho^S)$ of outcome x^n is given by

$$\Pr(x^m|\varrho^S) \equiv \mathrm{Tr}\{P^m U \varrho^S U^\dagger P^n\}, \tag{9.35}$$

where we use $P^n P^n = P^n$.

Now suppose we can find a set $\{M^n : n = 1, 2, \ldots, N\}$ of operators called *Kraus* (or *detection*) operators such that

$$\sum_{n=1}^{N} M^{n\dagger} M^n = I^S. \tag{9.36}$$

Then the generalization of (9.35) is to assert that

$$\Pr(y^n|\varrho^S) \equiv \mathrm{Tr}\{M^n \varrho^S M^{n\dagger}\} \tag{9.37}$$

is the probability associated with detection outcome associated with M^n, where now the number N of possible outcomes is not necessarily equal to $d \equiv \dim \mathcal{H}^S$.

We define the *POVM elements* $\{E^n : n = 1, 2, \ldots, N\}$ of the probability (or positive) operator-valued measure (POVM) by

$$E^n \equiv M^{n\dagger} M^n. \tag{9.38}$$

Then (9.36) gives

$$\sum_{n=1}^{N} E^n = I^{\mathcal{H}^S}. \tag{9.39}$$

The POVM operators E^n are positive.

The detection operators M^n are defined up to unitary transformations; that is, given a detection operator M^n, then for unitary operator V, $M'^n \equiv V^n M^n$ is a valid detection operator giving the same POVM element $E'^n \equiv M'^{n\dagger} M'^n$ as M^n, i.e.,

$$E'^n = E^n. \tag{9.40}$$

This scheme is called a *generalized measurement*.

9.12 Naimark's Theorem

This theorem is critical in the formal justification and rationalization of QDN. The theorem goes under the name of Naimark, but Naimark was not concerned with application to quantum theory per se.

The theorem has two parts (Paris, 2012). We write it out here in its QDN form, where S refers to the SUO and A refers to *apparatus* or **ancilla**.

Theorem 9.6 ***Part 1***
Suppose we are given a POVM set $\{E^n : n = 1, 2, \ldots, N\}$ on SUO Hilbert space $\mathcal{H}^S$ with finite dimension d. Then we know that the E^n are positive operators and

$$\sum_{n=1}^{N} E^n = I^S. \tag{9.41}$$

Then there exists a Hilbert space $\mathcal{H}^A$ of dimension at least N, and a pure state $|\omega^A\rangle$ in $\mathcal{H}^A$ such that

1. *The density operator $\varrho^A \equiv |\omega^A\rangle\langle\omega^A| \in L(\mathcal{H}^A)$, the Hilbert space of linear operator over $\mathcal{H}^A$.*
2. *There is some unitary evolution operator (in QDN a semi-unitary operator) $U \in L(\mathcal{H}^S \otimes \mathcal{H}^A)$ such that*

$$U^\dagger U = UU^\dagger = I^{SA}. \tag{9.42}$$

3. *There is a set $\{P^n\}$ of projectors over $\mathcal{H}^A$,*

such that

$$E^n = Tr_A\{I^S \otimes P^n U\ I^S \otimes \varrho^A U^\dagger\}. \tag{9.43}$$

Note that the U operator can encode any dynamical evolution in the combined system.

Part 2
In this part, we derive an expression for the detection operators $\{M^n\}$.

Suppose an initial mixed state is prepared, such that in $\mathcal{H}^S \otimes \mathcal{H}^A$ it is described by density operator

$$\varrho_0{}^{SA} \equiv \varrho_0{}^S \otimes |\omega_0{}^A\rangle\langle\omega_0{}^A| \tag{9.44}$$

and then allowed to evolve under unitary evolution U, giving the final density operator

$$\varrho^{SA} = U\varrho_0{}^{AB}U^\dagger. \tag{9.45}$$

Then a projective measurement to test for outcome $|x^n\rangle$ of observable X is made via the apparatus. The probability $Pr(x^n|\varrho_0{}^{SA})$ is given by the equivalent of the Born rule, in this case

$$\Pr(x^n|\varrho_0{}^{SA}) = \mathrm{Tr}_{SA}\{\varrho^{SA} I^S \otimes P^n\}, \tag{9.46}$$

where $P^n \equiv |x^n\rangle\langle x^n|$. *Then*

$$\begin{aligned}\Pr(x^n|\varrho_0{}^{SA}) &= \mathrm{Tr}_{SA}\{U\varrho_0{}^{SA}U^\dagger I^S \otimes P^n\}\\ &= \mathrm{Tr}_{SA}\{U\varrho_0{}^{S} \otimes |\omega_0{}^A\rangle\langle\omega_0{}^A|U^\dagger I^S \otimes |x^n\rangle\langle x^n|\}\\ &= \mathrm{Tr}_{SA}\{\varrho_0{}^{S} \otimes |\omega_0{}^A\rangle\langle\omega_0{}^A|U^\dagger I^S \otimes |x^n\rangle\langle x^n|U\}\\ &= \mathrm{Tr}_{S}\{\varrho_0{}^{S} \langle\omega_0{}^A|U^\dagger I^s \otimes |x^n\rangle\langle x^n|U|\omega_0{}^A\rangle\}\\ &= \mathrm{Tr}_{S}\{\varrho_0{}^{S} \langle\omega_0{}^A|U^\dagger |x^n\rangle\langle x^n|U|\omega_0{}^A\rangle\}\\ &= \mathrm{Tr}_{S}\{\underbrace{\langle x^n|U|\omega_0{}^A\rangle}_{M^n}\varrho_0{}^{S} \underbrace{\langle\omega_0{}^A|U^\dagger|x^n\rangle}_{M^{n\dagger}}\}\end{aligned} \tag{9.47}$$

i.e.

$$\Pr(x^n|\varrho_0{}^{SA}) = \mathrm{Tr}_{S}\{M^n\varrho_0{}^S M^{n\dagger}\} = \mathrm{Tr}_{S}\{E^n\varrho_0{}^S\}, \tag{9.48}$$

where $E^n \equiv M^{n\dagger}M^n$.

9.13 QDN and POVM Theory

We turn our attention now to the relationship between QDN and POVMs. This is important because a generalized QDN-POVM formalism is used extensively in our computer algebra (CA) implementation of QDN, program MAIN, discussed in Chapter 12 and used in all our specific calculations.

The first step is to recognize that the quantum registers we are concerned with model only the labstates, that is, the apparatus states, and say nothing per se about the imagined SUO states. As we stressed previously, QDN was not designed to give specific information about what happens in the information void. Therefore, the quantum physics of SUOs has to be appended "by hand". This is achieved by introducing a stage-dependent Hilbert space $\mathcal{H}_n^S$ at each stage Σ_n to contain the SUO "internal" states, and a separate quantum register $\mathcal{Q}_n$ to contain the signal state of the apparatus. Elements of $\mathcal{H}_n^S$ will be called *system states*, elements of $\mathcal{Q}_n$ are called *labstates*, and elements of $\mathcal{H}_n \equiv \mathcal{H}_n^S \otimes \mathcal{Q}_n$ will be called *total states*.

Notation

In the following, Dirac's bracket notation $|\psi_n\rangle$ is used for system states; our bold notation $\boldsymbol{i}_n$ denotes computational basis representation (CBR) of labstates; and modified Dirac brackets $|\Psi_n) = |\psi_n, \boldsymbol{i}_n) \equiv |\psi_n\rangle \otimes \boldsymbol{i}_n$ describe total states. We make an exception to our rule suppressing tensor product symbols in the case of total states, in order to separate out system states and labstates.

By definition, system states are unobservable directly and there is no natural preferred basis for $\mathcal{H}_n^S$.[4] Therefore, we are free to choose any convenient

[4] The principle of "general covariance," or independence of frame of reference, is meaningful as far as system states are involved, but vacuous as far as labstates are concerned.

orthonormal basis for $\mathcal{H}^S$. We shall denote elements of our choice by $|\alpha_n\rangle$, that is, with lowercase Greek labels, with the inner product rule $\langle\alpha_n|\beta_n\rangle \equiv \delta^{\alpha\beta}$, and assume that $\mathcal{H}_n^S$ has finite dimension d_n. For instance, $d_n = 2$ if $\mathcal{H}_n$ describes vertical and horizontal electromagnetic polarization eigenstates (ignoring momenta and other attributes).

Summations occur frequently in the following, so we make the following simplification:

$$\sum_{\alpha=1}^{d_n}\sum_{i=1}^{2^{r_n}-1} \to \sum_{\alpha i}^{[n]}, \tag{9.49}$$

where r_n is the rank of $\mathcal{Q}_n$.

Initial Total State

A pure initial total state $|\Psi_0)$ and its dual $(\Psi_0|$ will take the general form

$$|\Psi_0) \equiv \sum_{i}^{[0]} \psi_0^{\alpha i}|\alpha_0\rangle \otimes \boldsymbol{i}_0, \qquad (\Psi_0| = \sum_{\alpha i}^{[0]} \psi_0^{\alpha i*}\langle\alpha_0| \otimes \overline{\boldsymbol{i}_0}, \tag{9.50}$$

where $\psi_0^{\alpha i*}$ is the complex conjugate of $\psi_0^{\alpha i}$. If the initial total state is normalized to unity, then we have

$$(\Psi_0|\Psi_0) = \sum_{\alpha i}^{[0]} |\psi_0^{\alpha i}|^2 = 1. \tag{9.51}$$

Final Total State

Assuming semi-unitary evolution, the final total state $|\Psi_N)$ is given by

$$|\Psi_N) \equiv U_{N,0}|\Psi_0), \qquad N \geqslant 0, \tag{9.52}$$

where Σ_N is the final stage and the contextual evolution operator $U_{N,0}$ is given by

$$U_{N,0} \equiv \sum_{\alpha i}^{[N]}\sum_{\beta j}^{[0]} U_{N,0}^{\alpha i,\beta j}|\alpha_N\rangle\langle\beta_0| \otimes \boldsymbol{i}_N\overline{\boldsymbol{j}_0}. \tag{9.53}$$

The retraction operator $\overline{U}_{N,0}$ is given by

$$\overline{U}_{N,0} = \sum_{\alpha i}^{[N]}\sum_{\beta j}^{[0]} U_{N,0}^{\alpha i,\beta j*}|\beta_0\rangle\langle\alpha_N| \otimes \boldsymbol{j}_0\overline{\boldsymbol{i}_N}. \tag{9.54}$$

The semi-unitary condition $\overline{U}_{N,0}U_{N,0} = I_0$, where I_0 is the identity operator for the initial total space $\mathcal{H}_0$, requires the conditions

$$\sum_{\alpha i}^{[N]} U_{N,0}^{\alpha i,\beta j*}U_{N,0}^{\alpha i,\gamma k} = \delta^{\beta\gamma}\delta^{jk}. \tag{9.55}$$

QDN POVM Operators

The QDN Kraus operators are defined as

$$M^i_{N,0} \equiv \overline{\boldsymbol{i}_N} U_{N,0} = \sum_{\alpha}^{[N]} \sum_{\beta j}^{[0]} U^{\alpha i,\beta j}_{N,0} |\alpha_N\rangle\langle\beta_0| \otimes \overline{\boldsymbol{j}_0},$$

$$\overline{M}^i_{N,0} \equiv \overline{U}_{N,0} \boldsymbol{i}_N = \sum_{\alpha}^{[N]} \sum_{\beta j}^{[0]} U^{\alpha i,\beta j *}_{N,0} |\beta_0\rangle\langle\alpha_N| \otimes \boldsymbol{j}_0, \tag{9.56}$$

giving the POVM operators

$$E^i_{N,0} \equiv \overline{M}^i_{N,0} M^i_{N,0}$$
$$= \sum_{\beta j}^{[0]} \sum_{\delta k}^{[0]} \left\{ \sum_{\alpha}^{[N]} U^{\alpha i,\beta j *}_{N,0} U^{\alpha i,\delta k}_{N,0} \right\} |\beta_0\rangle\langle\delta_0| \otimes \boldsymbol{j}_0 \overline{\boldsymbol{k}_0}, \quad i = 0, 1, 2, \ldots, 2^{r_N} - 1. \tag{9.57}$$

Using the semi-unitary conditions (9.36), we readily find

$$\sum_{i}^{[0]} E^i_{N,0} = I_0, \tag{9.58}$$

which is the QDN analogue of the standard POVM condition (9.39).

Interpretation

We may readily understand the QDN POVM formalism if we consider the possible questions that the observer could ask. Those questions can be only asked about the signal status of the apparatus. If the final stage normalized total state is $|\Psi_N)$ and the final stage apparatus has rank r_N, then there is a grand total of 2^{r_N} maximal questions, each of them equivalent to a projector of the form $I^S_N \otimes \boldsymbol{i}_N \overline{\boldsymbol{i}_N}$, for $i = 0, 1, 2, \ldots, 2^{r_N} - 1$. The probability $P(\boldsymbol{i}_N|\Psi_0)$ that the final labstate is $\boldsymbol{i}_N$ is given by

$$P(\boldsymbol{i}_N|\Psi_0) \equiv (\Psi_N|\boldsymbol{i}_N\overline{\boldsymbol{i}_N}|\Psi_N) = (\Psi_0|\overline{U}_{N,0}\boldsymbol{i}_N\overline{\boldsymbol{i}_N}U_{N,0}|\Psi_0)$$
$$= (\Psi_0|\underbrace{\overline{U}_{N,0}\boldsymbol{i}_N}_{\overline{M}^i_{N,0}}\underbrace{\overline{\boldsymbol{i}_N}U_{N,0}}_{M^i_{N,0}}|\Psi_0) = (\Psi_0|\underbrace{\overline{M}^i_{N,0}M^i_{N,0}}_{E^i_{N,0}}|\Psi_0) \tag{9.59}$$
$$= \mathrm{Tr}\left\{E^i_{N,0}\varrho_0\right\}, \tag{9.60}$$

where $\varrho_0 \equiv |\Psi_0)(\Psi_0|$ is the initial stage density operator.

We saw in Chapter 8 that we can also ask partial questions, involving only a subset of all the final state detectors. We can answer these questions as well, because any partial question is equivalent to some combination of maximal questions.

Example 9.7 Consider a rank-three quantum register. There are $2^3 = 8$ maximal questions, corresponding to the eight projection operators associated with the CBR. Specifically, we have $\mathbb{Q}^0 \equiv \mathbb{P}^1\mathbb{P}^2\mathbb{P}^3 = \boldsymbol{0}\overline{\boldsymbol{0}}$, $\mathbb{Q}^1 \equiv \widehat{\mathbb{P}}^1\mathbb{P}^2\mathbb{P}^3 = \boldsymbol{1}\overline{\boldsymbol{1}}$, ..., $\mathbb{Q}^7 \equiv \widehat{\mathbb{P}}^1\widehat{\mathbb{P}}^2\widehat{\mathbb{P}}^3 = \boldsymbol{7}\overline{\boldsymbol{7}}$.

Suppose we wanted to ask the partial question $\widehat{\mathbb{P}}^2$, that is, look only at detector 2 and ask whether it is in its signal state. Then we use the rule that for i $= 1, 2, ..., $d, $\mathbb{P}^i + \widehat{\mathbb{P}}^i = \mathbb{I}$, the register identity, to write

$$\begin{aligned} \widehat{\mathbb{P}}^2 &= \mathbb{I}\widehat{\mathbb{P}}^2\mathbb{I} = (\mathbb{P}^1 + \widehat{\mathbb{P}}^1)\widehat{\mathbb{P}}^2(\mathbb{P}^3 + \widehat{\mathbb{P}}^3) \\ &= \mathbb{P}^1\widehat{\mathbb{P}}^2\mathbb{P}^3 + \widehat{\mathbb{P}}^1\widehat{\mathbb{P}}^2\mathbb{P}^3 + \mathbb{P}^1\widehat{\mathbb{P}}^2\widehat{\mathbb{P}}^3 + \widehat{\mathbb{P}}^1\widehat{\mathbb{P}}^2\widehat{\mathbb{P}}^3 \\ &= \mathbb{Q}^2 + \mathbb{Q}^3 + \mathbb{Q}^6 + \mathbb{Q}^7. \end{aligned} \tag{9.61}$$

We conclude from this that knowledge of all the maximal questions and their answers allows us to answer all partial questions.

Mixed Initial States

The above has assumed that the initial total state is a pure one. We can readily extend the formalism to the case of mixed initial states. All we need do is replace the pure density operator $\varrho_0 \equiv |\Psi_0)(\Psi_0|$ with the appropriate mixed density operator. This will have the generic form

$$\varrho_0 \equiv \sum_{\kappa} \omega^{\kappa} |\Psi_0^{\kappa})(\Psi_0^{\kappa}|, \tag{9.62}$$

where the ω^{κ} are probabilities summing to unity and $|\Psi_0^{\kappa})$ is the normalized total state occurring with probability ω^{κ} in the mixture.

There is an interesting possibility here: not only could the randomness be associated with the system states, but there could be uncertainty about the apparatus. In other words, we should be prepared for the possibility that the apparatus at any given stage is not determined beforehand but is created by random processes. This raises deeper questions to do with the general theory of observation that will certainly need to be addressed in the future. We discuss some aspects of this topic in Chapter 21, on self-intervening networks.

There are three important aspects to our modeling. First, in common with standard quantum modeling of pointer states, our initial apparatus register (at stage Σ_0) will be a rank-one quantum register $\mathcal{Q}_1$ referred to as a *preparation switch*. This models the logic of state preparation: if the observer knows that an SUO state $|\Psi\rangle$ has been prepared, then that state is tensored with element $\mathbf{1}_0$ of $\mathcal{Q}_0$, whereas if such a state has *not* been prepared, then that state is tensored with element $\mathbf{0}_0$ of $\mathcal{Q}_0$. We see from this that QDN attaches significance to what has *not* been done as much as to what has been done. This is reminiscent of Renniger's thought experiment, where an absence of observation has measurable consequences (Renniger, 1953).

The second aspect concerns the interpretation of what is going on. In QDN, only signal states of the apparatus are physically meaningful. The SUO and therefore the mathematical representation of its states is a convenient fiction encoding contextuality. Therefore, we do not need to treat $\mathcal{H}^S$ on the same footing as $\mathcal{Q}_n$. It may be convenient to employ the Schrödinger picture for evolution of

SUO states and treat $\mathcal{H}^S$ as stage independent.[5] That is certainly not the case for the quantum registers $\mathcal{Q}_n$.

The third aspect concerns the notion of *purification* that we discussed above for standard QM. There it was pointed out that the dimension of the ancilla Hilbert space $\mathcal{H}^A$ had to be at least as great as the number of independent states in the prepared mixture that was being purified, and that there was always an infinite number of ancillas that could be employed. In the case of QDN, neither of these comments applies. The analogue of the ancilla concept is the quantum register modeling the apparatus, and the rank of that register is determined by the number of detectors that the observer has constructed. That number is independent of the dimensionality of $\mathcal{H}^S$. Moreover, the concept of preferred basis does not apply to the standard ancilla concept, whereas it is relevant in QDN. The moral here is that standard QM treats states of SUO as the primary objects of interest, whereas QDN relegates them to auxiliary devices and elevates the apparatus to be the only thing that matters. This is exactly in line with Wheeler's participatory principle, mentioned previously.

We start our analysis therefore with a statement as to what is known at initial stage Σ_0. We will assume that a mixed SUO state has been prepared at stage Σ_0,[6] such that the initial density matrix ${\boldsymbol{\varrho}_0}^{SA}$ is an element of $L(\mathcal{H}^S \otimes \mathcal{Q}_0)$ given by

$$ {\boldsymbol{\varrho}_0}^{SA} \equiv \boldsymbol{\varrho}^S \otimes \mathbf{1}_0\overline{\mathbf{1}_0} = \boldsymbol{\varrho}^S \otimes \widehat{\mathbb{P}}_0^1, \tag{9.63} $$

where $\boldsymbol{\varrho}^S$ is a density operator, an element of $L(\mathcal{H}^S)$.

We now imagine that the combined SUO-apparatus state evolves from stage Σ_0 to some final stage, Σ_N, where $N > 0$. Consider quantum evolution from Hilbert space $\mathcal{H}^S \otimes \mathcal{Q}_n$ at stage Σ_n to Hilbert space $\mathcal{H}^S \otimes \mathcal{Q}_{n+1}$ at stage Σ_{n+1}, where $\mathcal{H}^S$ is the SUO Hilbert space of dimension d and $\mathcal{Q}_n, \mathcal{Q}_{n+1}$ are the quantum registers representing labstates of the apparatus. An ONB at Σ_n is given by $\{|A, \boldsymbol{i}_n) \equiv |A\rangle \otimes \boldsymbol{i}_n : i = 0, 1, \ldots, d_n\}$ with the orthonormalization $(A, \boldsymbol{i}_n | B, \boldsymbol{j}_n) = \delta^{AB}\delta^{ij}$, while at Σ_{n+1} we have ONB $\{|A, \boldsymbol{i}_{n+1}) \equiv |A\rangle \otimes \boldsymbol{i}_{n+1} : i = 0, 1, \ldots, d_{n+1}\}$ with the orthonormalization $(A, \boldsymbol{i}_{n+1} | B, \boldsymbol{j}_{n+1}) = \delta^{AB}\delta^{ij}$. Here $d_n \equiv 2^{r_n} - 1$.

Pure quantum evolution is given by

$$ U_{n+1,n}|A, \boldsymbol{i}_n) = \sum_{B=1}^{d} \sum_{j=0}^{d_{n+1}} U_{n+1,n}^{BA,ji} |B, \boldsymbol{j}_{n+1}) : \qquad i = 0, 2, \ldots, d_n. \tag{9.64} $$

[5] In our computer algebra program MAIN, discussed in Chapter 12, we take the system space to change from stage to stage.

[6] States of SUOs are treated in QDN as convenient repositories of context. By their actions in the state preparation process, an observer will in general be entitled to model some of that context in terms of a density operator representing a mixed state of an SUO.

Using completeness, this gives the dyadic representation of $U_{n+1,n}$

$$U_{n+1,n} = \sum_{A,B=1}^{d} \sum_{j=0}^{d_{n+1}} \sum_{i=0}^{d_n} |B, \boldsymbol{j}_{n+1}) U_{n+1,n}^{BA,ji} (A, \boldsymbol{i}_n|. \tag{9.65}$$

The retraction operator $\overline{U}_{n+1,n}$ is defined to be

$$\overline{U}_{n+1,n} = \sum_{A,B=1}^{d} \sum_{j=0}^{d_{n+1}} \sum_{i=0}^{d_n} |A, \boldsymbol{i}_n) U_{n+1,n}^{BA,ji*} (B, \boldsymbol{j}_{n+1}|. \tag{9.66}$$

By definition, we have

$$\overline{U}_{n+1,n} U_{n+1,n} = I^S \otimes I_1{}^A, \tag{9.67}$$

where I^S is the identity over $\mathcal{H}^S$ and $I_1{}^A$ is the identity over $\mathcal{Q}_1$. From this, we arrive at the semi-unitary conditions

$$\sum_{C=1}^{d} \sum_{j=0}^{d_{n+1}} U_{n+1,n}^{CA,ja*} U_{n+1,n}^{CB,jb} = \delta^{AB} \delta^{ab}. \tag{9.68}$$

Now consider a pure total state

$$|\Psi_0) \equiv \sum_{A=1}^{d} \Psi_0^A |A\rangle \otimes \mathbf{1}_0, \qquad \sum_{A=1}^{d} |\Psi_0^A|^2 = 1, \tag{9.69}$$

evolving from state Σ_0 to stage Σ_1, giving total state $|\Psi_1) \equiv U_{1,0}|\Psi_0)$.

Continuing this process, we find the state at stage Σ_N is given by $|\Psi_N) = U_{N,0}|\Psi_0)$, where $U_{N,0} \equiv U_{N,N-1} U_{N-1,N-2} \ldots U_{1,0}$.

Suppose now the observer asks a partial question $\mathbb{Q}_N^\theta$ of the state of the apparatus at stage Σ_N. The expectation value $E[\mathbb{Q}_N^\theta|\Psi_0]$ of the answer is given by

$$E[\mathbb{Q}_N^\theta|\Psi_0] \equiv (\Psi_N| I^S \otimes \mathbb{Q}_N^\theta |\Psi_N) = \sum_{A,B=1}^{d} \Psi_0^{A*} \Psi_0^B \langle A| E_N^\theta |B\rangle, \tag{9.70}$$

where

$$E_N^\theta \equiv \overline{\mathbf{1}_0 \overline{U}_{N,0}} I^S \otimes \mathbb{Q}_N^\theta U_{N,0} \mathbf{1}_0. \tag{9.71}$$

If now the initial state is a mixed state, such that the initial SUO state $|\Psi_0^\alpha\rangle$ is given with probability ω^α, then the expectation value is given by

$$E[\mathbb{Q}_N^\theta|\boldsymbol{\varrho}_0] = \mathrm{Tr}_S\{\boldsymbol{\varrho}_0^A E_N^\theta\}, \tag{9.72}$$

where

$$\boldsymbol{\varrho}_0^A \equiv \sum_{\alpha} \omega^\alpha |\Psi_0^\alpha\rangle\langle\Psi_0^\alpha|. \tag{9.73}$$

This is the QDN generalization of the standard POVM formalism.

We note that, provided we sum θ over all elements of an identity class Θ only,[7] then

$$\sum_{\theta \text{ in } \Theta} E_N^\theta = \overline{\mathbf{1}_0 U}_{N,0} I^S \otimes \sum_{\theta \text{ in } \Theta} \mathbb{Q}_N^\theta U_{N,0} \mathbf{1}_0 = \overline{\mathbf{1}_0 U}_{N,0} I^S \otimes \mathbb{I}_N U_{N,0} \mathbf{1}_0 = I^S. \quad (9.74)$$

Hence

$$\sum_{\theta \text{ in } \Theta} E[\mathbb{Q}_N{}^\theta | \boldsymbol{\varrho}_0] = \mathrm{Tr}_S \left\{ \boldsymbol{\varrho}_0{}^A \sum_{\theta \text{ in } \Theta} E_N{}^\theta \right\} = \mathrm{Tr}_S \{ \boldsymbol{\varrho}_0{}^A I^S \} = 1. \quad (9.75)$$

The interpretation of the expectation values $E[\mathbb{Q}_N{}^\theta | \boldsymbol{\varrho}_0]$ therefore is consistent with probability, provided we restrict θ to a single identity class.

[7] Identity classes are discussed in the previous chapter.

10
Double-Slit Experiments

10.1 Introduction

In this chapter we show how the quantized detector network (QDN) formalism describes the double-slit (DS) experiment. This is arguably the simplest experiment that demonstrates quantum affects such as wave–particle duality and quantum interference. It continues to be the focus of much debate and experiment (Mardari, 2005), because theoretical modeling of what is going on reflects current understanding of quantum physics and hence physical reality. We will apply QDN to two variants: the original DS experiment and the *monitored* DS experiment, where an attempt is made to determine the imagined path of the particle.

The DS experiment is widely acknowledged by physicists to be of importance to the understanding of quantum mechanics (QM). So much so that in 2002, the single electron version, first performed by Merli, Missiroli, and Pozzi (Merli et al., 1976), was voted by readers of *Physics World* to be "the most beautiful experiment in physics" (Rosa, 2012).

The DS experiment can be discussed in terms of three stages, shown in Figure 10.1. By the end of the preparation stage, Σ_0, a monochromatic beam of light or particles has been prepared by a source P, such as a laser. The beam emerges from point O and then passes through an information void V_1 to the first stage, Σ_1, which consists of a wall or barrier W. This wall has two openings denoted A and B that allow parts of the beam to pass through into another information void V_2 and onto the second and final stage Σ_2, which consists of a detecting screen S.

The screen S is in general some material that can absorb and record particle impacts. In reality, any screen will consist of a finite number of signal detectors, such as photosensitive molecules, but the typical QM modeling is done as if there were a continuum of sites on the screen, such as C, that could register particles.

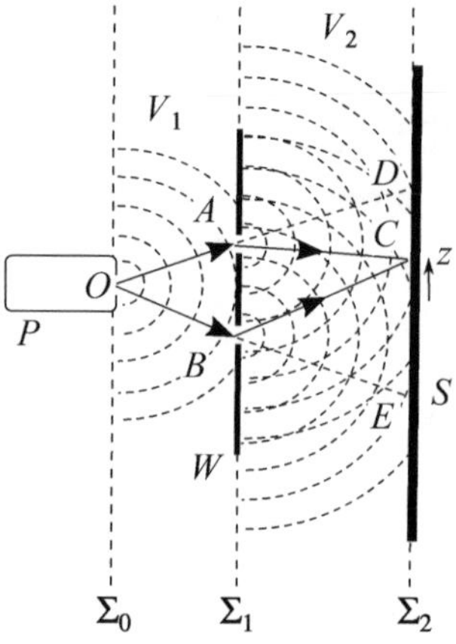

Figure 10.1. The DS experiment. C is a typical detector site in the detecting screen S.

10.2 Run Protocol

In addition to the above geometrical architecture, there are some important protocol features that need to be clarified.

Runs

Any DS experiment will consist of a large, possibly enormous, number of runs, or repetitions of a basic protocol.

Statistics

The conclusions of the experiment are based on a statistical analysis of the data averaged over all valid runs.

Preparation Stage

Each run consists of the observer establishing the start of that run, indicated by stage Σ_0 in Figure 10.1. This means that the observer will have reliable contextual information that some previously agreed-on procedure has been carried out at device P. If anything occurs to prevent confidence in that information, then that run is discarded.

Outcome Stage

At the end of each run, corresponding to stage Σ_2 in Figure 10.1, the observer looks at every accessible point/detector on the screen S and records whether each detector has a signal or not. The data from each run are entirely classical at this point, being in the form of a vast number of bits of yes/no values, each value coming from a given detector on the screen S.

Information Void

It is a critical feature of any quantum experiment, apart from the quantum Zeno-type experiments discussed in Chapter 15, that between stages Σ_0 and

Σ_2, no attempt is made by the observer to extract any information from either information void V_1 or V_2, or at stage Σ_1.

Reset

At the end of each run, the detectors in screen S are reset to their ground states, ready for the next run, which takes place as if no other run had ever occurred.

10.3 Baseball in the Dark

It is significant that in the above description of protocol, Section 10.2, no reference is made to *particles*, *beams*, *interactions*, *waves*, or any such classical mechanics (CM) imagery. What has been described is only what the observer actually does in the laboratory, not what theorists imagine they are doing. Indeed, we not pushing this point too far to remind the reader that even the above "objective" account is contextual, being human-centric. From any other species' point of view, nothing of importance would be going on in the laboratory. A dog, for example, would be more concerned where the observer kept their sandwiches. Of course, when we discuss such experiments in practice, it is most convenient to objectify procedures in familiar terms, so we will usually talk about preparing a beam of particles at P and allowing that beam to pass through a double slit. We may even be caught out referring to photons impacting on a screen. But all of that is to be read as a convenience, not as a statement of belief that there are objective things known as photons.

When discussing quantum processes, we should be wary of invoking undue mental imagery. In this respect, a helpful analogy is to imagine the DS experiment as "baseball played in the dark." Suppose we were asked to describe a game of baseball or cricket played not in broad daylight but at night in pitch black conditions and with no sound. Now the game would look very different from what it would look like during the day. The pitcher or bowler would be the analogue of the preparation device P; the batter or batsman would be the analogue of one of the openings, A or B, in the wall W; and the fielders would be the detectors in the screen S. P would have some contextual information that they should throw the ball in a certain direction. Suppose they did that. If the ball reached the batter, it would be struck in some random direction. The odds of any of the fielders catching that ball at night (the analogue of a quantum experiment) would be quite different from the odds during daytime (the analogue of a classical experiment), because in daylight, fielders would have constant visual information as to the current position of the ball. At night, they would be faced with a real information void.

The above description of baseball in the dark is a classically based attempt to convey some of the attributes of observation; it cannot adequately account for all the nonclassical attributes of quantum processes. Any experimentalist who has done real DS experiments would possibly find the description of protocol in Section 10.2 simplistic and perhaps misleading. What we have described is a highly idealized version of rather complex procedures that involve a great deal of beam calibration, noise suppression, timing protocols, shielding, detector physics, and complex electronics. Indeed, major experiments such as the search for the Higgs particle at the Large Hadron Collider take data during a small fraction of the duration of the experiment, the rest of the time being spent in planning, funding, construction, and calibration of the apparatus. It is a common feature of experimental physics doctorates that constructing the apparatus takes two or more years in preparation, and then there is a frantic race to take enough data to justify submitting a thesis.

Perhaps the best way to think of these issues is to accept that theory does not describe reality directly but deals with *equivalence classes* of processes (Kraus, 1983). For example, what we mean by a "double-slit experiment" is a theoretical model of the equivalence class of activities in the laboratory that each has the essential features outlined in Section 10.2, disregarding many contexts such as whether the observer is male or female, wears a hat or not, and so on.

What is truly remarkable is that when we overlook these issues in much in the same way as friction is overlooked in Newtonian mechanics, then there emerges from the overall complexity of any experiment some simple theoretical rules as to what is going on. That is really how QM was discovered. Moreover, these rules have great applicability and in the case of QM, work particularly well and far better than CM in experiments such as the DS experiment. QDN should be seen in this light: it will give the essential architecture of an experiment but not a detailed description of the "friction" encountered in quantum experiments, unless that is called for.

10.4 Observed Phenomena

The DS experiment reveals a number of phenomena that continue to puzzle physicists, because those phenomena cannot be fully explained according to the principles and ideology of CM.

Line-of-Sight Violation

With reference to Figure 10.1, suppose opening B is blocked off but otherwise the experiment runs as described above. According to CM, particles passing through opening A should pass more or less undisturbed onto position D on the screen S. Of course, the opening at A would deflect some of the particles, since the wall W consists of atoms, so it is to be expected that at the edges of opening A, forces will act on any beam particles passing nearby. However, the fraction of all the particles passing through either opening that comes close to the edges of the slits is expected to be relatively small, so we expect to find a relatively large distribution of particles around position D on the screen S with relatively few to either side.

What is observed is quite different. There is indeed a broad peak found centered on D, but it extends further over the screen than expected from naive CM expectations. Let us call the probability profile of this broad peak $P^A(z)$, where z is the position coordinate of points along the screen as shown in Figure 10.1. If now instead of B we blocked off opening A, then we would get a similar broad distribution $P^B(z)$, this time centered on E, on the line of sight from O to B.

Interference

What is astonishing is that if now we ran the experiment with both A and B open, then we would not find a simple distribution $P^A(z) + P^B(z)$ as expected from CM principles. Instead, we would find a distribution $P^A(z) + P^B(z) + I^{AB}(z)$, where I^{AB} is known as an interference term. It is this interference term that causes all the fuss.

Self-interference

It was originally believed that a theoretical particle–wave conflict could be avoided if the explanation of the interference term I^{AB} was that there were interactions of the particles coming from opening A that were somehow interacting with particles coming from opening B, thereby disrupting the basic addition of P^A to P^B on the detecting screen.

But the mystery was only deepened when DS experiments were done such that the rate of particles falling on the screen was extremely low (Taylor, 1909). So low in fact that there would be one (or even fewer than one)[1] on average landing on the screen S per run. It was found that the interference term occurred in the statistical analysis when a large number of runs was performed, even when only one particle could possibly have passed through at a time. The point is that the aforementioned picture of clouds of particles interfering with each other cannot be valid here when only one particle passes through at a time. The paradox is that it is then hard if not impossible to understand why the interference term should occur in the analysis of many separate, uncorrelated runs. When QM is interpreted in terms of particle–waves interfering with each other, the low-intensity interference pattern is generally referred to as *self-interference*, a term commonly attributed to Dirac in his famous book on QM (Dirac, 1958).

Not all theorists subscribe to this view (Mardari, 2005). From the QDN perspective, the term *self-interference* is a dangerous one to take seriously, as it invokes a confusing picture of a classical particle that behaves unlike a classical particle. We do *not* have to believe in particles: the only things that we can be sure of are signals in detectors.

[1] This is deep. An average of fewer than one particle per run would mean, of course, that during some runs, *no* particles were recorded as having landed on the screen. An average of fewer than one particle per run vindicates the view that what matters is what the observer does, not what we think they do. Observers push buttons and look at screens. They do *not* know for sure what is happening beyond that description.

Path Indeterminacy

The interference term I_{AB} requires both openings A and B to be present. If any attempt is made to block off one or another opening, then I^{AB} disappears. Now according to CM, every time a detector registers a signal at S, that is evidence for a particle having traveled from source P to that detector. According to CM principles, that particle had to have traveled along a continuous path from O through either A or B. If true, it should be possible to establish which of the two openings it was. But any attempt to do this appears to destroy I^{AB}.

It is as if the observer has two mutually exclusive choices: either have no knowledge about which path was taken, and then P^A, P^B, and the interference term occur, or know which path was taken, such as through A, but then the interference term and P^B disappear and only P^A is observed. In QM, the principle that there is this exclusive choice is generally referred to as *complementarity*.

It may be possible to trade off information, with the observer having only a probability estimate for each of these alternatives. In that case, the overall distribution on the screen would depend on that probability estimate in some way. However, predicting that distribution would undoubtedly require the most careful analysis of context. Experiments along such lines have been attempted, such as that of Afshar (Afshar, 2005). The results of such experiments remain controversial (Kastner, 2005).

Wave–Particle Duality

We shall see in the next section that physicists can get a good theoretical handle on the experiment by applying the particle–wave concept inherent to Schrödinger wave mechanics. The DS experiment touches on particle-like attributes because the detectors in the screen S respond in a discrete *yes/no* way characteristic of particle impacts, while on the other hand the distribution $P^A+P^B+I^{AB}$ is characteristic of wave dynamical processes. The mystery only arises when theorists suppose that there are objects with particle and wave properties simultaneously and fail to recognize empirical context as a critical factor in the experiment. In actuality, each aspect (particle or wave) is significant within its own particular empirical context, and there are no real paradoxes in the laboratory. The so-called wave–particle paradox occurs only because of the way humans generally choose to interpret their experiments.

10.5 A Wave-Mechanics Description

In this section we discuss the DS scenario using standard non-relativistic QM for a wave–particle of nonzero mass m propagating in three spatial dimensions. The standard theory assumes that apart from the production process in the source P, the interaction with the wall W, and the screen S, the quantum state representing the dynamics propagates in the information void regions V_1 and V_2 according to the Schrödinger–Dirac equation

$$i\hbar\frac{d}{dt}|\Psi,t\rangle = \widehat{H}|\Psi,t\rangle, \tag{10.1}$$

where $\widehat{H}$ is the free particle Hamiltonian given by

$$\widehat{H} = \frac{\widehat{\boldsymbol{p}}\cdot\widehat{\boldsymbol{p}}}{2m}. \tag{10.2}$$

Relative to a standard improperly normalized particle position basis $\{|\boldsymbol{x}\rangle, \boldsymbol{x} \in \mathbb{E}^3\}$, the state vector $|\Psi,t\rangle$ is given by

$$|\Psi,t\rangle = \int d^3\boldsymbol{x}\Psi(\boldsymbol{x},t)|\boldsymbol{x}\rangle. \tag{10.3}$$

The wave-function part $\Psi(\boldsymbol{x},t)$ of the solution to Eq. (10.1) is readily found by standard methods to be given by

$$\Psi(\boldsymbol{x},t) = \int d^3\boldsymbol{y}F(\boldsymbol{x},t;\boldsymbol{y},t_0)\Psi(\boldsymbol{y},t_0), \qquad t > t_0, \tag{10.4}$$

where the propagator $F(\boldsymbol{x},t;\boldsymbol{y},t_0) \equiv \langle\boldsymbol{x}|e^{-i\widehat{H}(t-t_0)/\hbar}|\boldsymbol{y}\rangle$ is given by (Feynman and Hibbs, 1965)

$$F(\boldsymbol{x},t;\boldsymbol{y},t_0) = \left[\frac{-im}{2\pi\hbar(t-t_0)}\right]^{3/2}\exp\left\{\frac{im(\boldsymbol{x}-\boldsymbol{y})^2}{2\hbar(t-t_0)}\right\}, \qquad t > t_0. \tag{10.5}$$

This propagator will be valid for events $(\boldsymbol{x},t)$ and $(\boldsymbol{y},t_0)$ in the same information void region. We shall consider what happens in region V_1 and then in region V_2.

In the following, we take standard Cartesian coordinates $\boldsymbol{x} \equiv (x,y,z)$ with origin at O in Figure 10.1, x-axis along the beam direction, y-axis transverse to the beam, and z-axis in the direction from B to A. For simplicity, we shall suppress any transverse effects and assume that openings A and B are almost point-like, with coordinates $\boldsymbol{x}_A = (d,0,a)$ and $\boldsymbol{y}_B = (d,0,-a)$, respectively, where d is the distance of the wall W from the opening O and a is positive. We shall consider what happens at a point C on the detecting screen S, with coordinates $\boldsymbol{x}_C = (d+D,0,c)$, where D is the distance of the screen S from the wall W.

For the rest of this section we use the notation $x_1 \equiv (\boldsymbol{x}_1,t_1)$, $y_2 \equiv (\boldsymbol{y}_2,t_2)$, and so on.

Region V_1

We imagine that at the first stage Σ_0, a normalized pulse is emitted from O, characterized by wave function $\Phi(x_0)$. At a given event with coordinates x_1 on the wall W, the wave function $\Psi(x_1)$ after propagation through V_1 is given by

$$\Psi(x_1) = \int d^3\boldsymbol{x}_0F(x_1;x_0)\Phi(x_0), \qquad t_1 > t_0. \tag{10.6}$$

Conservation of probability requires that

$$\int d^3\boldsymbol{x}_1|\Psi(x_1)|^2 = \int d^3\boldsymbol{x}_0|\Phi(x_0)|^2, \tag{10.7}$$

which is satisfied by virtue of the relation

$$\int d^3\boldsymbol{x}_1 F^*(x_1; y_0) F(x_1; x_0) = \delta^3(\boldsymbol{x}_0 - \boldsymbol{y}_0). \tag{10.8}$$

Region V_2

Equation (10.6) gives the wave function impacting on the wall W on the side facing the first information void V_1. We need an expression for the wave function on the other side of W that acts as an initial wave function propagating into information void V_2 and hitting the screen S. Stage Σ_1 can be thought of in this respect as a preparation stage.

To this end we introduce *shape functions* $G^A(x_1)$ and $G^B(x_1)$ that characterize the openings A and B. Feynman and Hibbs (1965) discuss this calculation where a Gaussian shape function is assumed. The prepared wave function on the V_2 side of the wall is then given by $\{G^A(x_1) + G^B(x_1)\}\,\Psi(x_1)$.

Propagation through V_2 follows the same pattern as through V_1. The final stage wave function $\Psi(x_1)$ for an event on S at stage Σ_2 is given by

$$\Psi(x_2) = \int d^3\boldsymbol{x}_1 F(x_2; x_1)\left\{G^A(x_1) + G^B(x_1)\right\}\Psi(x_1). \tag{10.9}$$

What matters here is the squared modulus $|\Psi(x_2)|^2$, which according to the Born interpretation (Born, 1926) is the probability density relevant to outcome detection on the screen S. From (10.9) we find

$$|\Psi(x_2)|^2 = P^A(x_2) + P^B(x_2) + I^{AB}(x_2), \tag{10.10}$$

where

$$\begin{aligned}
P^A(x_2) &= \int d^3\boldsymbol{x}_1 d^3\boldsymbol{y}_1 F(x_2; x_1) F^*(x_2; y_1) G^A(x_1) G^{*A}(y_1) \Psi(x_1) \Psi^*(y_1),\\
P^B(x_2) &= \int d^3\boldsymbol{x}_1 d^3\boldsymbol{y}_1 F(x_2; x_1) F^*(x_2; y_1) G^B(x_1) G^{*B}(y_1) \Psi(x_1) \Psi^*(y_1),\\
I^{AB}(x_2) &= \int d^3\boldsymbol{x}_1 d^3\boldsymbol{y}_1 F(x_2; x_1) F^*(x_2; y_1) \left\{\begin{array}{c} G^A(x_1) G^{*B}(y_1) + \\ G^B(x_1) G^{*A}(y_1) \end{array}\right\} \Psi(x_1) \Psi^*(y_1).
\end{aligned} \tag{10.11}$$

In these integrals, we may use (10.6) to work out the outcome probabilities from a knowledge of the detectors in the screen S and the characteristics of the preparation device P; that is, we should be able to specify the initial wave function $\Phi(x_0)$ reasonably well.

Blocking off opening B corresponds to setting G^B to zero, and then we see from (10.11) that the interference term and P^B vanish. A similar remark applies the blocking off of opening A.

10.6 The QDN Account of the Double-Slit Experiment

We are now in position to describe the DS experiment via QDN. Applying our bitification process, we introduce qubits at all those sites where significant

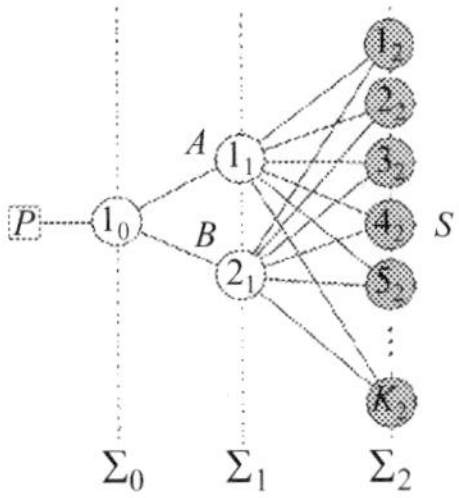

Figure 10.2. The DS experiment.

information could in principle be extracted. This excludes the information void regions V_1 and V_2, and all parts of the wall W apart from the two openings A and B. The QDN architecture is given by Figure 10.2. We model the detecting screen as a (possibly vast) number K of detectors and now discuss each run, stage by stage.

Stage Σ_0

The initial normalized labstate $\mathbf{\Psi}_0$ prepared by apparatus P by stage Σ_0 is denoted in QDN by

$$\mathbf{\Psi}_0 = \mathbf{1}_0, \tag{10.12}$$

where $\mathbf{1}_0$ is the signal state element of the one-qubit quantum computational basis representation (CBR) $\{\mathbf{0}_0, \mathbf{1}_0\}$. Essentially, $\mathbf{\Psi}_0$ carries the information that the run should proceed, analogous to the proposition "*go for burn*," validated by Mission Control prior to moon rocket engine ignition, *after* all safety checks had been completed (Apollo Program Office, 1969). We shall refer to such a one-qubit register as a *preparation switch*.

Note that we are ignoring the internal spin state of the electromagnetic radiation, as polarization is not a factor in this version of the experiment. However, such effects can be included easily if required.

Stage Σ_1

The QDN description of $\mathbf{\Psi}_1$, the labstate at stage Σ_1, has a dual role. On the one hand it represents what the observer could find (statistically) *if* they looked at either or both openings A and B at that stage. Actually, doing this would be part of the complex calibration processes involved in setting up the experiment in the first place. On the other hand, $\mathbf{\Psi}_1$ should be regarded as the initial labstate for propagation to the next stage, Σ_2, landing on the screen S. If this latter alternative is chosen, then the observer should make no attempt to extract information from $\mathbf{\Psi}_1$.

The rules of QM lead us to assert that

$$\mathbf{\Psi}_1 = \mathbb{U}_{1,0}\mathbf{\Psi}_0 = (\alpha^1\widehat{\mathbb{A}}_1^1 + \alpha^2\widehat{\mathbb{A}}_1^2)\mathbf{0}_1, \tag{10.13}$$

where α^1 and α^2 are complex numbers satisfying the normalization condition $|\alpha^1|^2 + |\alpha^2|^2 = 1$.

Here $\mathbb{U}_{2,1}$ is a semi-unitary operator taking normalized labstates from $\mathcal{Q}_0$ to $\mathcal{Q}_1 \equiv Q_1^1 Q_1^2$. We shall comment on the nature of this operator presently. At this point, we note that signality is preserved in the transition from stage Σ_0 to stage Σ_1, since the initial labstate $\boldsymbol{\Psi}_0 \equiv \widehat{\mathbb{A}}_0^1 \mathbf{0}_0$ has signality one, and by inspection of (10.13), $\boldsymbol{\Psi}_1$ also has signality one. We rule out (by hand) terms proportional to the signality-zero state, $\mathbf{0}_1$, or to the signality-two state $\widehat{\mathbb{A}}_1^1 \widehat{\mathbb{A}}_1^2 \mathbf{0}_1$, as these represent dynamics different from the one that we wish to explore here. When charged particles such as electrons are involved in the DS experiment, charge conservation rules out changes of signality. For bosons such as photons, it is quite possible to encounter signality nonconservation, such as in the case of parametric down-conversion (Burnham and Weinberg, 1970; Klyshko et al., 1970).

Stage Σ_2

The transition from stage Σ_1 to stage Σ_2 is handled as follows. We assert that there is a semi-unitary operator $\mathbb{U}_{2,1}$ such that for each term $\widehat{\mathbb{A}}_1^a \mathbf{0}_1$, $a = 1, 2$, on the right-hand side of equation (10.13),

$$\mathbb{U}_{2,1} \widehat{\mathbb{A}}_1^a \mathbf{0}_1 = \sum_{j=1}^{K} U_{2,1}^{j,a} \widehat{\mathbb{A}}_2^j \mathbf{0}_2, \quad a = 1, 2, \tag{10.14}$$

assuming there are K detector sites on the detecting screen S. This process preserves signality. We shall comment on the nature of $\mathbb{U}_{2,1}$ presently. Conservation of probability requires the complex coefficients $\left\{U_{2,1}^{j,a}\right\}$ to satisfy the semi-unitarity rule

$$\sum_{j=1}^{K} U_{2,1}^{j,b*} U_{2,1}^{j,a} = \delta^{ab}, \tag{10.15}$$

where $U_{2,1}^{j,b*}$ is the complex conjugate of $U_{2,1}^{j,b}$. We note that Eq. (10.15) is the QDN analogue of Eq. (10.8).

The linearity rules of QM now give the relationship between the initial labstate $\boldsymbol{\Psi}_0$ and the final labstate $\boldsymbol{\Psi}_2$ to be

$$\boldsymbol{\Psi}_2 = \mathbb{U}_{2,1} \mathbb{U}_{1,0} \boldsymbol{\Psi}_0 = \sum_{a=1}^{2} \sum_{j=1}^{K} \alpha^a U_{2,1}^{j,a} \widehat{\mathbb{A}}_2^j \mathbf{0}_2. \tag{10.16}$$

It can be readily checked, using the normalization condition and the above semi-unitarity rules that $\overline{\boldsymbol{\Psi}_2} \boldsymbol{\Psi}_2 = 1$.

We can now readily calculate all outcome probabilities, by choosing any of the 3^N maximal or partial questions. For example, the conditional probability $\Pr(k_2|\boldsymbol{\Psi}_0)$ that the kth detector at stage Σ_2 would be in its signal state is given by $\Pr(k_2|\boldsymbol{\Psi}_0) = \overline{\boldsymbol{\Psi}_2} \widehat{\mathbb{P}}_2^k \boldsymbol{\Psi}_2$, which readily evaluates to the value

$$\Pr(k_2|\Psi_0) = \sum_{a,b=1}^{2} \alpha^{a*}\alpha^b U_{2,1}^{k,a*} U_{2,1}^{k,b}, \quad k = 1, 2, \ldots, K. \tag{10.17}$$

Exercise 10.1 Prove (10.17) and show that total probability is conserved, that is,

$$\sum_{k=1}^{K} \Pr(k_2|\Psi_0) = 1. \tag{10.18}$$

10.7 Contextual Subspaces

Suppose $\mathbb{U}_{n+1,n}$ is the semi-unitary evolution operator from quantum register $\mathcal{Q}_n$ at stage Σ_n to quantum register $\mathcal{Q}_{n+1}$ at stage Σ_{n+1}. Suppose further that we have a complete specification of the action of $\mathbb{U}_{n+1,n}$, in the form of the rules for CBR element evolution given by

$$\mathbb{U}_{n+1,n}\boldsymbol{i}_n = \sum_{j=0}^{2^{r_{n+1}}-1} U_{n+1,n}^{j,i}\boldsymbol{j}_{n+1}, \qquad r_{n+1} = rank\ \mathcal{Q}_{n+1}. \tag{10.19}$$

Then using completeness, we find the dyadic representation

$$\mathbb{U}_{n+1,n} = \sum_{i=0}^{2^{r_n}-1}\sum_{j=0}^{2^{r_{n+1}}-1} \boldsymbol{j}_{n+1} U_{n+1,n}^{j,i} \overline{\boldsymbol{i}}^n. \tag{10.20}$$

Then we have the rule

$$\overline{\mathbb{U}}_{n+1,n}\mathbb{U}_{n+1,n} = \mathbb{I}_n, \tag{10.21}$$

where $\overline{\mathbb{U}}_{n+1,n}$ is the retraction of $\mathbb{U}_{n+1,n}$ and $\mathbb{I}_n$ is the identity operator over $\mathcal{Q}_n$.

At this point, we are confronted with what appears to be a serious problem; we do not have all the information that allows us to construct the full evolution operator $\mathbb{U}_{2,1}$ in the *DS* experiment. The number of elements in the initial preferred basis B_1 for $\mathcal{Q}_1$ is four, but of these, only two have signality one, that is, $\widehat{\mathbb{A}}_1^1\mathbf{0}_1$ and $\widehat{\mathbb{A}}_1^2\mathbf{0}_1$. The only specific information we have is given by the relations (10.14) for those two elements of the preferred basis.

Fortunately, this problem is easily circumvented by the observation that for a DS experiment, an observer will not in general be interested in the complete quantum registers $\mathcal{Q}_n$ and $\mathcal{Q}_{n+1}$ but only in the subspaces spanned by the signality-one elements of their respective preferred bases. This is really the meaning of the term *self-interference*.

This leads us to define the notion of *contextual basis* and *contextual subspace*.

Definition 10.2 In a given experiment, a *contextual basis* B_n^c is a subset of the preferred basis B_n for the quantum register $\mathcal{Q}_n$ at stage Σ_n, the elements of B_n^c being dictated by the context of the experiment.

Definition 10.3 A *contextual subspace* is a subspace $\mathcal{Q}_n^c$ of a quantum register $\mathcal{Q}_n$, the preferred basis for $\mathcal{Q}_n^c$ being a given contextual subset B_n^c of the preferred basis B_n for $\mathcal{Q}_n$

In the case of the DS experiment, the contextual bases B_1^c, B_2^c are given by all the respective signality-one states, so we have

$$B_1^c \equiv \left\{ \widehat{\mathbb{A}}_1^1 \mathbf{0}_1, \widehat{\mathbb{A}}_1^2 \mathbf{0}_1 \right\}, \quad B_2^c \equiv \left\{ \widehat{\mathbb{A}}_2^i \mathbf{0}_2 : i = 1, 2, \ldots, K \right\}. \tag{10.22}$$

These define the contextual subspaces $\mathcal{Q}_1^c$ and $\mathcal{Q}_2^c$. From (10.14), we can construct the contextual evolution operator $\mathbb{U}_{2,1}^c$ and its retraction $\overline{\mathbb{U}}_{2,1}^c$. These are given in the CBR by

$$\mathbb{U}_{2,1}^c \equiv \sum_{a=1}^{2} \sum_{j=1}^{K} U_{2,1}^{j,a} \mathbf{2}_2^{j-1} \overline{\mathbf{2}_1^{a-1}}, \quad \overline{\mathbb{U}}_{2,1}^c \equiv \sum_{a=1}^{2} \sum_{j=1}^{K} U_{2,1}^{j,a*} \mathbf{2}_2^{a-1} \overline{\mathbf{2}_1^{j-1}} \tag{10.23}$$

and satisfy the required relation $\overline{\mathbb{U}}_{2,1}^c \mathbb{U}_{2,1}^c = \mathbb{I}_1^c$, where $\mathbb{I}_1^c$ is the stage-Σ_1 *contextual identity*

$$\mathbb{I}_1^c = \mathbf{1}_1 \overline{\mathbf{1}_1} + \mathbf{2}_1 \overline{\mathbf{2}_1}. \tag{10.24}$$

Remark 10.4 The real world of experience and the world of the theorist's imagination are each far too complex to understand fully. Experiments attempt to limit the complexity of the former by focussing on a limited number of detectors. Contextual subspaces implement that strategy as closely as possible, limiting the amount of complexity that the theorist needs to face.

Usually it will be clear by context when we are dealing with contextual subspaces, so we shall usually drop the superscript c in our notation.

10.8 The Sillitto–Wykes Variant

The DS experiment was identified by Feynman, Leighton, and Sands (FLS) as *the* fundamental experiment to understand. They wrote that "it contains the only mystery" (Feynman et al., 1966). From the beginning of QM, experimentalists sought to probe this mystery deeper, such as greatly reducing the light intensity (Taylor, 1909) in the case of electromagnetic waves. Technology finally made possible experiments where one electron came through the device at a time (Merli et al., 1976).

Another variant was performed by Sillitto and Wykes, who arranged for the two openings, A and B, to be opened and closed so that only one was open at a time (Sillitto and Wykes, 1972). Nevertheless, an interference pattern was observed. This seems at first sight impossible to understand if we think in terms of particles.

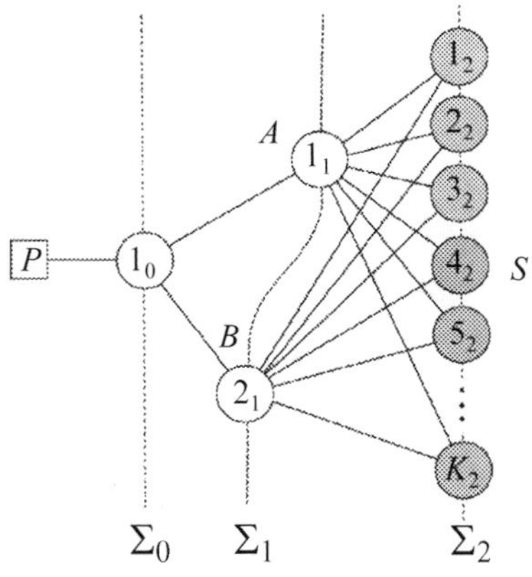

Figure 10.3. The QDN explanation of the Sillitto-Wykes experiment result is that stage Σ_1 need not coincide with a hyperplane of simultaneity in the laboratory, or indeed, in any other frame of reference.

We can readily explain the result of the Sillitto–Wykes experiment in QDN by pointing out that time is handled in QDN in terms of *stages*. In Figure 10.2, we note that the openings A, B are associated with *stage* Σ_1, not a specific laboratory time. We have stressed previously that stages are analogues of space-like hypersurfaces in conventional relativity, but with an important proviso: shielding can play a role. In the case of the Sillitto–Wykes experiment, a more appropriate diagram is Figure 10.3. In that representation, stage Σ_1 is not necessarily physically space-like everywhere. What matters is that the optical paths (in the case of photons) or trajectory channels (in the case of electrons) from P through openings A and B are such that wave trains from the two slits subsequently "intersect and interfere" at the screen S. Whether A and B are open "at the same labtime" turns out to be irrelevant. Indeed, that is as it should be, because it is not possible anyway to determine when any "particle" passes through either hole without destroying the interference pattern on S.

10.9 The Monitored Double-Slit Experiment

A fundamental question raised by the DS experiment concerns the particle interpretation of quanta. Speaking classically, if a particle such as an electron is released from source P and lands on screen S, then "it stands to reason" that that particle must have passed through either opening A or opening B. That is, after all, what is meant by a "particle."

But this is just an appeal to intuition. We have emphasized that our interpretation of QM is that it is really a theory of *entitlement*: QM tells us what we are entitled to say in a given context, and no more and no less. Therefore, if we have not monitored through which slit an electron has gone, we are not entitled to say it had to have gone through one opening for sure. We do not even have to think of that as having happened in any way; physicists are not in the business of believing in unverified propositions.

It is that lack of entitlement that really upsets people, because it destroys the comfortable belief that real things are going on, even when we are not observing them. Physicists, being people, felt compelled to find out whether the path of electron could be determined when an interference pattern on screen S was found. The *monitored DS experiment* was devised in order to investigate this issue.

The Feynman–Leighton–Sands Discussion

We shall first discuss the treatment given by FLS in Feynman et al. (1966). The apparatus is given in Figure 10.4.

The apparatus is the same as for the original DS experiment except for three items. L is a source of light and D^A and D^B are photon detectors. The idea is that if an electron passes through opening A, then there is a possibility that it will interact with light from the source L, causing a signal in D^A. A signal in D^A may therefore be an indicator that the electron has passed through opening A and not B. A similar remark applies if the electron passes through opening B.

The analysis goes as follows. First, consider the case when the light source is absent. If opening B is closed off and an electron passes through opening A and lands on the screen at C, then the amplitude is ϕ_1. Conversely, if opening A is closed off and an electron passes through opening B and lands on C, then the amplitude is ϕ_2. If A and B are both open, then the amplitude at C is $\phi_1 + \phi_2$. This is just the QM description of the original DS experiment.

When the light source L is present and both slits are open, then the situation needs some careful analysis. Suppose an electron has been detected at C *and*

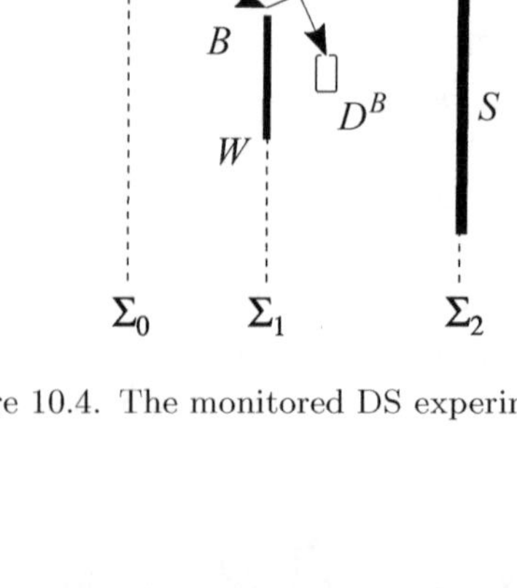

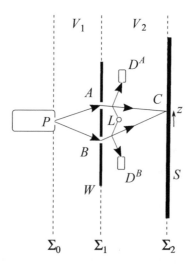

Figure 10.4. The monitored DS experiment.

a photon detected at D^A. Then two alternative paths contribute to the overall amplitude $\mathcal{A}(C|P)$.

Path A

The electron passes through slit A and interacts with the light, with an amplitude at C written as $a\phi_1$, where a is a factor representing the electron–electromagnetic field interaction.

Path B

The electron passes through slit B and interacts with the light, with an amplitude at C written as $b\phi_2$, where b is a factor representing the electron–electromagnetic field interaction.

The coefficients a and b are expected to be different, from the geometry of the situation. The overall amplitude at C, according to Feynman's sum of paths prescription, is therefore the sum of the two contributions, and so given by $\mathcal{A}(C, D^A|P) = a\phi_1 + b\phi_2$. Assuming suitable normalization, the probability $\Pr(C, D^A|P)$ for an electron to land on C and a photon to be registered at D^A, conditional on the electron being fired from P, is therefore given by

$$\Pr(C, D^A|P) = |\mathcal{A}(C, D^A|P)|^2 = |a\phi_1 + b\phi_2|^2. \tag{10.25}$$

If, on the other hand, an electron has been detected at C *and* a photon detected at D^B, then a similar argument (using symmetry) gives $\Pr(C, D^B|P) = |a\phi_2 + b\phi_1|^2$.

The probability $\Pr(C, D^A \text{ or } D^B|P)$ of an electron landing at C and a photon being detected at either D^A or D^B is the sum of the two probabilities, not the squared modulus of the sum of the two amplitudes (a point stressed by FLS), and is therefore given by

$$\Pr(C, D^A \text{ or } D^B|P) = |a\phi_1 + b\phi_2|^2 + |a\phi_2 + b\phi_1|^2, \tag{10.26}$$

which is Eq. (3.10) in Feynman et al. (1966).

The QDN Discussion

The relevant QDN version of Figure 10.4 is Figure 10.5. In the latter figure, we have two extra nodes, labeled A_2 and B_2, added to stage Σ_2 detectors in the screen S. These new nodes represent the detectors D^A and D^B shown in Figure 10.4.

The QDN analysis in this case goes by the three stages discussed in the original scenario. Nothing is different by stage Σ_1, so Eq. (10.13) is still valid. However, the jump to stage Σ_2 has to treated more carefully. We deal with each opening separately.

In the following, $a = 1$ corresponds to opening A and $a = 2$ corresponds to opening B. We expect the following dynamics, which is more general than that considered in Feynman et al. (1966):

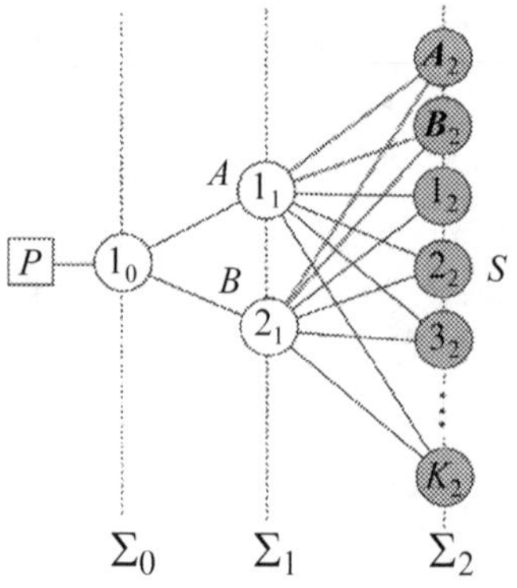

Figure 10.5. The QDN stage diagram for the monitored DS experiment.

$$\mathbb{U}_{2,1}\widehat{\mathbb{A}}_1^a\mathbf{0}_1 = \sum_{j=1}^{K}(\underbrace{U^{j,a}\widehat{\mathbb{A}}_2^j}_{(a)} + \underbrace{V^{j,a}\widehat{\mathbb{A}}_2^j\widehat{\mathbb{A}}_2^A}_{(b)} + \underbrace{W^{j,a}\widehat{\mathbb{A}}_2^j\widehat{\mathbb{A}}_2^B}_{(c)} + \underbrace{X^{j,a}\widehat{\mathbb{A}}_2^j\widehat{\mathbb{A}}_2^A\widehat{\mathbb{A}}_2^B}_{(d)})\mathbf{0}_2. \quad (10.27)$$

The four terms on the right-hand side represent the following alternatives:

(*a*) $U^{j,a}$ is the signality-one amplitude that an electron has passed through opening a and has landed on the screen at site j and no signal is detected in either D^A or D^B.

(*b*) $V^{j,a}$ is the signality-two amplitude that an electron has passed through opening a and has landed on the screen at site j and a signal is detected in D^A and not in D^B.

(*c*) $W^{j,a}$ is the signality-two amplitude that an electron has passed through opening a and has landed on the screen at site j and a signal is detected in D^B and not in D^A.

(*d*) $X^{j,a}$ is the signality-three amplitude that an electron has passed through opening a and has landed on the screen at site j and a signal is detected in D^A and a signal is detected in D^B. This possibility is not considered in the calculation of FLS.

The labstate $\mathbf{\Psi}_2$ is given by

$$\mathbf{\Psi}_2 = \sum_{a=1}^{2}\sum_{j=1}^{K}\alpha^a(U^{j,a}\widehat{\mathbb{A}}_2^j + V^{j,a}\widehat{\mathbb{A}}_2^j\widehat{\mathbb{A}}_2^A + W^{j,a}\widehat{\mathbb{A}}_2^j\widehat{\mathbb{A}}_2^B + X^{j,a}\widehat{\mathbb{A}}_2^j\widehat{\mathbb{A}}_2^A\widehat{\mathbb{A}}_2^B)\mathbf{0}_2. \quad (10.28)$$

From this we can readily determine all the various probabilities of interest by asking the appropriate questions. For example, if we ask for the probability that the electron has landed on the kth detector site, and there is a signal in D^A, and there is no signal in D^B, we need to calculate the expectation value $\overline{\mathbf{\Psi}}_2\widehat{\mathbb{P}}_2^k\widehat{\mathbb{P}}_2^A\mathbb{P}_2^B\mathbf{\Psi}_2$. To evaluate this, we first note that the signal algebra gives

$$\widehat{\mathbb{P}}_2^k\widehat{\mathbb{P}}_2^A\mathbb{P}_2^B\mathbf{\Psi}_2 = \sum_{a=1}^{2}\alpha^a V^{k,a}\widehat{\mathbb{A}}_2^k\widehat{\mathbb{A}}_2^A\mathbf{0}_2, \quad (10.29)$$

and then we readily find

$$\overline{\mathbf{\Psi}}_2 \widehat{\mathbb{P}}^k_2 \widehat{\mathbb{P}}^A_2 \mathbb{P}^B_2 \mathbf{\Psi}_2 = \left| \sum_{a=1}^{2} \alpha^a V^{k,a} \right|^2 = |\alpha^1 V^{k,1} + \alpha^2 V^{k,2}|^2. \tag{10.30}$$

Similarly, the probability $\overline{\mathbf{\Psi}}_2 \widehat{\mathbb{P}}^k_2 \mathbb{P}^A_2 \widehat{\mathbb{P}}^B_2 \mathbf{\Psi}_2$, that the electron landed on the kth screen site and there was no signal in D^A and there was a signal in D^B, is given by

$$\overline{\mathbf{\Psi}}_2 \widehat{\mathbb{P}}^k_2 \mathbb{P}^A_2 \widehat{\mathbb{P}}^B_2 \mathbf{\Psi}_2 = |\alpha^1 W^{k,1} + \alpha^2 W^{k,2}|^2. \tag{10.31}$$

We recover the FLS result in their notation if we use symmetry, setting $\alpha^1 = \alpha^2$ and

$$V^{k,1} = a\phi_1, \quad V^{k,2} = b\phi_2, \quad W^{k,1} = a\phi_2, \quad W^{k,2} = b\phi_1. \tag{10.32}$$

The QDN calculation allows greater flexibility in the architecture of the experiment than that assumed in the FLS calculation, but the price is that more assumptions have to be made about the various amplitudes.

11
Modules

11.1 Modules

In this chapter, we discuss important *modules* such as Wollaston prisms, Newtonian prisms, beam splitters, phase shifters, mirrors, and null modules. Modules are pieces of apparatus placed at various strategic positions in the information void and are designed to influence labstate amplitudes at signal detectors. Modules are therefore a critical component in the quantized detector network (QDN) approach to quantum mechanics (QM).

Modules are not classified as either real or virtual detectors, because no information is extracted from them. Their function is solely to influence quantum amplitude propagation *between* stages in specific ways. Even empty space (the vacuum) can be regarded as a module, for the propagation of signals through nominally "empty" space is a fundamental subject in its own right. A particularly important example illustrating how nontrivial that can be is the Hubble–Doppler red shift of light from distant galaxies.

In the QDN description of the double-slit (DS) experiment, shown in Figure 10.2, the stage Σ_1 detectors labeled 1_1 and 2_1 may be considered as on the V_2 side of the wall W, where V_2 is the information void between stages Σ_1 and Σ_2.

Viewed in this way, the wall may be interpreted as a part of the apparatus that is positioned in the information void region V_1 so as to influence the signals obtained at detectors 1_1 and 2_1. Virtually all experiments have similar components of apparatus in the information void. We shall call any such piece of apparatus a *module*. Modules are necessary to the architecture of the experiment, are situated in the information void, and therefore are not detectors.

Examples of modules are Stern–Gerlach (SG) inhomogeneous field magnets, Wollaston prisms, mirrors, phase changers, and so on. Even empty space, otherwise known as the *vacuum*, should be regarded as a module.

Each module has its own physical properties. These include the number of detectors feeding amplitudes into that module and the number of detectors that amplitudes are fed out from. We shall discuss a number of modules relevant to the experiments discussed in this book, starting with the vacuum.

11.2 The Vacuum

In this book, we make a distinction between the *information void* concept and the *vacuum* of elementary particle physics. The differences are subtle.

Each concept has the characteristic that no signal detectors are associated with it. While it is commonplace to talk about "observers *in* empty space" or "detectors *in* the vacuum," such expressions are manifestly inconsistent if taken literally, even in classical mechanics (CM). What distinguishes the information void concept from the vacuum concept is that the latter is a mathematical objectivization of the former, based on some input context. For instance, if we believe that the information void has some structure or physical properties associated with a vacuum, then we may choose to believe in or set up some sort of mathematical model of empty space, such as a three-dimensional manifold with a Euclidean metric, as in Newtonian mechanics, or a general relativistic (GR) spacetime with a metric, or think of it in operator terms, as in Snyder's noncommuting spacetime (Snyder, 1947a,b), or even assign a quantum state vector to it, as in Fock space and relativistic quantum field theory (RQFT).

We are here faced with an important question as to the status of the vacuum: is it part of the relative external context (REC) that defines the observer, or is it part of the relative internal context (RIC) that defines the apparatus?

On the one hand, we have argued in earlier chapters that real observers are always *endophysical*, that is, are sitting inside the physical Universe. According to GR, the physical space that we image observers to be sitting in will have some physical attributes, such as metrical structure, curvature, mass, and energy densities. In this context, empty space plays a classical, auxiliary role as part of REC.

On the other hand, the vacuum concept will play a fundamental role in the calculation of quantum signal propagation amplitudes, such as black hole physics, where the classical background spacetime structure plays an important, even crucial, role (Birrell and Davies, 1982; DeWitt, 1975). In that context, the vacuum contributes to the RIC.

Elementary particle physics has been very successful in modeling the vacuum as having special relativistic (SR) symmetries and certain physical properties such as zero electric charge density, and so on. The standard model of particle physics appears not to need or use any concepts associated with the so-called quantum gravity program, a conjectural program based on the belief that the equations of GR and those of QM should be unified. The jury is out on all programs of research that attempt to give a detailed model of the vacuum, such as quantum gravity, string theory, and noncommutative geometry such as that of Snyder (Snyder, 1947a,b).

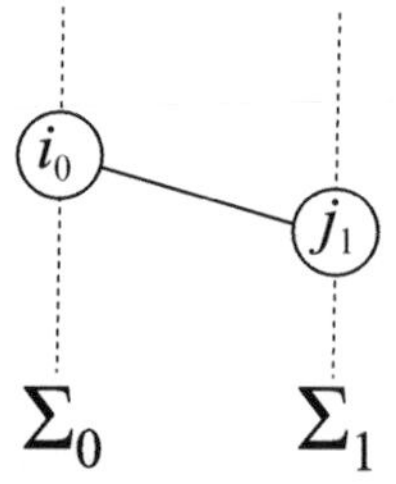

Figure 11.1. Amplitude propagation across the information void.

The question of what the best model of the "internal" vacuum of the RIC could be is a crucial one, because there are many contexts where the answer will influence the calculation of the quantum signal amplitudes of relevance to QDN. For example, we may well need to take spacetime curvature into account if we are sending signals across vast distances. Such a scenario is found in all observations in astronomy and cosmology.

We have indicated previously that QDN has nothing to say about the information void or the vacuum, precisely because QDN models what is happening at the signal detectors. Therefore, we will need to bring in standard physics theory, such as relativistic quantum field theory, to help us write down signal amplitudes between those detectors. Should that appear to make QDN rather limited, we point out that standard theories do have their limitations as well, such as the appearance of infinite renormalization constants in quantum field theory. We interpret that as the same problem seen from the opposite direction. It will probably be only by a judicious combination of QDN (or whatever should replace it) on the apparatus side of the physics coin and of standard relativistic quantum field theory on the reductionist side that we will find a better approach to empirical physics than with just either alone.

In QDN notation, signal amplitude propagation across the information void is represented by featureless lines, as in Figure 11.1.

11.3 The Wollaston Prism

A Wollaston prism is a quantum optics module that splits up a beam of light into two orthogonally polarized beams. Figure 11.2 is a schematic QDN diagram of such a device.

Consider an initial total state

$$|\mathbf{\Psi}_0) \equiv \left\{\alpha|s_0^1\rangle + \beta|s_0^2\rangle\right\} \otimes \mathbf{1}_0, \tag{11.1}$$

where $\mathbf{1}_0 \equiv \widehat{\mathbb{A}}_0^1\mathbf{0}_0$ and$|s_0^i\rangle$ is a stage-Σ_0 polarization state vector in two-dimensional photon polarization Hilbert space $\mathcal{H}_0$. Here $i = 1$ or 2 represents either of two orthogonal polarizations such as *horizontal* and *vertical.* The coefficients α and β are complex numbers. If the initial state is normalized to unity, then $|\alpha|^2 + |\beta|^2 = 1$.

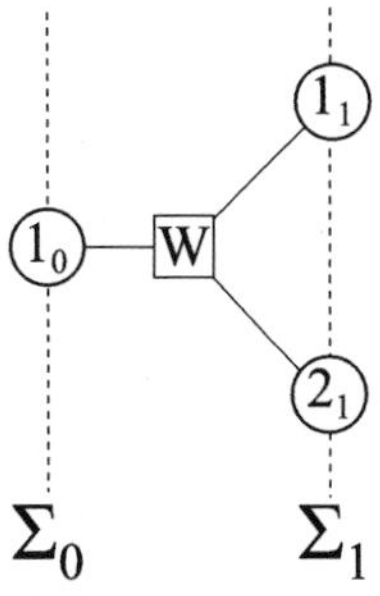

Figure 11.2. The Wollaston prism.

The contextual evolution operator[1] $\mathbb{U}_{1,0}$ from stage Σ_0 to stage Σ_1 is assumed semi-unitary, mapping from a two-dimensional contextual subspace of the initial four-dimensional Hilbert space $\mathcal{H}_0 \otimes Q_0^1$ to the final eight-dimensional one $\mathcal{H}_1 \otimes \mathcal{Q}_1$, where $\mathcal{H}_1$ is a copy of $\mathcal{H}_0$ and $\mathcal{Q}_1 \equiv Q_1^1 Q_1^2$. The assumed rules for the Wollaston prism are

$$\mathbb{U}_{1,0}\{|s_0^1\rangle \otimes \mathbf{1}_0\} = |s_1^1\rangle \otimes \mathbf{1}_1, \quad \mathbb{U}_{1,0}\{|s_0^2\rangle \otimes \mathbf{1}_0\} = |s_1^2\rangle \otimes \mathbf{2}_1, \tag{11.2}$$

where $\mathbf{1}_1 \equiv \widehat{\mathbb{A}}_1^1 \mathbf{0}_1$ and $\mathbf{2}_1 \equiv \widehat{\mathbb{A}}_1^2 \mathbf{0}_1$.

Using contextual completeness we have

$$\mathbb{U}_{1,0} = |s_1^1\rangle\langle s_0^1| \otimes \mathbf{1}_1\overline{\mathbf{1}_0} + |s_1^2\rangle\langle s_0^2| \otimes \mathbf{2}_1\overline{\mathbf{1}_0} \tag{11.3}$$

and its retraction

$$\overline{\mathbb{U}}_{1,0} = |s_0^1\rangle\langle s_1^1| \otimes \mathbf{1}_0\overline{\mathbf{1}_1} + |s_0^2\rangle\langle s_1^2| \otimes \mathbf{1}_0\overline{\mathbf{2}_1}. \tag{11.4}$$

These operators satisfy the semi-unitary relation

$$\overline{\mathbb{U}}_{1,0}\mathbb{U}_{1,0} = \{|s_0^1\rangle\langle s_0^1| + |s_0^2\rangle\langle s_0^2|\} \otimes \mathbf{1}_0\overline{\mathbf{1}_0} = I_0^{\mathcal{H}} \otimes \mathbb{I}_0^c, \tag{11.5}$$

the identity operator for the initial contextual total Hilbert space $\mathcal{H}_0 \otimes \mathcal{Q}_0^c$.

There are two generalized Kraus matrices associated with stage Σ_1, given by

$$\begin{aligned} M_{1,0}^1 &\equiv \overline{\mathbf{1}_1}\mathbb{U}_{1,0} = |s_1^1\rangle\langle s_0^1| \otimes \overline{\mathbf{1}_0}, \\ M_{1,0}^2 &\equiv \overline{\mathbf{2}_1}\mathbb{U}_{1,0} = |s_1^2\rangle\langle s_0^2| \otimes \overline{\mathbf{1}_0}, \end{aligned} \tag{11.6}$$

with retractions

$$\begin{aligned} \overline{M}_{1,0}^1 &= |s_0^1\rangle\langle s_1^1| \otimes \mathbf{1}_0, \\ \overline{M}_{1,0}^2 &= |s_0^2\rangle\langle s_1^2| \otimes \mathbf{1}_0. \end{aligned} \tag{11.7}$$

From these, the generalized POVM operators associated with Σ_1 are given by

$$\begin{aligned} E_{1,0}^1 &\equiv \overline{M}_{1,0}^1 M_{1,0}^1 = |s_0^1\rangle\langle s_0^1| \otimes \mathbf{1}_0\overline{\mathbf{1}_0}, \\ E_{1,0}^2 &\equiv \overline{M}_{1,0}^2 M_{1,0}^2 = |s_0^2\rangle\langle s_0^2| \otimes \mathbf{1}_0\overline{\mathbf{1}_0}. \end{aligned} \tag{11.8}$$

[1] As stated in a previous chapter, we drop the superscript "c" denoting "contextual."

In this particular case, these POVMs satisfy the relations $E^i_{1,0}E^j_{1,0} = \delta^{ij}E^i_{1,0}$ (no sum over i) and

$$E^1_{1,0} + E^2_{1,0} = \left\{|s^1_0\rangle\langle s^1_0| + |s^2_0\rangle\langle s^2_0|\right\} \otimes \mathbf{1}_0\overline{\mathbf{1}_0} = I_{\mathcal{H}} \otimes \mathbb{I}^c_0, \tag{11.9}$$

the identity operator for the initial contextual Hilbert space $\mathcal{H} \otimes \mathcal{Q}^c_0$. From these operators we find the conditional outcome rates

$$\Pr(\mathbf{1}_1|\mathbf{\Psi}_0) \equiv Tr\left\{E^1_{1,0}\varrho_0\right\} = |\alpha|^2, \quad \Pr(\mathbf{2}_1|\mathbf{\Psi}_0) \equiv Tr\left\{E^2_{1,0}\varrho_0\right\} = |\beta|^2, \tag{11.10}$$

assuming complete efficiency. Here $\varrho_0 \equiv |\mathbf{\Psi}_0)(\mathbf{\Psi}_0|$.

11.4 The Newtonian Prism

Newton's researches in optics revealed features of light that demonstrate fundamental properties of relevance to us here (Newton, 1704). In this section we discuss two of his observations with prisms.

The Splitting of Light

Newton found that a beam of white light incident on a prism P^1 would be split into a *spectrum*, a set of emerging rays each of a different color, according to anyone looking at it.[2] If any one of those single-color component subbeams were in turn passed through another prism P^2, no further splitting occurred; Figure 11.3(a). From this, Newton concluded that the primary colors in white light were associated with properties of that light, and not induced by the prism. Figure 11.3(b) shows the same process as described by QDN.

Significantly, Newton took the view that the perception of color itself was a sensation induced in the mind by the processes of visual observation. His reasoning is based on empirical evidence: he noticed that superposing certain primary colors in the spectrum emerging from a prism created colors such as purple, in the mind of the observer, that were not contained in the original spectrum. In that respect he could reasonably be regarded as having a deeper view of observation than that generally associated with classical mechanics (CM), which pays no lip service to the observer and their subjective perceptions.

Newton's idea was vindicated subsequently by the development of the theory of color vision (the colors that humans believe they "see"), principally by Young and Helmholtz. They proposed that the human eye uses three distinct types of receptor (that is, detector). The signal information from these detectors is then processed to trigger the color sensations that we think we see.

The QDN modeling is based on the tensor product total Hilbert space $\mathcal{H}_n$ at stage Σ_n defined as $\mathcal{H}_n \equiv \mathcal{H}^{EM}_n \otimes \mathcal{Q}_n$, where $\mathcal{H}^{EM}_n$ is the standard RQFT

[2] Note that this expression is based on the ancient and misleading paradigm that observers look "*at*" objects.

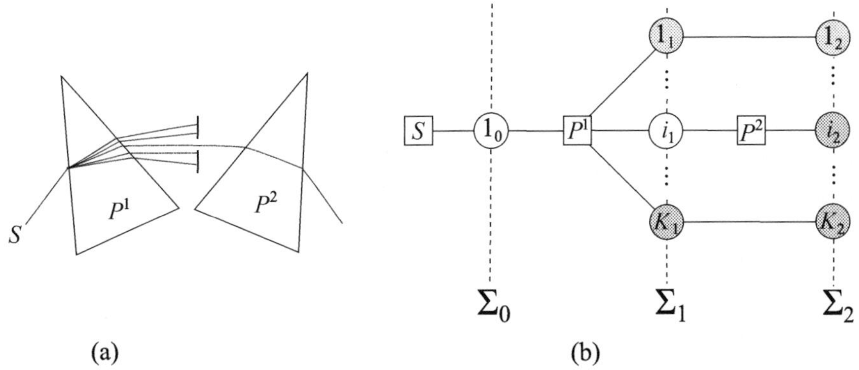

Figure 11.3. (a) Newtonian prism P^1 splits an incoming beam of white light from source S (the Sun) into subbeams associated with primary colors. Prism P^2 does not split any one of those subbeams further. (b) The QDN schematic of the same process.

Hilbert space containing free photon states[3] and $\mathcal{Q}_n$ is the apparatus quantum register at that stage. Relevant states in $\mathcal{H}_n^{EM}$ are constructed in terms of a suitable orthonormal set $\{|i_n\rangle : i = 1, 2, \dots K\}$, with orthonormality condition $\langle i_n | j_n\rangle = \delta^{ij}$.

With reference to Figure 11.3(b), the stage-Σ_0 normalized incident state $|\mathbf{\Psi}_0)$ is defined as

$$|\mathbf{\Psi}_0) \equiv \sum_{i=1}^{K} \psi_0^i |i_0\rangle \otimes \mathbf{1}_0, \qquad \sum_{i=1}^{K} |\psi_0^i|^2 = 1. \tag{11.11}$$

The action of the first prism P^1 in Figure 11.3 is taken to be

$$|\mathbf{\Psi}_0) \underset{P^1}{\rightarrow} |\mathbf{\Psi}_1) = \sum_{i=1}^{K} \psi_0^i |i_1\rangle \otimes \widehat{\mathbb{A}}_1^i \mathbf{0}_1, \tag{11.12}$$

which models the splitting up of the original beam into its primary colour constituents.

Three points to note here are the following.

Index Labels

The electromagnetic state indices i_0 and i_1 have a temporal subscript that labels stages only and does not affect the "value" of the index. So, for example, we take $i_1 = i_0 \equiv i$.

Discreteness

Contrary to what we might anticipate from experience with RQFT, the electromagnetic index i_n in the above is discrete, not continuous, because we can

[3] Any reference to *photons* is for convenience. The "basis" states here form a discrete set and therefore are not plane wave solutions but approximate them suitably.

only ever have a discrete set of detectors. This overturns the usual approach to RQFT, which presupposes a spectrum of photon states with a continuous frequency range. We do not need to deal with any supposed continuity in $\mathcal{H}_n^{EM}$, because the only parts of it that we can ever deal with are found at the finite, discrete detector end. Everything done here is contextual: it is the presupposition that an unobserved (and unobservable) continuum of states should play a role in, for example, Feynman graph integrals that contributes to the divergence of renormalization constants in RQFT. In the laboratory, all signals are finite.

The effect of the first prism P^1 on the incident beam is modeled by the action of a contextual semi-unitary evolution operator $U_{1,0}$, given by

$$U_{1,0} \equiv \sum_{i=1}^{K} |i_1\rangle\langle i_0| \otimes \widehat{\mathbb{A}}_1^i \mathbf{0}_1 \overline{\mathbf{1}_0}, \qquad (11.13)$$

where K is the number of subbeams in the emerging spectrum that is relevant to the discussion. Newton chose K to be 7, referring to the subbeams as *red, orange, yellow, green, blue, indigo*, and *violet* (Newton, 1704). This choice is dependent on the observer, for if we had equipment that could detect infrared or ultraviolet light, we would have a different value for K.

Coherence versus Incoherence

When physical process are affected by either constructive or destructive interference of amplitude waves, there are two factors to take into account: the amplitudes should have the same frequency (more or less), and if so, they should be coherent. In the case of white light, neither of these factors is in play because by definition, white light consists of waves of many different frequencies, and these are necessarily incoherent. That does not mean that we cannot describe such situations with our QDN formalism. Newton's recombination experiment is a demonstration of incoherent superposition but not of constructive or destructive interference.

With reference to Figure 11.3(a), we see that only one chosen subbeam, labeled i in Figure 11.3(b), enters the second prism P^2, all the other subbeams being blocked off. Then we may write

$$U_{2,1}\left\{|i_1\rangle \otimes \widehat{\mathbb{A}}_1^i \mathbf{0}_1\right\} \equiv |i_2\rangle \otimes \widehat{\mathbb{A}}_2^i \mathbf{0}_2. \qquad (11.14)$$

But what about the other, blocked subbeams?

There is a significant point here that we will be able to address more carefully in Chapter 25 and that has to do with information extraction versus *decommissioning*. By the term *decommissioning*, we mean the blocking off of signals and the elimination of the corresponding detectors from further consideration. We can see this in Figure 11.3(a), where all except one of the spectral subbeams emerging from P^1 are blocked off, with the remaining one allowed to pass through prism P^2.

The QDN approach to this issue is two-fold. In Figure 11.3(b), all of the stage-Σ_1 detectors except i_1 are shown shaded. This indicates that those components are either blocked off or observed irreversibly, with consequent information extraction. This blocking off/information extraction plays no further role in the experiment, and so is represented in the diagram by null tests taken from stage Σ_1 to stage Σ_2. On the other hand, Figure 11.3(b) indicates that i_1 sends a signal through the prism P^2 that is subsequently observed at i_2, which is now shown shaded in the figure.

Since the evolution from stage Σ_1 to stage Σ_2 is essentially trivial (which is what null tests are in practical terms), we may also write

$$U_{2,1}\left\{|j_1\rangle \otimes \widehat{\mathbb{A}}_1^j \mathbf{0}_1\right\} \equiv |j_2\rangle \otimes \widehat{\mathbb{A}}_2^j \mathbf{0}_2, \qquad j \neq i. \tag{11.15}$$

Hence we deduce

$$U_{2,1} = \sum_{\substack{j=1,\\ j\neq i}}^{K} \underbrace{|j_2\rangle\langle j_1| \otimes \widehat{\mathbb{A}}_2^j \mathbf{0}_2 \overline{\mathbf{0}}_1 \mathbb{A}_1^j}_{\text{null tests}} + \underbrace{|i_2\rangle\langle i_1| \otimes \widehat{\mathbb{A}}_2^i \mathbf{0}_2 \overline{\mathbf{0}}_1 \mathbb{A}_1^i}_{\text{through } P^2}. \tag{11.16}$$

Although the two terms on the right-hand side of this expression look similar, they are very different contextually. A practical difference would be that $|j_2\rangle$ is an exact copy (in $\mathcal{H}_2^{EM}$) of $|j_1\rangle$ (in $\mathcal{H}_1^{EM}$), for $j \neq i$, whereas $|i_2\rangle$ would not be an exact copy of $|i_1\rangle$ because of, say, attenuation effects as that subbeam passed through P^2.

The point here is that context underpins the significance of everything, and that context is determined by the observer. If they decide to block off all subbeams except the one labeled i, and then actually do that, then that action (not the decision alone) automatically brings in empirically significant context that involves separate dynamics and apparatus. It is the essence of Wheeler's participatory principle quoted on Chapter 1 that action be actually carried out when discussing quantum processes.

We see from this that the freedom that the observer has to choose how to rearrange apparatus creates a contextuality. This is emphasized in the next part of our description of the Newtonian prism, the recombination of a split spectrum back into the original beam.

The Recombination of Light

In his remarkable book *Opticks*, Newton discusses an extraordinary variant of the prism experiment shown in Figure 11.3(a) (Newton, 1704). He inverted prism P^2 and placed a lens between the two prisms. The lens refocused the spectrum emerging from P^1 onto P^2, as shown in Figure 11.4(a). The result was to recombine the spectrum, back into the original form of a beam of white light emerging from P^2.

Certainly, the recombined beam would not be precisely the same as the original, in that there would be some attenuation due to passing through glass, but as in the case of friction in mechanics, such effects can be regarded as secondary and not relevant to the main discussion.

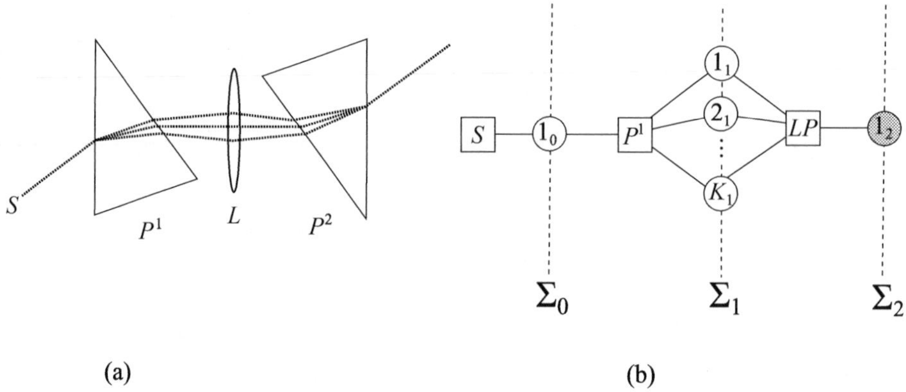

Figure 11.4. (a) Newton used lens L and prism P^2 to recombine the spectrum from prism P^1 back into a beam of white light. (b) In the QDN schematic of the same process, the lens and second prism constitute module LP.

The QDN stage diagram for Newton's recombination experiment is shown in Figure 11.4(b), with the following analysis.

Stage $\boldsymbol{\Sigma_0}$

The initial labstate is

$$|\boldsymbol{\Psi}_0) \equiv \sum_{i=1}^{K} \psi_0^i |i_0\rangle \otimes \mathbf{1}_0, \qquad \sum_{i=1}^{K} |\psi_0^i|^2 = 1. \tag{11.17}$$

Stage $\boldsymbol{\Sigma_0 \to \Sigma_1}$

The evolution operator $U_{1,0}$ that models the action of prism P^1 is given by

$$U_{1,0}^c = \sum_{i=1}^{K} |i_1\rangle\langle i_0| \otimes \widehat{\mathbb{A}}_1^i \mathbf{0}_1 \overline{\mathbf{1}_0}. \tag{11.18}$$

Hence the labstate $|\boldsymbol{\Psi}_1)$ emerging from prism P^1 is given by

$$|\boldsymbol{\Psi}_1) \equiv U_{1,0}|\boldsymbol{\Psi}_0) = \sum_{i=1}^{K} \psi_0^i |i_1\rangle \otimes \widehat{\mathbb{A}}_1^i \mathbf{0}_1. \tag{11.19}$$

Stage $\boldsymbol{\Sigma_1 \to \Sigma_2}$

The evolution operator $U_{2,1}$ that models the undoing action of the module LP consisting of the lens L and the prism P^2 is given, by inspection, by

$$U_{2,1} = \sum_{i=1}^{K} |i_2\rangle\langle i_1| \otimes \widehat{\mathbb{A}}_2^1 \mathbf{0}_2 \overline{\mathbf{0}_1} \mathbb{A}_1^i. \tag{11.20}$$

Here, we have chosen to represent the action of LP as refocusing the spectrum from P^1 onto labstate $\mathbf{1}_2 \equiv \widehat{\mathbb{A}}_2^1 \mathbf{0}_2$. The final outcome total state $|\boldsymbol{\Psi}_2)$ is therefore given by

$$|\mathbf{\Psi}_2) \equiv U_{2,1}|\mathbf{\Psi}_1) = \sum_{i=1}^{K} \psi_0^i |i_2\rangle \otimes \mathbf{1}_2. \tag{11.21}$$

Comparing this with the original labstate (11.17), we see that the final total state $|\mathbf{\Psi}_2)$ is a persistent image of $|\mathbf{\Psi}_0)$. Therefore, the combination of prism P^1 followed by prism P^2 is equivalent to a null test of the original beam.

It might be believed that we have encountered an example of semi-unitary evolution where the initial quantum register had a larger rank than the final quantum register. In fact, that is not the case. The action of the lens between the prisms is designed to focus on a one-dimensional subspace of the stage-Σ_1 register, a subspace containing the image of the original beam. It is that subspace alone that is then mapped by the lens and second prism into the one-dimensional subspace representing the outgoing recombined beam.

When Newton's recombination experiment is examined in fine detail, only then can it be appreciated how extraordinarily difficult it is to model what happens with some degree of correctness. The reader is invited to model this experiment using only quantum electrodynamics (QED) in its standard formulation, and they will then see what we mean.

11.5 Nonpolarizing Beam Splitters

Despite the implication in its name, a beam splitter will in general have two input channels and two outcome channels, as shown in Figure 11.5. In fact, it is perhaps best to think of a beam splitter as a specific example in quantum optics of a more general two-two particle scattering process that satisfies certain properties, such as unitarity and various conservation laws, such as conservation of energy, momentum, and charge.

In applications, we shall be interested in coherent, signality-one, normalized initial labstates $|\mathbf{\Psi}_0)$ involving both channels 1_0 and 2_0, of the generic form

$$|\mathbf{\Psi}_0) \equiv \psi^1 |s_0^1\rangle \otimes \widehat{\mathbb{A}}_0^1 \mathbf{0}_0 + \psi^2 |s_0^2\rangle \otimes \widehat{\mathbb{A}}_0^2 \mathbf{0}_0, \tag{11.22}$$

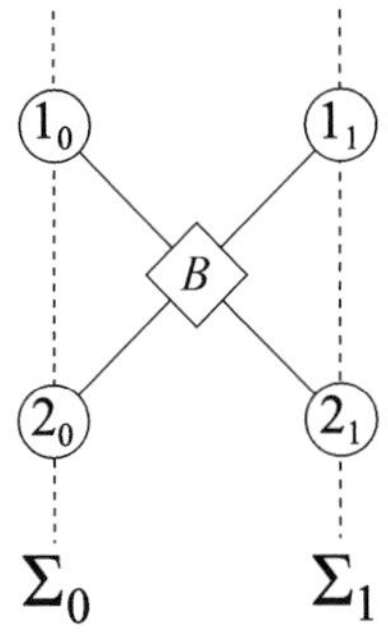

Figure 11.5. The beam splitter.

where $|\psi^1|^2 + |\psi^2|^2 = 1$, and $|s_0^i\rangle$, $i = 1, 2$, are initial, normalized one-photon states, not necessarily orthogonal. In this scenario, the polarization properties associated with each input channel are unspecified, so the discussion here is rather general in that respect.

The dynamics is specified by stating the beam splitter rules for each input channel. In general, semi-unitary evolution applies (if we ignore inefficiency and dissipation), so we write

$$U_{1,0}\left\{|s_0^1\rangle \otimes \widehat{\mathbb{A}}_0^1\mathbf{0}_0\right\} = \alpha|s_1^1\rangle \otimes \widehat{\mathbb{A}}_1^1\mathbf{0}_1 + \beta|s_1^1\rangle \otimes \widehat{\mathbb{A}}_1^2\mathbf{0}_1,$$
$$U_{1,0}\left\{|s_0^2\rangle \otimes \widehat{\mathbb{A}}_0^2\mathbf{0}_0\right\} = \gamma|s_1^2\rangle \otimes \widehat{\mathbb{A}}_1^1\mathbf{0}_1 + \delta|s_1^2\rangle \otimes \widehat{\mathbb{A}}_1^2\mathbf{0}_1, \tag{11.23}$$

where the beam splitter coefficients satisfy the semi-unitary relations

$$|\alpha|^2 + |\beta|^2 = |\gamma|^2 + |\delta|^2 = 1, \qquad \alpha^*\gamma + \beta^*\delta = 0. \tag{11.24}$$

It will be useful now and later to define the matrix of coefficients

$$B \equiv \begin{bmatrix} \alpha & \beta \\ \gamma & \delta \end{bmatrix}. \tag{11.25}$$

Then the semi-unitary relations (11.24) tell us that B is in fact a unitary matrix.

It is easy to prove from (11.24) that $|\alpha| = |\delta|$ and $|\beta| = |\gamma|$, so we may write $\alpha \equiv ue^{iA}$, $\beta \equiv ve^{iB}$, $\gamma \equiv ve^{iC}$, and $\delta \equiv ue^{iD}$, where $|u|^2 + |v|^2 = 1$ and the phases A, B, C, and D are real. Now define the phase E by $2E \equiv A + D$. Then it is straightforward to show that B can be written in the form

$$B \equiv e^{iE}\begin{bmatrix} a & -b^* \\ b & a^* \end{bmatrix}, \tag{11.26}$$

where $|a|^2 + |b|^2 = 1$ and the phase arguments of a and b are linear combinations of A, B, and C. We shall refer to the right-hand side of equation (11.26) as the *standard form* of a 2×2 unitary matrix.

Usually, the phase E in the standard form may be ignored. As for the complex coefficients a and b, these are related by Fresnel's equations for the reflection and transmission of light through optical media. Suppose a monochromatic beam of light passes through medium μ^1 and is incident on a plane surface boundary of medium μ^2. Generally, it will split into a reflected part that goes back into medium μ^1 and a transmitted part that moves into μ^2. If the speed of light c^1 in μ^1 is greater than the speed of light c^2 in μ^2, then the reflected part will be out of phase with the incident beam by π, that is, equivalent to a sign change. There is no such change in the phase of the transmitted part. If, conversely, c^1 is less than c^2, then there are no phase changes in either reflected or transmitted parts.

With this information there are two forms we may choose for B:

Form 1

$$B = \begin{bmatrix} t & -r \\ r & t \end{bmatrix}, \tag{11.27}$$

where t and r are real and satisfy the equation $t^2 + r^2 = 1$.

Form 2

$$B = \begin{bmatrix} t & ir \\ ir & t \end{bmatrix}, \tag{11.28}$$

where t and r are real and satisfy the equation $t^2 + r^2 = 1$.

In stage diagrams such as Figure 11.5, where the square representing the beam splitter is rotated as shown to face the two input ports 1_0 and 2_0, we shall adopt the convention that 2_1 and 1_1 are the transmitted and reflected beams, respectively, associated with input port 1_0, while they are the reflected and transmitted beams, respectively, associated with input port 2_0. This convention is used with Form 2 above throughout this book in the encoding of our computer algebra program MAIN, discussed in the next chapter.

In many experiments discussed in the literature, Form 2 is assumed, with t and r taken equal, that is, $t = r = 1/\sqrt{2}$. In our computer algebra program MAIN discussed in the next chapter, we do not do this. We use Form 2, but it is generally most instructive to leave the transmission and reflection coefficients arbitrary up to the required constraints. For instance, t^i and r^i will be the transmission and reflection coefficients for beam splitter B^i and will be arbitrary, apart from the condition that $(t^i)^2 + (r^i)^2 = 1$.

Signality-Two Input

On occasion, the possibility arises that a signality-two labstate or its equivalent is incident on a beam splitter. We shall refer to this as *beam splitter saturation*. The dynamics for such a scenario has to be decided contextually. One possibility is that the two photons (speaking loosely) do not emerge and are absorbed by the device. In such a case, the output channels of the beam splitter remain in their ground states. On the other hand, we could decide to model what happens as a case of *transparency*, such that the output labstate is a signality-two state with both output detectors registering a signal.

Exercise 11.1 Suppose a calibrated beam splitter satisfies the semi-unitary dynamics

$$\begin{aligned} \mathbb{U}_{1,0}\mathbf{0}_0 &= \mathbf{0}_1, \\ \mathbb{U}_{1,0}\widehat{\mathbb{A}}_0^1\mathbf{0}_0 &= \alpha\widehat{\mathbb{A}}_1^1\mathbf{0}_1 + \beta\widehat{\mathbb{A}}_1^2\mathbf{0}_1, \\ \mathbb{U}_{1,0}\widehat{\mathbb{A}}_0^2\mathbf{0}_0 &= \gamma\widehat{\mathbb{A}}_1^1\mathbf{0}_1 + \delta\widehat{\mathbb{A}}_1^2\mathbf{0}_1, \end{aligned} \tag{11.29}$$

where α, β, γ, and δ satisfy the semi-unitary relations (11.24).

Show that if

$$\mathbb{U}_{1,0}\widehat{\mathbb{A}}_0^1\widehat{\mathbb{A}}_0^2\mathbf{0}_0 = a\mathbf{0}_1 + b\widehat{\mathbb{A}}_1^1\mathbf{0}_1 + c\widehat{\mathbb{A}}_1^2\mathbf{0}_1 + d\widehat{\mathbb{A}}_1^1\widehat{\mathbb{A}}_1^2\mathbf{0}_1, \tag{11.30}$$

where $|a|^2 + |b|^2 + |c|^2 + |d|^2 = 1$ and $\mathbb{U}_{1,0}$ is semi-unitary, then $a = b = c = 0$ and $|d|^2 = 1$.

11.6 Mirrors

Mirrors are important modules in many experiments. The basic action of a mirror is to deflect electromagnetic radiation, that is, change path direction.

Depending on the physics, mirrors can also change the phase of electromagnetic waves. According to Fresnel's laws of optics, light incident from air on a mirror undergoes a phase shift of π, if reflecting from the front surface of a mirror, while there is no phase shift on rear surface reflection.

Naturally occurring mirrors can have a severe influence on signal detection amplitudes. For example, television signals received at an aerial are built up from the superposition of all the signals that have followed separate paths from the transmitting station to that aerial. If a transmitted signal has one path that goes in line of sight from transmitter to detector and another path that goes from transmitter, is reflected from the surface of the sea, and then arrives at the detector, then destructive interference can occur, thereby degrading the overall detected signal (Laven et al., 1970). Moreover, if the sea level changes due to tidal action, then the interference varies during the day and cannot be eliminated easily.

In stage diagrams, mirrors are labeled M.

11.7 Phase Changers

Mirrors are a specific example of more general modules referred to as *phase changers*. Such modules are fundamental to experiments where quantum interference is used to explore material properties, such as in interferometry.

Although phase changers can be used in any context, an important scenario involves electromagnetic signals, because in classical optics, Maxwell's equations lead to the conclusion that light is a wave process involving transversely oscillating electric and magnetic waves. Phase changer modules other than mirrors will be labeled by the phase angle involved.

11.8 Polarization Rotators

Electromagnetic waves have transverse electric and magnetic field polarization degrees of freedom, and this plays a significant role in many experiments. Generally, the convention is to define the polarization of a plane polarized electromagnetic wave as that of the electric field component. By default, the magnetic polarization is orthogonal to the electric field polarization.

We shall encounter some experiments where a polarized electromagnetic wave passes through a module that turns the wave's polarization plane by a known angle. Such a module will be labeled R^{θ} in stage diagrams, where θ is the angle of rotation.

11.9 Null Modules

A null module is any process that essentially does nothing to a signal. Because of this, a signal that is passed through a succession of null modules can be thought of as "on hold" until such time as the observer decides to look.

Null modules are used to model the concept of *persistence*, discussed in Chapter 18. On that account, null modules should not be considered trivial in the sense of identity operators, the action of which does nothing observable. In contrast, null modules require the right context and reflect a fundamental property of physics: that structures can persist over time in an observable sense.

Given a rank-r quantum register $\mathcal{Q}_n$ at stage Σ_n, a computational basis representation of a complete null operator (one that acts on the whole register) $\mathbb{N}_{n+1,n}$ that maps into stage Σ_{n+1} register $\mathcal{Q}_{n+1}$ of the same rank is

$$\mathbb{N}_{n+1,n} = \sum_{i=0}^{2^r-1} \boldsymbol{i}_{n+1}\overline{\boldsymbol{i}_n}. \tag{11.31}$$

12

Computerization and Computer Algebra

12.1 Introduction

The quantized detector network (QDN) formalism was designed from the outset to deal with detector networks of arbitrary complexity and rank. However, there is a price to be paid. A qubit register of rank r is a Hilbert space of dimension 2^r, so the dimensionality of quantized detector networks grows exponentially with rank. For example, a relatively small system of, say, 10 detectors involves a Hilbert space of dimension $2^{10} = 1,024$.

Even such a relatively small system cannot be dealt with easily by manual calculation, because quantum mechanics (QM) involves entangled states. Labstates in QDN are far more complex (in the sense of having far more mathematical structure) than the corresponding states in a classical register of the same rank. Some of the mathematical entanglement and separability properties of quantum labstates are discussed in Chapters 22 and 23.

The quantum entanglement structure of labstates poses a ubiquitous and serious problem, for both theorists and experimentalists. At this time, there is significant interest in quantum entanglement, particularly regarding its use in quantum computing, but theoretical understanding of entanglement is still surprisingly limited.

On the experimental side, quantum computers with 2000 qubits are currently being developed (D-Wave Systems, 2016). There is scope here for the application of QDN to networks of rank going into the many thousands. For those sorts of systems under observation (SUOs), quantum entanglement makes calculations by hand far too laborious to be of any practical use.

Fortunately, three factors come to our aid here, making the application of QDN to large-rank networks a potentially viable proposition.

Modularization
We saw in previous chapters that QDN deals with discrete aspects of observation, despite the fact that standard QM and relativistic quantum field theory (RQFT) deal with continuous degrees of freedom. This discreteness comes in three forms, referred to us as stages, nodes, and modules, and all of that is due to three inescapable physical facts. First, all real observations take time and involve discrete signals in finite numbers of detectors.[1] Second, real apparatus is constructed from atoms, not continua. Third, these atoms form finite numbers of well-characterized modules, as discussed in the previous chapter.

Contextuality
A critically helpful fact here is that in any particular experiment, the contextual subspaces that the observer needs to deal with will usually be of significantly lower dimensions than that of the full quantum registers involved. For example, while the register ground state $\mathbf{0}_n$ at stage Σ_n is an indispensable component of the QDN formalism, there are few situations that will involve the completely saturated labstate $\underline{\mathbf{2^r - 1}}_n \equiv \widehat{\mathbb{A}}_n^1 \widehat{\mathbb{A}}_n^2 \dots \widehat{\mathbb{A}}_n^r \mathbf{0}_n$. This will certainly be the case when the rank r is large, say, of the order of hundreds or even possibly thousands.

Computerization
While QDN calculations can often be done by hand for small rank calculations involving a small number of stages, the use of computer algebra (CA) software, such as Maple, Mathlab, and Mathematica, allows problems with much greater rank registers running over many more stages to be dealt with. All but the simplest network calculations in this book were done using a computer algebra program developed by us called Program MAIN, for example.

12.2 Program MAIN

In this chapter we discuss the application of CA to QDN. We illustrate this approach to QDN by describing a typical experiment, referred to here as the *Wollaston interferometer* (WI), shown in Figure 12.1.

The aim in this experiment is to investigate the polarization structure of an unpolarized monochromatic beam of light. From source S, the beam is first passed through a Wollaston prism W that splits it into two orthogonally polarized components 1_1 and 2_1 as shown. Component 1_1 has "internal" polarization represented by ket $|s_1^1\rangle$ while component 2_1 has polarization state $|s_1^2\rangle$, orthogonal to $|s_1^1\rangle$. Component 1_1 is then deflected by mirror M to 1_2 and then toward one input channel of beam splitter B. Component 2_1, meanwhile, is passed through polarization rotator R^θ that rotates $|s_1^2\rangle$ into $\cos(\theta)|s_2^2\rangle+\sin(\theta)|s_2^1\rangle$, before passing

[1] Technically, the signals are not discrete or continuous. It is the observer who interprets whatever they find at a detector as a discrete bit of information.

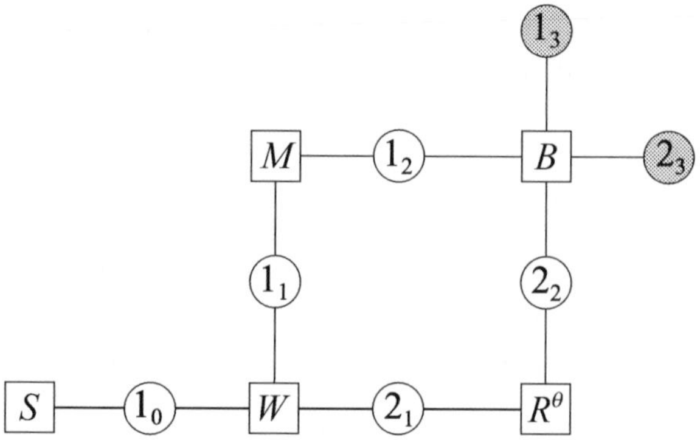

Figure 12.1. The Wollaston interferometer. A monochromatic, unpolarized beam of light from source S is passed through Wollaston prism W. Emerging polarized beam 1_1 is deflected by mirror M onto beam splitter B, while emerging polarized beam 2_1 has its polarization rotated by angle θ through module R^θ, before hitting B. Signal detection is at detectors 1_3 and 2_3.

into the other input channel of B. The two beam splitter outcome channels are monitored by final stage detectors 1_3 and 2_3. The angle θ is chosen by the observer before the preparation switch 1_0 is thrown.

If the transmission t and reflection r coefficients of the beam splitter B are known, then the observed outcome frequencies at the final stage detectors will depend on θ, and this gives information about the relative magnitudes of the two polarizations in the initial beam.

For applying QDN to the WI and all other experiments we have a single CA program, referred to here as *MAIN*. This program consists of two sections, A and B. In Section A, we input the details of the particular experiment we want to calculate, such as how many stages the experiment involves, the rank of each stage quantum register, and so on.

In practice, MAIN carries full details of all the experiments we have investigated, with a path parameter that allows us to select any experiment of current interest. For example, the path parameter for WI is 44.

Given the path parameter, MAIN jumps to Section A, initializing the variables from the relevant data there. Then MAIN jumps to Section B, automatically calculating everything of interest, such as the outcome probabilities at each final labstate detector.

The power of this approach is that all the work is done in encoding Section A. In general, this does not take long and is straightforward, given a stage diagram for the process and the module rules discussed in the previous chapters. Section B is universal, that is, is the same for all experiments. This gives a very economical and powerful approach indeed, allowing a great range of questions to be posed and answered rapidly. The only real theoretical problem is in deciding the stage-to-stage evolution operators $U_{n+1,n}$ in the encoding of Section A. This part of the process is currently where the computer does not help and the theorist has to decide what goes in. However, it should be possible to automate this part as well, if stage diagrams could be encoded first. We comment on this further, below.

12.3 The Wollaston Interferometer

We now describe the steps followed for a typical experiment, in this case the WI. This example illustrates the fact that MAIN can deal with *total* states, that is, entangled quantum states of the form

$$|\Psi_n) \equiv \sum_a |s_n^a\rangle \otimes \boldsymbol{i}_n^a, \tag{12.1}$$

where $|s_n^a\rangle$ is an "internal" SUO state, carrying information such as polarization, and $\boldsymbol{i}_n^a$ is a computational basis representation (CBR) labstate element carrying information about the signal state of the detectors.

Encoding

Our QDN notation is transcribed into program MAIN as follows. Although the observer's focus is the signality of the signal detector amplitudes, the internal calculations of MAIN are written in terms of the CBR. The reason is that labstates of signality greater than one are nonlinear in the signal operators $\widehat{\mathbb{A}}_n^i$, whereas the CBR is transparent to signality.

Our encoding takes the following form:

$$\begin{aligned} \boldsymbol{i}_n &\to a[i,n], \quad \overline{\boldsymbol{i}_n} \to A[i,n], \\ |s_n^p\rangle &\to s[p,n], \quad \langle s_n^p| \to S[p,n]. \end{aligned} \tag{12.2}$$

Here the symbols a, A, s, and S were arbitrarily chosen, with the rule that lowercase letters represent quantum "ket" states and capital letters represent their duals.

The power of the CA approach comes in at this point in that we can use differentiation to implement "inner product" rules. For example,

$$\overline{\boldsymbol{i}_n}\boldsymbol{j}_n \to \frac{\partial a[j,n]}{\partial a[i,n]} = \delta^{ij}, \quad \langle s_n^p|s_n^q\rangle \to \frac{\partial s[q,n]}{\partial s[p,n]} = \delta^{pq}, \tag{12.3}$$

which is readily encoded in CA.[2]

In the following, the asterisk symbol $*$ denotes ordinary commutative multiplication in the CA program. It is an important and helpful fact that our encoding does not need any noncommutative product operation, although CA is capable of handling those when required.

[2] Differentiation is a good example of the dichotomy between the continuous and the discrete. Mathematicians frequently deal with continuous variables, but mathematicians themselves operate on discrete principles: all of mathematics is done line by line, symbol by symbol, in a discrete way, very much like the way experimentalists have to operate in the real world.

If r_n is the rank of the stage-Σ_n qubit register and d_n is the dimension of the internal SUO Hilbert space, also at that stage, then total states will be of the form

$$\begin{aligned}|\Psi_n) &\equiv \sum_{i=0}^{2^{r_n}-1}\sum_{p=1}^{d_n} c_n^{p,i}|s_n^p\rangle \otimes \boldsymbol{i}_n \\ \to \Psi[n] &\equiv \sum_{i=0}^{2^{r_n}-1}\sum_{p=1}^{d_n} c[p,i,n] * s[p,n] * a[i,n],\end{aligned} \tag{12.4}$$

while their duals are of the form

$$\begin{aligned}(\Phi_n| &\equiv \sum_{i=0}^{2^{r_n}-1}\sum_{p=1}^{d_n} D_n^{p,i}\langle s_n^p| \otimes \overline{\boldsymbol{i}_n} \\ \to \Phi[n] &\equiv \sum_{i=0}^{2^{r_n}-1}\sum_{p=1}^{d_n} D[p,i,n] * S[p,n] * A[i,n],\end{aligned} \tag{12.5}$$

where the $c_n^{p,i}$ and $D_n^{p,i}$ are complex coefficients (readily handled in CA).

Inner products are evaluated by the CA prescription

$$\begin{aligned}(\Phi_n|\Psi_n) &\to \sum_{i=0}^{2^{r_n}-1}\sum_{p=1}^{d_n} \frac{\partial^2\Phi[n]}{\partial S[p,n]\partial A[i,n]} * \frac{\partial^2\Psi[n]}{\partial s[p,n]\partial a[i,n]} \\ &= \sum_{i=0}^{2^{r_n}-1}\sum_{p=1}^{d_n} D[p,i,n] * c[p,i,n].\end{aligned} \tag{12.6}$$

Draw a Stage Diagram

Having decided to model a particular experiment, we need to have a clear overview of the experiment's space-time architecture. That is best done graphically. Any quantum experiment can be described diagrammatically by a *stage diagram*, that is, a representation of the detectors and modules at each stage, with different stages linked by amplitude transmission lines.

Currently, MAIN has no facility for automatically transcribing such diagrams into code, but we envisage the development of graphical interface software that would make this part of the programming straightforward. The reason for this optimism is that there is only a finite variety of modules, and their properties can be well specified and catalogued mathematically, as discussed in Chapter 11. Figure 12.1 is the relevant diagram for the WI calculation.

Transcription Phase

Given the stage diagram, the next step is to read from it the following information and encode it into MAIN:

1. N, the number of stages, not counting the initial stage, which is labeled Σ_0. There is no limit to the size of N apart from the capacity limits of the

computer. In our case, a personal computer with 32 gigabytes random access memory and a 240 gigabyte hard disc was quite sufficient to deal with all the experiments discussed in this book. For our WI calculation, $N = 3$.

2. $r[0], r[1], \ldots, r[N]$, the rank at each stage. From Figure 12.1 we have $r[0] = 1$, $r[1] = r[2] = r[3] = 2$.
3. $d[0], d[1], \ldots, d[N]$, the dimensions of the "internal" SUO Hilbert space at each stage. In the case of the WI, we have two orthonormal polarizations of light, $|s_n^1\rangle, |s_n^2\rangle$, to factor in. These are often taken as horizontal and vertical polarizations, but in principle could be circular polarizations if required. Hence we take $d[0] = d[1] = d[2] = 2$.
4. The initial labstate: for the WI being discussed, we take $|\Psi_0) \equiv \{c^1|s_0^1\rangle + c^2|s_0^2\rangle\} \otimes \mathbf{1}_0$, and its dual $(\Psi_0| \equiv \{c^{1*}\langle s_0^1| + c^{2*}\langle s_0^2|\} \otimes \overline{\mathbf{1}_0}$, where c^{i*} is the complex conjugate of c^i. The coefficients c^i characterize the two incident beam polarizations and satisfy $|c^1|^2 + |c^2|^2 = 1$.

   These states are encoded into MAIN as

$$|\Psi_0) \rightarrow \psi[0] \equiv (c[1] * s[1, 0] + c[2] * s[2, 0]) * a[1, 0],$$
$$(\Psi_0| \rightarrow \Psi[0] \equiv (C[1]) * S[1, 0] + C[2] * S[2, 0]) * A[1, 0]. \quad (12.7)$$

   MAIN checks at each stage that normalization is preserved, so that at any stage Σ_n, we have

$$(\Psi_n|\Psi_n) = (\Psi_0|\Psi_0) \rightarrow C[1] * c[1] + C[2] * c[2]. \quad (12.8)$$

5. From Figure 12.1 and the rules of the modules concerned (covered in the previous chapter) we write (on paper)

Stage $\Sigma_0 \rightarrow \Sigma_1$

$$U_{1,0}\left\{|s_0^1\rangle \otimes \widehat{\mathbb{A}}_0^1 \mathbf{0}_0\right\} = |s_1^1\rangle \otimes \widehat{\mathbb{A}}_1^1 \mathbf{0}_1,$$
$$U_{1,0}\left\{|s_0^2\rangle \otimes \widehat{\mathbb{A}}_0^1 \mathbf{0}_0\right\} = |s_1^2\rangle \otimes \widehat{\mathbb{A}}_1^2 \mathbf{0}_1. \quad (12.9)$$

Stage $\Sigma_1 \rightarrow \Sigma_2$

$$U_{2,1}\left\{|s_1^1\rangle \otimes \widehat{\mathbb{A}}_1^1 \mathbf{0}_1\right\} = |s_2^1\rangle \otimes \widehat{\mathbb{A}}_2^1 \mathbf{0}_2, \quad (12.10)$$
$$U_{2,1}\left\{|s_1^2\rangle \otimes \widehat{\mathbb{A}}_1^2 \mathbf{0}_1\right\} = (\cos\theta|s_2^2\rangle + \sin\theta|s_2^1\rangle) \otimes \widehat{\mathbb{A}}_2^2 \mathbf{0}_2. \quad (12.11)$$

Stage $\Sigma_2 \rightarrow \Sigma_3$

$$U_{3,2}\left\{|s_2^1\rangle \otimes \widehat{\mathbb{A}}_2^1 \mathbf{0}_2\right\} = |s_2^1\rangle \otimes \left(t\widehat{\mathbb{A}}_3^2 \mathbf{0}_3 + ir\widehat{\mathbb{A}}_3^1 \mathbf{0}_3\right),$$
$$U_{3,2}\left\{|s_2^1\rangle \otimes \widehat{\mathbb{A}}_2^2 \mathbf{0}_2\right\} = |s_2^1\rangle \otimes \left(t\widehat{\mathbb{A}}_3^1 \mathbf{0}_3 + ir\widehat{\mathbb{A}}_3^2 \mathbf{0}_3\right),$$
$$U_{3,2}\left\{|s_2^2\rangle \otimes \widehat{\mathbb{A}}_2^2 \mathbf{0}_2\right\} = |s_2^2\rangle \otimes \left(t\widehat{\mathbb{A}}_3^1 \mathbf{0}_3 + ir\widehat{\mathbb{A}}_3^2 \mathbf{0}_3\right). \quad (12.12)$$

Here t and r are beam splitter parameters, and we use (11.28).

That is all the theoretical input needed. The next step is to transcribe it into Section A code.

6. To transcribe the above information, we first use contextual completeness to write down $U_{1,0}$, $U_{2,1}$, and so on, and then use the rules (12.2). For instance, from (12.9) this gives

$$\begin{aligned} U_{1,0} &= |s_1^1\rangle \otimes \widehat{\mathbb{A}}_1^1 \mathbf{0}_1 \langle s_0^1| \otimes \overline{\mathbf{0}_1}\mathbb{A}_0^1 + |s_1^2\rangle \otimes \widehat{\mathbb{A}}_1^2 \mathbf{0}_1 \langle s_0^2| \otimes \overline{\mathbf{0}_1}\mathbb{A}_0^1 \\ &= |s_1^1\rangle\langle s_0^1| \otimes \mathbf{1}_1 \overline{\mathbf{1}_0} + |s_1^2\rangle\langle s_0^2| \otimes \mathbf{2}_1 \otimes \overline{\mathbf{1}_0} \\ \to U[1,0] &\equiv s[1,1] * S[1,0] * a[1,1] * A[1,0] + \\ & \quad\; s[2,1] * S[2,0] * a[2,1] * A[1,0], \end{aligned} \tag{12.13}$$

and similarly for $U_{2,1}$ and $U_{3,2}$.

This is all the input we need to provide program MAIN for it to answer all possible maximal questions about this particular experiment.

Evaluation Phase

Once Section A has been initialized, Section B goes into action, employing several key procedures (subroutines):

1. Given $U[n+1,n]$ and $U[n+2,n+1]$, then the operator $U_{n+2,n} \equiv U_{n+1,n+1} U_{n+1,n}$ is given by the rule $U_{n+2,n} \to U[n+2,n]$, where

$$U[n+2,n] \equiv \sum_{i=0}^{2^{r[n+1]}-1} \sum_{p=1}^{d[n+1]} \frac{\partial^2 U[n+2,n+1]}{\partial S[p,n+1]\partial A[i,n+1]} * \frac{\partial^2 U[n+1,n]}{\partial s[p,n+1]\partial a[i,n+1]}. \tag{12.14}$$

The overall contextual evolution operator $U[N,0]$ is evaluated by iterating this process.

2. The retraction operator $\overline{U}_{N,0}$ is calculated from $U[N,0]$ by complex conjugation of coefficients and the interchange $s \leftrightarrow S$, $a \leftrightarrow A$. Complex conjugation is readily handled in CA.
3. The detector POVM operators $E_{N,0}^i \to E[i,N]$ discussed in Chapter 9 are calculated by the rule

$$E[i,N] \equiv \sum_{p=1}^{d[N]} \frac{\partial^2 \overline{U}[N,0]}{\partial S[p,N]\partial A[i,N]} * \frac{\partial^2 U[N,0]}{\partial s[p,N]\partial a[i,N]}. \tag{12.15}$$

These should satisfy the rule

$$\sum_{i=0}^{2^{r[N]}-1} E[i,N] = \sum_{p=1}^{d[0]} s[p,N] * S[p,N] \sum_{i=0}^{2^{r[0]}-1} a[i,0] * A[i,0], \tag{12.16}$$

which is a representation of I_0, the contextual identity operator at stage Σ_0. In MAIN, this is used as a useful check on the evolution operators $U[n+1,n]$ written down in Section A.

4. The outcome probabilities $P(\boldsymbol{i}_N|\Psi_0)$, $i = 0, 1, 2, \ldots, 2^{r[N]} - 1$ are given by the rule (9.59) and are readily evaluated in MAIN, again using differentiation in the taking of inner products and operator action on state vectors. This gives the answers to all the 2^{r_N} maximal questions. Because large-rank stages may have excessive dimensions, the program gives a listing only of the nonzero answers to the maximal questions. This listing will generally be significantly smaller than the full potential set of answers, reflecting the fact that experiments generally deal with contextual subspaces of quantum registers, not the complete registers.

Points to note are the following.

Answers

Answers can be either in numerical form or in the far more useful algebraic form. For example, the output listing for the Wollaston interferometer is

$$\begin{aligned} P(\widehat{\mathbb{A}}_3^1\mathbf{0}_3|\Psi_0) &= |c^1|^2r^2 + |c^2|^2t^2 + i(c^{2*}c^1 - c^{1*}c^2)rt\sin\theta, \\ P(\widehat{\mathbb{A}}_3^2\mathbf{0}_3|\Psi_0) &= |c^1|^2t^2 + |c^2|^2r^2 - i(c^{2*}c^1 - c^{1*}c^2)rt\sin\theta. \end{aligned} \tag{12.17}$$

These sum to unity, as expected. The implication here is that MAIN has found the other two potential probabilities $P(\mathbf{0}_3|\Psi_0)$ and $P(\widehat{\mathbb{A}}_3^1\widehat{\mathbb{A}}_3^2\mathbf{0}_3|\Psi_0)$ to be zero.

Signal Decomposition

The CBR is excellent for CA but not so helpful with understanding signality. To assist in the interpretation, MAIN converts CBR probabilities into a listing of total signal probability for individual final-stage detectors, excluding any that have probabilities of zero. This gives valuable insight into the patterns of information flow. With some experience, it becomes clear why the photon concept has become so popular: signality is conserved in many quantum optics experiments unless some specific module such as a nonlinear crystal that creates photons is used, as in parametric down-conversion.

Partial Questions

MAIN answers the full set of maximal questions. We have stated before that this set of answers allows all partial questions to be answered. Therefore, CA gives us the possibility of answering all possible questions in the context of QDN. This becomes important in several complex experiments, such as the two-photon interferometer of Horne, Shimony, and Zeilinger (Horne et al., 1989) and the double-slit quantum eraser of Walborn et al. (Walborn et al., 2002), experiments covered in detail in Chapter 14.

12.4 Going to the Large Scale

There is little doubt in our mind that the greatest challenge facing quantum physics lies not with the reductionist laws of physics, which have been determined empirically to be excellent, but with the relatively unexplored laws of complexity and emergence. Even the simplest of experiments, such as the double-slit (DS) experiment discussed in Chapter 10, poses a serious challenge to QDN, simply because the dimensions of quantum registers grow exponentially with rank. With reference to Figure 10.2, suppose we wanted to model a DS experiment with K detector sites on screen. Three possible approaches that we could take are the following.

Approach 1: The *Theoretical* Approach
Here K is regarded as fixed but undetermined, and treated algebraically by the theorist on paper.

Approach 2: The *Brute Force* CA Approach
Here K is given a specific value, such as $K = 20$ in a CA program such as MAIN, and a relatively laborious calculation is done by a computer.

Approach 3: The *Symbolic* Approach
Here K is encoded symbolically and treated as an algebraic parameter characterizing the experiment. CA does in principle have the potential to handle this approach, but it requires more sophisticated programming, because that would essentially be an attempt to simulate the way the human theorist handles abstract modeling. A good example is that theorists can with training readily discuss infinite-dimensional Hilbert spaces on paper, but CA cannot easily handle infinities directly.

Approach 1 works for the relatively simple DS experiment but is impractical for other large-rank networks, prompting the development of the CA approach.

Approach 2 has definite limitations. A CA simulation of the DS experiment using MAIN was done with a range of values of K and timed. The results are given in the following table:

Rank r	T_r (seconds)
10	0.137
12	0.515
14	2.609
16	11.73
18	40.92
20	192.3
22	1284

Analysis of this table suggests that the time T_r for MAIN to complete a DS simulation with a rank-r screen register is approximately of the form $T_r = A + B2^r$, where A and B are constants. It is clear that even a rank-100 simulation could not be completed on this basis.

Approach 3 should be made to work if contextual subspaces are incorporated adequately (which they are not, currently, in our program MAIN).

Three conclusions can be drawn from this analysis:

1. Computerization seems inevitable.
2. Computerization will be very useful in a broad range of problems, but some problems may remain intractable.
3. Intelligent use of contextual subspaces will be necessary for problems where the rank is too large for the brute force approach to work.

12.5 Prospects

Our current CA program, MAIN, requires the user to transcribe "by hand" the information contained in a stage diagram into CA form. This process could be, and perhaps should be, fully automated, opening the door to potentially vast applications. Ideally, this would utilize the currently accessible technology of touch screens, whereby the user used electronic graphics pens to sketch stage diagrams in freehand. These would then be automatically transcribed into the information currently fed in by hand in Section A of MAIN.

We envisage the development of user-friendly graphical software that would allow the user to select all the necessary modules that were of interest, using palettes of symbols representing real and virtual detectors, modules of all sorts, and lines representing transmission through various media. With the right information, it should be possible to encode different material properties, so that quantized detector networks representing crystals, glasses, and other complex systems such as neural networks and retinas could be dealt with in a systematic and coherent way.

There is little doubt in our mind that such an approach to complex problems involving quantum mechanical effects in engineering, materials science, and medical science will one day become commonplace.

13

Interferometers

13.1 Introduction

The discussion of modules and computerization in previous chapters gives us the means to consider empirically useful quantized detector networks (QDN). In this chapter we shall focus on a particular class of network known as *interferometers*.

Interferometers play a crucial role in quantum mechanics (QM) because they demonstrate the "paradox" of wave–particle duality in a direct way. On the one hand, discrete signals are detected by the observer and those signals are usually interpreted as particles or quanta. On the other hand, the observed frequencies built up over many runs show effects interpreted as due to the interference of waves.

In the previous chapter, we used computer algebra (CA) to discuss the Wollaston interferometer (WI), a relatively simple interferometer. We shall use the same approach in the following discussion of the Mach–Zehnder interferometer (MZI), apparatus that allows the observer to investigate optical transmission through various materials.

13.2 The Mach–Zehnder Interferometer

The basic structure of an MZI is given in Figure 13.1. A source S of light sends a monochromatic, unpolarized beam 1_0 into one input channel of beam splitter B^1. Output channel 1_1 is deflected by mirror M^1 onto a module labeled ϕ that contains some medium under investigation, such as a crystal or liquid. The net effect is to create a phase change in that deflected beam by an amount ϕ, and then that modified beam, 1_2, is passed into beam splitter B^2. The second output channel, 2_1, from B^1, meanwhile, is deflected by mirror M^2 into beam splitter B^2. The two deflected beams, 1_2 and 2_2, that pass into B^2 interfere and are finally monitored by detectors 1_3 and 2_3.

Our analysis of the MZI follows the CA approach used for the WI in the previous chapter.

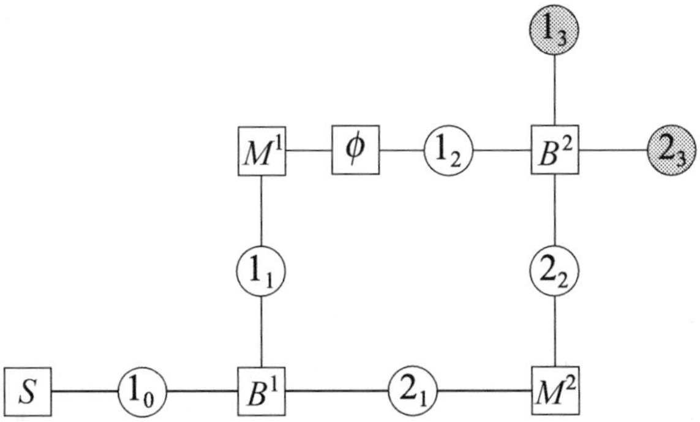

Figure 13.1. The Mach–Zehnder interferometer: S is a source of monochromatic, unpolarized light, B^1 and B^2 are beam splitters, M^1 and M^2 are mirrors, and ϕ contains a medium that changes the phase of the light by ϕ.

Parameters
For this experiment, the number of stages N is $N = 3$. From Figure 13.1, the rank at each stage is seen to be $r_0 = 1$, $r_1 = r_2 = r_3 = 2$. Since polarization is not a factor in this version of the experiment, we take the dimensions of the internal (photonic) space at the four stages to be given by $d_0 = d_1 = d_2 = d_3 = 1$.

The Initial State
Since we are not interested here with polarization, the initial total state $|\Psi_0)$ is given by $|\Psi_0) \equiv |s_0\rangle \otimes \widehat{\mathbb{A}}_0^1\mathbf{0}_0 = |s_0\rangle \otimes \mathbf{1}_0$, where $|s_0\rangle$ is a normalized photon state.

From Figure 13.1 and the known properties of the modules, we write

Stage Σ_0 to Stage Σ_1

$$U_{1,0}\left\{|s_0\rangle \otimes \widehat{\mathbb{A}}_0^1\mathbf{0}_0\right\} = t^1|s_1\rangle \otimes \widehat{\mathbb{A}}_1^2\mathbf{0}_1 + ir^1|s_1\rangle \otimes \widehat{\mathbb{A}}_1^1\mathbf{0}_1$$
$$\equiv |s_1\rangle \otimes (t^1\mathbf{2}_1 + ir^1\mathbf{1}_1). \tag{13.1}$$

Stage Σ_1 to Stage Σ_2

$$U_{2,1}\left\{|s_1\rangle \otimes \widehat{\mathbb{A}}_1^1\mathbf{0}_1\right\} = e^{i\phi}|s_2\rangle \otimes \widehat{\mathbb{A}}_2^1\mathbf{0}_2 \equiv e^{i\phi}|s_2\rangle \otimes \mathbf{1}_2,$$
$$U_{2,1}\left\{|s_1\rangle \otimes \widehat{\mathbb{A}}_1^2\mathbf{0}_1\right\} = |s_2\rangle \otimes \widehat{\mathbb{A}}_2^2\mathbf{0}_2 \equiv |s_2\rangle \otimes \mathbf{2}_2. \tag{13.2}$$

Stage Σ_2 to Stage Σ_3

$$U_{3,2}\left\{|s_2\rangle \otimes \widehat{\mathbb{A}}_2^1\mathbf{0}_2\right\} = t^2|s_3\rangle \otimes \widehat{\mathbb{A}}_3^2\mathbf{0}_3 + ir^2|s_3\rangle \otimes \widehat{\mathbb{A}}_3^1\mathbf{0}_0 \equiv |s_3\rangle \otimes (t^2\mathbf{2}_3 + ir^2\mathbf{1}_3),$$
$$U_{3,2}\left\{|s_2\rangle \otimes \widehat{\mathbb{A}}_2^2\mathbf{0}_2\right\} = ir^2|s_3\rangle \otimes \widehat{\mathbb{A}}_3^2\mathbf{0}_3 + t|s_3\rangle \otimes \widehat{\mathbb{A}}_3^1\mathbf{0}_0 \equiv |s_3\rangle \otimes (ir^2\mathbf{2}_3 + t\mathbf{1}_3). \tag{13.3}$$

Here we have ignored any phase changes at the mirrors and characterized each beam splitter separately. In the above, all superscripts are labels, not powers.

This is all the input information required for our CA program MAIN, as discussed in the previous chapter. The nonzero conditional outcome probabilities are found to be

$$\begin{aligned}\Pr(\widehat{\mathbb{A}}_3^1\mathbf{0}_3|\Psi_0) &= -2r^1r^2t^1t^2\cos(\phi) + (r^1r^2)^2 + (t^1t^2)^2,\\ \Pr(\widehat{\mathbb{A}}_3^2\mathbf{0}_3|\Psi_0) &= +2r^1r^2t^1t^2\cos(\phi) + (r^1t^2)^2 + (t^1r^2)^2.\end{aligned} \tag{13.4}$$

These sum up to unity as required, given that $(t^i)^2 + (r^i)^2 = 1$, for $i = 1, 2$.

The significance here is that the outcome probabilities are affected by the phase change module. By altering the path length in that module, and other parameters such as temperature and density of the medium in that module, significant information about that medium can be extracted.

13.3 Brandt's Network

The next example is a quantum optics network discussed by Brandt (Brandt, 1999) in terms of conventional positive operator-valued measure operators (POVMs) and shown in Figure 13.2. A source S prepares a monochromatic unpolarized beam of light 1_0 that is split by Wollaston prism W into two orthogonally polarized components 1_1 and 2_1. One component 1_1 is then passed into beam splitter B^1 and thereby split into two components 1_2 and 2_2 with no change in polarization. Component 1_2 is subsequently observed at detector 1_3, while component 2_2 is passed into beam splitter B^2. Meanwhile, component 2_1 emerging from the Wollaston prism W has its polarization turned by $\pi/2$ at module R. The resulting beam 3_2 is then passed into beam splitter B^2, where it interferes with 2_2, with subsequent detection at detectors 2_3 and 3_3.

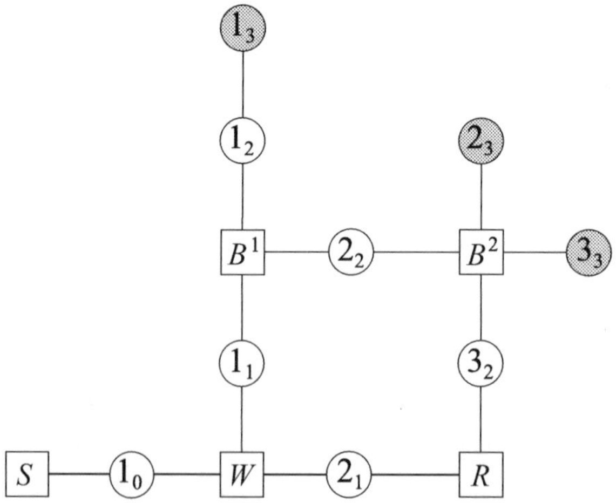

Figure 13.2. Brandt's network: source S prepares an unpolarized, monochromatic beam of light that passes through Wollaston prism W. This splits the beam into two orthogonally polarized components. These are passed through beam splitters B^1 and B^2 as shown. Module R rotates the polarization of one of the polarized beams into that of the other, prior to it being passed through B^2.

Brandt's analysis was in terms of nonorthogonal system under observation (SUO) state vectors. Our analysis avoids nonorthogonality issues directly. The initial state is given by

$$|\Psi_0\rangle \equiv (\alpha|s_0^1\rangle + \beta|s_0^2\rangle) \otimes \widehat{\mathbb{A}}_0^1 \mathbf{0}_0 = (\alpha|s_0^1\rangle + \beta|s_0^2\rangle) \otimes \mathbf{1}_0, \tag{13.5}$$

where $|s_0^1\rangle, |s_0^2\rangle$ denote orthogonal photon polarization states, and α and β are complex coefficients satisfying $|\alpha|^2 + |\beta|^2 = 1$. By inspection of Figure 13.2 we take $N = 3$, $r_0 = 1$, $r_1 = 2$, $r_2 = r_3 = 3$, $d_0 = d_1 = d_2 = d_3 = 2$. The dynamics is given by the following rules.

Stage Σ_0 to Stage Σ_1

$$\begin{aligned} U_{1,0}\left\{|s_0^1\rangle \otimes \widehat{\mathbb{A}}_0^1 \mathbf{0}_0\right\} &\equiv |s_1^1\rangle \otimes \widehat{\mathbb{A}}_1^1 \mathbf{0}_1 = |s_1^1\rangle \otimes \mathbf{1}_1, \\ U_{1,0}\left\{|s_0^2\rangle \otimes \widehat{\mathbb{A}}_0^1 \mathbf{0}_0\right\} &\equiv |s_1^2\rangle \otimes \widehat{\mathbb{A}}_1^2 \mathbf{0}_1 = |s_1^2\rangle \otimes \mathbf{2}_1. \end{aligned} \tag{13.6}$$

Stage Σ_1 to Stage Σ_2

$$\begin{aligned} U_{2,1}\left\{|s_1^1\rangle \otimes \widehat{\mathbb{A}}_1^1 \mathbf{0}_1\right\} &\equiv t^1|s_2^1\rangle \otimes \widehat{\mathbb{A}}_2^1 \mathbf{0}_1 + ir^1|s_2^1\rangle \otimes \widehat{\mathbb{A}}_2^2 \mathbf{0}_2 = |s_2^1\rangle \otimes \left\{t^1 \mathbf{1}_1 + ir^1 \mathbf{2}_2\right\}, \\ U_{2,1}\left\{|s_1^2\rangle \otimes \widehat{\mathbb{A}}_1^2 \mathbf{0}_1\right\} &\equiv -|s_2^1\rangle \otimes \widehat{\mathbb{A}}_2^3 \mathbf{0}_2 = -|s_2^1\rangle \otimes \mathbf{4}_2, \end{aligned} \tag{13.7}$$

where $(t^1)^2 + (r^1)^2 = 1$. The second equation in (13.7) represents a rotation of the photon polarization vector $|s_1^2\rangle$ by $-\frac{1}{2}\pi$ into $-|s_2^1\rangle$ as it passes through the module labeled R in Figure 13.2 (the sign change follows the convention used in Brandt (1999)).

Stage Σ_2 to Stage Σ_3

$$\begin{aligned} U_{3,2}\left\{|s_2^1\rangle \otimes \widehat{\mathbb{A}}_2^2 \mathbf{0}_2\right\} &\equiv |s_3^1\rangle \otimes \widehat{\mathbb{A}}_3^2 \mathbf{0}_3 = |s_3^1\rangle \otimes \mathbf{1}_3, \\ U_{3,2}\left\{|s_2^1\rangle \otimes \widehat{\mathbb{A}}_2^2 \mathbf{0}_2\right\} &\equiv t^2|s_3^1\rangle \otimes \widehat{\mathbb{A}}_3^3 \mathbf{0}_3 + ir^2|s_3^1\rangle \otimes \widehat{\mathbb{A}}_3^2 \mathbf{0}_3 \\ &= t^2|s_3^1\rangle \otimes \mathbf{4}_3 + ir^2|s_3^1\rangle \otimes \mathbf{2}_3, \\ U_{3,2}\left\{|s_2^1\rangle \otimes \widehat{\mathbb{A}}_2^3 \mathbf{0}_2\right\} &\equiv ir^2|s_3^1\rangle \otimes \widehat{\mathbb{A}}_3^3 \mathbf{0}_3 + t^2|s_3^1\rangle \otimes \widehat{\mathbb{A}}_3^2 \mathbf{0}_3 \\ &= ir^2|s_3^1\rangle \otimes \mathbf{4}_3 + t^2|s_3^1\rangle \otimes \mathbf{2}_3, \end{aligned} \tag{13.8}$$

where $(t^2)^2 + (r^2)^2 = 1$.

With this information transcribed into Section A of the CA program MAIN, we find the following nonzero outcome probabilities for the Brandt network:

$$\begin{aligned} \Pr(\widehat{\mathbb{A}}_3^1 \mathbf{0}_3|\Psi_0) &= |\alpha|^2(t^1)^2, \\ \Pr(\widehat{\mathbb{A}}_3^2 \mathbf{0}_3|\Psi_0) &= |\alpha|^2(r^1 r^2)^2 - (\alpha^*\beta + \alpha\beta^*)r^1 r^2 t^2 + |\beta|^2(t^2)^2, \\ \Pr(\widehat{\mathbb{A}}_3^3 \mathbf{0}_3|\Psi_0) &= |\alpha|^2(r^1 t^2)^2 + (\alpha^*\beta + \alpha\beta^*)r^1 r^2 t^2 + |\beta|^2(r^2)^2, \end{aligned} \tag{13.9}$$

assuming perfect efficiency and wave-train overlap. When the reflection and transmission coefficients are chosen as by Brandt (Brandt, 1999), these rates agree with his precisely.

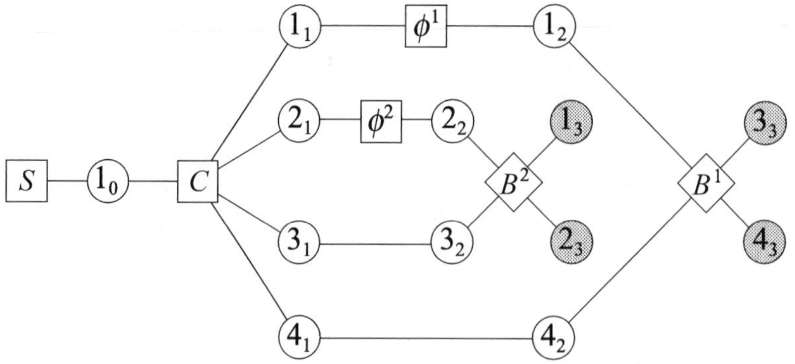

Figure 13.3. The HSZ two-photon interferometer.

13.4 The Two-Photon Interferometer

Up to now, we have restricted attention to signality-one processes. It is important to test our QDN approach on more subtle situations. With this in mind, in this section we discuss the application of QDN to a signality-two process discussed by Horne, Shimony, and Zeilinger (HSZ) (Horne et al., 1989). The relevant stage diagram is given in Figure 13.3.

A source S creates a preparation switch 1_0 that by stage Σ_1 has generated a two-photon entangled state via module C. As discussed in Horne et al. (1989), module C creates four separate components that are processed as follows. Component 1_1 passes through phase changer ϕ^1 and then enters beam splitter B^1, while component 2_1 passes through phase changer ϕ^2 and then enters beam splitter B^2. Components 3_1 and 4_1 pass directly on to beam splitters B^2 and B^1, respectively, as shown. The observer monitors signals at stage Σ_3 as shown. The aim of the experiment is to investigate photon signal pair correlations in the detectors.

Parameters

We take $N = 3$, $d[0] = d[1] = d[2] = d[3] = 1$, $r[0] = 1$, $r[1] = r[2] = r[3] = 4$.

Initial State

The preparation switch is given by

$$|\Psi, 0) = |s_0^1\rangle \otimes \widehat{\mathbb{A}}_0^1 \mathbf{0}_0. \tag{13.10}$$

Evolution

Stage Σ_0 to Σ_1

$$U_{1,0}\left\{|s_0^1\rangle \otimes \widehat{\mathbb{A}}_0^1 \mathbf{0}_0\right\} \equiv \frac{1}{\sqrt{2}}|s_1^1\rangle \otimes \left\{\widehat{\mathbb{A}}_1^1 \widehat{\mathbb{A}}_1^3 \mathbf{0}_1 + \widehat{\mathbb{A}}_1^2 \widehat{\mathbb{A}}_1^4 \mathbf{0}_1\right\} = \frac{1}{\sqrt{2}}|s_1^1\rangle \otimes \{\mathbf{5}_1 + \underline{\mathbf{10}}_1\}. \tag{13.11}$$

Stage Σ_1 to Σ_2

$$U_{2,1}\left\{|s_1^1\rangle \otimes \widehat{\mathbb{A}}_1^1\widehat{\mathbb{A}}_1^3\mathbf{0}_1\right\} \equiv e^{i\phi^1}|s_2^1\rangle \otimes \widehat{\mathbb{A}}_2^1\widehat{\mathbb{A}}_2^3\mathbf{0}_2 = e^{i\phi^1}|s_2^1\rangle \otimes \mathbf{5}_2, \tag{13.12}$$

$$U_{2,1}\left\{|s_1^1\rangle \otimes \widehat{\mathbb{A}}_1^2\widehat{\mathbb{A}}_1^4\mathbf{0}_1\right\} \equiv e^{i\phi^2}|s_2^1\rangle \otimes \widehat{\mathbb{A}}_2^2\widehat{\mathbb{A}}_2^4\mathbf{0}_2 = e^{i\phi^2}|s_2^1\rangle \otimes \underline{\mathbf{10}}_2. \tag{13.13}$$

Stage Σ_2 to Σ_3

$$\begin{aligned} U_{3,2}\left\{|s_2^1\rangle \otimes \widehat{\mathbb{A}}_2^1\widehat{\mathbb{A}}_2^3\mathbf{0}_2\right\} &= |s_3^1\rangle \otimes \left\{t^1\widehat{\mathbb{A}}_3^4 + ir^1\widehat{\mathbb{A}}_3^3\right\}\left\{t^2\widehat{\mathbb{A}}_3^1 + ir^2\widehat{\mathbb{A}}_3^2\right\} \\ &= |s_3^1\rangle \otimes \left\{t^1t^2\mathbf{9}_3 + ir^1t^2\mathbf{5}_3 + ir^2t^1\underline{\mathbf{10}}_3 - r^1r^2\mathbf{6}_3\right\}, \\ U_{3,2}\left\{|s_2^1\rangle \otimes \widehat{\mathbb{A}}_2^2\widehat{\mathbb{A}}_2^4\mathbf{0}_2\right\} &= |s_3^1\rangle \otimes \left\{t^2\widehat{\mathbb{A}}_3^2 + ir^2\widehat{\mathbb{A}}_3^1\right\}\left\{t^1\widehat{\mathbb{A}}_3^3 + ir^1\widehat{\mathbb{A}}_3^4\right\} \\ &= |s_3^1\rangle \otimes \left\{t^1t^2\mathbf{6}_3 + ir^2t^1\mathbf{5}_3 + ir^1t^2\underline{\mathbf{10}}_3 - r^1r^2\mathbf{9}_3\right\}. \end{aligned} \tag{13.14}$$

With this information, program MAIN gives the following nonzero correlation probabilities:

$$\begin{aligned} \Pr(\widehat{\mathbb{A}}_3^1\widehat{\mathbb{A}}_3^3\mathbf{0}_3|\Psi_0) = \Pr(\widehat{\mathbb{A}}_3^2\widehat{\mathbb{A}}_3^4\mathbf{0}_3|\Psi_0) &= \frac{1}{2}(r^1t^2)^2 + \frac{1}{2}(r^2t^1)^2 + r^1r^2t^1t^2\cos(\phi^1 - \phi^2), \\ \Pr(\widehat{\mathbb{A}}_3^2\widehat{\mathbb{A}}_3^3\mathbf{0}_3|\Psi_0) = \Pr(\widehat{\mathbb{A}}_3^1\widehat{\mathbb{A}}_3^4\mathbf{0}_3|\Psi_0) &= \frac{1}{2}(r^1r^2)^2 + \frac{1}{2}(t^2t^1)^2 - r^1r^2t^1t^2\cos(\phi^1 - \phi^2). \end{aligned} \tag{13.15}$$

These agree with the results given in Horne et al. (1989), assuming no losses and taking $t^1 = r^1 = t^2 = r^2 = 1/\sqrt{2}$.

The following comments are relevant.

Partial Questions
The above signal pair correlations show dependency on the phases ϕ^1, ϕ^2, whereas as pointed out by HSZ, the answers to the partial questions involving single detectors only have no such dependence. For instance, the probability $\Pr(1_3|\Psi_0)$ that stage Σ_3 detector 1_3 has fired is given by

$$\Pr(1_3|\Psi_0) = \Pr(\widehat{\mathbb{A}}_3^1\widehat{\mathbb{A}}_3^3\mathbf{0}_3|\Psi_0) + \Pr(\widehat{\mathbb{A}}_3^1\widehat{\mathbb{A}}_3^4\mathbf{0}_3|\Psi_0) = \frac{1}{2}, \tag{13.16}$$

which is independent of the phases. The same result holds for the other single detectors.

Nonlocality
The QDN formalism shows that the phase change ϕ^1 applied to component 1_1 could be applied to component 3_1 instead with no change in physical predictions; this is evident from the fact that the phase factor $\exp(i\phi^1)$ multiplies the product $\widehat{\mathbb{A}}_2^1\widehat{\mathbb{A}}_2^3$ in Eq. (13.12). A similar remark applies to the other phase change ϕ^2. Since the component beams are spatially separated when this phase change is applied, this means that quantum states are inherently nonlocal in character.

From such considerations, we conclude that physical space is not as simple as it seems from a classical perspective. Perhaps a better way of saying this is that the classical model of physical space as a three-dimensional continuum of spatial position parameters is a good classical model but contextually incomplete as far as quantum mechanics is concerned.

Probabilities versus Rates

As with all calculations in QDN, normalization to unity is an idealization that avoids empirically significant but theoretically marginal considerations to do with flux factors, particle production rates, and such like. Perhaps the best way to deal with these issues is to recognize that what is calculated represents idealized situations. The more empirically useful interpretation of stated probabilities is that they are *best case rates*, that is, predicted relative average signal rates during the time when incoming wave trains are long enough to intersect and interfere, with no inefficiencies or extraneous losses in detection.

14
Quantum Eraser Experiments

14.1 Introduction

In recent years, numerous quantum optics experiments have demonstrated novel quantum effects based on quantum amplitude superposition. Some of these empirical observations raise disturbing questions about the nature of reality, particularly concerning quantum nonlocality. In this chapter we discuss some experiments that suggest (to us, at least) that the information void, that uncharted regime between signal preparation and outcome detection, is not well modeled by classical spacetime. Something strange seems to be lurking there.

For the purposes of exposition, the sequence of experiments we discuss in this chapter does not follow the historical sequence of those experiments, We discuss first the *delayed-choice quantum eraser* experiment of Kim et al. (Kim et al., 2000), referred to here as KIM.[1] This leads us to define a heuristic measure of *which-path* information that is central to the theme of this chapter. This leads to a discussion of Wheeler's thought experiments on the conundrum of *delayed choice* in physics, particularly evident in the phenomenon of galactic lensing (Wheeler, 1983), and its empirical implementation, the *delayed-choice interferometer* experiment of Jacques et al., referred to as JACQUES. Our final experiment in this sequence is the *double-slit quantum eraser* of Walborn et al. (Walborn et al., 2002), referred to as WALBORN. That experiment gives a significant challenge for quantized detector networks (QDN) to reproduce its empirical results.

Such experiments have led to suggestions that interference patterns formed by particles impacting on a screen may be influenced in some way by decisions made long after those particles had landed on that screen. Our objective in this chapter is to show by a detailed QDN analysis of those experiments that

[1] With profound apologies to co-authors, we will employ the convention that several experiments are referred to by the capitalized surname of the first named author.

quantum principles do not support these suggestions. Nevertheless, something bizarre seems to be there.

There are two complementary aspects in these experiments that have deeply worried physicists. The first involves *spatial nonlocality* and is the point of Wheeler's concerns. Historically, it has been a great source of debate among physicists. It motivated Einstein to write to Born thus:[2]

> So clearly that you consider my attitude towards statistical quantum mechanics to be strange and archaic ...
>
> ... I cannot make a case for my attitude in physics which you would consider at all reasonable. I admit, of course, that there is a considerable amount of validity in the statistical approach which you were the first to recognise clearly as necessary given the framework of the existing formalism. I cannot seriously believe in it because the theory cannot be reconciled with the idea that physics should represent a reality in time and space, free from spooky action at a distance.
>
> (*Born–Einstein Letters*, 1971, p. 158)

The second aspect involves *temporal nonlocality*: the temporal sequence of actions taken in the laboratory at different places seems to be irrelevant in certain experiments. We regard that as a vindication of the *stages* concept. The empirical evidence from experiments such as WALBORN, who specifically investigated the temporal aspects, is that the passage of "detector time" is synonymous with quantum information acquisition (QIA) occurring in a sequence of *stages* (Eakins and Jaroszkiewicz, 2005). Stages have rules that are different in some aspects from those of classical information acquisition, and this accounts for some of the apparent strangeness of quantum mechanics.

The rules governing QIA are those of quantum mechanics (QM) and indeed conform with all known physics. For example, QIA never violates the light-cone constraint of relativity that classical information cannot be acquired between space-like intervals. Underpinning this is the so-called *no-communication theorem* in QM, discussed in Section 16.7. This states that the actions taken by an observer on a substate of a total state of a system under observation (SUO) cannot be detected by another observer of another substate of that total state.

One way of understanding this is to note that quantum correlations appearing to violate the principle of Einstein locality always require observations to be completed *before* those correlations can be defined by observers, and this completion necessarily always takes place in a classically consistent matter. Currently,

[2] The specific English translation of what Einstein wrote here has been disputed: *mysterious* rather than *spooky* has been suggested by some authorities as closer to what Einstein intended to say.

there is no empirical evidence for, or theoretical necessity of, the contextually incomplete notion that information can flow backward in time, as suggested by Cramer's transactional interpretation of QM (Cramer, 1986), or for the existence of closed time-like curves (CTCs), as found in some relativistic spacetimes such as that of Gödel (Gödel, 1949).

On the other hand, quantum information *can* be shielded against the effects of decoherence and preserved in a state of stasis for arbitrarily long periods of laboratory time. QDN uses the concept of null test to encode this phenomenon, as demonstrated in our discussion of particle decay experiments in Chapter 15. In the delayed-choice experiments discussed in the present chapter, suitably shielded observations involving separate detectors can be taken in apparently random order relative to laboratory time without affecting correlations.

14.2 Delayed-Choice Quantum Eraser

We turn first to the *delayed-choice quantum eraser*. The architecture of an experiment proposed in Kim et al. (2000), and referred to here as KIM, is shown in Figure 14.1. In that stage diagram, we leave out dotted lines indicating stages and use subscripts for this purpose. In the following description, we use the term *photon* for convenience only.

A source S produces an initial total state 1_0 that is a superposition of two coherent photon pairs. By stage Σ_1 these are split by suitable apparatus into four components 1_1, 2_1, 3_1, and 4_1 as shown. Components 1_1 and 3_1 are from different initial pairs and are passed onto a detecting screen DS, consisting of detectors $5_2, 6_2, \ldots, K_2$, where it is supposed that they interfere, just as in the original double-slit experiment. Our analysis will show that the situation is more subtle than that.

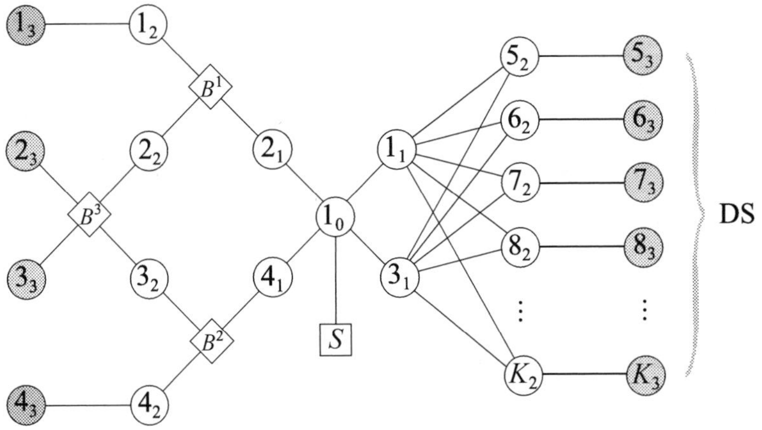

Figure 14.1. KIM, a proposed delayed-choice quantum eraser experiment (Kim et al., 2000).

Meanwhile, components 2_1 and 4_1 are each passed through beam splitters B^1, B^2 as shown. The transmitted components from these beam splitters are sent on to detectors 1_3 and 4_3 as shown, while the reflected components 2_2 and 3_2 are passed through beam splitter B^3. The outcome channels of this beam splitter are monitored by detectors 2_3 and 3_3.

Analysis of this arrangement suggests that choices made by the experimentalist at B^3 can influence the signal patterns seen on DS, even though the signals in that screen had been captured earlier at stage Σ_2. Kim et al.'s argument is that correlations between signals in 1_3 and DS or between signals in 4_3 and DS give *which-path* information, whereas correlations between signals in either 2_3 and DS, and 3_3 and DS do not. Therefore, there should be no interference terms on DS when there are signals in 1_3 or 4_4, whereas interference terms are expected when 1_3 or 3_3 register a signal.

We proceed with our stage analysis as follows. From Figure 14.1 we read off the following parameters for our CA program MAIN: $N = 3$, $r_0 = 1$, $r_1 = 4$, $r_2 = r_3 = K, d_0 = d_1 = d_2 = d_3 = 1$, where K is chosen to be large enough so as to show what happens on the screen DS. In MAIN a value $K = 6$ was sufficient to show all the essential features of the experiment.

As with the double-slit experiment discussed in Chapter 10, we represent our initial preparation switch at stage Σ_0 by the total state $|\Psi_0) = |s_0\rangle \otimes \widehat{\mathbb{A}}_0^1\mathbf{0}_0$.

Stage Σ_0 to Stage Σ_1

The jump to stage Σ_1 is defined by the production of two correlated photon pairs. Photon polarization is ignored here, but the formalism can readily deal with any situation where this is not the case. Therefore, MAIN assumes a one-dimensional internal Hilbert space in this experiment with normalized basis $|s_n\rangle$ for the internal degrees of freedom at any given stage Σ_n.

The semi-unitary transformation from $\Sigma_0 \to \Sigma_1$ is given by

$$\mathbb{U}_{1,0}\left\{|s_0\rangle \otimes \widehat{\mathbb{A}}_0^1\mathbf{0}_0\right\} = |s_1\rangle \otimes \left\{\alpha\widehat{\mathbb{A}}_1^1\widehat{\mathbb{A}}_1^2 + \beta\widehat{\mathbb{A}}_1^3\widehat{\mathbb{A}}_1^4\right\}\mathbf{0}_1, \tag{14.1}$$

where $|\alpha|^2 + |\beta|^2 = 1$. As we stated above we assume that photon spin is not relevant to the issues being explored here.

Stage Σ_1 to Stage Σ_2

For this jump we have

$$\begin{aligned}
\mathbb{U}_{2,1}\left\{|s_1\rangle \otimes \widehat{\mathbb{A}}_1^1\widehat{\mathbb{A}}_1^2\mathbf{0}_1\right\} &= |s_2\rangle \otimes \sum_{i=5}^{K} C^i\widehat{\mathbb{A}}_2^i\{t^1\widehat{\mathbb{A}}_2^1 + ir^1\widehat{\mathbb{A}}_2^2\}\mathbf{0}_2, \\
\mathbb{U}_{2,1}\left\{|s_1\rangle \otimes \widehat{\mathbb{A}}_1^3\widehat{\mathbb{A}}_1^4\mathbf{0}_1\right\} &= |s_2\rangle \otimes \sum_{i=5}^{K} D^i\widehat{\mathbb{A}}_2^i\{t^2\widehat{\mathbb{A}}_2^4 + ir^2\widehat{\mathbb{A}}_2^3\}\mathbf{0}_2.
\end{aligned} \tag{14.2}$$

Here, one component beam from each pair is focused onto the detecting screen DS, while the other component is channeled onto either beam-splitter B^1 or B^2, as shown. The $\{C^i\}$ coefficients represent the amplitudes for landing on the screen DS from 1_1, while the $\{D^i\}$ are from 3_1. The coefficients t^i, r^i are characteristic transmission and reflection parameters associated with beam splitter B^i, $i = 1, 2, 3$, and satisfy the rule $(t^i)^2+(r^i)^2 = 1$. It is useful not to set these parameters to the conventional value $1/\sqrt{2}$ but to keep them open and available to be changed. It is in these parameters that we encode the observer's freedom of choice in this particular experiment.

Stage Σ_2 to Stage Σ_3

The final transition from stage Σ_2 to Σ_3 involves four terms, each of which involves null tests in one way or another. Most significantly for this discussion, it is supposed that the DS screen detectors have registered their signals irreversibly by stage Σ_2. Therefore, this information is carried to stage Σ_3 by null tests, shown by the horizontal lines on the right-hand side of Figure 14.1.

We write, for $i = 5, 6, \dots, K$:

$$\begin{aligned}
\mathbb{U}_{3,2}\left\{|s_2\rangle \otimes \widehat{\mathbb{A}}_2^i\widehat{\mathbb{A}}_2^1\mathbf{0}_2\right\} &= |s_3\rangle \otimes \widehat{\mathbb{A}}_3^i\widehat{\mathbb{A}}_3^1\mathbf{0}_3,\\
\mathbb{U}_{3,2}\left\{|s_2\rangle \otimes \widehat{\mathbb{A}}_2^i\widehat{\mathbb{A}}_2^2\mathbf{0}_2\right\} &= |s_3\rangle \otimes \widehat{\mathbb{A}}_3^i\{t^3\widehat{\mathbb{A}}_3^3 + ir^3\widehat{\mathbb{A}}_3^2\}\mathbf{0}_3\\
\mathbb{U}_{3,2}\left\{|s_2\rangle \otimes \widehat{\mathbb{A}}_2^i\widehat{\mathbb{A}}_2^3\mathbf{0}_2\right\} &= |s_3\rangle \otimes \widehat{\mathbb{A}}_3^i\{t^3\widehat{\mathbb{A}}_3^2 + ir^3\widehat{\mathbb{A}}_3^3\}\mathbf{0}_3\\
\mathbb{U}_{3,2}\left\{|s_2\rangle \otimes \widehat{\mathbb{A}}_2^i\widehat{\mathbb{A}}_2^4\mathbf{0}_2\right\} &= |s_3\rangle \otimes \widehat{\mathbb{A}}_3^i\widehat{\mathbb{A}}_3^4\mathbf{0}_3.
\end{aligned} \tag{14.3}$$

This is all the information needed for our CA program MAIN to answer all possible maximal questions. MAIN gives four sets of nonzero maximal question answers, equivalent to two-site correlations:

$$\begin{aligned}
\Pr(\widehat{\mathbb{A}}_3^1\widehat{\mathbb{A}}_3^i\mathbf{0}_3|\Psi_0) &= |t^1C^i\alpha|^2,\\
\Pr(\widehat{\mathbb{A}}_3^2\widehat{\mathbb{A}}_3^i\mathbf{0}_3|\Psi_0) &= |r^1r^3C^i\alpha - ir^2t^3D^i\beta|^2,\\
\Pr(\widehat{\mathbb{A}}_3^3\widehat{\mathbb{A}}_3^i\mathbf{0}_3|\Psi_0) &= |r^1t^3C^i\alpha + ir^2r^3D^i\beta|^2,\\
\Pr(\widehat{\mathbb{A}}_3^4\widehat{\mathbb{A}}_3^i\mathbf{0}_3|\Psi_0) &= |t^2D^i\beta|^2, \qquad i = 5, 6, \dots, K.
\end{aligned} \tag{14.4}$$

These give the total sum

$$\sum_{i=1}^{4}\sum_{j=5}^{K}\Pr(\widehat{\mathbb{A}}_3^i\widehat{\mathbb{A}}_3^j\mathbf{0}_3|\Psi_0) = |\alpha|^2\sum_{j=5}^{K}|C^i|^2 + |\beta|^2\sum_{j=5}^{K}|D^i|^2, \tag{14.5}$$

which is consistent with probability conservation if we take the semi-unitarity conditions $\sum_{j=5}^{K}|C^i|^2 = \sum_{j=5}^{K}|D^i|^2 = 1$ into account. Note also that semi-unitarity requires $\sum_{j=5}^{K}C^{i*}D^i = 0$, where C^{i*} is the complex conjugate of C^i.

There are several observations to be made about these results.

1. The only genuine *interference* found in our analysis occurs in the two-signal correlation probabilities $\Pr(\widehat{\mathbb{A}}_3^2\widehat{\mathbb{A}}_3^i\mathbf{0}_3|\Psi_0)$ and $\Pr(\widehat{\mathbb{A}}_3^3\widehat{\mathbb{A}}_3^i\mathbf{0}_3|\Psi_0)$.
2. The parameters t^i, r^i for beam splitter B^i represent places in the apparatus where the experimentalist could make changes, either before or after signals

have been registered on the screen DS during any given run of the experiment. In other words, choices can be made at B^1, B^2, and B^3 that affect various incidence rates. The question is, does any change made by the experimentalist at any beam splitter affect anything that has been measured *before* that change was made? In particular, can any change in B^3 affect what has already happened on the screen?

By inspection of (14.3), we see that no change in t^3 or r^3, subject to $(t^3)^2 + (r^3)^2 = 1$, has any effect whatsoever on $\Pr(\widehat{\mathbb{A}}_3^1\widehat{\mathbb{A}}_3^i\mathbf{0}_3|\Psi_0)$ or $\Pr(\widehat{\mathbb{A}}_3^4\widehat{\mathbb{A}}_3^i\mathbf{0}_3|\Psi_0)$. These coincidence rates actually involve signal detection completed during *earlier* stages. The conclusion therefore is that any suggestion that delayed choice can erase information irreversibly acquired in the past is incorrect and misleading.

3. By inspection, it is true that changes in t^3 and r^3 affect $\Pr(\widehat{\mathbb{A}}_3^2\widehat{\mathbb{A}}_3^i\mathbf{0}_3|\Psi_0)$ and $\Pr(\widehat{\mathbb{A}}_3^3\widehat{\mathbb{A}}_3^i\mathbf{0}_3|\Psi_0)$. However no acausality is involved, because a coincidence rate is undefined until signals from both detectors involved have been counted. $\Pr(\widehat{\mathbb{A}}_3^2\widehat{\mathbb{A}}_3^i\mathbf{0}_3|\Psi_0)$ and $\Pr(\widehat{\mathbb{A}}_3^3\widehat{\mathbb{A}}_3^i\mathbf{0}_3|\Psi_0)$ cannot be measured until *after* the choice of t^3 and r^3 has been made.

   Suggestions that events in stage Σ_3 could influence events in earlier stages do not take into account the crucial role of *post-selection*[3] in such experiments. The proper way to understand what is happening is to view the role of the four detectors $1_3, 2_3, 3_3$, and 4_3 as post-selection apparatus for the processing of data already accumulated on the screen DS.

4. Program MAIN allows for partial questions to be answered. If we look at the individual total counting rates at each of the four detectors i_3, $i = 1, 2, 3, 4$, we find

$$
\begin{aligned}
\Pr(\widehat{\mathbb{A}}_3^1\mathbf{0}_3|\Psi_0) &\equiv \sum_{i=5}^{K} \Pr(\widehat{\mathbb{A}}_3^1\widehat{\mathbb{A}}_3^i\mathbf{0}_3|\Psi_0) = |t^1\alpha|^2, \\
\Pr(\widehat{\mathbb{A}}_3^2\mathbf{0}_3|\Psi_0) &\equiv \sum_{i=5}^{K} \Pr(\widehat{\mathbb{A}}_3^2\widehat{\mathbb{A}}_3^i\mathbf{0}_3|\Psi_0) = |r^1r^3\alpha|^2 + |t^3r^2\beta|^2, \\
\Pr(\widehat{\mathbb{A}}_3^3\mathbf{0}_3|\Psi_0) &\equiv \sum_{i=5}^{K} \Pr(\widehat{\mathbb{A}}_3^3\widehat{\mathbb{A}}_3^i\mathbf{0}_3|\Psi_0) = |r^1t^3\alpha|^2 + |r^2r^3\beta|^2, \\
\Pr(\widehat{\mathbb{A}}_3^4\mathbf{0}_3|\Psi_0) &\equiv \sum_{i=5}^{K} \Pr(\widehat{\mathbb{A}}_3^4\widehat{\mathbb{A}}_3^i\mathbf{0}_3|\Psi_0) = |t_2\beta|^2,
\end{aligned} \tag{14.6}
$$

using the semi-unitarity of the $\{C^i\}$ and $\{D^i\}$ coefficients. Again, changes made at B^3 would have no effect on $\Pr(\widehat{\mathbb{A}}_3^1\mathbf{0}_3|\Psi_0)$ or $\Pr(\widehat{\mathbb{A}}_3^4\mathbf{0}_3|\Psi_0)$ but would

[3] In ordinary usage, the term *post-selection* means selection occurring *after* some given process. In probability theory, given an event A, then the probability of some event B conditional on that given is a conditional probability, denoted $P(B|A)$. Either way, the concern is with what happens *after* something has been done or occurred.

affect $\Pr(\widehat{\mathbb{A}}_3^2\mathbf{0}_3|\Psi_0)$ and $\Pr(\widehat{\mathbb{A}}_3^3\mathbf{0}_3|\Psi_0)$. These probabilities sum to unity as expected.

5. If we look at the individual relative probabilities for a given detector i_3, $i = 5, 6, \ldots, K$, on the screen DS, we find

$$\Pr(\widehat{\mathbb{A}}_3^i\mathbf{0}_3|\Psi_0) \equiv \Pr\sum_{j=1}^{4}\Pr(\widehat{\mathbb{A}}_3^j\widehat{\mathbb{A}}_3^i\mathbf{0}_3|\Psi_0) = |\alpha C^i|^2 + |\beta D^i|^2, \tag{14.7}$$

which shows no interference on DS. No change at any of the beam splitters affects the single detector rates observed on the screen DS. This is essentially the same phenomenon, discussed in Section 13.4, found in the two-photon interferometer experiment of Horne, Shimony, and Zeilinger (Horne et al., 1989).

We should ask: *Given that signals on the detector screen DS came from the double slits, why is there no interference on that screen? Surely there should be such interference.*

The answer is deep. This experiment is not just a double-slit experiment. The mere fact that a photon has gone off toward the beam splitters B^1 or B^2 *and* the observer could in principle find out which beam splitter it was is enough to provide which-path information. It is that contextual information that destroys any possibility of an interference pattern on DS. In fact, the beam splitters are not needed for this. It is enough to have 2_1 and 4_1 as identifiable potential sites for signal detection to destroy any chance of interference on DS.

14.3 Which-Path Measure

The double-slit and delayed-choice eraser experiments discussed up to this point in this book belong to an important class of experiment that, to use colloquial terminology, provide partial or complete information about which path a photon had taken in its journey from initial to final stages. We shall discuss in the next section another example, Wheeler's delayed choice experiment.

Each of these experiments carries with it contextual attributes arising from the experimental setup that determine the extent to which photon paths can be determined from the data or not. For example, the double-slit experiment with both slits open gives zero information about which slit a particular detected photon came from. On the other hand, the same setup with one of the slits blocked up gives total information as to where any of the detected photons originated.

It is of interest therefore to find some measure or parameter Φ that is characteristic of any given experimental setup and that gives us an indication as to how much which-path information we could extract from the experiment. There is no precise theory for this known to us at present, so in the absence of any deeper analysis, our choice is to define Φ heuristically as the total probability of determining with certainty full path information from a single detected "indicator" photon, that is, from a single signal in a specific set of detectors. We define

$$\Phi_{DS} \equiv Prob(\text{full path information}|\text{any single indicator detector fires}). \quad (14.8)$$

For example, in the case of the double-slit experiment discussed in Chapter 10, the detecting screen contains the indicator detectors. Then with both slits open, we find $\Phi_{DS} = 0$. When one of the slits is open, then $\Phi_{DS} = 1$. In the case of the delayed choice quantum eraser discussed above, the indicator detectors are 1_3 and 4_3, so we define

$$\Phi_{DC} \equiv \Pr(\widehat{\mathbb{A}}_3^1\mathbf{0}_3|\Psi_0) + \Pr(\widehat{\mathbb{A}}_3^4\mathbf{0}_3|\Psi_0) = |t^1\alpha|^2 + |t_2\beta|^2.$$

In the conventional symmetric situation, $|\alpha| = |\beta| = t^1 = t^2 1/\sqrt{2}$, giving $\Phi_{DC} = 1/2$, as we should expect. When $t^1 = t^2 = 0$, transmission to 1_3 or 4_3 cannot occur, so there is normally interference for sure, giving $\Phi_{DC} = 0$, meaning no path information can be established.

There is a pathology in this last case, because $t^1 = t^2 = 0$ gives $\Pr(\widehat{\mathbb{A}}_3^2\mathbf{0}_3|\Psi_0) = |r^3\alpha|^2 + |t^3\beta|^2$ and $\Pr(\widehat{\mathbb{A}}_3^3\mathbf{0}_3|\Psi_0) = |t^3\alpha|^2 + |r^3\beta|^2$. If now, in addition to setting $t^1 = t^2 = 0$, the experimentalist had set $r^3 = 0$, then a single photon would be detected at 2_3 or 3_3 and it would clear which path had been taken by the photons. But overall, this is equivalent to having no beam splitters, so this scenario is of no value here.

14.4 Wheeler's Double-Slit Delayed Choice Experiment

The physicist John Wheeler gave a theoretical discussion of a quantum interference experiment that has stimulated a great deal of puzzlement and controversy concerning the nature of reality (Wheeler, 1983). It is our considered judgment that his concern reflects the inadequacy of our classical views about reality.

There are two forms of his experiment that are commonly discussed: one is a modified Mach–Zehnder interferometer (MZI) and the other is a modified double-slit (DS) experiment. We shall refer to these as WHEELER-1 and WHEELER-2, respectively.

WHEELER-1

Figure 13.1 is relevant to WHEELER-1. Specifically, WHEELER-1 is concerned with stage-Σ_1 nodes 1_1 and 2_1 and their relationship to stage-Σ_2 nodes 1_2 and 2_2. Suppose the laboratory time between stage Σ_1 and stage Σ_2 is significantly long enough for anything done to one of the beams, at say module ϕ, to be long *after* the action of beam splitter B^1 in any reasonable sense of the word. Then there appears to be a clash of particle and wave concepts.

To be specific, consider a pulse of light 1_0 coming from a distant galaxy billions of years ago at stage Σ_0 being split by gravitational lensing (equivalent to B^1)

into components 1_1 and 2_1. These two components follow very different paths across the Universe until an observer here on Earth observes stage-Σ_3 signals at detectors 1_3 or 2_3. Now two kinds of process could affect outcome probabilities at those detectors: (1) there could be some galactic cloud affecting, say, the optical path from 1_1 to 1_2, equivalent to module ϕ, and (2) the observer could choose to do something significant at B^2 so that either an interference pattern is built up or no such pattern is built up.

The conventional picture here is of a particle known as a photon traveling through space. Intuitively, we would like to think of such objects. But it seems incredible that the presence of some disruptive process ϕ on the potential path $1_1 \to 1_2$ could affect a photon traveling from 2_1 to 2_2. Yet that is just what the photon paradigm requires us to contemplate, particularly if the observer decides to arrange the apparatus at B^2 to detect an interference pattern at the detectors 1_3 and 2_3, or not.

Of this scenario, Wheeler wrote:

> We get up in the morning and spend the day in meditation whether to observe by "which route" or to observe interference between "both routes." When night comes and the telescope is at last usable we leave the half-silvered mirror out or put it in, according to our choice. The monochromatizing filter placed over the telescope makes the counting rate low. We may have to wait an hour for the first photon. When it triggers a counter, we discover "by which route" it came with the one arrangement; or by the other, what the relative phase is of the waves associated with the passage of the photon from source to receptor "by both routes" – perhaps 50,000 light years apart as they pass the lensing galaxy. But the photon has already *passed* that galaxy billions of years before we made our decision. This is the sense in which, in a loose way of speaking, what the photon *shall have done* after it has *already* done it. In actuality it is wrong to talk about the "route" of the photon. For a proper way of speaking we recall once more that it makes no sense to talk of the phenomenon until it has been brought to a close by an irreversible act of amplification. "No elementary phenomenon is a phenomenon until it is a registered (observed) phenomenon." (Wheeler, 1983)

The problem as we see it lies not with the quantum physics but with mental imagery associated with the particle-wave concept. There is, for instance, a classically conditioned vacuous assertion glaringly obvious in Wheeler's statement above: that "the photon has already passed that galaxy billions of years before we made our decision." Proof? There can be none, but of course, we are all strongly conditioned to believe this counterfactual, vacuous statement and many like it. According to quantum physics principles, as we see them in this book, we have no direct empirical entitlement to make that assertion. In quantum physics, we have to resist the constant temptation to think classically in those places where such thinking is unwarranted.

WHEELER-2

Wheeler's delayed choice experiment can be discussed as a double-slit experiment with a modified screen. Some of the detectors on the screen can receive quantum signals from both slits, while the others can receive a signal from only one of the slits. The interest in this arrangement comes from the possibility that the observer can decide in principle which detectors receive which signal(s) long *after* light has left the two slits.

An idealized version of WHEELER-2 is shown in Figure 14.2. The details are much the same as the double-slit experiment studied in Chapter 10, but with the difference that now there are three groups of detectors on the screen. Detectors 1_2 to R_2 can receive a quantum amplitude from 1_1 only, $(R+1)_2$ to $(R+S)_2$ can receive quantum amplitudes from 1_1 and from 2_1, and $(R+S+1)_2$ to K_2 can receive an amplitude from 2_1 only. Here, $K = R + S + T$.

Following Wheeler, we imagine that the experimentalist can shuffle the values of R, S, and T during any given run, *after* they are sure that light has left the two slits 1_1 and 2_2 and *before* any impact on the final detecting screen at stage Σ_2. Any actual experiment would require analysis of the final data, post-selecting signals corresponding to equivalent values of R, S, and T.

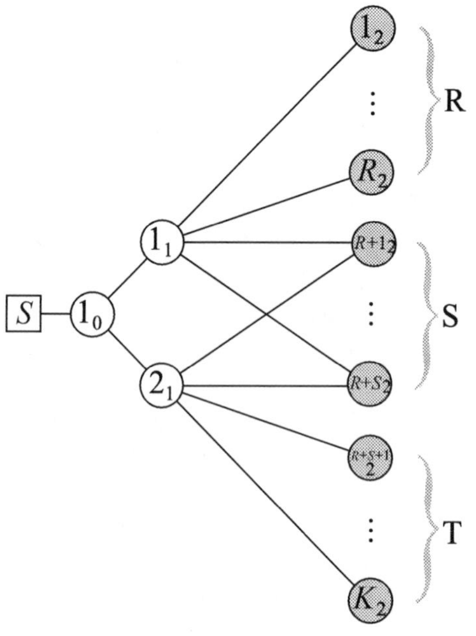

Figure 14.2. Idealization of Wheeler's delayed-choice experiment, WHEELER-2.

For this discussion, we do not need CA assistance. Photon polarization is not an issue in this particular instance either, so the initial labstate $|\Psi_0)$ is given by $|\Psi_0) \equiv |s_0\rangle \otimes \widehat{\mathbb{A}}_0^1 \mathbf{0}_0$. The stage dynamics goes as follows.

Stage Σ_0 to Stage Σ_1
The evolution of the preparation switch is given by

$$U_{1,0}\left\{|s_0\rangle \otimes \widehat{\mathbb{A}}_0^1 \mathbf{0}_0\right\} = |s_1\rangle \otimes \left\{\alpha \widehat{\mathbb{A}}_1^1 + \beta \widehat{\mathbb{A}}_1^2\right\} \mathbf{0}_1, \tag{14.9}$$

where $|\alpha|^2 + |\beta|^2 = 1$.

Stage Σ_1 to Stage Σ_2
The evolution of the two separate components at stage Σ_1 is given by

$$U_{2,1}\left\{|s_1\rangle \otimes \widehat{\mathbb{A}}_1^1 \mathbf{0}_1\right\} = |s_2\rangle \otimes \left\{\underbrace{\sum_{i=1}^{R} R^i \widehat{\mathbb{A}}_2^i \mathbf{0}_2}_{\text{Region } R} + \underbrace{\sum_{i=R+1}^{R+S} S^{i,1} \widehat{\mathbb{A}}_2^i \mathbf{0}_2}_{\text{Region S}}\right\},$$

$$U_{2,1}\left\{|s_1\rangle \otimes \widehat{\mathbb{A}}_1^2 \mathbf{0}_1\right\} = |s_2\rangle \otimes \left\{\underbrace{\sum_{i=R+1}^{R+S} S^{i,2} \widehat{\mathbb{A}}_2^i \mathbf{0}_2}_{\text{Region } S} + \underbrace{\sum_{i=R+S+1}^{K} T^i \widehat{\mathbb{A}}_2^i \mathbf{0}_2}_{\text{Region T}}\right\}. \tag{14.10}$$

Here the coefficients $\{R^i\}, \{S^{i,1}, S^{i,2}\}$, and $\{T^i\}$ satisfy the semi-unitary conditions

$$\sum_{i=1}^{R} |R^i|^2 + \sum_{i=R+1}^{R+S} |S^{i,1}|^2 = \sum_{i=R+S+1}^{K} |T^i|^2 + \sum_{i=R+1}^{R+S} |S^{i,2}|^2 = 1, \quad \sum_{i=R+1}^{R+S} S^{i,1*} S^{i,2} = 0. \tag{14.11}$$

The outcome probabilities are found to be

$$\begin{aligned} \Pr(\mathbb{A}_2^i \mathbf{0}_2 | \Psi_0) &= |\alpha R^i|^2, & 1 \leqslant i \leqslant R \\ &= |\alpha S^{i,1} + \beta S^{i,2}|^2, & R < i \leqslant R + S \\ &= |\beta T^i|^2 & R + S < i \leqslant K \equiv R + S + T. \end{aligned} \tag{14.12}$$

These probabilities sum to unity as required.

From this we find the which-path parameter Φ_{W2} for $W2$ to be the sum of the probabilities over regions R and T, that is,

$$\Phi_{W2} = |\alpha|^2 \sum_{i=1}^{R} |R^i|^2 + |\beta|^2 \sum_{i=R+S+1}^{K} |T^i|^2. \tag{14.13}$$

This reduces to unity when $S = 0$ as expected and to zero when both R and T are zero.

14.5 The Delayed Choice Interferometer

A delayed choice experiment that confirms the predictions of QM was done by Jacques et al. with a Mach–Zehnder interferometer (Jacques et al., 2007), the relevant stage diagram being the same as Figure 13.1. In this experiment, referred to here as JACQUES, the final beam-splitter B^2 could be removed while the light was on its way from the first beam-splitter B^1.

Although this experiment is of type WHEELER-1, it is also equivalent to the above WHEELER-2 scenario with $K = 2$, because a Mach–Zehnder experiment represented by stage diagram Figure 13.1 is equivalent to a double-slit experiment with just two detectors in the final stage-detecting screen. In JACQUES, the configuration with the second beam splitter B^2 removed corresponds to taking $R = T = 1$, $S = 0$ in WHEELER-2, while that with B^2 in operation corresponds to $R = T = 0$, $S = 2$.

14.6 The Double-Slit Quantum Eraser

The above discussed experiments do not involved photon spin specifically. The experiment we discuss next requires a careful analysis of spin.

Prior to the delayed-choice quantum eraser experiment of Jacques et al. (Jacques et al., 2007), the double-slit quantum eraser experiment of Walborn et al. (Walborn et al., 2002) had demonstrated the empirical validity of the stage concept in quantum mechanics. The Walborn et al. experiment consists of three subexperiments, referred to here as WALBORN-1, WALBORN-2, and WALBORN-3.

WALBORN-1: No Which-Way Information

The first subexperiment, WALBORN-1, is shown in Figure 14.3. Source S prepares a spinless photon pair 1_0, which is then split, with photon 2_1 passing onto a double-slit (2_2, 3_2), and then onto a screen with final stage Σ_3 detectors $2_3, 3_4, \ldots, K$. Meanwhile, 1_1 is passed onto a detector 1_3. Coincidence measurements are taken involving screen impacts and 1_3 detection, with no polarization involved.

The initial total state $|\Psi_0)$ is given by $|\Psi_0) \equiv |s_0\rangle \otimes \widehat{\mathbb{A}}_0^1 \mathbf{0}_0$.

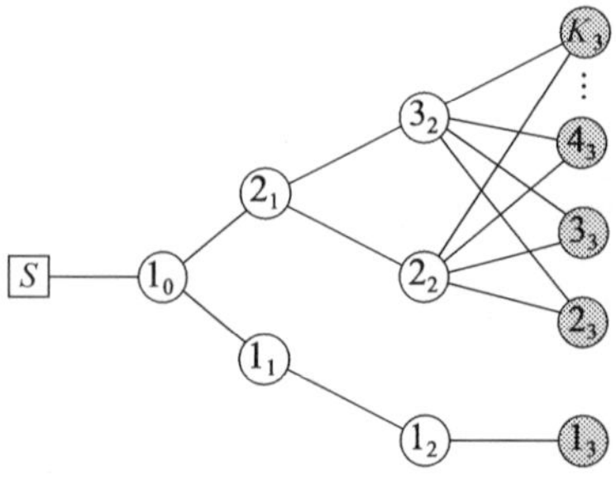

Figure 14.3. WALBORN-1: zero which-path information.

Stage Σ_0 to Stage Σ_1
Evolution from $\Sigma_0 \to \Sigma_1$ is given by

$$U_{1,0}\left\{|s_0\rangle \otimes \widehat{\mathbb{A}}_0^1 \mathbf{0}_0\right\} = |s_1\rangle \otimes \widehat{\mathbb{A}}_1^1 \widehat{\mathbb{A}}_1^2 \mathbf{0}_1. \tag{14.14}$$

Stage Σ_1 to Stage Σ_2
Evolution from $\Sigma_1 \to \Sigma_2$ is given by

$$U_{2,1}\left\{|s_1\rangle \otimes \widehat{\mathbb{A}}_1^1 \widehat{\mathbb{A}}_1^2 \mathbf{0}_1\right\} = |s_2\rangle \otimes \left\{\alpha \widehat{\mathbb{A}}_2^1 \widehat{\mathbb{A}}_2^2 + \beta \widehat{\mathbb{A}}_2^1 \widehat{\mathbb{A}}_2^3\right\} \mathbf{0}_2, \tag{14.15}$$

where $|\alpha|^2 + |\beta|^2 = 1$.

Stage Σ_2 to Stage Σ_3
The final stage transition $\Sigma_2 \to \Sigma_3$ is given by

$$\begin{aligned} U_{3,2}\left\{|s_2\rangle \otimes \widehat{\mathbb{A}}_2^1 \widehat{\mathbb{A}}_2^2 \mathbf{0}_2\right\} &= |s_3\rangle \otimes \sum_{i=2}^{K} C^i \widehat{\mathbb{A}}_3^1 \widehat{\mathbb{A}}_3^i \mathbf{0}_3, \\ U_{3,2}\left\{|s_2\rangle \otimes \widehat{\mathbb{A}}_2^1 \widehat{\mathbb{A}}_2^3 \mathbf{0}_2\right\} &= |s_3\rangle \otimes \sum_{i=2}^{K} D^i \widehat{\mathbb{A}}_3^1 \widehat{\mathbb{A}}_3^i \mathbf{0}_3, \end{aligned} \tag{14.16}$$

where the screen consists of $K-1$ detectors and the $\{C^i\}, \{D^i\}$ coefficients satisfy the usual semi-unitarity conditions.

The coincidence rates $\Pr(\widehat{\mathbb{A}}_3^1 \widehat{\mathbb{A}}_3^i \mathbf{0}_3 | \Psi_0)$ are given by

$$\Pr(\widehat{\mathbb{A}}_3^1 \widehat{\mathbb{A}}_3^i \mathbf{0}_3 | \Psi_0) = |\alpha C^i + \beta D^i|^2, \quad i = 2, 3, \ldots, K, \tag{14.17}$$

and these are the same as the single site detection rates $\Pr(\widehat{\mathbb{A}}_3^i \mathbf{0}_3 | \Psi_0)$. These results demonstrate double-slit interference because the detection of the 1_3 signal provides no which-way information, so that this version of the experiment is equivalent to a standard double-slit experiment.

WALBORN-2: Creation of Which-Path Information

WALBORN-1 is now reconsidered with some modifications, shown schematically in Figure 14.4. Walborn et al. placed two quarter-wavelength polarizers P^2 and P^3 in front of 2_3 and 3_3 as shown (Walborn et al., 2002). Each polarizer alters the beam it acts on in a way that distinguishes it from the other beam. The consequence is that each signal observed on the detecting screen contains information about the path taken. Hence no interference should be observed on the screen.

To understand the action of the P^2 and P^3 modules, we need to consider three sets of orthonormalized photon polarization bases, and the fact that we have a two-spin photonic internal space.

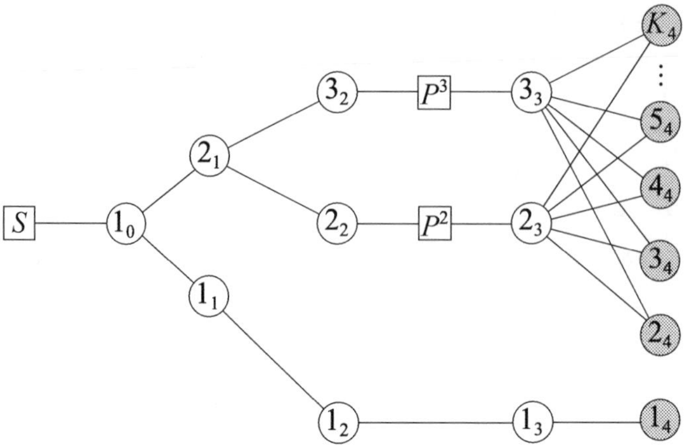

Figure 14.4. WALBORN-2: creation of which-path information.

Spin Bases

The first orthonormal photon spin basis $B^{HV} \equiv \{|H\rangle, |V\rangle\}$ consists of a horizontal polarization state $|H\rangle$ and a vertical polarization state $|V\rangle$; the second photon spin basis $B^{LR} \equiv \{|R\rangle, |L\rangle\}$ consists of a right circularly polarized state $|R\rangle$ and a left circularly polarized state $|L\rangle$; and the third photon spin basis is $B^{WP} \equiv \{|P\rangle, |N\rangle\}$, a conceptual "which-path" basis defined by Walborn et al. (Walborn et al., 2002). For clarity, our $|P\rangle$ and $|N\rangle$ states are the same as $|+\rangle$ and $|-\rangle$ states, respectively, used in Walborn et al. (2002).

These bases are related as follows:

$$\begin{aligned} |H\rangle &= \frac{1}{\sqrt{2}}\{|P\rangle + |N\rangle\}, \qquad & |R\rangle &= \frac{1-i}{2}\{|P\rangle + i|N\rangle\}, \\ |V\rangle &= \frac{1}{\sqrt{2}}\{|P\rangle - |N\rangle\}, \qquad & |L\rangle &= \frac{1-i}{2}\{i|P\rangle + |N\rangle\}. \end{aligned} \tag{14.18}$$

Photonic Space Basis

The photonic internal spin space involves symmetrized spin states of two photons denoted p and s by Walborn et al., where the p beam is passed into module X while the s beam is passed onto the double slit. Therefore, we have to consider tensor products of the form $|H^p\rangle|H^s\rangle$, $|H^p\rangle|V^s\rangle$, and so on, where we drop the usual tensor product symbol $\otimes$. For clarity, we keep a tensor product symbol between "inner" (SUO) states and "outer" (detector) labstates. In our CA program MAIN, we use the following four states, $\{s[i,n] : i = 1,2,3,4\}$, to serve as an orthonormal basis for the internal photonic degrees of freedom, defined at stage Σ_n by

$$\begin{aligned} |H_n^p\rangle|H_n^s\rangle &\to s[1,n], \quad & |H_n^p\rangle|V_n^s\rangle &\to s[2,n], \\ |V_n^p\rangle|H_n^s\rangle &\to s[3,n], \quad & |V_n^p\rangle|V_n^s\rangle &\to s[4,n]. \end{aligned} \tag{14.19}$$

Double-Slit Polarizers

The two modules P^2 and P^3 have the following active[4] actions on their respective input beams:

$$\text{beam } 2_2 \left\{ \begin{array}{l} P^2|H_2^s\rangle \underset{P^2}{\rightarrow} \quad |L_3^s\rangle = \frac{1}{\sqrt{2}}\left\{|H_3^s\rangle + i|V_3^s\rangle\right\}, \\ P^2|V_2^s\rangle \underset{P^2}{\rightarrow} \quad i|R_3^s\rangle = \frac{1}{\sqrt{2}}\left\{i|H_3^s\rangle + |V_3^s\rangle\right\}, \end{array} \right. \tag{14.20}$$

$$\text{beam } 3_2 \left\{ \begin{array}{l} P^3|H_2^s\rangle \underset{P^3}{\rightarrow} \quad |R_3^s\rangle = \frac{1}{\sqrt{2}}\left\{|H_3^s\rangle - i|V_3^s\rangle\right\}, \\ P^3|V_2^s\rangle \underset{P^3}{\rightarrow} -i|L_3^s\rangle = \frac{1}{\sqrt{2}}\left\{-i|H_3^s\rangle + |V_3^s\rangle\right\}. \end{array} \right. \tag{14.21}$$

The stage dynamics is as follows.

Stage Σ_0 to Stage Σ_1

$$U_{1,0}|\Psi_0\rangle = \frac{1}{\sqrt{2}}\left\{ \underbrace{|H_1^p\rangle|V_1^s\rangle}_{s[2,1]} + \underbrace{|V_1^p\rangle|H_1^s\rangle}_{s[3,1]} \right\} \otimes \underbrace{\widehat{\mathbb{A}}_1^1\widehat{\mathbb{A}}_1^2\mathbf{0}_1}_{a[2^0+2^1,1]\equiv a[3,1]} . \tag{14.22}$$

We show in (14.22) how the various terms are transcribed into the notation used in program MAIN.

Stage Σ_1 to Stage Σ_2

$$U_{2,1}\left\{|H_1^p\rangle|V_1^s\rangle \otimes \widehat{\mathbb{A}}_1^1\widehat{\mathbb{A}}_1^2\mathbf{0}_1\right\} = \frac{1}{\sqrt{2}}|H_2^p\rangle|V_2^s\rangle \otimes \left\{\widehat{\mathbb{A}}_2^1\widehat{\mathbb{A}}_2^2\mathbf{0}_2 + \widehat{\mathbb{A}}_2^1\widehat{\mathbb{A}}_2^3\mathbf{0}_2\right\},$$

$$U_{2,1}\left\{|V_1^p\rangle|H_1^s\rangle \otimes \widehat{\mathbb{A}}_1^1\widehat{\mathbb{A}}_1^2\mathbf{0}_1\right\} = \frac{1}{\sqrt{2}}|V_2^p\rangle|H_2^s\rangle \otimes \left\{\widehat{\mathbb{A}}_2^1\widehat{\mathbb{A}}_2^2\mathbf{0}_2 + \widehat{\mathbb{A}}_2^1\widehat{\mathbb{A}}_2^3\mathbf{0}_2\right\}. \tag{14.23}$$

Stage Σ_2 to Stage Σ_3

Here the modules P^2 and P^3 take effect:

$$U_{3,2}\left\{|H_2^p\rangle|V_2^s\rangle \otimes \widehat{\mathbb{A}}_2^1\widehat{\mathbb{A}}_2^2\mathbf{0}_2\right\} = \frac{1}{\sqrt{2}}\left\{i|H_3^p\rangle|H_3^s\rangle + |H_3^p\rangle|V_3^s\rangle\right\}\widehat{\mathbb{A}}_3^1\widehat{\mathbb{A}}_3^2\mathbf{0}_3,$$

$$U_{3,2}\left\{|V_2^p\rangle|H_2^s\rangle \otimes \widehat{\mathbb{A}}_2^1\widehat{\mathbb{A}}_2^2\mathbf{0}_2\right\} = \frac{1}{\sqrt{2}}\left\{|V_3^p\rangle|H_3^s\rangle + i|V_3^p\rangle \otimes |V_3^s\rangle\right\}\widehat{\mathbb{A}}_3^1\widehat{\mathbb{A}}_3^2\mathbf{0}_3,$$

$$U_{3,2}\left\{|H_2^p\rangle|V_2^s\rangle \otimes \widehat{\mathbb{A}}_2^1\widehat{\mathbb{A}}_2^3\mathbf{0}_2\right\} = \frac{1}{\sqrt{2}}\left\{-i|H_3^p\rangle|H_3^s\rangle + |H_3^p\rangle \otimes |V_3^s\rangle\right\}\widehat{\mathbb{A}}_3^1\widehat{\mathbb{A}}_3^3\mathbf{0}_3,$$

$$U_{3,2}\left\{|V_2^p\rangle|H_2^s\rangle \otimes \widehat{\mathbb{A}}_2^1\widehat{\mathbb{A}}_2^3\mathbf{0}_2\right\} = \frac{1}{\sqrt{2}}\left\{|V_3^p\rangle|H_3^s\rangle - i|V_3^p\rangle \otimes |V_3^s\rangle\right\}\widehat{\mathbb{A}}_3^1\widehat{\mathbb{A}}_3^3\mathbf{0}_3. \tag{14.24}$$

The point here, as stressed by Walborn et al., is that all four photon polarization states on the right-hand side of (14.24) are mutually orthogonal. Therefore,

[4] Here "*active*" means that the changes are physically observable.

no interference is to be expected in any subsequent pattern of signals. To confirm that QDN gives such a conclusion, we need to evolve the total state to the final stage.

Stage Σ_3 to Stage Σ_4

No polarizations are affected on passage through the double slit, so we have

$$U_{4,3}\left\{|s_3^i\rangle \otimes \widehat{\mathbb{A}}_3^1\widehat{\mathbb{A}}_3^2\mathbf{0}_3\right\} = \sum_{j=2}^{K} C^j|s_4^i\rangle \otimes \widehat{\mathbb{A}}_4^1\widehat{\mathbb{A}}_4^j\mathbf{0}_4,$$
$$U_{4,3}\left\{|s_3^i\rangle \otimes \widehat{\mathbb{A}}_3^1\widehat{\mathbb{A}}_3^3\mathbf{0}_3\right\} = \sum_{j=2}^{K} D^j|s_4^i\rangle \otimes \widehat{\mathbb{A}}_4^1\widehat{\mathbb{A}}_4^j\mathbf{0}_4, \qquad (14.25)$$

for $i = 1, 2, 3, 4$. Here the $\{C^j\}$, $\{D^j\}$ coefficients satisfy the usual semi-unitarity relations.

The above information was encoded into program MAIN, with the following results for the nonzero maximal question answers:

$$\Pr\left(\widehat{\mathbb{A}}_4^1\widehat{\mathbb{A}}_4^j\mathbf{0}_4|\Psi_0\right) = \frac{1}{2}|C^j|^2 + \frac{1}{2}|D^j|^2, \qquad j = 2, 3, \ldots, K. \qquad (14.26)$$

This confirms that QDN reproduces the empirical results of Walborn et al. (Walborn et al., 2002). Essentially, placing P^2 and P^3 in front of their respective slits destroys the lack of which-way information observed in the unpolarized double-slit experiment WALBORN-1 discussed above.

MAIN also confirms that $\Pr\left(\widehat{\mathbb{A}}_4^j\mathbf{0}_4|\Psi_0\right) = \Pr\left(\widehat{\mathbb{A}}_4^1\widehat{\mathbb{A}}_4^j\mathbf{0}_4|\Psi_0\right)$, $j = 2, 3, \ldots, K$, that is, that the single signal pattern on the screen shows no interference.

WALBORN-3: Erasure of Which-Path Information

The variants WALBORN-1 and WALBORN-2 confirm standard QM expectations: interference occurs in WALBORN-1 because no which-path information is available, while the modules P^2 and P^3 in WALBORN-2 provide such information by labeling the two beams passing through the double slit, and so there is no interference. The essence of WALBORN-3 is that the module labeled X in Figure 14.5 counteracts the effects of P^2 and P^3 so that interference in correlations is now observed. What is incomprehensible from a classical perspective is that the action of X is nonlocal relative to P^2 and P^3. This is the point made by Wheeler: P^1 and P^2 could be operating on one side of the Universe and X on the other, but the interference destroyed in WALBORN-2 is restored by the action of module X.

The QDN analysis of the total state evolution for WALBORN-3 is identical to that for WALBORN-2 up to stage Σ_2. However, at stage Σ_3, the most suitable internal polarization basis to use is B^{WP} rather than B^{HV}, because X essentially filters out the two elements $|P\rangle$ and $|N\rangle$ of that basis.

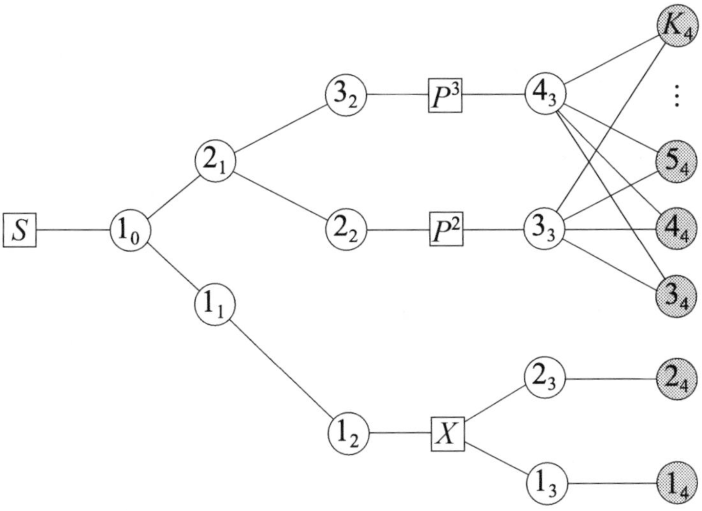

Figure 14.5. WALBORN-3: erasure of which-path information.

The combined actions of P^2, P^3, and X are now given by

$$\text{beam } 2_2 \begin{cases} P^2|H_2^s\rangle \underset{P^2}{\rightarrow} \; |L_3^s\rangle = \frac{(1-i)}{2}\{i|P_3^s\rangle + |N_3^s\rangle\}, \\ P^2|V_2^s\rangle \underset{P^2}{\rightarrow} \; i|R_3^s\rangle = \frac{(i+1)}{2}\{|P_3^s\rangle + i|N_3^s\rangle\}, \end{cases} \tag{14.27}$$

$$\text{beam } 3_2 \begin{cases} P^3|H_2^s\rangle \underset{P^3}{\rightarrow} \; |R_3^s\rangle = \frac{(1-i)}{2}\{|P_3^s\rangle + i|N_3^s\rangle\}, \\ P^3|V_2^s\rangle \underset{P^3}{\rightarrow} -i|L_3^s\rangle = -\frac{(i+1)}{2}\{i|P_3^s\rangle + |N_3^s\rangle\}. \end{cases} \tag{14.28}$$

Module X has the following effects on the beam from 1_2:

$$\text{beam } 1_2 \begin{cases} X|H_2^p\rangle \underset{X}{\rightarrow} \; \frac{1}{\sqrt{2}}\{|P_3^p\rangle + |N_3^p\rangle\} \\ X|V_2^p\rangle \underset{X}{\rightarrow} \; \frac{1}{\sqrt{2}}\{|P_3^p\rangle - |N_3^p\rangle\}. \end{cases} \tag{14.29}$$

Although this looks like a passive basis change, module X has the property of splitting the beam into two separately observable components and this is critical to the experiment. Module X is really an active transformation, with stage Σ_3 output beams denoted 1_3 and 2_3.

Stage Σ_2 to Stage Σ_3

Here the pattern of information flow is intricate, requiring great care in the programming. The difficulty of correctly encoding what the experimentalists had done turned out to generate the most significant test of the QDN formalism: a single sign error would easily invalidate the whole calculation. This underlines what we said at the end of Chapter 12, that there is an obvious need for a more sophisticated process of transcribing stage diagrams into CA code. We find the following transition rules:

$$U_{3,2}\left\{|H_2^p\rangle|V_2^s\rangle\otimes\widehat{\mathbb{A}}_2^1\widehat{\mathbb{A}}_2^2\mathbf{0}_2\right\}=\tfrac{(1+i)}{2\sqrt{2}}\begin{bmatrix}\{|P_3^p\rangle|P_3^s\rangle+i|P_3^p\rangle|N_3^s\rangle\}\widehat{\mathbb{A}}_3^1\widehat{\mathbb{A}}_3^3\mathbf{0}_3+\\ \{|N_3^p\rangle|P_3^p\rangle+i|N_3^p\rangle|N_3^s\rangle\}\widehat{\mathbb{A}}_3^2\widehat{\mathbb{A}}_3^3\mathbf{0}_3\end{bmatrix}, \tag{14.30}$$

$$U_{3,2}\left\{|H_2^p\rangle|V_2^s\rangle\otimes\widehat{\mathbb{A}}_2^1\widehat{\mathbb{A}}_2^3\mathbf{0}_2\right\}=-\tfrac{(1+i)}{2\sqrt{2}}\begin{bmatrix}\{i|P_3^p\rangle|P_3^s\rangle+|P_3^p\rangle|N_3^s\rangle\}\widehat{\mathbb{A}}_3^1\widehat{\mathbb{A}}_3^4\mathbf{0}_3+\\ \{i|N_3^p\rangle|P_3^p\rangle+|N_3^p\rangle|N_3^s\rangle\}\widehat{\mathbb{A}}_3^2\widehat{\mathbb{A}}_3^4\mathbf{0}_3\end{bmatrix}, \tag{14.31}$$

$$U_{3,2}\left\{|V_2^p\rangle|H_2^s\rangle\otimes\widehat{\mathbb{A}}_2^1\widehat{\mathbb{A}}_2^2\mathbf{0}_2\right\}=\tfrac{(1-i)}{2\sqrt{2}}\begin{bmatrix}\{i|P_3^p\rangle|P_3^s\rangle+|P_3^p\rangle|N_3^s\rangle\}\widehat{\mathbb{A}}_3^1\widehat{\mathbb{A}}_3^3\mathbf{0}_3-\\ \{i|N_3^p\rangle|P_3^p\rangle+|N_3^p\rangle|N_3^s\rangle\}\widehat{\mathbb{A}}_3^2\widehat{\mathbb{A}}_3^3\mathbf{0}_3\end{bmatrix}, \tag{14.32}$$

$$U_{3,2}\left\{|V_2^p\rangle|H_2^s\rangle\otimes\widehat{\mathbb{A}}_2^1\widehat{\mathbb{A}}_2^3\mathbf{0}_2\right\}=\tfrac{(1-i)}{2\sqrt{2}}\begin{bmatrix}\{|P_3^p\rangle|P_3^s\rangle+i|P_3^p\rangle|N_3^s\rangle\}\widehat{\mathbb{A}}_3^1\widehat{\mathbb{A}}_3^4\mathbf{0}_3-\\ \{|N_3^p\rangle|P_3^p\rangle+i|N_3^p\rangle|N_3^s\rangle\}\widehat{\mathbb{A}}_3^2\widehat{\mathbb{A}}_3^4\mathbf{0}_3\end{bmatrix}. \tag{14.33}$$

Stage $\boldsymbol{\Sigma_3}$ to Stage $\boldsymbol{\Sigma_4}$

The evolution is given by (14.25), taking into account that detector 2_4 is not associated now with the double-slit detecting screen but is part of the X module output channel detectors.

The above information when fed into program MAIN gives the results:

$$\begin{aligned}\Pr\left(\widehat{\mathbb{A}}_4^1\widehat{\mathbb{A}}_4^j\mathbf{0}_4\right)&=\frac{1}{4}|C^j|^2+\frac{1}{4}|D^j|^2+\frac{i}{4}\left(C^jD^{j*}-C^{j*}D^j\right),\\ \Pr\left(\widehat{\mathbb{A}}_4^2\widehat{\mathbb{A}}_4^j\mathbf{0}_4\right)&=\frac{1}{4}|C^j|^2+\frac{1}{4}|D^j|^2-\frac{i}{4}\left(C^jD^{j*}-C^{j*}D^j\right),\qquad j=3,4,\ldots,K,\end{aligned} \tag{14.34}$$

demonstrating the interference observed by Walborn et al. Moreover, the two alternative output channels, 1_4 and 2_4, show the out-of-phase interference referred to by Walborn et al. as *fringe* and *antifringe*, respectively.

QDN does indeed simulate the empirical results of Walborn et al. The following comments are relevant.

The No-Signaling Theorem Is Vindicated

The action of module X is entirely local, being applied only to beam 1_2. The question is, could anything done by X be observed at the possibly remote screen consisting of detectors $3_4, 5_4, \ldots, K_4$ alone? The answer is emphatically *no*. Specifically, we find the single detector outcome probabilities to be given as

$$\begin{aligned}\Pr\left(\widehat{\mathbb{A}}_4^j\mathbf{0}_4\right)&\equiv\Pr\left(\widehat{\mathbb{A}}_4^1\widehat{\mathbb{A}}_4^j\mathbf{0}_4\right)+\Pr\left(\widehat{\mathbb{A}}_4^2\widehat{\mathbb{A}}_4^j\mathbf{0}_4\right)\\ &=\frac{1}{4}|C^j|^2+\frac{1}{4}|D^j|^2,\ j=3,4,\ldots,K,\end{aligned} \tag{14.35}$$

that is, showing no interference terms.

The *Stage* Concept Is a Valid Model of Empirical Time
Most significantly, Walborn et al. repeated their experiment with the screen and p photon detection order reversed with significant time differences and found no change in the results (Walborn et al., 2002). This is strong evidence for the validity of the stages concept in such quantum process.

14.7 Concluding Remarks

Our analysis supports the notion that QM never actually needs to involve any acausality in order to account for empirical data. We should be worried if it did, for then our entire view of what probability and information represent would need drastic revision.

Detailed QDN analysis reveals the basic fact that interference phenomena involve a lack of information about quantum *states*, and not specifically about particles per se. Conceptual problems arise when our classical conditioning is relied on too much. We would like to believe in photons as particles and we would like to believe that time runs continuously. Both concepts have their uses, but QM seems to require a generalization of both. In the case of the former, experiments tell us that we have to deal with interference of amplitudes, not particles. In the case of the latter, we cannot expect quantum processes to evolve strictly according to an integrable timetable, such as coordinate time, or even the physical time in a laboratory. What is important is whether or not quantum information has been extracted. If it has been placed "on hold," as can be seen in our analysis of the delayed-choice eraser and the double-slit eraser, then it can remain in a stage that could in principle persist until the end of the Universe.

15
Particle Decays

15.1 Introduction

In this chapter we discuss experiments where the run architecture is significantly different from that of standard *in-out* experiments such as particle scattering. We apply the quantized detector network (QDN) formalism to particle decays, the ammonium molecular system, Kaon-type regeneration decay experiments, and quantum Zeno experiments. In all of these experiments, the problem is the modeling of time, which conventionally is taken to be continuous. In QDN, time is treated in terms of stages, which are discrete. We show how the QDN formalism deals with such experiments.

In standard quantum mechanics (QM), time is assumed to be continuous. That is a legacy from classical mechanics (CM), which does not concern itself in general with the processes of observation. CM assumes systems under observation (SUOs) "have" physical properties that are independent of how they are observed. In contrast, QM cannot be considered without a discussion of the processes of observation. On close inspection of any process of observation, *as it is actually carried out in the laboratory and not how it is modeled theoretically*, the continuity of time does not look quite so obvious.

The problem is that there are two mutually exclusive views about the nature of observation in physics. These were discussed in detail by Misra and Sudarshan (MS) in an influential paper on the quantum Zeno effect (Misra and Sudarshan, 1977). On the one hand, no known principle forbids the continuity of time, so the axioms of QM are stated implicitly in terms of continuous time. When the Schrödinger equation is postulated to be one of them (Peres, 1995), temporal continuity is assumed explicitly. On the other hand, it is an empirical fact that no experiment can actually monitor any SUO in a truly continuous way. All references to continuous time measurements are invariably based on statistical modeling of complex processes, with the continuity of time having much the same status as that of temperature. Such effective parameters are extremely useful in physics, but their status as model-dependent, emergent attributes of SUOs, and the apparatus used to observe them, should always be kept in mind.

The best that could be done toward simulating temporal continuity in physics would be to perform a sequence of experiments with a carefully prescribed decreasing measurement time scale, such as occurs in experiments investigating the phenomenon known as the quantum Zeno effect (Itano et al., 1990).

MS analyzed particle decay processes and asked certain questions about them not normally investigated in quantum mechanics. Three of these questions were referred to as P, Q, and R and this convention will be followed here.

$P(t|\Psi)$

This question asks for the probability that an unstable system prepared at time zero in state Ψ has decayed sometime during the interval $[0, t]$.

$Q(t|\Psi)$

This question asks for the probability that the prepared state has *not* decayed during this interval.

$R(t_1, t|\Psi)$

This question asks for the probability that the state has not decayed during the interval $[0, t_1]$, where $0 < t_1 < t$, and has decayed during the interval $[t_1, t]$.

Here we come across an example where mathematics and logic cannot be used to explore physics. We pointed out in Section 2.10 that the validation[1] of the negation $\neg P$ of a physical proposition P cannot always be undertaken by the same apparatus that is used to validate P. The point is that suppose we had used apparatus A_P to answer MS's question $P(t|\Psi)$ by looking for decay products of an unstable SUO and had found no such decay products over any given interval of time. We could not conclude that the SUO was absolutely stable; there could be decay products that our apparatus could not detect. At best we could only say that the SUO was stable *relative* to A_P.

In physics, therefore, we cannot simply assert $Q(t|\Psi) = 1 - P(t|\Psi)$ as an empirical fact, because as we stated in Chapter 2, what are important in physics are generalized propositions, and these require full specification of apparatus. In order to answer $Q(t|\Psi)$, we would have to use apparatus A_Q, which could be very different from A_P.

Likewise, in order to answer MS's question $R(t_1, t|\Psi)$, we would have to use apparatus A_R.

The point made by MS is not quite the same as what we have just made. Our concern is about apparatus, theirs was about time. MS stressed that $P(t|\Psi)$, $Q(t|\Psi)$, and $R(t_1, t|\Psi)$ are not what quantum mechanics normally calculates,

[1] Our convention is that *validation* means the *attempt* to establish the truth of a proposition.

which is the probability distribution of the time at which decay occurs, denoted by T. The difference as they saw it is that the P, Q, and R questions involve a continuous set of observations (according to the standard paradigm), or the nearest practical equivalent of it, during each run of the experiment, whereas T involves a set of repeated runs, each with a one-off observation at a different time to determine whether the particle has decayed or not by that time. Because P, Q, and R involve an experimental architecture different from T, it should be expected that empirical differences might be observed.

Note that the observations referred to by MS to can have negative outcomes; i.e., a failure to detect an expected particle decay in an experiments counts as an observation. The correct statement of such an observation is not that the particle is stable, but that that particular experiment has failed to detect any decay.

MS emphasized the limitations of QM, stressing that although it works excellently in many situations, QM does not readily give a complete picture of experiments probing questions such as P, Q, and R. They concluded that "there is no standard and detailed theory for the actual coupling between quantum systems and the *classical measuring apparatus*" (Misra and Sudarshan, 1977). We fully agree. QDN is a relatively simplified attempt to move toward such a theory.

Our first task in this chapter is to apply QDN to the simplest idealized decay process, a particle decaying via a single channel. The quantum Zeno effect is then discussed. That effect demonstrates that the answer as to whether a system decays while it is being monitored or whether it remains in its initial state depends on the experimental context, i.e., the details of the apparatus and the measurement protocol involved. We follow this by applying QDN to more complex phenomena such as the ammonium molecule and neutral Kaon decay. We show how QDN can readily provide the empirical architecture to describe the spectacular phenomenon of Kaon decay regeneration, originally discussed by Gell-Mann and Pais in standard QM (Gell-Mann and Pais, 1955).

It will be shown that for all of these phenomena, QDN incorporates probability conservation at all levels of the discussion and therefore does not require the introduction of any ad hoc imaginary terms in any energies or the use of non-Hermitian Hamiltonians.

15.2 One Species Decays

In this subsection, we apply QDN to the description of what in standard terminology would be called the decay of an unstable particle, the initial state X of which can decay into some multiparticle state Y. Our aim is to show that the QDN account of such processes readily conserves total probability at all stages. Because the essence of such processes lies in the temporal architecture, the momenta and other attributes of the particles involved will be ignored here, the discussion being designed to illuminate the basic principles of the formalism only. Should such aspects be required, the formalism can readily deal with them by introducing an "internal" Hilbert space associated with the SUO states.

The run architecture follows the pattern used throughout this book, with labstate preparation for each run being completed by an initial stage denoted as Σ_0, and referred to as stage zero. All subsequent stages in that run are counted from stage zero, so Σ_1 is the first stage (after stage zero).

By stage Σ_0 of any given run, the observer will have contextual evidence that they have prepared an X-particle state, in the language of standard QM. This is represented in QDN by the normalized labstate $\mathbf{\Psi}_0 \equiv \widehat{\mathbb{A}}_0^X \mathbf{0}_0$, which we have previously designated a *preparation switch.*

Now consider the first stage, Σ_1, at which the observer has the means to detect any decay. Suppose by that stage, the labstate is now represented by $\mathbf{\Psi}_1$ and given by

$$\mathbf{\Psi}_1 \equiv \mathbb{U}_{1,0}\mathbf{\Psi}_0 = (\alpha\widehat{\mathbb{A}}_1^X + \beta\widehat{\mathbb{A}}_1^{Y_1})\mathbf{0}_1, \quad |\alpha|^2 + |\beta|^2 = 1. \tag{15.1}$$

Here the first term on the right-hand side (RHS) represents the possibility that the particle has *not* decayed, whereas the second term, involving Y, represents the possibility that a decay has occurred.

It is part of the underlying philosophy of QDN that the term in Y in Eq. (15.1) does not model specific details of the Y state. It models a *yes/no* possibility that something has happened. It is an example of a virtual detector rather than a real detector. A virtual detector is informational in character, not necessarily directly identifiable with a specific, real detector in the laboratory, although such things are necessary to establish the context for the labstate Y.

To clarify this point further, suppose that the multiparticle state Y consisted of N identifiable particles. We have *not* modeled here the stage Σ_1 labstate by a term in (15.1) such as $\beta\widehat{\mathbb{A}}_1^{Y_1^1}\widehat{\mathbb{A}}_1^{Y_1^2}\widehat{\mathbb{A}}_1^{Y_1^3}\dots\widehat{\mathbb{A}}_1^{Y_1^N}\mathbf{0}_1$, where for example $\widehat{\mathbb{A}}_1^{Y_1^i}$ would create a signal in a specific detector for decay component particle Y^i at stage Σ_1. We could do that, if we wanted to, however. That would undoubtedly add to the complexity of a problem that already has some degree of complexity in its architecture, so that is a scenario where a computer algebra approach to QDN would be most suitable.

The modeling of the labstate $\mathbf{\Psi}_1$ given by (15.1) does however include some desirable features that we have put in "by hand." We exclude from the RHS of (15.1) the possibility that we find no signals whatsoever at stage Σ_1; that is, we exclude the signal ground state $\mathbf{0}_1$. This means that the apparatus is what we have referred to before as *calibrated*.

In the same spirit, we exclude the signality-two state $\widehat{\mathbb{A}}_1^X\widehat{\mathbb{A}}_1^{Y_1}\mathbf{0}_1$, on the grounds that any run with a labstate consisting of the original particle *and* its decay product would be discounted by the observer as contaminated by external influences (as happens in real experiments).

From (15.1), the amplitude $\mathcal{A}(X_1|X_0)$ for the particle *not* to have decayed by stage Σ_1 is given by

$$\mathcal{A}(X_1|X_0) \equiv \overline{\mathbf{0}_1}\mathbb{A}_1^X \mathbf{\Psi}_1 = \alpha, \tag{15.2}$$

while the amplitude $\mathcal{A}(Y_1|X_0)$ for the particle to have made the transition to state Y by stage Σ_1 is given by

$$\mathcal{A}(Y_1|X_0) \equiv \overline{\mathbf{0}_1}\mathbb{A}_1^{Y_1} \mathbf{\Psi}_1 = \beta. \tag{15.3}$$

Here we have used the fact that $[\widehat{\mathbb{A}}_n^X, \mathbb{A}_n^Y] = 0$, and so on.

From the above, we see that

$$|\mathcal{A}(X_1|X_0)|^2 + |\mathcal{A}(Y_1|X_0)|^2 = 1, \tag{15.4}$$

so total probability is conserved.

The above probabilities can also be calculated directly as expectation values of partial questions. For the probability $\Pr(X_1|X_0)$ of no decay by stage Σ_1, we find $\Pr(X_1|X_0) \equiv \overline{\mathbf{\Psi}_1}\widehat{\mathbb{P}}_1^X \mathbf{\Psi}_1 = |\alpha|^2$, while the probability $\Pr(Y_1|X_0)$ of decay into Y by stage Σ_1 is given by $\Pr(Y_1|X_0) \equiv \overline{\mathbf{\Psi}_1}\, \widehat{\mathbb{P}}_1^{Y_1} \mathbf{\Psi}_1 = |\beta|^2$. Note that these are contextual probabilities: we noted at the previous section that these are statements valid only relative to the detectors used, which are assumed suitable for what they are supposed to detect.

On the RHS of (15.3), the label is Y_1; that is, the decay state label is itself labeled by a temporal subscript, in this case the number 1, which is the stage Σ_1 at which the amplitude is calculated for. This label of a label is significant. It registers the fact that when a detector is triggered, it does so irreversibly. The stage at which this happens is a crucial feature of the analysis, being directly related to the measurement issues discussed by MS (Misra and Sudarshan, 1977). Our architecture is based on monitoring the state of the SUO as much as possible, that is, attempting to perform as good an approximation to continuous-in-time monitoring as our equipment allows.

The above process conserves signality one, so the dynamics can be discussed economically in terms of the evolution of the signal operators rather than the labstates. For instance, evolution from stage Σ_0 to stage Σ_1 can be given in the form

$$\widehat{\mathbb{A}}_0^X \to \mathbb{U}_{1,0}\widehat{\mathbb{A}}_0^X\, \overline{\mathbb{U}}_{1,0} = \alpha\widehat{\mathbb{A}}_1^X + \beta\widehat{\mathbb{A}}_1^{Y_1}, \tag{15.5}$$

where $\mathbb{U}_{1,0}$ is a semi-unitary operator satisfying the rule $\overline{\mathbb{U}}_{1,0}\mathbb{U}_{1,0} = \mathbb{I}_0$, with $\mathbb{I}_0$ being the identity for the initial lab register $\mathcal{Q}_0 \equiv Q_0^X$ and $\overline{\mathbb{U}}_{1,0}$ being the retraction of $\mathbb{U}_{1,0}$.

Process (15.5) involves a change in rank, since $\mathcal{Q}_1 \equiv Q_1^X Q_1^{Y_1}$, but not in signality. Because $\dim \mathcal{Q}_1 > \dim \mathcal{Q}_0$, the evolution operator is properly semi-unitary, that is, satisfies the condition $\mathbb{U}_{1,0}\overline{\mathbb{U}}_{1,0} \neq \mathbb{I}_1$, which is a statement of irreversibility relative to the observer. This is a critical feature of the experiments discussed in this chapter, apart for the ammonium molecule, and is the reason for the apparent loss of probability in conventional Schrödinger wave mechanics descriptions of unstable particles. In those descriptions, a common strategy is to consider only the Hilbert space of the original SUO and add an imaginary term $-i\Gamma$ to energies, thereby forcing wave functions to fall off with time, with the interpretation that this represents particle decay. Where this probability loss goes is left unstated.

In relativistic quantum field theory, such as quantum electrodynamics (QED), the architecture is usually different. There, the Hilbert space is big enough to accommodate particles (such as muons) and their decay products. Decays are treated as scattering problems, with initial undecayed particles coming in at remote negative (infinite past) time and decay products going out at remote positive (infinite future) time. The extraction of decay lifetimes then is usually done in a heuristic manner, usually involving manipulation with symbols that are nominally divergent, such as dividing an amplitude by the four-volume measure of Minkowski spacetime in order to determine a flux decay rate. Our ambition in QDN is to avoid such manipulations while retaining probability conservation.

The QDN description of the next stage of the process, from stage Σ_1 to stage Σ_2, is more subtle and involves a *null test*. Considering the labstate of the above decay process at stage Σ_1, there are now two terms to consider.

No Decay

The first term on the RHS in (15.1), $\alpha\widehat{\mathbb{A}}_1^X\mathbf{0}_1$, corresponds to a *no decay* outcome by stage Σ_1. This potential outcome can now be regarded as a preparation, at stage Σ_1, of an initial X state that could subsequently decay into a Y state or not, with the same dynamical characteristics as for the first temporal link of the run, held between stages Σ_0 and Σ_1. If the measured laboratory time interval $\tau_{10} \equiv t_1 - t_0$ between stages Σ_0 and Σ_1 is the same, within experimental uncertainty, as the measured laboratory time interval $\tau_{21} \equiv t_2 - t_1$ between stages Σ_1 and Σ_2, and so on for subsequent links, then spatial and temporal homogeneity may be assumed, if the apparatus has been set up in the laboratory carefully enough. This will be a physically reasonable assumption in the absence of gravitational fields and in the presence of suitable apparatus.

Decay

The second term, $\beta\widehat{\mathbb{A}}_1^{Y_1}\mathbf{0}_1$, in (15.1) corresponds to *decay having occurred during the first time interval.* Such an outcome is irreversible in this example, but this is not an inevitable assumption in general. Situations where the Y state could revert back to the X state are more complicated but of empirical interest, such as in the ammonium maser and Kaon and B meson decay. These scenarios are discussed later sections in this chapter.

Assuming homogeneity, the next stage of the evolution is given by

$$\widehat{\mathbb{A}}_1^X \to \mathbb{U}_{2,1}\widehat{\mathbb{A}}_1^X\,\overline{\mathbb{U}}_{2,1} = \alpha\widehat{\mathbb{A}}_2^X + \beta\widehat{\mathbb{A}}_2^{Y_2}, \tag{15.6}$$

$$\widehat{\mathbb{A}}_1^{Y_1} \to \mathbb{U}_{2,1}\widehat{\mathbb{A}}_1^{Y_1}\,\overline{\mathbb{U}}_{2,1} = \widehat{\mathbb{A}}_2^{Y_1}. \tag{15.7}$$

Equation (15.7) is justified as follows. The decay term in (15.1), proportional to $\widehat{\mathbb{A}}_1^{Y_1}$ at stage Σ_1, corresponds to the possibility of detecting a decay product state Y at that time. Now there is nothing that requires this information to be extracted precisely at that stage. The experimentalist could choose, or indeed be required, to delay information extraction until some later stage, effectively placing the decay product observation "on hold." The inherent irreversibility of signal detectors means that, as a rule, such signals on hold are not lost.

As stated above, this may be represented in QM by passing a state through a null test, which does not alter it. In QDN this is represented by Eq. (15.7). Essentially, quantum information about a decay is isolated from the rest of the experiment and passed forward in time until it is physically extracted.

The lab register $\mathcal{Q}_2$ at stage Σ_2 has rank three, being the tensor product $\mathcal{Q}_2 \equiv Q_2^X Q_2^{Y_1} Q_2^{Y_2}$. Semi-unitary evolution from stage Σ_0 to stage Σ_2 is still of signality one and is given by

$$\widehat{\mathbb{A}}_0^X \to \mathbb{U}_{2,1}\mathbb{U}_{1,0}\widehat{\mathbb{A}}_0^X\overline{\mathbb{U}}_{1,0}\overline{\mathbb{U}}_{2,1} = \alpha^2\widehat{\mathbb{A}}_2^X + \alpha\beta\widehat{\mathbb{A}}_2^{Y_2} + \beta\widehat{\mathbb{A}}_2^{Y_1}, \tag{15.8}$$

with the various probabilities being read off as the squared moduli of the corresponding terms.

The temporal architecture of this process is given in Figure 15.1. It will be apparent from a close inspection of (15.8) that what appears to look like a space-time description with a specific arrow of time is being built up, with a memory of the change of rank of the lab register at stage Σ_1 being propagated forward in time to stage Σ_2. This is represented by the contribution involving $\widehat{\mathbb{A}}_2^{Y_1}$, which is interpreted as a potential decay process that may have occurred by stage Σ_1 and contributing to the overall labstate amplitude at stage Σ_2.

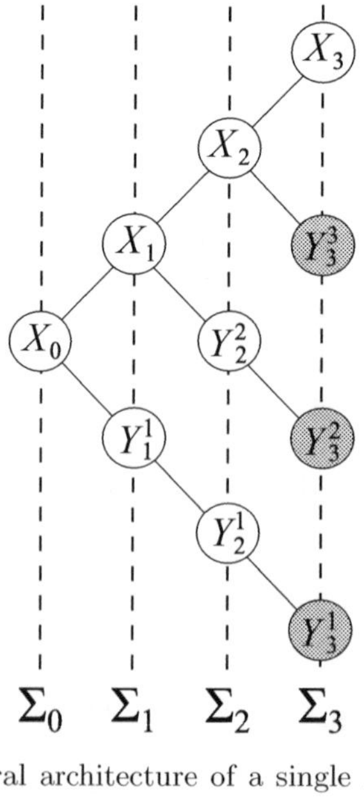

Figure 15.1. The temporal architecture of a single X particle decay experiment.

Subsequently the process continues in an analogous fashion, with the rank of the lab register increasing by one at each time step. By stage Σ_n, assuming homogeneity, the dynamics is given by

$$\widehat{\mathbb{A}}_0^X \to \mathbb{U}_{n,0}\widehat{\mathbb{A}}_0^X\,\overline{\mathbb{U}}_{n,0} = \alpha^n \widehat{\mathbb{A}}_n^X + \beta \sum_{i=1}^{n} \alpha^{i-1} \widehat{\mathbb{A}}_n^{Y_i}, \tag{15.9}$$

where $\mathbb{U}_{n,0} \equiv \mathbb{U}_{n,n-1}\mathbb{U}_{n-1,n-2}\dots\mathbb{U}_{1,0}$ is semi-unitary and satisfies the constraint $\overline{\mathbb{U}}_{n,0}\mathbb{U}_{n,0} = \mathbb{I}_0$.

The amplitude $\mathcal{A}(X, n|X, 0)$ that the original state has *not* decayed by stage Σ_n can be immediately read off the RHS of (15.9) and is given by

$$\mathcal{A}(X, n|X, 0) = \overline{\mathbf{0}}_n \mathbb{A}_n^X \mathbb{U}_{n,0} \widehat{\mathbb{A}}_0^X \mathbf{0}_0 = \alpha^n. \tag{15.10}$$

The probability $\Pr(X, n|X, 0)$ of no decay by stage Σ_n is the squared modulus of this amplitude, so provided $\beta \neq 0$, this probability falls monotonically with increasing n, consistent with our expectations of particle decay. Specifically, if we write $\alpha = e^{i\theta - \frac{1}{2}\Gamma\tau}$, where θ is some real phase, Γ is a characteristic lifetime associated with the decay, and τ is the effective time between successive stages, then we have

$$\Pr(X, n|X, 0) = e^{-\Gamma n \tau}, \tag{15.11}$$

which is the exponential decay form expected with such phenomena.

Commentary

Figure 15.1 does not reveal the full complexity of what is going on. That will be appreciated by the observation that the labstate has signality one at every stage. This means that at any stage, either the original state has not decayed or it has decayed *once*, either at the stage being examined or prior to that stage.

This stage diagram reinforces the message that the Block Universe picture of reality is too simplistic, because that picture is a classical record of what was actually observed and cannot include the future of whatever "now" is being discussed (which is stage Σ_0 here), unless the vacuous assumption is made that the future is single valued and predetermined. That would not be compatible with quantum principles as we know them, however. Figure 15.1 refers to the future of stage Σ_0 and the probability outcomes predicted for the observer by QM, for that stage only; it is not a valid stage diagram for any process time stage after stage Σ_0. The contextuality of stage diagrams underlines the message that physics is contextual, never absolute.

15.3 The Quantum Zeno Effect

The discussion at this point calls for some care with limits, because there arises the theoretical possibility of encountering the so-called *quantum Zeno effect*, in which a carefully monitored state of an unstable SUO appears not to change.

In the following, we will assume that the parameter α in the one-particle decay discussed above satisfies $|\alpha| < 1$, because the case $|\alpha| = 1$ corresponds to a stable particle, which is of no interest here.

Consider the physics of particle decay. Calculated probabilities should be functions of labtime, the clock time used by the observer in the laboratory. Labtime is not assumed here to be a continuous variable on the microscopic level. Instead, it is linked to the scale of time associated with successive stages, and this is determined by the apparatus used. The temporal subscript n in our concept of stages will, when it is so arranged, correspond to a physical interval of time τ, where τ is some reasonably well-defined time scale characteristic of the apparatus.

Certainly, stages need not be strictly regulated in terms of being equally spaced out in time. But in the sort of experiments relevant to this chapter, there will be such an interval τ, and it will be typically a minute fraction of a second, but certainly nowhere near the Planck time scale of 10^{-44} second. Indeed, the conjecture that there is such a Planck time scale remains conjectural and has received some meaningful criticism (Meschini, 2007). The smallest interval currently that has been measured empirically is of the order 10^{-23} second, which is on the shortest hadronic resonance scale, comparable with the time light takes to cross a proton diameter. More realistic measurement scales that could be involved in our discussion directly would probably be electromagnetic in origin, in the range of 10^{-9} to 10^{-18} second. For instance, the shortest controllable time is about 10 attoseconds, that is, about 10^{-17} second (Koke et al., 2010). Experimentalists would generally have a good understanding of what their relevant τ is.

Suppose first that we have some reason to believe that we can relate the transition amplitude α to the characteristic time τ by the rule $|\alpha|^2 \equiv e^{-\Gamma\tau}$, where Γ is a characteristic inverse time introduced to satisfy this relation. Then the survival probability $P(t_n)$ is given by $P(t) \equiv \Pr(X, n|X, 0) = e^{-\Gamma t}$, which is the usual exponential decay formula. No imaginary term proportional to Γ in any supposed Hamiltonian or energy has been introduced in order to obtain exponential decay.

A subtlety may arise here, however. Exponential decay implies that $|\alpha|^2$ is an analytic function of τ with a Taylor expansion of the form

$$|\alpha|^2 = 1 - \Gamma\tau + O\left(\tau^2\right), \tag{15.12}$$

i.e., one with a nonzero linear term. Under such circumstances, the standard result $\lim_{n\to\infty}(1 - x/n)^n = e^{-x}$ leads to the exponential decay law. The possibility remains, however, that the dynamics of the apparatus is such that the linear term in (15.12) is zero, so that the actual expansion is of the form

$$|\alpha|^2 = 1 - \gamma\tau^2 + O\left(\tau^3\right), \tag{15.13}$$

where γ is a positive constant (Itano et al., 1990). Then in the limit $n \to \infty$, where $n\tau \equiv t$ is held fixed, the result is given by

$$\lim_{n\to\infty,\ n\tau=t\text{ fixed}} \left(1-\gamma\tau^2+O\left(\tau^3\right)\right)^n = 1, \tag{15.14}$$

which gives rise to the quantum Zeno effect scenario. An expansion of the amplitude of the form $a = 1+i\mu\tau+\nu\tau^2+O(\tau^3)$ is consistent with (15.13), for example, if μ is real and $\mu^2+\nu+\nu^* < 0$.

To understand properly what is going on, it is necessary to appreciate that there are two competing limits being considered: one where an SUO is being repeatedly observed over an increasingly large macroscopic laboratory time scale $t \equiv n\tau$, and another one where more and more observations are being taken in succession, each separated on a time scale τ that is being brought as close to zero as possible by the experimentalist. In each case, the limit cannot be achieved in the laboratory. The result is that in such experiments, the specific properties of the apparatus and the experimental protocol may play a decisive role in determining the results. If the apparatus is such that (15.12) holds, then exponential decay will be observed, whereas if the apparatus behaves according to the rule (15.13), or any reasonable variant of it, then approximations to the quantum Zeno effect should be observed.

From the QDN perspective, the quantum Zeno effect can be understood from the architecture of decay observation as follows. Looking at Figure 15.1, we see that there is one channel, denoted by circles with an X, that runs across all stages. That channel is the "no decay" channel. If during a run involving a great number of stages the net probability of any of the other outcomes being detected is sufficiently low, then it would appear that the original system was stable. However, given enough stages with a fixed duration τ between each, the decay outcomes would eventually win out. The quantum Zeno effect therefore relies on having as brief a duration τ as possible and finding the critical time scale over which the apparent effect could be observed.

Another way of understanding the quantum Zeno effect is in terms of *environment*. For instance, a free neutron will decay with a mean lifetime of about 880 seconds, whereas inside a nucleus, neutrons are generally stable. We can understand the quantum Zeno effect as the effect of the detection environment on an otherwise unstable system.

15.4 Matrix Analysis

The single-particle decay scenario discussed above can be discussed efficiently in terms of semi-unitary matrices.

Definition 15.1 A semi-unitary matrix M is an $m \times n$ complex matrix such that $M^\dagger M = I_n$, where I_n is the $n \times n$ identity matrix.

Exercise 15.2 Prove that no semi-unitary matrix exists if $m < n$.

Consider the X decay scenario discussed above. If the initial labstate $\boldsymbol{\Psi}_0$ is represented by the 1×1 column vector $[\boldsymbol{\Psi}_0] \equiv [1]$, then the action of $\mathbb{U}_{1,0}$ acting on that labstate $\boldsymbol{\Psi}_0$ given by (15.1) may be represented by the action of the 2×1 semi-unitary matrix $U_{1,0} \equiv \begin{bmatrix} \alpha \\ \beta \end{bmatrix}$ acting on $[\boldsymbol{\Psi}_0]$, giving

$$[\boldsymbol{\Psi}_1] \equiv U_{1,0}[\boldsymbol{\Psi}_0] = \begin{bmatrix} \alpha \\ \beta \end{bmatrix} [1] = \begin{bmatrix} \alpha \\ \beta \end{bmatrix}. \tag{15.15}$$

The two required transition amplitudes at stage Σ_1 are just the two components of this vector.

Continuing this process to the next stage, we deduce that the labstate $[\boldsymbol{\Psi}_2]$ at stage Σ_2 is represented by the action of the semi-unitary matrix $U_{2,1} \equiv \begin{bmatrix} \alpha & 0 \\ \beta & 0 \\ 0 & 1 \end{bmatrix}$ on $[\boldsymbol{\Psi}_1]$, giving

$$[\boldsymbol{\Psi}_2] \equiv U_{2,1}[\boldsymbol{\Psi}_1] = \begin{bmatrix} \alpha & 0 \\ \beta & 0 \\ 0 & 1 \end{bmatrix} \begin{bmatrix} \alpha \\ \beta \end{bmatrix} = \begin{bmatrix} \alpha^2 \\ \alpha\beta \\ \beta \end{bmatrix}. \tag{15.16}$$

For $n > 1$, the relevant semi-unitary matrix is an $(n+1) \times n$ matrix given by

$$U_{n+1,n} = \begin{bmatrix} \alpha & \boldsymbol{\theta}_{n-1}^T \\ \beta & \boldsymbol{\theta}_{n-1}^T \\ \boldsymbol{\theta}_{n-1} & I_{n-1} \end{bmatrix}, \tag{15.17}$$

where $\boldsymbol{\theta}_n$ is a column of n zeros, $\boldsymbol{\theta}_n^T$ is its transpose, and I_n is the $n \times n$ identity matrix. This leads to the final state $[\boldsymbol{\Psi}_n]$ at stage Σ_n:

$$[\boldsymbol{\Psi}_n] = U_{n,n-1}U_{n-1,n-2}\dots U_{1,0}[\boldsymbol{\Psi}_0] = \begin{bmatrix} \alpha^n \\ \beta\alpha^{n-1} \\ \vdots \\ \beta\alpha \\ \beta \end{bmatrix}. \tag{15.18}$$

The squared modulus of the first component of this column vector gives the same survival probability $|\alpha|^{2n}$ as before. It is also easy to read off all the other transition amplitudes and from them determine discrete time versions of the P, Q, and R functions discussed by MS (Misra and Sudarshan, 1977).

Although the QDN analysis gives results that look formally like the standard decay result, the scenario involved is equivalent to that discussed by MS; namely, there is a constant questioning (or its discrete equivalent) by the apparatus as to whether decay has taken place or not. In this case the results are simple. For Kaon and B meson decays, the results are more complicated.

15.5 The Ammonium System

In order to understand the QDN approach to neutral Kaon decay, discussed in the next section, it will be necessary to review first how stable systems such as the ammonium molecule are dealt with in our formalism.

The ammonium molecule consists of three hydrogen atoms and one nitrogen atom. If we ignore molecular rotation and translation as inessential to this argument, then we can think of the three hydrogen atoms as defining a plane in three dimensions. Then the nitrogen can be found on either side of this plane. What is observed is consistent with the classical explanation that the nitrogen oscillates from one side of this plane to the other and back with a characteristic frequency. It is this behavior that is the focus of our discussion here.

The Standard QM Account

With the above assumptions about neglecting rotation and translation, a simple but effective model of the ammonium molecule is described quantum mechanically as a superposition of two orthonormal states, $|X\rangle$ and $|Y\rangle$, each of which represents one of the two possible position states of the single nitrogen atom relative to the plane defined by the three hydrogen atoms. These two states form a basis for a two-dimensional Hilbert space $\mathcal{H}_{AM}$, in other words, a qubit. It is interesting to note that we have encountered here a naturally occurring preferred basis for a qubit, that is, one that is dictated not by detector apparatus but by the assumed geometry of the SUO.

An account of the nonrelativistic quantum theory for ammonium is given by Feynman et al. (1966) so we give only a simplified brief resume here to set the scene.

It is most convenient to use a matrix representation for the states and the Hamiltonian operator. We define the preferred basis representation

$$|X\rangle \underset{R}{=} \begin{bmatrix} 1 \\ 0 \end{bmatrix}, \quad |Y\rangle \underset{R}{=} \begin{bmatrix} 0 \\ 1 \end{bmatrix}. \tag{15.19}$$

Then relative to this representation, the Hamiltonian for the system is given by the Hermitian matrix

$$H \underset{R}{=} \begin{bmatrix} e & f \\ f^* & e \end{bmatrix}, \tag{15.20}$$

where e is real and f can be complex. When f is zero, the two states are degenerate energy eigenstates and so are stable. This possibility is of no in interest here, so we shall assume that $f = |f|e^{i\phi}$, where $|f|$ is nonzero and ϕ is a constant phase.

The two eigenvalues of H are $E^{\pm} \equiv e \pm |f|$ with corresponding normalized energy eigenstates

$$|\pm\rangle \underset{R}{=} \frac{1}{\sqrt{2}} \begin{bmatrix} e^{i\phi} \\ \pm 1 \end{bmatrix}, \tag{15.21}$$

where we have set the arbitrary phases to zero for convenience. Hence an arbitrary normalized solution to the Schrödinger equation

$$i\hbar \frac{d}{dt}|\Psi, t\rangle = H|\Psi, t\rangle \tag{15.22}$$

has matrix representation

$$|\Psi, t\rangle \underset{R}{=} ae^{-iE^+t/\hbar}|+\rangle + be^{-iE^-t/\hbar}|-\rangle, \tag{15.23}$$

where $|a|^2 + |b|^2 = 1$.

If now we calculate the probability $\Pr(X, t|\Psi, 0) \equiv \langle X|\Psi, t\rangle$ that the SUO be found in state $|X\rangle$ at time $t > 0$ we find

$$\Pr(X, t|\Psi, 0) = \tfrac{1}{2} + |ab|\cos(\beta - \alpha + 2|f|t/\hbar), \tag{15.24}$$

where $a = |a|e^{i\alpha}$, $b = |b|e^{i\beta}$. Similarly, we find

$$\Pr(Y, t|\Psi, 0) = \tfrac{1}{2} - |ab|\cos(\beta - \alpha + 2|f|t/\hbar). \tag{15.25}$$

These probabilities successfully model our expectations. First, if f is zero, then these probabilities are fixed. Second, if the initial state is prepared in an energy eigenstate to begin with, which means setting either a or b to zero, then the probability of finding the SUO in state $|X\rangle$ is the same as that of finding it in state $|Y\rangle$.

The QDN Account

In the QDN description, the temporal architecture is given by Figure 15.2. It is assumed that there are two different detectable signal states, X, Y, with signal operators $\widehat{\mathbb{A}}_n^X$, $\widehat{\mathbb{A}}_n^Y$, respectively, evolving from stage Σ_n to stage Σ_{n+1} according to the rule

$$\begin{aligned} \mathbb{U}_{n+1,n}\widehat{\mathbb{A}}_n^X \mathbf{0}_n &= \{A\widehat{\mathbb{A}}_{n+1}^X + B\widehat{\mathbb{A}}_{n+1}^Y\}\mathbf{0}_{n+1}, \\ \mathbb{U}_{n+1,n}\widehat{\mathbb{A}}_n^Y \mathbf{0}_n &= \{C\widehat{\mathbb{A}}_{n+1}^X + D\widehat{\mathbb{A}}_{n+1}^Y\}\mathbf{0}_{n+1}, \end{aligned} \tag{15.26}$$

where $\mathbb{U}_{n+1,n}$ is a semi-unitary operator and A, B, C, and D are constants determined by the dynamics of the situation. Semi-unitarity requires the constraints

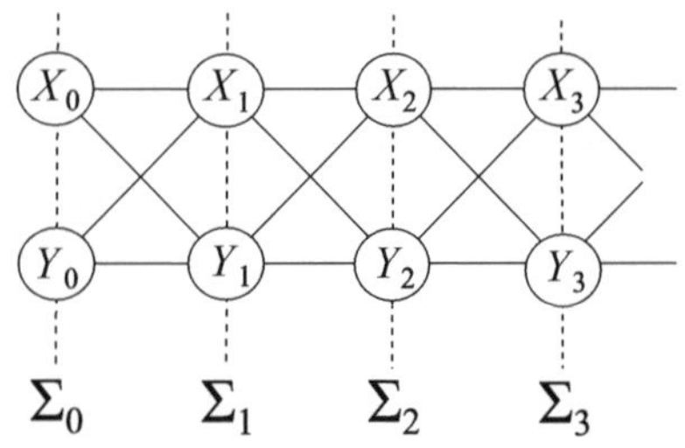

Figure 15.2. The temporal architecture of the ammonium molecule, with the two orthogonal states denoted X and Y.

$|A|^2+|B|^2=|C|^2+|D|^2=1$, $A^*C+B^*D=0$, which is equivalent to standard unitarity in quantum mechanics in this case, because the rank of the quantum register is constant from stage to stage. All other states will be disregarded on the basis that there are no dynamical channels between them and states X, Y.

From (15.26) we can find a dyadic representation for $\mathbb{U}_{n+1,n}$ in the form

$$\mathbb{U}_{n+1,n} = \mathbf{\Phi}^T_{n+1} U \overline{\mathbf{\Phi}_n}, \tag{15.27}$$

where

$$\mathbf{\Phi}^T_{n+1} \equiv [\widehat{\mathbb{A}}^X_{n+1}\mathbf{0}_{n+1}, \widehat{\mathbb{A}}^Y_{n+1}\mathbf{0}_{n+1}], \qquad \overline{\mathbf{\Phi}_n} \equiv \begin{bmatrix} \overline{\mathbf{0}_n}\mathbb{A}^X_n \\ \overline{\mathbf{0}_n}\mathbb{A}^Y_n \end{bmatrix}, \tag{15.28}$$

and U is the unitary matrix

$$U \equiv \begin{bmatrix} A & C \\ B & D \end{bmatrix}. \tag{15.29}$$

The retraction $\overline{\mathbb{U}_{n+1,n}}$ of $\mathbb{U}_{n+1,n}$ is given by

$$\overline{\mathbb{U}_{n+1,n}} = \mathbf{\Phi}^T_n U^\dagger \overline{\mathbf{\Phi}_{n+1}} \tag{15.30}$$

and satisfies the relation $\overline{\mathbb{U}_{n+1,n}}\mathbb{U}_{n+1,n} = \mathbb{I}^c_n$, where $\mathbb{I}^c_n$ is the contextual identity operator

$$\mathbb{I}^c_n \equiv \widehat{\mathbb{A}}^X_n \mathbf{0}_n \overline{\mathbf{0}_n} \mathbb{A}^X_n + \widehat{\mathbb{A}}^Y_n \mathbf{0}_n \overline{\mathbf{0}_n} \mathbb{A}^Y_n \tag{15.31}$$

for the contextual subspace $\mathcal{Q}^c_n$ with orthonormal basis $\{\widehat{\mathbb{A}}^X_n\mathbf{0}_n, \widehat{\mathbb{A}}^Y_n\mathbf{0}_n\}$.

The form (15.27) is particularly suitable for finding the evolution operator $\mathbb{U}_{N,0}$ taking states of the SUO from stage Σ_0 to some final stage Σ_N. We find

$$\mathbb{U}_{N,0} = \mathbf{\Phi}^T_N U^N \overline{\mathbf{\Phi}_0}, \qquad N = 0, 1, 2, \ldots \tag{15.32}$$

The problem therefore reduces to finding U^N, which we do as follows.

As discussed in Section 11.5, a unitary matrix U such as (15.29) can always be put in standard form, that is,

$$U = e^{i\eta} \begin{bmatrix} a & -b^* \\ b & a^* \end{bmatrix}, \tag{15.33}$$

where η is real and a and b satisfy the condition $|a|^2+|b|^2=1$.

Exercise 15.3 Find expressions for η, a, and b in terms of A, B, C, and D.

We now state without proof that matrix U can be written in the form

$$U = e^{i\eta} V \Lambda V^\dagger, \tag{15.34}$$

where matrix V is a unitary matrix given by

$$V \equiv \begin{bmatrix} u & -v^* \\ v & u^* \end{bmatrix}, \tag{15.35}$$

where $|u|^2 + |v|^2| = 1$ and Λ is the diagonal matrix

$$\Lambda \equiv \begin{bmatrix} \lambda^+ & 0 \\ 0 & \lambda^- \end{bmatrix}. \tag{15.36}$$

Here λ^+, λ^- are the two eigenvalues of U and so are the roots of the equation $\lambda^2 - (a + a^*)\lambda + 1 = 0$. These roots are complex conjugates of each other and have magnitude one, so we can write $\lambda^+ = e^{i\theta}$, $\lambda^- = e^{-i\theta}$ for some real angle θ. The relations between u, v and a, b are

$$a = |u|^2 e^{i\theta} + |v|^2 e^{-i\theta}, \quad b = u^* v(e^{i\theta} - e^{-i\theta}), \tag{15.37}$$

noting that these are nonlinear in u and v. Indeed, u and v are not unique: we can multiply each by an element of the unit circle without changing relations (15.37).

The significance of the form (15.34) is that it is now easy to evaluate powers of U. Specifically, we find

$$\mathbb{U}_{n,0} = e^{i\eta n} \mathbf{\Phi}_n^T V \Lambda^n V^\dagger \overline{\mathbf{\Phi}_0}. \tag{15.38}$$

Applying this to the evolution of the signal operators then gives, modulo a phase factor,

$$\begin{aligned} \widehat{\mathbb{A}}_0^X &\to \{|u|^2 e^{in\theta} + |v|^2 e^{-in\theta}\} \widehat{\mathbb{A}}_n^X + u^* v\{e^{in\theta} - e^{-in\theta}\} \widehat{\mathbb{A}}_n^Y, \\ \widehat{\mathbb{A}}_0^Y &\to uv^*\{e^{in\theta} - e^{-in\theta}\} \widehat{\mathbb{A}}_n^X + \left\{|u|^2 e^{-in\theta} + |v|^2 e^{in\theta}\right\} \widehat{\mathbb{A}}_n^Y. \end{aligned} \tag{15.39}$$

Hence we find the conditional probabilities

$$\begin{aligned} \Pr(X, n|X, 0) = \Pr(Y, n|Y, 0) &= |u|^4 + |v|^4 + 2|u|^2|v|^2 \cos(2n\theta), \\ \Pr(Y, n|X, 0) = \Pr(X, n|Y, 0) &= 4|u|^2|v|^2 \sin^2(n\theta), \end{aligned} \tag{15.40}$$

which agrees with the QM expressions (15.24) and (15.25) when the parameters u, v, and θ are chosen suitably.

It was noted by Itano et al. (1990) that a survival probability of the form $P(\tau) \sim 1 - \gamma\tau^2 + O(\tau^3)$ would be needed to make observations of the quantum Zeno effect viable. The above calculation of the ammonium survival probabilities is compatible with this, as can be seen from the expansion

$$\begin{aligned} \Pr(X, n|X, 0) &= |u|^4 + |v|^4 + 2|u|^2|v|^2 \cos(2n\theta) \\ &\sim 1 - 4|u|^2|v|^2 n^2\theta^2 + O\left(n^4\theta^4\right). \end{aligned} \tag{15.41}$$

Therefore, it is predicted that the quantum Zeno effect (or at least behavior analogous to it) should be observable in the ammonium system, if the right experimental conditions are set up. As with the particle decays discussed in the previous section, it would be necessary to ensure that the two limits, $t \to \infty$, $\tau \to 0$, were carefully balanced.

15.6 Kaon-type Decays

The explanation by Gell-Mann and Pais (Gell-Mann and Pais, 1955) of the phenomenon of regeneration in neutral Kaon decay was a successful application of

QM to particle physics. In the standard calculation (Feynman et al., 1966), a non-Hermitian Hamiltonian is used to introduce the two decay parameters needed to describe the observations. We will show that QDN readily reproduces the results of the Gell-Mann and Pais calculation while conserving total probability and without the introduction of any complex energies.

The analysis of the Kaon system is more complicated than the single-particle decay process discussed above, involving the interplay of two distinct neutral Kaons, the K^0 and its antiparticle, the $\bar{K}^0$. In that respect, our discussion in the previous section of the ammonium molecule is useful. The QDN discussion of neutral Kaon decay goes as follows (Jaroszkiewicz, 2008b).

Consider three different particle states, X, Y, and Z, making transitions between each other in the specific way described below. An important example of such behavior in particle physics involves the neutral Kaons, with X representing a K^0 meson, Y representing a $\overline{K}^0$ meson, and Z representing their various decay products. Kaon decay is remarkable for the phenomenon of regeneration, in which the Kaon survival probabilities fall and then rise with time. A similar phenomenon has been observed in B meson decay (Karyotakis and de Monchenault, 2002).

As before, attention can be restricted to signality-one states. The dynamics is described by the transition rules

$$\widehat{\mathbb{A}}_n^X \to \alpha\widehat{\mathbb{A}}_{n+1}^X + \beta\widehat{\mathbb{A}}_{n+1}^Y + \gamma\widehat{\mathbb{A}}_{n+1}^{Z^{n+1}}, \tag{15.42}$$

$$\widehat{\mathbb{A}}_n^Y \to u\widehat{\mathbb{A}}_{n+1}^X + v\widehat{\mathbb{A}}_{n+1}^Y + w\widehat{\mathbb{A}}_{n+1}^{Z^{n+1}}, \tag{15.43}$$

$$\widehat{\mathbb{A}}_n^{Z^a} \to \widehat{\mathbb{A}}_{n+1}^{Z^a}, \quad a = 1, 2, \ldots, n \tag{15.44}$$

where semi-unitarity requires the transition coefficients to satisfy the constraints

$$|\alpha|^2 + |\beta|^2 + \gamma|^2 = |u|^2 + |v|^2 + |w|^2 = 1, \quad \alpha^* u + \beta^* v + \gamma^* w = 0. \tag{15.45}$$

The above process is a combination of the decay and oscillation processes discussed in previous sections.

The temporal architecture is given by Figure 15.3: the dynamics given by (15.42)–(15.44) rules out transitions from Z states to either X or Y states. Therefore, once a Z state is created, it remains a Z state, so there is an irreversible

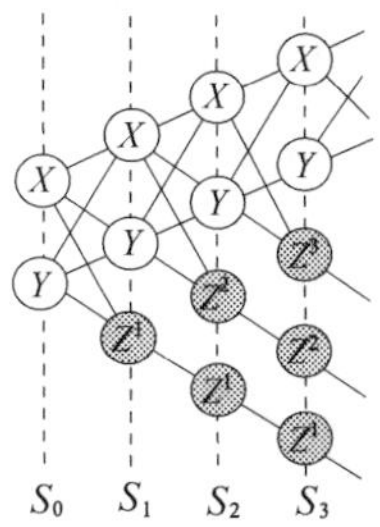

Figure 15.3. The temporal architecture of Kaon decay.

flow from the X and Y states and so these eventually disappear. Before that occurs, however, there will be back-and-forth transitions between the X and Y states that give rise to the phenomenon of regeneration.

In actual Kaon decay experiments, pure K^0 states can be prepared via the strong interaction process $\pi^- + p \rightarrow K^0 + \Lambda$, while pure $\overline{K}^0$ states can be prepared via the process $\pi^+ + p \rightarrow K^+ + \overline{K}^0 + p$. In our notation, these preparations correspond to initial labstates $\widehat{\mathbb{A}}_0^X \mathbf{0}_n$ and $\widehat{\mathbb{A}}_0^Y \mathbf{0}_0$, respectively. In practice, superpositions of K^0 and $\overline{K}^0$ states may be difficult to prepare directly, but the analysis of Gell-Mann and Pais shows that such states could in principle be prepared indirectly (Gell-Mann and Pais, 1955). Therefore, labstates corresponding to X and Y superpositions are physically meaningful and will be used in the following analysis.

Consider an initial labstate of the form $\mathbf{\Psi}_0 \equiv \left\{ x_0 \widehat{\mathbb{A}}_0^X + y_0 \widehat{\mathbb{A}}_0^Y \right\} \mathbf{0}_0$, where $|x_0|^2 + |y_0|^2 = 1$. Matrix methods are appropriate here. The dynamics of the system will be discussed in terms of the initial column vector $\underline{\Psi}_0 \equiv [\ x_0 \quad y_0\ ]^T$, equivalent to the statement that each run of the experiment starts with the rank-two lab register $\mathcal{Q}_0 \equiv Q_0^X Q_0^Y$. The dynamical rules (15.42)–(15.44) map labstates in $\mathcal{Q}_0$ into $\mathcal{Q}_1 \equiv Q_1^X Q_1^Y Q_1^{Z^1}$, so there is a change of rank from two to three. The transition is represented by the semi-unitary matrix

$$\mathsf{U}_{1,0} \equiv \begin{bmatrix} \alpha & u \\ \beta & v \\ \gamma & w \end{bmatrix}, \tag{15.46}$$

which subsequently generalizes to

$$\mathsf{U}_{n+1,n} \equiv \begin{bmatrix} \alpha & u & \underline{0}_n^T \\ \beta & v & \underline{0}_n^T \\ \gamma & w & \underline{0}_n^T \\ \underline{0}_n & \underline{0}_n & I_n \end{bmatrix}, \quad n > 0, \tag{15.47}$$

where I_n is the $n \times n$ identity matrix and $\underline{0}_n$ is a column of n zeros. The observer's detector array increases rank by one over each time step. The state at stage Σ_n is represented by a column vector $\underline{\Psi}_n$ with $n+2$ components, given by $\underline{\Psi}_n = \mathsf{U}_{n,n-1}\mathsf{U}_{n-1,n-2}\dots\mathsf{U}_{2,1}\mathsf{U}_{1,0}\underline{\Psi}_0$. Overall probability is conserved, because of the semi-unitarity of the transition operators.

As before, the key to unraveling the dynamics is linearity, which is guaranteed by the use of semi-unitary evolution operators. Suppose the state $\underline{\Psi}_n$ at time n is represented by

$$\underline{\Psi}_n = [x_n, y_n, z_n^n, \dots, z_n^1]^T, \tag{15.48}$$

where the components x_n and y_n are such that $x_n = \lambda^n x_0$ and $y_n = \lambda^n y_0$, where λ is some complex number to be determined. Such states will be referred to as *eigenmodes*. They are not eigenstates of any physical operator, but their first two components, x_n and y_n, behave as if they were. The dynamics gives the relations $x_{n+1} = \alpha x_n + u y_n = \lambda x_n$, $y_{n+1} = \beta x_n + v y_n = \lambda y_n$, and $z_{n+1}^{n+1} = \gamma x_n + w y_n$.

Experimentalists will be interested principally in survival probabilities for the X and Y states, so the dynamics of Z states will be ignored here; i.e., the behavior of the components z_n^k for $k < n$ will not be discussed. Clearly, however, the QDN formalism is capable of giving much more specific details about the process than just the X and Y survival probabilities.

It will be seen from the above that λ is an eigenvalue of the matrix

$$\begin{bmatrix} \alpha & u \\ \beta & v \end{bmatrix},$$

which means that in principle there are two solutions, λ^+ and λ^-, for the eigenmode values, given by $\lambda^\pm = \frac{1}{2}\{\alpha + v \pm \sqrt{(\alpha - v)^2 + 4\beta u}\}$. It is expected that these will not be mutual complex conjugates in actual experiments, because if they were, the analysis could not explain observed Kaon physics. Therefore, the coefficients α, β, u, and v will be such that the above two eigenmode values are complex and of different magnitude and phase, giving rise to two decay channels with different lifetimes, as happens in neutral Kaon decay. In the quantum mechanics analysis of neutral Kaon decays, Gell-Mann and Pais described the neutral Kaons as superpositions of two hypothetical particles known as K_1^0 and K_2^0, which are charge-parity eigenstates and have different decay lifetimes (Gell-Mann and Pais, 1955). The K_1^0 decays to a two-pion state with a lifetime of about 0.9×10^{-10} second, while the K_2^0 decays to a three-pion state with a lifetime of about 0.5×10^{-7} second.

Semi-unitarity guarantees that

$$|x_{n+1}|^2 + |y_{n+1}|^2 + |z_{n+1}^{n+1}|^2 = |x_n|^2 + |y_n|^2, \tag{15.49}$$

and so it can be deduced that

$$|\lambda|^2 = 1 - \frac{|z_{n+1}^{n+1}|^2}{|x_n|^2 + |y_n|^2} < 1, \qquad n = 0, 1, 2, \ldots, \tag{15.50}$$

given $|x_n|^2 + |y_n|^2 > 0$. From this and the conditions $x_n = \lambda^n x_0$, $y_n = \lambda^n y_0$, the eigenmode values can be written in the form $\lambda_2 \equiv \lambda^+ = \rho_1 e^{i\theta_1}$, $\lambda_2 \equiv \lambda^- = \rho_2 e^{i\theta_2}$, where $0 < \rho_1, \rho_2 < 1$ and θ_1 and θ_2 are real. The eigenmodes at time $t = 0$ corresponding to λ_1 and λ_2 will be denoted by $\underline{\Lambda}_{1,0}$ and $\underline{\Lambda}_{2,0}$ respectively, i.e. $\underline{\Lambda}_{1,0} = [\ a_1 \quad b_1\ ]^T$, $\underline{\Lambda}_{2,0} = [\ a_2 \quad b_2\ ]^T$, and then the evolution rules give

$$\underline{\Lambda}_{1,n} = \begin{bmatrix} \lambda_1^n a_1 \\ \lambda_1^n b_1 \\ c_{n,n} \\ \vdots \\ c_{1,n} \end{bmatrix}, \quad \underline{\Lambda}_{2,n} = \begin{bmatrix} \lambda_2^n a_2 \\ \lambda_2^n b_2 \\ d_{n,n} \\ \vdots \\ d_{1,n} \end{bmatrix}, \tag{15.51}$$

where the coefficients $\{c_{k,n}\}, \{d_{k,n}\}$ can be determined from the dynamics. The initial modes $\underline{\Lambda}_{1,0}$ and $\underline{\Lambda}_{2,0}$ are linearly independent provided λ_1 and λ_2 are different. Given that, then any initial labstate $\underline{\Psi}_0$ can be expressed uniquely as a normalized linear combination of $\underline{\Lambda}_{1,0}$ and $\underline{\Lambda}_{2,0}$, i.e., $\underline{\Psi}_0 = \mu_1\underline{\Lambda}_{1,0} + \mu_2\underline{\Lambda}_{2,0}$, for some coefficients μ_1 and μ_2. This is the analogue of the decompositions

$$
\begin{aligned}
|K^0\rangle &= \{|K_1^0\rangle + |K_2^0\rangle\}/\sqrt{2}, \\
|\overline{K}^0\rangle &= \{|K_1^0\rangle - |K_2^0\rangle\}/\sqrt{2}
\end{aligned} \tag{15.52}
$$

in the Gell-Mann and Pais approach.

From this, the amplitude $\mathcal{A}(X, n|\Psi, 0)$ to find an X signal at time n is given by

$$
\mathcal{A}(X, n|\Psi, 0) = \mu_1 a_1 \lambda_1^n + \mu_2 a_2 \lambda_2^n, \tag{15.53}
$$

so that the survival probability for X is given by

$$
\begin{aligned}
\Pr(X, n|\Psi, 0) = {} & |\mu_1|^2 |a_1|^2 \rho_1^{2n} + |\mu_2|^2 |a_2|^2 \rho_2^{2n} \\
& + 2\rho_1^n \rho_2^n Re\{\mu_1^* \mu_2 a_1^* a_2 e^{-i(\theta_1 - \theta_2)}\},
\end{aligned} \tag{15.54}
$$

and similarly for $\Pr(Y, n|\Psi, 0)$.

There is scope here for various limits to be considered, as discussed in the single-channel decay analysis, such that either particle decay is seen or the quantum Zeno effect appears to hold over limited time spans. If we are justified on empirical grounds in writing $\rho_1^n \equiv e^{-\Gamma_1 t/2}$, $\rho_2^n \equiv e^{-\Gamma_2 t/2}$, where $t \equiv n\tau$ and Γ_1, Γ_2 correspond to long and short lifetime decay parameters, respectively, then the various constants can always be chosen to get full agreement with the standard Kaon survival intensity functions

$$
\begin{aligned}
I(K^0) &= (e^{-\Gamma_1 t} + e^{-\Gamma_2 t} + 2e^{-(\Gamma_1+\Gamma_2)t/2} \cos \Delta m t)/4, \\
I(\overline{K}^0) &= (e^{-\Gamma_1 t} + e^{-\Gamma_2 t} - 2e^{-(\Gamma_1+\Gamma_2)t/2} \cos \Delta m t)/4,
\end{aligned} \tag{15.55}
$$

for pure K^0 decays. Here Δm is proportional to the proposed mass difference between the hypothetical K_1^0 and K_2^0 "particles," which are each charge-parity eigenstates and are supposed to have charge-parity–conserving decay channels. From the QDN approach, such objects need not exist. Instead, they are regarded as manifestations of different possible superpositions of K^0 and $\overline{K}^0$ labstates, each of which is physically realizable via the strong interactions, as mentioned above. Conversely, the apparatus dynamics may be such that quantum Zeno-type effects are observed instead of long-term decays. Again, this will depend on the details of the experiment chosen.

16

Nonlocality

16.1 Introduction

Our concern in this chapter is *locality* in quantum mechanics (QM). Locality is a heuristic physics principle based on the following propositions.

No Action-at-a-Distance

All the evidence points to the principle that physical actions, taken within restricted (localized) regions of space and time by observers or other agencies such as systems under observation (SUOs), do not cause instantly observable effects on other SUOs at large distances. This does not apply to mathematical/metaphysical concepts such as quantum wave functions or correlations, as these are conceptual objects (Scarani et al., 2000). Statements about instantaneous wave function collapse are vacuous (have no empirical significance) and are therefore *not* an issue of significance in physics. Such statements *are* an issue to theorists who objectivize wave functions, as in Hidden Variables (HV) theory.

Action-at-a-distance is generally regarded as anathema by most physicists. For example, Newton's law of universal gravitation is well known for mathematically encoding action-at-a-distance. There is direct evidence, however, in the form of a letter written by Newton to Bentley, that Newton believed that gravity acting "at a distance through a vacuum without the mediation of anything else" was an absurdity (Newton, 2006).

The *no action-at-a-distance* principle is encoded in quantized detector networks (QDN) by the requirement that labstate preparation and consequent signal detection never occur at the same stage.

Causal Transmission

All physically observable consequences of local actions taken by an observer or SUO are transmitted by identifiable physical processes, such as electromagnetic waves or neutrinos. There is no such thing as magic or *action-at-a-distance.*

In QDN this proposition is taken into account implicitly in the labstate outcome amplitudes at each stage, as these model how information is propagated from stage to stage. When necessary, the information void can be modeled as if there were fields and/or particles propagating through it, giving scope for different mathematical models, such as Euclidean space, curved spacetime, noncommuting spacetimes, and so on. The structure of the information void is essentially a discussion of whatever modules have to be taken into account between labstate preparation and signal detection.

No Superluminosity

According to the standard principles of relativity, the speed of transmission of any observed physical effect is never greater than the local speed of light, as measured in a standard localized laboratory. To date, no particles or signals that travel faster than the speed of light (tachyons) have been observed.

A necessary, but not sufficient, condition for QM to respect the no-superluminosity principle is that the *no-communication theorem* holds. This theorem in standard QM and in QDN is discussed at the end of this chapter.

The no-communication theorem is insufficient in relativistic QM because it does not mention the speed of light. In relativistic quantum field theory (RQFT), the theorem is replaced by the above-mentioned no-superluminosity principle directly, which involves *lightcones*. These have a special place in physics, having both emergent and reductionist aspects.

Lightcones are obviously emergent structures because they define macroscopic subsets of spacetime consisting of events that are all either time-like or space-like relative to a given event (identified with the vertex of a given lightcone). On the other hand, lightcones are intimately involved in the reductionist formulation of RQFT, such as the postulate that operators representing observables at relatively spacelike intervals have to commute. Interestingly, RQFT places no such restriction on unobservables such as Dirac fields, which obey anticommutation relations. In Chapter 24, we discuss the construction of fermionic quantum fields from a QDN perspective, based on Jordan and Wigner's nonlocal quantum register approach, a manifestly emergent formulation (Jordan and Wigner, 1928).

QDN is not a reductionist approach to QM, as it deals principally with apparatus sitting on the interface between relative internal context (the world of SUO states) and relative external context (the environment in which that apparatus and the observer are situated). The precise relationship between QDN and lightcone structure is not clear at this time. This is consistent with the general situation at this time that a proper quantum theory of spacetime, so-called quantum gravity, has not been established. All attempts to do so from a reductionist approach have failed, to date.[1] It is not even clear at this time what has to be "quantized."

[1] Failure from the point of view of proving empirically vacuous.

We shall show in the last section of this chapter that QDN does obey the no-communication theorem, which is certainly necessary for lightcone physics to work. However, that theorem alone is not sufficient to establish lightcone structure.

Shielding

QDN actually goes further than the no-superluminosity principle, in that it allows for *shielding*. This is the empirical possibility of bypassing *lightcone causality*, the standard relativistic assumption that if event B is inside the forward lightcone of event A, then A can be the location of processes that "cause" effects to be observed at B. In practice it is possible, and often necessary, to materially isolate detectors so that whatever happens at one detector does not affect others, even when they are time-like separated and could in principle interfere with each other. Lest this be thought contrary to standard physics, we point out that all experiments are done on this basis. For example, neutrino detectors are located in deep mines, to filter out noise.

The shielding concept is really what underpins the stages concept. Our discussion of the double-slit experiments in Chapter 10 illustrates the point well: when looking at a screen for signals, an observer can do so over extended periods of real laboratory time, provided the screen is not interfered with by processes external to the experiment. The final stage of such a run, then, need not be identified with a definite instant of labtime. This is a form of loss of absolute simultaneity, different from the one discussed in special relativity (SR).

The above criteria are based on current empirical evidence and are subject to constant empirical reexamination: experimentalists continue to search for tachyons, for example. The hard fact is, however, that there have been no observed violations of the locality principle to date. Therefore, since QDN is greatly concerned with signal preparation and outcome detection, the locality principle is one that should be respected by QDN. There are several aspects to be discussed here regarding this.

Observers and Their Apparatus Are Essential

Any discussion of locality or its breakdown (referred as *nonlocality*) is meaningless if there is no mention of the observers involved, their apparatus, and the protocols of observation employed. All of these are necessary in order to define what is meant by the phrase "instantly observable" in the above.

The Information Void Cannot Be an Absolute Void

The information void refers to an absence of detectors, relative to a given observer, but makes no claims about the reality, structure, or otherwise of "empty space." Indeed, what an information void is to one observer may be a seething mass of detectors to another. For example, an observer looking to detect neutrinos faces immense technical difficulties, while an observer looking to detect sunlight need only open their eyes.

Apparatus Nonlocality

A critical attribute of observation that cannot be ignored here concerns the spatial distribution of the observer's apparatus. No apparatus is perfectly localized in space or in time. All equipment consists of vast numbers of atoms, which have spatial extent, and all observations take time. We have in previous chapters formalized this latter fact, that observations take time, into the stage concept.

Correlations

Apparatus nonlocality plays a critical role in quantum correlations. The superluminal transmission of certain types of information, interpreted as correlations, has been investigated and speeds in excess of 10,000 times the speed of light reported (Scarani et al., 2000). One of our aims in this book is to demystify such phenomena. QDN interprets quantum nonlocality as originating from the fact that apparatus is invariably nonlocal, as are the processes of extracting information from it, rather than reflecting strange, nonclassical properties of SUOs. Since apparatus has to be constructed *before* a quantum state or wave function can be given any meaning, or a correlation measured, it is then obvious that nonlocality is built into quantum physics from the word go, in the form of correlations arising from empirical context.

The Interpretation of Relativity

A particular problem with nonlocality arises from the principles of physics embedded in special and general relativity (GR). Relativity holds a strategic position in physics. It has passed all its tests and its principles cannot be dismissed, according to all current empirical evidence. It is as good in its domain of applicability as QM is in its domain.

Before we go further, we should clarify what we mean by *relativity*. There really are two different discussions going on here: the *gravitational* and the *relational*. The former concerns spacetime curvature, while the latter concerns relationships between observers.

Gravitational Side of Relativity

Einstein's field equations in GR couple spacetime curvature to the stress energy tensor. Although phrased in local differential form (thereby giving GR a reductionist flavor), Einstein's equations are really aspects of emergent physics that belong to the relative external context side of any Heisenberg cut: observers will usually think of themselves as moving along time-like worldlines in a GR spacetime, and that is an emergent concept.

Relational Side of Relativity

This side of relativity can be seen as an attempt to formulate a theory of observ*ers* in classical mechanics. However, it is emphatically not a theory of observ*ation* per se: there are no prescriptions in SR or GR about apparatus, for instance, beyond how it is related to relative external context. This aspect of relativity gives the rules relating the classical data held by one observer with the classical data held by another, and no more. Put in these terms, we can appreciate why relativity and QM are not contradictory or incompatible. They are frameworks discussing different aspects of observation.

We may summarize these comments by saying that relativity deals with relative external context (REC), whereas QM discusses relative internal context (RIC). Two historical category errors, the quantum gravity program and the Multiverse paradigm, appear to have been made here. Quantum gravity attempts to extend GR into RIC, while Multiverse attempts to extend QM into REC. Both attempts appear to be empirically vacuous at this time.

Our division of empirical context into REC and RIC is not clear cut but is identified in QDN with *Heisenberg's cut*:

> *The dividing line between the system to be observed and the measuring apparatus is immediately defined by the nature of the problem but it obviously signifies no discontinuity of the physical process. For this reason there must, within certain limits, exist complete freedom in choosing the position of the dividing line.* (Heisenberg, 1952)

Relativity impacts on QM because there is a principle in relativity, known as *Einstein locality* or *the principle of local causes* (Peres, 1995), that crosses the line between REC and RIC and has a direct impact on the sort of information that an observer can extract from their apparatus (which we remind the reader has nonlocality built into it even before an experiment starts). This principle asserts that

> events occurring in a given spacetime region are independent of external parameters that may be controlled, at the same moment, by agents located in distant spacetime regions. (Peres, 1995)

The Einstein principle affects the QDN formalism because relativity asserts that detectors that are outside each other's light cones cannot causally influence each other, yet QDN may give amplitude effects between those detectors. Explaining how the classical Einstein locality principle can survive in QM, and indeed in QDN, is a major challenge.

16.2 Active and Passive Transformations

Before we go further, we need to clarify a fundamental point about transformations, as it concerns the relationship between mathematics and physics.

In general, transformations come in two types, called *active* and *passive*. We illustrate the difference between these by considering some set $\Theta \equiv \{\theta^a\}$ of objects, labeled by an index a.

Active Transformations

An active transformation is something done to the original set Θ, replacing it with a new set, $\Theta' \equiv \{\theta'^a\}$. We represent an active transformation by the rule

$$\Theta \to \Theta' \neq \Theta, \tag{16.1}$$

where the arrow $\to$ means "*is replaced by*." An active transformation implies the existence of some observer (that is, mathematician, experimentalist, or theorist) who is making a specific change in, or of, a set of objects. Moreover, there is an implicit assumption that some observer (who may be a "superobserver" playing the role of a god) has a memory of the original set and can compare it with the new set.

Passive Transformations

On the other hand, a passive transformation is merely a relabeling of the elements in a set, with no actual change in the contextually significant properties of those elements. For such a transformation, we write

$$\Theta \to \Theta^{''} = \Theta, \tag{16.2}$$

where $\Theta'' \equiv \{\theta^{a'}\}$. A passive transformation therefore concerns how a set is *described*, which, again, implies the existence of some observer (that is, mathematician, experimentalist, or theorist) who is changing their way of describing a given set.

Example 16.1 Consider a d-dimensional real vector space V with orthonormal basis $\{\boldsymbol{e}^i : i = 1, 2, \ldots, d\}$. Then an arbitrary vector $\boldsymbol{v}$ in V may be written in the form $\boldsymbol{v} = v^i\boldsymbol{e}^i$, where the components $\{v^i\}$ are real and the summation convention is used.

An active transformation of $\boldsymbol{v}$ is defined by a change in the components of $\boldsymbol{v}$ but not in the basis: v^i is replaced by an arbitrary v'^i, with the basis vectors remaining the same. Hence for an active transformation, we write $\boldsymbol{v} \to \boldsymbol{v}' \equiv v'^i\boldsymbol{e}^i \neq \boldsymbol{v}$.

On the other hand, a passive transformation leaves the vector $\boldsymbol{v}$ unchanged but the basis vectors are changed for a new set. For example, a change of basis from $\{\boldsymbol{e}^i\}$ to $\{\boldsymbol{e}'^i\}$, defined by the invertible linear relations $\boldsymbol{e}^i \equiv C^{ij}\boldsymbol{e}'^j$, gives $\boldsymbol{v} \to \boldsymbol{v}' \equiv v'^j\boldsymbol{e}'^j = \boldsymbol{v}$. In this case, the new coefficients $\{v'^j\}$ are given by $v'^j \equiv v^iC^{ij}$.

Active transformations are generally the only ones of interest to physicists, because they concern the real, physical world, whereas passive transformations are done in the mind. There are several kinds of active transformation in physics.

Construction of Apparatus

The construction of apparatus is the severest form of active transformation, as this creates physical context. We discuss this form of active transformation in Chapter 25.

Changes in State Preparation

A common form of investigation involves an observer sitting in a fixed laboratory and making active changes in prepared states, keeping the detectors unchanged. For example, switching on an electric field will often be associated with the acceleration of a charged particle in a scattering experiment. In terms of generalized propositions, we may describe such a change by the rule

$$(S, D|\Omega, F) \to (S', D|\Omega, F), \tag{16.3}$$

where S is a proposition about a prepared state, D is the detection apparatus, Ω is the observer, and F is the relative external context (the frame of reference) that defines the observer.

Changes in Outcome Detection

Another common form of investigation involves active changes in detection apparatus, keeping prepared states and relative external context unchanged. For example, the main magnetic field axis of a Stern–Gerlach experiment may be rotated to a new orientation.

We may describe such changes by the rule

$$(S, D|\Omega, F) \to (S, D'|\Omega, F). \tag{16.4}$$

It is commonly assumed that in some cases such as spatial rotations, transformations (16.3) and (16.4) are related by a change in sign in the parameters involved. That is a matter for empirical validation and **not** an absolute truth. The overthrow of parity in 1956 in the Wu–Ambler experiment demonstrated emphatically that symmetry and logic must not be treated as equivalent to empirical truths (Wu et al., 1957).

Interframe Experiments

An *interframe* experiment is essentially an experiment involving two observers: one of these is associated with state preparation and the other is involved with outcome detection. Such experiments explore perhaps the most spectacular and deepest issues in physics. Examples are the Doppler shift, the Unruh effect (Unruh, 1976), and indeed, the notion that more than one observer is a meaningful topic in physics. Questions about standardization of physics protocols (units and such like), observation of observers, the constancy of physical "constants," Early Universe physics, and so on, come flying at us immediately.

Such experiments may be represented symbolically by

$$(S, A|\Omega, F) \to (S', D'|\Omega', F'), \tag{16.5}$$

where A is the apparatus that creates the state S, relative to observer Ω and D' is the detecting apparatus relative to observer Ω'.

We place a prime on the transformed state/proposition, S', on the right-hand side in (16.5), for two reasons. First, what the detecting observer Ω' detects will appear to have properties different from those of the state that the preparing observer Ω believes they have prepared. Second, that a quantum state means the same thing to different observers is a notion that needs to be questioned in several respects, concerning the standardization of physics and the exchange of context. In QDN, for instance, we do not accept that quantum states are objective "things."

To illustrate our concerns, consider the detection by observer Ω' of a single photon signal. On what basis could Ω' believe that they had detected a signal sent from a distant galaxy? The only plausible scenario is that the observer had already received sufficient contextual information about that galaxy to formulate and justify such a belief.

We could take the view that such an exchange of sufficient contextual information from the source observer to detecting observer is equivalent to having a single observer, encompassing both source and detectors. While reasonable in most cases, that point of view seems bizarre as far as intergalactic processes are concerned.

Active transformations play a role in mathematics, where they are associated with functions, maps, and operations. A function can be regarded as a form of active transformation: given a set Θ, a function f maps elements of that set into some other set, Θ'. We can even think of this as defining a "mathematical arrow of time."

Passive transformations have played a role in physics in some important situations.

Symmetries in the Void

By the fact that there are no detectors in the information void, it is permissible to model signal propagation through it in any theoretically useful way, such as through Minkowski spacetime. It may then be useful to consider passive transformations in that spacetime and explore the empirical consequences. For example, relativistic quantum field theory, which is used to calculate the dynamics of quantum fields in the information void, is generally assumed to Lorentz covariant, meaning that it does not matter which inertial frame is used to do the calculations.

Space-Time Symmetries

Passive coordinate transformations played a critical role in the development of modern physics. Aristotelian physics was firmly based on the proposition that the Earth is an absolute frame of reference. After the works of Galileo and Newton, observers discussed physics more carefully. The observer's frame of reference now became identified as an important ingredient. For convenience, the important frames were usually taken to be inertial frames. It was pointed out by Bishop Berkeley early on in the development of Newtonian mechanics that Newton's laws of motion are invariant to (unchanged by) any passive Galilean transformation of an observer's inertial frame of reference (Berkeley, 1721). Later, this idea was extended to special relativity, leading to the notion that the laws of physics (excluding gravitation) are invariant to passive Lorentz and Poincaré transformations of inertial frames of reference.

We describe this concept in generalized proposition form by

$$(L, \emptyset|\Omega, F) \equiv (L, \emptyset|\Omega', F'), \tag{16.6}$$

where L are Newton's laws of motion in the nonrelativistic case or the laws of relativistic mechanics in the relativistic case, and frames F and F' are related by either a Galilean or Poincaré transformation, as appropriate.

We note the absence of any relative internal context in (16.6), typical of both nonrelativistic classical mechanics and relativistic mechanics. This contextual incompleteness gives the generalized proposition classification of such theories as two.

16.3 Local Operations

In this and following sections, we pin down our meaning of *local operation.* In line with the discussion in the previous section, such an operation is always an active one, taking place over at least one stage and possibly more. We consider an active physical operation $\mathbb{L}_{n+1,n}$ on a rank-r apparatus $\mathcal{A}_n^{[r]}$ at stage Σ_n, an operation that affects a number p of detectors in $\mathcal{A}_n^{[r]}$ and leaves the remaining $q \equiv r - p$ detectors unaffected in a specific way, to be explained below. The affected detectors and their corresponding signal qubits will be called *relatively local,* while the unaffected detectors and their corresponding signal qubits will be called *relatively remote*. By unaffected, we mean that no possible partial measurements on the remote detectors alone would detect any changes due to $\mathbb{L}_{n+1,n}$.[2]

Our approach is to split the quantum register $\mathcal{Q}_n$ at stage Σ_n into two subregisters $\mathcal{Q}_n^{[L]}$ and $\mathcal{Q}_n^{[R]}$, such that $\mathcal{Q}_n = \mathcal{Q}_n^{[L]}\mathcal{Q}_n^{[R]}$. Splitting a quantum register is a purely mathematical operation, discussed in detail in Chapter 22, although the motivation for doing this comes entirely from the physics of the experiment.

Here $\mathcal{Q}_n^{[L]}$ is the rank-p tensor product $Q_n^1 Q_n^2 \dots Q_n^p$ of the local signal qubits and is therefore referred to as the *local subregister*, while $\mathcal{Q}_n^{[R]}$ is the rank $q \equiv r-p$

[2] Note that the language here is imprecise. Experiments to detect changes in the remote detectors would actually involve two ensembles of runs, comparing partial measurements on apparatus evolving with the action of $\mathbb{L}_{n+1,n}$ with partial measurements on apparatus evolving without it.

tensor product $Q_n^{p+1}Q_n^{p+2}\dots Q_n^r$ of the remote signal qubits and is therefore referred to as the *remote subregister*.

Note that the labeling of detectors is arbitrary in principle, so we are always entitled to order local and remote qubits in the way given above.

Example 16.2 An observer has prepared a labstate $\mathbf{\Psi}$ in a rank-five quantum register $\mathcal{Q}^{[5]} \equiv Q^1Q^2Q^3Q^4Q^5$, given in the computational basis representation (CBR) by

$$\mathbf{\Psi} = ac\underline{\mathbf{24}} + ad\underline{\mathbf{29}} + ae\underline{\mathbf{20}} + bc\underline{\mathbf{10}} + bd\underline{\mathbf{15}} + be\mathbf{6}, \tag{16.7}$$

where each underlined number in bold, such as $\underline{\mathbf{24}}$, represents a single element of the CBR, and a, b, c, d, and e are complex coefficients. Show that this state is separable relative to the split $\mathcal{Q}^{[5]} = \mathcal{Q}^{[L]}\mathcal{Q}^{[R]}$, where $\mathcal{Q}^{[L]} \equiv Q^2Q^5$ and $\mathcal{Q}^{[R]} \equiv Q^1Q^3Q^4$.

Solution

The CBR is not best suited to discuss splits, so we need to find the equivalent of the signal basis representation (SBR) of the state. We translate the above CBR basis vectors into SBR counterparts as follows. By inspection, the binary decomposition of the integers 6, 10, 15, 20, 24, and 29 is

$$6 = 2{+}4, 10 = 2{+}8, 15 = 1{+}2{+}4{+}8, 20 = 4{+}16, 24 = 8{+}16, 29 = 1{+}4{+}8{+}16. \tag{16.8}$$

Then for example we have $\mathbf{6} = \widehat{\mathbb{A}}^2\widehat{\mathbb{A}}^3\mathbf{0}$, and so on. Hence we can write

$$\begin{aligned}\mathbf{\Psi} = {} & ac\widehat{\mathbb{A}}^4\widehat{\mathbb{A}}^5\mathbf{0} + ad\widehat{\mathbb{A}}^1\widehat{\mathbb{A}}^3\widehat{\mathbb{A}}^4\widehat{\mathbb{A}}^5\mathbf{0} + ae\widehat{\mathbb{A}}^3\widehat{\mathbb{A}}^5\,\mathbf{0} + \\ & + bc\widehat{\mathbb{A}}^2\widehat{\mathbb{A}}^4\mathbf{0} + bd\widehat{\mathbb{A}}^1\widehat{\mathbb{A}}^2\widehat{\mathbb{A}}^3\widehat{\mathbb{A}}^4\mathbf{0} + be\widehat{\mathbb{A}}^2\widehat{\mathbb{A}}^3\mathbf{0}.\end{aligned} \tag{16.9}$$

By inspection, this can be operator factorized into the form

$$\mathbf{\Psi} = (a\widehat{\mathbb{A}}^5 + b\widehat{\mathbb{A}}^2)(c\widehat{\mathbb{A}}^4 + d\widehat{\mathbb{A}}^1\widehat{\mathbb{A}}^3\widehat{\mathbb{A}}^4 + e\widehat{\mathbb{A}}^3)\mathbf{0}. \tag{16.10}$$

This can now be interpreted as the tensor product of two subregister states; that is, we may write $\mathbf{\Psi} = \mathbf{\Psi}^{[L]}\mathbf{\Psi}^{[R]}$, where $\mathbf{\Psi}^{[L]} \equiv (a\widehat{\mathbb{A}}^5{+}b\widehat{\mathbb{A}}^2)\mathbf{0}^{[L]}$ is in $\mathcal{Q}^{[L]} \equiv Q^2Q^5$, $\mathbf{\Psi}^{[R]} \equiv (c\widehat{\mathbb{A}}^4{+}d\widehat{\mathbb{A}}^1\widehat{\mathbb{A}}^3\widehat{\mathbb{A}}^4{+}e\widehat{\mathbb{A}}^3)\mathbf{0}^{[R]}$ is in $\mathcal{Q}^{[R]} \equiv Q^1Q^3Q^4$, $\mathbf{0} = \mathbf{0}^{[L]}\mathbf{0}^{[R]}$, and we define signal creation operators for each subregister accordingly.

16.4 Primary and Secondary Observers

There is an issue here, the physical implications of which are deep to say the least and that underpins many debates about the nature of reality. In the above section we considered splitting a quantum register into a local subregister and a remote subregister. What we have in mind, of course, is to associate different "observers" with each of these subregisters, as this is of interest in various branches of physics and information theory. Typically we would call the local observer *Alice*, and the remote observer *Bob*.

What needs to be addressed is the following: if Alice and Bob have no connection, meaning that they have no channels of communication between them, then for whom is the combined scenario *Alice and Bob* meaningful? We have argued elsewhere in this book that truth values are contextual. So if we want to discuss Alice and Bob together, we have to specify the context in which we are doing so.

The only answer that makes empirical sense is that there must be (implicitly, if not explicitly) some third observer *Carol* who has the contextual information to know about Alice and Bob and what they are observing and the outcomes that they have found.

In CM, such an overseeing observer is generally not specified, a factor that contributes to the essential contextual incompleteness of that discipline. In QM, we cannot allow such contextual incompleteness. Whatever is asserted must have some empirical basis for its truth values. Quantum entanglement runs directly into this issue.

This chain of reasoning leads us to the *primary observer* concept. A primary observer is the overseer and custodian of all relevant context: the buck stops with a primary observer, there is nothing behind them, in the given context. So when we discuss Alice and Bob as local and remote observers respectively, they are by implication **not** primary observers. We will refer to them as *secondary observers*, or *subobservers*.

A fundamental difference in QM between a primary observer and any secondary observers is that dimensions of Hilbert spaces do not follow an additive rule: if Alice thinks she is dealing with a p-dimensional Hilbert space and Bob thinks he is dealing with a q-dimensional space, then primary observer Carol, who is overseeing Alice and Bob, is dealing with a pq-dimensional space, not a $p+q$-dimensional Hilbert space. What is additive is qubit register *rank*.

Exercise 16.3 If Alice models her experiment with a rank-a qubit quantum register and Bob models his with a rank-b qubit quantum register, prove that if Carol models both experiments by a rank $c = a + b$ qubit quantum register such that the dimension of Carol's quantum register equals the sum of the dimensions of Alice and Bob's registers, then $a = b = 1$.

There is an interesting question here: *is it possible for a primary observer to observe themselves completely, that is, know everything about themselves*? Our resolution of this is that it is not possible to do this in a complete way: real empirical information always comes in discrete form and real observers have finite capacity to store information. According to Sen (Sen, 2010), a finite set cannot be mapped bijectively to any proper subset (but an infinite set can be so mapped). Therefore, a real observer cannot faithfully observe themselves in a complete way.

An observer can observe themselves partially, however: we do this every time we look in a mirror. What we see is only part of ourselves (our face, clothes, and so on), and that information is stored not in those parts but in other parts that serve as memory (our brains). In essence, part of us plays the role of an SUO and other parts play the role of a primary observer.

16.5 Subregister Bases

Contextuality is equivalent in many respects to information. Given the split of the quantum register $\mathcal{Q}_n$ into the tensor product of a local subregister $\mathcal{Q}_n^{[L]}$ and a remote subregister $\mathcal{Q}_n^{[R]}$, discussed in Section 16.3, the question of basis set for each subregister in the split arises. It turns out that QDN gives an immediate answer as follows.

As we saw in previous chapters, context gives a preferred basis set for every signal qubit component of a quantum register. Factoring such a register into two subregisters, each consisting of an integral number of signal qubits, does not change this. A preferred basis is determined by physical context, whereas a split is a purely mathematical operation. Therefore, each signal qubit retains its preferred basis during a split.

Suppose we have a detector array consisting of $r = p + q$ detectors, where p and q are at least one and we intend to split it into the tensor product of a local subregister $\mathcal{Q}_n^{[L]}$ consisting of p detector qubits and a remote subregister $\mathcal{Q}_n^{[R]}$ of q detector qubits. Now in principle there is no natural way to order any array of detectors, because they are located in three-dimensional space, and there is no natural ordering in such a set. We are free to label the original array in any way that we want. In the context of a split such as we envisage, it seems sensible to do it as follows. The p detectors that we will include in the local subregister will be labeled 1 to p and the remote detectors will be labeled $p + 1$ to r when we discuss the original register, but from 1 to q when we are discussing the remote subregister as a separate entity.

We now have an obvious way of organizing each of our subregister bases. We can define CBRs and SBRs for each. Moreover, we can define signal projection operators $\mathbb{P}_n^i, \widehat{\mathbb{P}}_n^i, \mathbb{A}_n^i, \widehat{\mathbb{A}}_n^i$ for each separately, just as if they were independent registers (as indeed, they could be).

A natural question is: given a basis element $\boldsymbol{i}_n^{[L]}$ in the CBR for the local register, where $0 \leq i < 2^p$ and a basis element $\boldsymbol{j}_n^{[R]}$ in the CBR for the remote register, where $0 \leq j < 2^q$, to what element $\boldsymbol{k}_n$ of the CBR for the original register does the tensor product $\boldsymbol{i}_n^{[L]} \otimes \boldsymbol{j}_n^{[R]}$ correspond? It is easy to show that with the ordering described above, we have the rule $k = i + 2^p j$.

We may readily construct operators that act on elements of a given subregister and not on elements in another subregister. For example, if $\mathbb{U}^{[L]}$ is a subregister operator that acts over $\mathcal{Q}^{[L]}$, then relative to the original register $\mathcal{Q} \equiv \mathcal{Q}^{[L]}\mathcal{Q}^{[R]}$, its action is equivalent to that of the operator $\mathbb{U}^{[L]} \otimes \mathbb{I}^{[R]}$, where $\mathbb{I}^{[R]}$ is the identity operator for $\mathcal{Q}^{[R]}$.

16.6 Local and Remote Evolution

When a primary observer discusses operations performed by local and remote secondary observers, such actions are never implemented instantaneously but take laboratory time, so should be discussed in terms of stage to stage dynamics.

Suppose we split a register $\mathcal{Q}_n$ into local and remote subregisters as above, with their respective CBR bases. Consider further a process of evolution from stage Σ_m to stage Σ_n, for $n > m$, with the local secondary observer Alice performing an action $\boldsymbol{U}^{[L]}_{n,m}$ on her apparatus and remote secondary observer Bob performing an action $\boldsymbol{U}^{[R]}_{n,m}$ on his apparatus. The question is, how does Carol, the primary observer, describe both of these processes?

For simplicity, we shall assume that the rank of each subregister is stage independent. Following our discussion above on subregister bases, we can write

$$\boldsymbol{U}^{[L]}_{n,m} \equiv \sum_{i,j=0}^{2^{r_L}-1} \boldsymbol{i}^{[L]}_n U^{[L]i,j}_{n,m} \overline{\boldsymbol{j}^{[L]}_m}, \quad \boldsymbol{U}^{[R]}_{n,m} \equiv \sum_{i,j=0}^{2^{r_R}-1} \boldsymbol{i}^{[R]}_n U^{[R]i,j}_{n,m} \overline{\boldsymbol{j}^{[R]}_m}. \tag{16.11}$$

Then Carol can describe what Alice and Bob are doing by the evolution operator

$$\mathbb{U}_{n,m} \equiv \boldsymbol{U}^{[L]}_{n,m} \otimes \boldsymbol{U}^{[R]}_{n,m} = \sum_{i,j=0}^{2^{r_L}-1} \sum_{k,l=0}^{2^{r_R}-1} \boldsymbol{i}^{[L]}_n \otimes \boldsymbol{k}^{[R]}_n U^{[L]i,j}_{n,m} U^{[R]k,l}_{n,m} \overline{\boldsymbol{j}^{[L]}_m \otimes \boldsymbol{l}^{[R]}_m}. \tag{16.12}$$

It is straightforward to rewrite this expression in terms of the CBR for Carol's quantum register. Likewise, all subregister operators acting on local or remote labstates can be readily rewritten in terms of the global register (that is, from Carol's perspective).

Two important conclusions can be drawn from this analysis. (1) It is consistent to apply QM to parts of the universe, while ignoring the rest, even though all of it is subject to the laws of QM, and (2) it is the possibility of isolating apparatus that gives rise to the SUO concept in the first place

16.7 The No-Communication Theorem

We have stressed the critical significance of the no-communication concept in QM. We shall discuss it in simplistic terms in both standard QM and in QDN.

The QM Account

Consider two observers, Alice and Bob. Alice will be our *local* observer, conducting active transformations on locally prepared signal states, while Bob will be our *remote* observer, conducting observations at remote detectors. The aim is to see if anything Alice does can affect what Bob observes.

First, we put ourselves in the position of a primary observer, Carol, who has an overview of what Alice and Bob do. They are now to be regarded as secondary observers, relative to Carol, although each of them believes themself to be a primary observer. Suppose Alice models the states she can prepare by elements of a Hilbert space denoted $\mathcal{H}^A$, and suppose Bob models the states he observes by elements of a Hilbert space $\mathcal{H}^B$. We will suppose that Carol has enough information to model the combined system by the tensor product space $\mathcal{H}^C \equiv \mathcal{H}^A \otimes \mathcal{H}^B$.

In the following we shall deal with pure states only, assuming that mixed states present no exceptional concerns. This is on account of the fact that mixed states involve no more than the extra complication of classical probabilities, and these do not interfere in the way that entangled quantum states do.

Suppose Carol arranges for an entangled state to be prepared, described by a normalized element in $\mathcal{H}^C$ given by

$$|\Psi^C) \equiv \alpha|\psi^A\rangle \otimes |\psi^B\rangle + \beta|\phi^A\rangle \otimes |\phi^B\rangle, \tag{16.13}$$

where $|\psi^A\rangle$ and $|\phi^A\rangle$ are normalized elements in $\mathcal{H}^A$, $|\psi^B\rangle$ and $|\phi^B\rangle$ are normalized elements in $\mathcal{H}^B$, and $|\alpha|^2 + |\beta|^2 = 1$. The actions of Alice and Bob are respectively as follows.

Alice

Alice performs a local active operation U^A on states in $\mathcal{H}^A$ that she has access to, changing them according to the prescription

$$|\psi^A\rangle \to |\psi'^A\rangle \equiv U^A|\psi^A\rangle, \quad |\phi^A\rangle \to |\phi'^A\rangle \equiv U^A|\phi^A\rangle. \tag{16.14}$$

Bob

Bob performs a measurement of an observable O^B on states in $\mathcal{H}^B$ to which he has access.

Carol

From Carol's perspective, a holistic account has to be given, in terms of states in, and operators over, the total Hilbert space $\mathcal{H}^C$. From her perspective, Bob's observable corresponds to the operator $O^C \equiv I^A \otimes O^B$, where I^A is the identity operator over $\mathcal{H}^A$, while Alice's local operator corresponds to the operator $U^C \equiv U^A \otimes I^B$, where I^B is the identity operator over $\mathcal{H}^B$.

Before Alice performs her operation, Carol calculates that Bob's expectation value $\langle O^B\rangle$ will be given by

$$\begin{aligned}\langle O^B\rangle &\equiv (\Psi^C|O^C|\Psi^C) \\ &= |\alpha|^2\langle\psi^B|O^B|\psi^B\rangle + |\beta|^2\langle\phi^B|O|\phi^B\rangle \\ &\quad + \alpha^*\beta\langle\psi^A|\phi^A\rangle\langle\psi^B|O|\phi^B\rangle + \alpha\beta^*\langle\phi^A|\psi^A\rangle\langle\phi^B|O^B|\psi^B\rangle,\end{aligned} \tag{16.15}$$

where we note the presence of interference terms.

After Alice performs her operation, Carol calculates that Bob's expectation value $\langle O^B\rangle'$ will be given by $\langle O^B\rangle' \equiv (\Psi'^C|O^C|\Psi'^C)$, where now

$$|\Psi'^C) \equiv U^C|\Psi^C) = \alpha|\psi'^A\rangle \otimes |\psi^B\rangle + \beta|\phi'^A\rangle \otimes |\phi^B\rangle. \tag{16.16}$$

Then we find

$$\begin{aligned}\langle O^B\rangle' = |\alpha|^2\langle\psi^B|O^B|\psi^B\rangle + |\beta|^2\langle\phi^B|O|\phi^B\rangle \\ +\alpha^*\beta\langle\psi'^A|\phi'^A\rangle\langle\psi^B|O|\phi^B\rangle + \alpha\beta^*\langle\phi'^A|\psi'^A\rangle\langle\phi^B|O^B|\psi^B\rangle.\end{aligned} \quad (16.17)$$

Comparing (16.15) and (16.17), it is easy to see that $\langle O^B\rangle' = \langle O^B\rangle$, because $\langle\psi'^A|\phi'^A\rangle = \langle\psi^A|U^{A\dagger}U^A|\phi^A\rangle = \langle\psi^A|\phi^A\rangle$.

The prediction, therefore, is that no active unitary transformations performed by Alice would affect Carol's calculation of Bob's measured expectation value. In other words, Alice could not transmit any signals to Bob using entanglement.

The QDN Account

QDN has no problem in fully accommodating the no-communication theorem, because as discussed in previous sections, it is straightforward to define concepts of local and remote subregisters that can model all the necessary requirements for the no-communication theorem to hold. Indeed, we encounter such processes in our discussion of the quantum eraser experiments in Chapter 14.

17
Bell Inequalities

17.1 Introduction

From the dawn of the quantum age in 1900, there has been an ongoing debate, often fierce and hostile, between the supporters of the classical world view and those of the quantum world view. This debate eventually led to remarkable experiments, the convincing and repeated results of which are claimed by quantum theorists to support the predictions of quantum mechanics (QM) and not those of classical mechanics (CM). However, old paradigms tend to linger on the shelves of science long past their sell-by dates and there are extant schools of theorists that strive to find loopholes in the above-mentioned experiments. That is a legitimate activity up to a point, but the divisive nature of the debate requires commentary, which is the subject matter of this chapter.

QM was discovered only by advanced technology, so it stands to reason that any test of QM will require even more advanced technology. In the previous chapter we stressed the differences between active and passive transformations. The experiments discussed here rely on such technically difficult active transformations that the Hidden Variables (HV) theorists (for that is really what they are) opposing QM are frequently able to think of objections, known as *loopholes*, to the empirical protocols employed by the experimentalists. Some of these loopholes are reasonable but many are not. Closing those loopholes convincingly is an ongoing important activity in experimental quantum physics.

A significant feature of the experiments discussed in this chapter is that they go beyond a certain point of complexity in terms of the number of classical and quantum degrees of freedom involved. Historically, before that point had been reached, classical interpretations of quantum wave functions appeared viable and perhaps even attractive, such as that proposed by Bohm (Bohm, 1952). We shall refer to the collective of such interpretations as the *hidden variables* (HV) paradigm and the point in question as the *Heisenberg point*.

It seems obvious with hindsight that the Heisenberg point was reached almost immediately after Schrödinger introduced his wave mechanics formulation of QM in 1926 (Schrödinger, 1926). Certainly, the wave function $\Psi(\boldsymbol{x}, t)$ for a single-particle system under observation (SUO) can be visualized as an objective wave of sorts in physical space. By the term *physical space*, we mean the three-dimensional space $\mathbb{P}^3(\Lambda)$ of extension, position, and distance, associated with a real laboratory Λ,[1] a concept that would surely have been understandable to Aristotle, Galileo, Newton, Hamilton, and Lagrange. But the Schrödinger wave function $\Psi(\boldsymbol{x}^1, \boldsymbol{x}^2, \ldots, \boldsymbol{x}^N; t)$ for an SUO consisting of N particles is actually a time-dependent complex function over a real $3N$-dimensional space $\mathcal{C}^{3N}$, known as *configuration space*, which is quite different conceptually from $\mathbb{P}^3(\Lambda)$.

Humans are remarkably stable creatures, both physically and mentally. Good health is measured in years and mental outlook is measured in decades. The phenomenon of *persistence*, which we have attributed as the origin of objectivization, seems to us responsible for the classical world view. We see objects apparently unchanged over significant stretches of time, and we come to believe that those objects have real identities. In fact, we are generally strongly conditioned mentally to think in such classical terms. We imagine ourselves as sitting in physical space $\mathbb{P}^3(\Lambda)$ and we try to relate all experience to it. We may refer to this as *mental persistence*, or equivalently, *classical conditioning*.

Absolute physical space and absolute time (Newton, 1687)[2] form the conceptual foundations on which Newtonian mechanics was built. We shall refer to this mathematical model as *space-time*, noting the hyphen between "space" and "time." This hyphen is important: it marks the recognition that observers operate in their physical space with a *process* view of time rather than a *manifold* or Block Universe perspective. Indeed, QM seems to us empirically meaningful only in the space-time perspective, simply because probability, as we know it, makes no sense otherwise.[3]

Not only does the space-time model give a remarkably good account of many physical phenomena such as planetary orbits, but it conforms excellently to our inherited classical conditioning: we think in terms of objects moving around physical space with the passage of absolute time.

The advent of relativity did little to change this in practice, for the basic reason that the speed of light is so great in relative terms that the space-time model is a good one for all practical purposes. Indeed, its replacement, the Block

[1] We follow here Schwinger's statement, quoted in Chapter 24, that space and time are contextually defined by apparatus.

[2] We refer the reader to the *Principia* for Newton's defining comments on what he meant by "absolute space" and "absolute time".

[3] There will be theorists who interpret QM and probability in terms of abstract mathematical structures over Block Universe manifolds, with operator norms, C^* algebras, and such like. Those approaches to QM equate empirical physics with mathematical physics and generally neglect observers and apparatus, thereby usually having a generalized propositional classification of one.

Universe manifold $\mathbb{M}^4$ (four-dimensional Minkowski spacetime), came into the orbit of theorists' attention only relatively recently, in 1908 (Minkowski, 1908). It is too soon for human conditioning to have evolved (if it ever will) to the point where the world around us is interpreted naturally according to the $\mathbb{M}^4$ spacetime paradigm rather than the Newtonian space-time paradigm.

When Schrödinger introduced his wave function in 1926, therefore, classical conditioning ensured that attempts would be made to understand his theory in classical mechanical terms. Indeed, Schrödinger himself originally interpreted his wave function in realist terms, but soon realized that the configuration space argument required a revision in that interpretation.

Such attempts have persisted: there remain groups of theorists who have the agendas of either refuting relativity (Dingle, 1967) or refuting QM (Bohm, 1952), or refuting both (Kracklauer, 2002). A theme common to these agendas is contextual incompleteness, with little or no attention being paid to the details of relative internal context (discussed in Section 2.12). A significant point to make here is that none of these theorists has reported experiments that *they* have actually done, so *nullius in verba*[4] and *Hitchens' razor*[5] can legitimately be applied here.

Fortunately, the matter goes beyond talking-shop physics because of the remarkable contribution of the theorist J. S. Bell. In 1964, Bell discussed an HV model of a two spin-half SUO such as a two-electron state (Bell, 1964). He based his model on a number of reasonable assumptions about the nature of classical reality and discovered the possibility of testing those assumptions. The significance of this is that Bell opened the door to empirical tests of quantum principles, because these give predictions different from those based on classical principles.

To understand his ideas, we need to review some concepts.

Parameters

Parameters are used in the description of apparatus and observers, and they are to be found in relative internal context and relative external context. Every classical model, including that of Bell, relies on parameters, such as the masses of the particles being observed, orientation of apparatus relative to the laboratory, and so on. Parameters may be physical constants determined by previous experiments, such as particle masses, or they may be classical degrees of freedom under the control of the observer, such as orientation angles of a Stern–Gerlach (SG) main magnetization field.

Parameters should not be confused with dynamical variables, which are theoretical constructs related to states of SUOs. Parameters are generally expressed in terms of classical real numbers, associated with agreed systems of units. Some parameters are specific, meaning that they are assumed to be exact to within

[4] Take no one's word for it.

[5] That which is asserted without proof can be dismissed without proof.

measurement errors, or statistical in nature. An example of a specific parameter is the mass of the free electron in vacuo, $9.1938291\ldots \times 10^{-31}$ kg. An example of a statistical parameter would be the temperature of the laboratory in which an experiment was being carried out.

Variables

Variables have to do with states of SUOs. As the word suggests, variables change over the course of an experimental run. The point of doing experiments is to determine to what extent observers understand those changes.

Model Limits

A good model will make predictions about the range over which the variables can go, for a fixed set Θ of parameters. For example, Newtonian mechanics and Newton's law of gravitation predicts how far the Earth could go from the Sun in its annual orbit, given the known parameters, such as the mass of the Earth and of the Sun, and the current position and velocity of the Earth relative to the Sun.

Bell showed, in a CM model of two spin-half particles based on "reasonable" classical assumptions such as locality, that there was a limit Λ to a certain empirically measurable function F of the parameters Θ,[6] given in the form of an inequality $F(\Theta) \leqslant \Lambda$. Such inequalities are now universally referred to as Bell inequalities (Bell, 1988). They are in focus in this chapter, because they provide an empirical test of CM predictions versus those of QM.

In the following sections we shall first discuss the SG experiment from a classical perspective. Then we shall show how a classical Bell-type inequality can be derived. Then we shall show how standard QM predicts the possibility of a nonclassical violation of this inequality. Then we shall give the quantized detector network (QDN) account of the violation.

17.2 The Stern–Gerlach Experiment

Along with the double-slit experiment, the SG experiment serves as a standard test of QM principles. In this section we discuss the latter experiment from the perspective of HV theorists applying standard CM principles.

The stage diagram Figure 17.1 shows the SG architecture. Observer Ted operates a preparation device T that directs a beam of particles[7] 1_0 toward an SG module S^a containing a strong inhomogeneous magnetic field aligned principally

[6] By *empirically measurable*, we mean fixing the parameters and then performing an experiment to calculate a value for F.

[7] We emphasize again that such descriptions express a classical interpretation of what happens in the laboratory: Ted pushes buttons in preparation device T and Alice looks at signals on the detector screen. Neither Ted nor Alice see "particles" in the way spectators observe baseballs or cricket balls.

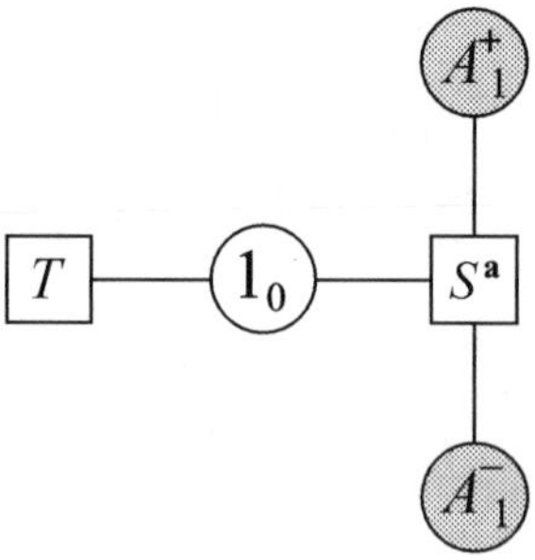

Figure 17.1. Stage diagram of the SG experiment.

along a fixed direction $\boldsymbol{a}$. This direction is the significant parameter in this experiment. Interestingly, while the classical explanation of what is going on requires a knowledge of the magnetic field in module $S^{\boldsymbol{a}}$, that turns out not to be directly relevant to the outcome of the experiment; what is important here is the forest, not the trees.

It was observed by Stern and Gerlach (Gerlach and Stern, 1922a,b) that on a detecting screen on the opposite side of $S^{\boldsymbol{a}}$ to the source T, signals were observed in two principal areas, or spots labeled A_1^+ and A_1^- in Figure 17.1. These spots will henceforth be assumed to be disjoint, that is, not overlapping, because experience suggests that this can always be arranged to a good approximation in real experiments.

The standard CM interpretation of these empirical observations is that particles from T have entered the magnetic field in $S^{\boldsymbol{a}}$ and that, by virtue of interaction with that field, some of them have been deflected into region A_1^+ while the rest have been deflected into region A_1^-. The standard CM theory is based on the following assumptions.

Extended Charged Particles

Each particle passing through $S^{\boldsymbol{a}}$ has nontrivial electromagnetic structure, meaning that it is not a point but an extended system of electric charge density that is swirling around in such a way as to create a time-dependent magnetic dipole moment, denoted $\boldsymbol{\mu}$.

Electrodynamic Forces

As it passes through the SG module $S^{\boldsymbol{a}}$, each particle's instantaneous magnetic dipole moment $\boldsymbol{\mu}$ interacts with the inhomogeneous magnetic field $\boldsymbol{B}$ in $S^{\boldsymbol{a}}$, an interaction modeled by a term proportional to $\boldsymbol{\mu} \cdot \boldsymbol{B}$ in the classical Hamiltonian. This generates an additional contribution to the standard Lorentz force $q\boldsymbol{v} \times \boldsymbol{B}$ on that particle, and it is this additional force that is interpreted classically as responsible for what Stern and Gerlach observed. Here q is the electric charge of the particle, $\boldsymbol{v}$ is its instantaneous velocity, and $\boldsymbol{B}$ is the effective magnetic field in which the particle is moving.

The fact that Stern and Gerlach observed two regions, labeled by us A_1^+ and A_1^- in Figure 17.1, is indisputable. Therefore, a CM calculation should account for that splitting. We assume that that can be done. There are good examples of such bifurcations in CM, such as comets either falling into elliptic captured orbits around the sun or entering the solar system once and then leaving forever on hyperbolic orbits. Another example is from the statistics of single car accidents on icy roads: cars will veer off a road either to the left or to the right.

Lacking more detailed information, particularly about any hidden variables that could contribute additional forces in the SG experiment, we simply assume that if we could complete such calculations, they would show the necessary bifurcation into either A_1^+ or A_1^-.[8]

Deterministic Outcomes

The classical electromagnetic forces guiding each particle through module S^a are deterministic, meaning the following. Suppose that at the start of a run, one of the particles, #1, in 1_0 has an initial position $\boldsymbol{r}_0$ and initial velocity $\boldsymbol{v}_0$. Subsequently, it enters S^a and ends up somewhere on the detecting screen. Now a typical beam will consist of a vast number of particles. Suppose another one of the particles, #2, started off in 1_0 at exactly the same initial position, the same initial velocity, and with the same internal degrees of freedom (HVs), as #1, although at some other time. Then it would subsequently follow exactly the same spatial path and would end up in exactly the same spatial position on the screen as #1.

Consider now a beam of many particles from 1_0 passing through the same apparatus. There will be a spread of initial positions and initial momenta, with most particle velocities approximately along the same common direction toward S^a. There may also be some additional hidden variables, such as those involved with the internal charge structure of the particles. Suppose there are N particles in such a beam. Then the ith particle starts in 1_0 with a set of initial variables denoted $\boldsymbol{\theta}^i$. This includes any hidden variables.

Now according to the deterministic principle outlined above, that particle will certainly end up in A_1^+ or else certainly in A_1^-, assuming perfect transmission, that is, with no outside interference and with completely efficient detection. It is important to note that for given $\boldsymbol{\theta}^i$, which of the two outcomes occurs is not the question; what matters is the asserted fact that one of them will be definitely forced to occur by the deterministic nature of the mechanics. Moreover, this is not a random outcome: which of the two sites it will be is predetermined by θ^i.

[8] The assumption is made that cases where the particle would end up in the middle of the detecting screen between A_1^+ and A_1^- constitute a tiny proportion of the whole and can be neglected. There must be, for instance, relatively few, if any, comets that are on genuinely parabolic trajectories.

It is also important to note that the question of whether we could calculate that outcome is irrelevant to the discussion.[9]

The classical deterministic paradigm outlined above leads naturally to the following assertion. Consider the set $\Theta \equiv \{\boldsymbol{\theta}^1, \boldsymbol{\theta}^2, \ldots, \boldsymbol{\theta}^N\}$ of all the initial HVs associated with a beam in a given run. Then this set can be regarded as the union $\Theta = \mathcal{A}^+ \cup \mathcal{A}^-$ of two disjoint subsets $\mathcal{A}^+$ and $\mathcal{A}^-$, where $\mathcal{A}^+$ is the subset of HVs that send a beam particle into A_1^+ and $\mathcal{A}^-$ is the subset of HVs that send a beam particle into A_1^-.

Classical Counterfactuality

So far, nothing controversial has been written. To understand the next step, we need to make a small diversion. There is a piece of logic (or metaphysics, if you prefer) that is at the heart of the problem in "understanding" QM. It is the essential point on which Einstein, Podolsky, and Rosen pitched their famous argument against standard QM (Einstein et al., 1935). It goes by the name of *counterfactuality*. A basic definition is the following.

Definition 17.1 If $\boldsymbol{P}$ and $\boldsymbol{Q}$ are two propositions (that is, statements) and $\boldsymbol{P}$ is known to be false, then the statement $\boldsymbol{P}$ implies $\boldsymbol{Q}$, written $\boldsymbol{P} \Rightarrow \boldsymbol{Q}$, is a counterfactual (statement) if it is logically true, despite the fact that $\boldsymbol{P}$ is false.

Example 17.2 Let $\boldsymbol{P}$ be the proposition **My computer is broken today** and $\boldsymbol{Q}$ is the proposition **I cannot do any typing today**. Then it is true that $\boldsymbol{P} \Rightarrow \boldsymbol{Q}$. But actually, at the time of writing this, $\boldsymbol{P}$ is false: my computer is working and I have just typed out these words.

There are serious questions about the above definition, particularly in view of the fact that we cannot avoid employing counterfactual reasoning in physics. The most obvious problem is that Definition 17.1 is contextually incomplete: there is no reference in it to any observer for whom the "truth" concept is valid, nor is there any statement of a method by which a "truth" value could be established. This is not hair-splitting: it matters in physics. That is why we have qualified the word *truth* in that definition with the adjective *logical.* Logic is not physics.

When it comes to physics, CM adopts a modified form of counterfactuality, referred to here as *the principle of classical counterfactuality*, also known

[9] It is remarkable that QM does not even attempt any such calculation, because not only is it regarded as a vacuous enterprise, but there is also no mechanism in QM to deal with individual outcomes. QM is after all a theory about the statistics of observation. HV theorists, on the other hand, do not consider it contradictory to assert (1) that there are variables they cannot know anything about, but (2) if they did know about them, they could predict everything about their behaviour, in principle.

as *counterfactual definiteness.* In brief, this principle states that counterfactual statements can be empirically meaningful in physics, that it is meaningful to assume SUOs exist and have properties that are independent of actual observation.

Classical counterfactuality is used by observers in all experiments, even those that are described by QM. This is because state preparation is based on prior established context: how do we *know* that a beam of particles is entering an SG apparatus? Such knowledge arises only because we believe in the constancy, internal consistency, and reliability of the laws of physics that we have established over centuries, and because we have previously prepared such a beam many times and tested it for its properties, a process called *calibration.*

It is just a fact of life that when it comes to actually using such a beam in a complete run in an experiment, we can no longer test it. If we were to interrupt the beam before it entered the apparatus, in order to check that the beam is what we think it is, then the rest of the run could not subsequently take place. We literally cannot have our cake and eat it.

This is perhaps the only place in science where metaphysics is critical. It is a self-evidently vacuous assertion to say that we cannot complete an uninterrupted run if we constantly interrupt it to check on the state of the apparatus. It is a quasi-religious belief throughout all of science that calibration allows us to make certain assumptions without the need to check them: *I have pressed this button and I am confident that a beam of electrons is now on its way into an SG module.*

It is on such a basis that when each run of the real experiment starts, we do not check the beam any more but rely on classical counterfactuality: that what we are doing in the beam preparation stage means what we think it means.

Classical counterfactuality is an article of faith in the laws of persistence and consistency, and in a universe that does not play tricks on us by arbitrarily changing its laws. It is implicit in the protocol of *every* experiment. The point about Bell inequality experiments is that they show that on the quantum level, classical counterfactuality is a false principle. That surely is enough to worry anyone, because it means we do not really "understand" physical reality: it is not completely described by a classical world view, only most of the time.

The relevance of classical counterfactuality to the SG experiment is as follows. Suppose that the same, identical (at the HV level) beam had been sent into the given SG apparatus but with one crucial difference: the direction of the magnetic field had previously been altered by the observer and was now along some different direction $\boldsymbol{b}$ and not $\boldsymbol{a}$. According to classical counterfactuality, it is legitimate to discuss both scenarios "simultaneously." Of course, that is fanciful, for it is not possible to go back and alter the past, as far as we know.

Classical counterfactuality and the principle of CM, however, allow us to *imagine* the possibility of a different past. We are permitted to decompose Θ as the union $\Theta = \mathcal{B}^+ \cup \mathcal{B}^-$, where $\mathcal{B}^+$ is the subset of HVs that would have sent a beam particle into B_1^+ and $\mathcal{B}^-$ is the subset of HVs that would have sent a beam particle into B_1^-, **if** the magnetic field axis had been $\boldsymbol{b}$ and not $\boldsymbol{a}$. Here B_1^+ and B_1^- are the two regions on the detecting screen observed with module $S^{\boldsymbol{b}}$.

Classical counterfactuality allows us to go further and to make the decomposition

$$\Theta = (\mathcal{A}^+ \cap \mathcal{B}^+) \cup (\mathcal{A}^+ \cap \mathcal{B}^-) \cup (\mathcal{A}^- \cap \mathcal{B}^+) \cup (\mathcal{A}^- \cap \mathcal{B}^-). \tag{17.1}$$

This is a proposition rather different from the assertion $\Theta = \mathcal{A}^+ \cup \mathcal{A}^- = \mathcal{B}^+ \cup \mathcal{B}^-$: we could at least perform each experiment $S^{\boldsymbol{a}}$ and $S^{\boldsymbol{b}}$ separately, at different times, with magnetization direction well defined each time, whereas none of the four terms on the right-hand side of (17.1) could be tested directly. For instance, the first term, $\mathcal{A}^+ \cap \mathcal{B}^+$, is the set of HV that would send a beam particle into A_1^+ **if** the magnetization axis were along direction $\boldsymbol{a}$, and would send the same beam particle into B_1^+ **if** the magnetization axis were along direction $\boldsymbol{b}$ instead of $\boldsymbol{a}$.

In the real world it is physically not possible to have both directions $\boldsymbol{a}$ *and* $\boldsymbol{b}$ *in the same run.*

That is not considered a problem by theorists who accept the CM paradigm, but is a point of view inconsistent with Wheeler's participatory principle, stated in Chapter 1. It is on such points of interpretation and natural philosophy that rest the irreconcilable differences between CM and QM. These *very* different, incompatible visions of physical reality are what this debate is all about.

Counterfactual Probabilities

HV theorists address the problem of the meaning of the counterfactual intersection $\mathcal{A}^+ \cap \mathcal{B}^+$ by the following argument. Consider a very large number $N^{\boldsymbol{a}}$ of particles, prepared by Ted in a standard way, passed through $S^{\boldsymbol{a}}$. By counting the number $N(A_1^+)$ that land in A_1^+, we can estimate the probability $P_{CM}(\mathcal{A}^+) \simeq N(A_1^+)/N^{\boldsymbol{a}}$ that a particle would land in A_1^+ and not in A_1^-. Likewise, if the axis were $\boldsymbol{b}$, then we can estimate the probability $P_{CM}(\mathcal{B}^+) \simeq N(B_1^+)/N^{\boldsymbol{b}}$ that a particle would land in B_1^+ and not in B_1^-.

Note that we are now discussing a probability measure on the set Θ of hidden variables associated with a beam prepared in a standard way by Ted.

According to CM, the rules of classical probability apply, so the counterfactual probability $P_{CM}(\mathcal{A}^+ \cap \mathcal{B}^+)$ that a particle would land in A_1^+ *if* the magnetization axis were $\boldsymbol{a}$, but would have landed in B_1^+ *if* the magnetization axis were $\boldsymbol{b}$, is given by the rule

$$P_{CM}(\mathcal{A}^+ \cap \mathcal{B}^+) = P_{CM}(\mathcal{A}^+)P_{CM}(\mathcal{B}^+). \tag{17.2}$$

HV theorists simply cannot escape this assertion, unless they are prepared to introduce novel possibilities such as nonlocality, contextual effects, and such like, and these generally make their line of argument unappealing.

The problem is, we cannot do any experiments to measure $P_{CM}(\mathcal{A}^+ \cap \mathcal{B}^+)$ directly, or any of the other counterfactual probabilities directly, because as stated, only one magnetization axis can be set up at a time. We shall call this the *simultaneity problem.*

17.3 Circumventing the Simultaneity Problem

An ingenious way of getting around the simultaneity problem is to use two electrons or photons at a time, as follows.

Spin-Zero Two-Electron States

In both CM and QM, electrons are imagined as having an internal degree of freedom called *spin*, associated with angular momentum. There is great evidence for the empirical validity of this idea.

CM and QM treat spin differently, however: the former puts no restriction on the direction of an electron's spin axis or the magnitude of the electron's angular momentum, whereas the latter requires external context (that is, the parameters associated with the apparatus being used) and predicts the quantization of angular momentum in units of $\hbar$, the reduced Planck's constant.[10] Despite their fundamental differences, however, both CM and QM theorists are happy to discuss a two-electron state with total angular momentum of zero.

This means the following in both paradigms: if a spin-zero state of two electrons is prepared and subsequently one of the electrons is passed through $S^{\boldsymbol{a}}$ and observed to land in region A_1^+, then if the other electron was passed through another, identical[11] apparatus $\overline{S}^{\boldsymbol{a}}$, then that second electron would land in region $\overline{A}_1^-$ and not in $\overline{A}_1^+$. Likewise, for every electron that landed in A_1^-, its partner electron would land in $\overline{A}_1^+$.

There is a useful fact about electrons that we can exploit: electrons repel each other. If we fired a beam of two-electron states along a given direction, we should expect that beam to spread out due to this repulsion. This would then allow us to make observations on separate electrons.

We are in position now to circumvent the simultaneity problem as follows.

Alice, Bob, and Ted

Imagine now three experimentalists, Alice, Bob, and Ted, with apparatus shown schematically in Figure 17.2(a). We will call this the *enhanced Stern–Gerlach* (eSG) experiment. In eSG, Bob has an identical copy $\overline{S}^{\boldsymbol{a}}$ of Alice's SG module $S^{\boldsymbol{a}}$, except Bob's module is displaced in the laboratory so as not to overlap Alice's module in their common physical space $\mathbb{P}^3(\Lambda)$.

By stage Σ_0, Ted has prepared a beam of two-electron, spin-zero states. By stage Σ_1, the beam has split by electric repulsion into subbeams 1_1 and 2_1.

[10] The angular momentum of an electron state is generally discussed in terms of $\hbar/2$.

[11] Empirically identical, apart from being spatially displaced, so not identical in the sense of Leibniz.

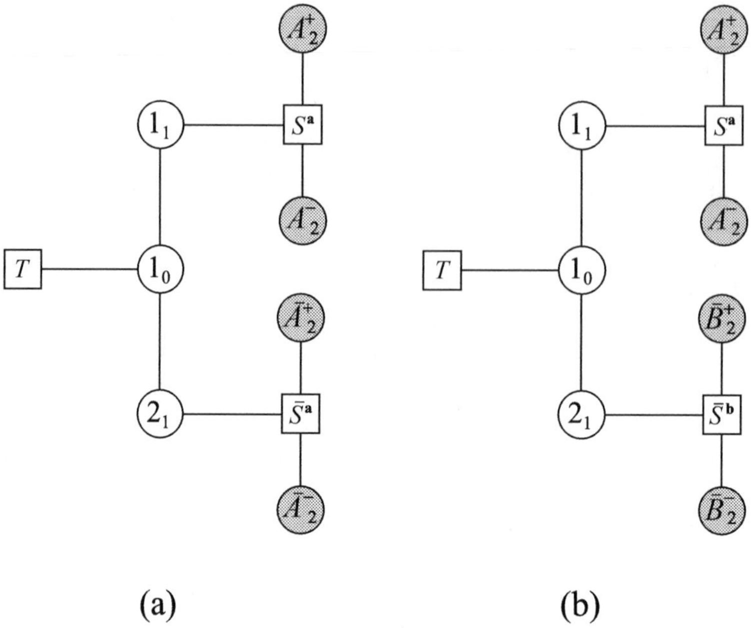

Figure 17.2. The enhanced SG experiment on two-electron, spinless states.

Subbeam 1_1 then enters Alice's SG module $S^{\boldsymbol{a}}$, while subbeam 2_1 enters Bob's module $\overline{S}^{\boldsymbol{a}}$. By conservation of angular momentum (which holds in both CM and QM), whenever Bob observes an electron in region $\overline{A}_2^+$ he can be sure that Alice has observed her electron in region A_2^-, and so on.

It is important to appreciate that this statement is not about statistics; it is about what happens in each individual run. The conservation of angular momentum (as well as that of electric charge, energy, linear momentum, and other classically motivated conserved quantities) takes place at the emergent, process time level of observation. It is an extraordinary and deep fact that in QM, there are classical concepts that can be relied on. The neutrino was discovered precisely because the conservation of energy applies at the individual-run level in beta decay, as proposed by Pauli in 1930, not at the statistical level, as proposed by Bohr.[12]

Now we come to the circumvention of simultaneity problem. The trick here is that Bob need not set the main magnetic field axis in his module to be in the same direction $\boldsymbol{a}$ as that set by Alice in her module $S^{\boldsymbol{a}}$. Suppose Bob sets his

[12] At the time, beta decay experiments could not account for a discrepancy between the energy going into a beta decay process and that going out. Pauli proposed that there was an unobserved particle, now known as the neutrino and subsequently detected by Cowan, Reines, and collaborators in 1956 (Cowan et al., 1956), carrying off the energy discrepancy.

field axis to some different direction $\boldsymbol{b}$ (we will now refer to his module as $\overline{S}^{\boldsymbol{b}}$ rather than $\overline{S}^{\boldsymbol{a}}$). The relevant stage diagram is now Figure 17.2(b). Then we can give an operational definition of $\Pr_{CM}(\mathcal{A}^+ \cap \mathcal{B}^+)$ and the other counterfactual probabilities as follows.

Consider a large number N of runs in the eSG experiment. Suppose Alice and Bob count their respective outcomes, recording the time of each. Afterward, they come together and compare data sets and times. Then for every outcome in which Alice found her electron in A_2^+ *and* Bob found his electron in $\overline{B}_2^-$, in the same run, then that is counted as originating from the subset $\mathcal{A}^+ \cap \mathcal{B}^+$, as if all that information came from the original DS experiment, with a single electron. Essentially, the second electron observed by Bob is like a ghost version of the electron observed by Alice, and both are observed in the same run. Of course, care is taken to take the opposite spins into account. Given the total count, an estimate for the counterfactual probability $\Pr_{CM}(\mathcal{A}^+ \cap \mathcal{B}^+)$ follows immediately.

With this procedure, it is clear that total probability is conserved, that is, we have

$$\Pr_{CM}(\mathcal{A}^+ \cap \mathcal{B}^+) + \Pr_{CM}(\mathcal{A}^+ \cap \mathcal{B}^-) + \Pr_{CM}(\mathcal{A}^- \cap \mathcal{B}^+) + \Pr_{CM}(\mathcal{A}^- \cap \mathcal{B}^-) = 1. \tag{17.3}$$

Bob & Carol & Ted & Alice

To derive the appropriate Bell inequality, the analysis requires the above discussion to be extended to a third direction, $\boldsymbol{c}$, with observer Carol, giving the decomposition $\Theta = \mathcal{C}^+ \cap \mathcal{C}^-$, where $\mathcal{C}^+$ is that subset of Θ that would send an electron into C^+, and similarly for C^-. Classical counterfactuality also entitles us to make the decomposition

$$\begin{aligned}\Theta = {} & (\mathcal{A}^+ \cap \mathcal{B}^+ \cap \mathcal{C}^+) \cup (\mathcal{A}^+ \cap \mathcal{B}^+ \cap \mathcal{C}^-) \cup (\mathcal{A}^+ \cap \mathcal{B}^- \cap \mathcal{C}^+) \cup \\ & (\mathcal{A}^+ \cap \mathcal{B}^- \cap \mathcal{C}^-) \cup (\mathcal{A}^- \cap \mathcal{B}^+ \cap \mathcal{C}^+) \cup (\mathcal{A}^- \cap \mathcal{B}^+ \cap \mathcal{C}^-) \cup \\ & (\mathcal{A}^- \cap \mathcal{B}^- \cap \mathcal{C}^+) \cup (\mathcal{A}^- \cap \mathcal{B}^- \cap \mathcal{C}^-).\end{aligned} \tag{17.4}$$

A Bell Inequality

We are now in position to construct a Bell inequality for this experiment. The first step is to direct the discussion from the set Θ of HV (which by definition we have no knowledge about), to outcome probabilities, for which we have empirical data. Suppose we passed a beam with a large number N of electrons through our SG device, and repeated that run a large number of times. By counting electron impacts on the detector screen, we would then determine outcome frequencies, and finally we would be in position to discuss probabilities.

Given a probability measure P over Θ, then for any subsets A, B of Θ, we have the rule

$$P(A \cup B) = P(A) + P(B) - P(A \cap B). \tag{17.5}$$

With this rule and what we know about the subsets involved, we can deduce the following relations:

$$P_{CM}(\mathcal{A}^+ \cap \mathcal{B}^-) = P_{CM}(\mathcal{A}^+ \cap \mathcal{B}^- \cap \mathcal{C}^+) + P_{CM}(\mathcal{A}^+ \cap \mathcal{B}^- \cap \mathcal{C}^-), \tag{17.6}$$

$$P_{CM}(\mathcal{B}^+ \cap \mathcal{C}^-) = P_{CM}(\mathcal{A}^+ \cap \mathcal{B}^+ \cap \mathcal{C}^-) + P_{CM}(\mathcal{A}^- \cap \mathcal{B}^+ \cap \mathcal{C}^-), \tag{17.7}$$

$$P_{CM}(\mathcal{A}^+ \cap \mathcal{C}^-) = P_{CM}(\mathcal{A}^+ \cap \mathcal{B}^+ \cap \mathcal{C}^-) + P_{CM}(\mathcal{A}^+ \cap \mathcal{B}^- \cap \mathcal{C}^-). \tag{17.8}$$

Every term in these equations is a probability, so is nonnegative. Adding (17.6) to (17.7) and subtracting (17.8), we exploit this nonnegativity directly to find the Bell inequality

$$\boxed{P_{CM}(\mathcal{A}^+ \cap \mathcal{B}^-) + P_{CM}(\mathcal{B}^+ \cap \mathcal{C}^-) \geqslant P_{CM}(\mathcal{A}^+ \cap \mathcal{C}^-).} \tag{17.9}$$

This and similar inequalities are the focus of interest in numerous experiments.

Exercise 17.3 Use (17.5) and other relevant information to prove (17.6), (17.7), and (17.8). Hence prove (17.9).

It has to be pointed out that there are more serious questions about this inequality than those we raised about $P_{CM}(\mathcal{A}^+ \cap \mathcal{B}^+)$. We used two-spin states to circumvent that latter problem, but that approach cannot deal with *three* simultaneous magnetization axes. Essentially, we have to perform separate subexperiments to determine the empirical values $P_{EMP}(\mathcal{A}^+ \cap \mathcal{B}^-)$, $P_{EMP}(\mathcal{B}^+ \cap \mathcal{C}^-)$, and $P_{EMP}(\mathcal{A}^+ \cap \mathcal{C}^-)$ separately, ensuring that standardization across all three subexperiments makes the test of the inequality (17.9) beyond reasonable doubt.

We note in passing that (17.9) looks like the triangle inequality $d(a,b) + d(b,c) \geqslant d(a,c)$ defined for a metric space, where $d(a,b)$ is the "distance" between elements a and b of the metric space.

17.4 The Standard Quantum Calculation

We give now a brief account of the standard QM calculation used to test the inequality (17.9). We will use the nonrelativistic Pauli electron theory, but taking account only of the internal spin degree of freedom of each particle, assumed to be an electron. This is modeled by a qubit Hilbert space $\mathcal{Q}$, no different in mathematical structure from the qubits used to construct QDN quantum registers.

Calibration of Single Electron States

With reference to Figure 17.1, consider an uncalibrated beam of electrons sent by Ted into an SG calibration module $S^{\boldsymbol{k}}$ where $\boldsymbol{k} = (0,0,1)$ in standard Cartesian coordinates defined previously by Ted. As discussed above, this beam will split into two subbeams, denoted K_1^+ and K_1^-. An electron state entering K_1^+ will be denoted $|+\boldsymbol{k}\rangle$, while one entering K_1^+ will be denoted $|-\boldsymbol{k}\rangle$. These two states form a *calibration basis* for $\mathcal{Q}$ and satisfy the orthonormality conditions

$$\langle +\boldsymbol{k}|+\boldsymbol{k}\rangle = \langle -\boldsymbol{k}|-\boldsymbol{k}\rangle = 1, \quad \langle +\boldsymbol{k}|-\boldsymbol{k}\rangle = 0. \tag{17.10}$$

In contrast to QDN, standard QM tends to work with a fixed Hilbert space over any given run. Neither of these approaches is incorrect: they encode the same information differently. We will use the calibration basis as a reference to describe all state vectors and other bases in the following discussion.

Having calibrated his module T, Ted now uses it to create a beam of normalized particle states, $|\Psi\rangle$, at stage Σ_0, given by

$$|\Psi\rangle = u|+\boldsymbol{k}\rangle + v|-\boldsymbol{k}\rangle, \tag{17.11}$$

where $|u|^2 + |v|^2 = 1$.[13]

With reference to Figure 17.1, now suppose a beam of particles represented by such a state is subsequently sent through SG module $S^{\boldsymbol{a}}$, where the main magnetic field direction $\boldsymbol{a}$ is given by $\boldsymbol{a} = (\sin\theta\cos\psi, \sin\theta\sin\psi, \cos\theta)$, where θ and ψ are standard spherical polar coordinates relative to the standard Cartesians referred to hitherto. Any beam passing through $S^{\boldsymbol{a}}$ will in turn be split into two components, denoted A_1^+ and A_1^-, as discussed above. In standard QM, each of these components will be associated with orthogonal, normalized quantum outcome states $|+\boldsymbol{a}\rangle$ and $|-\boldsymbol{a}\rangle$, respectively, and these can be used to form an orthonormal preferred basis for $\mathcal{Q}$, associated with $S^{\boldsymbol{a}}$.

Standard quantum theory gives the following relations between the calibration basis $\{|+\boldsymbol{k}\rangle, |-\boldsymbol{k}\rangle\}$ and the preferred basis $\{|+\boldsymbol{a}\rangle, |-\boldsymbol{a}\rangle\}$ (the outcome states), up to inessential arbitrary overall phase factors:

$$\begin{aligned} |+\boldsymbol{k}\rangle &= \frac{\sin\theta}{\sqrt{2-2\cos\theta}}|+\boldsymbol{a}\rangle + \frac{\sin\theta}{\sqrt{2+2\cos\theta}}|-\boldsymbol{a}\rangle, \\ |-\boldsymbol{k}\rangle &= \frac{(1-\cos\theta)}{\sqrt{2-2\cos\theta}}|+\boldsymbol{a}\rangle - \frac{(1+\cos\theta)}{\sqrt{2+2\cos\theta}}|-\boldsymbol{a}\rangle. \end{aligned} \tag{17.12}$$

From this we can find the amplitude $\mathcal{A}(+\boldsymbol{a}|+\boldsymbol{k}) \equiv \langle +\boldsymbol{a}|+\boldsymbol{k}\rangle$ for outcome state $|+\boldsymbol{a}\rangle$ given initial state $|+\boldsymbol{k}\rangle$, and so on. Hence we can find the conditional probabilities. We find, for example,

$$\Pr(+\boldsymbol{a}|+\boldsymbol{k}) \equiv |\mathcal{A}(+\boldsymbol{a}|+\boldsymbol{k})|^2 = \cos^2(\tfrac{1}{2}\theta). \tag{17.13}$$

Two-Spin States

Disregarding all factors inessential to the present discussion, a normalized two half-spin state $|\Phi\rangle$ of zero total spin zero is given by

$$|\Phi\rangle = \tfrac{1}{\sqrt{2}}|+\boldsymbol{k}^a\rangle \otimes |-\boldsymbol{k}^b\rangle - \tfrac{1}{\sqrt{2}}|-\boldsymbol{k}^a\rangle \otimes |+\boldsymbol{k}^b\rangle, \tag{17.14}$$

[13] The reader will appreciate by now how difficult it is to describe such a process without using suggestive, misleading language.

where the superscripts label the two spin-half particles.[14] We now imagine that particle a is sent through Alice's module $S^{\boldsymbol{a}}$ and particle b is sent through Bob's module $S^{\boldsymbol{b}}$.

Using (17.12), we may write

$$\begin{aligned} |+\boldsymbol{k}^a\rangle &= \alpha^a|+\boldsymbol{a}\rangle + \beta^a|-\boldsymbol{a}\rangle, \quad &|+\boldsymbol{k}^b\rangle = \alpha^b|+\boldsymbol{b}\rangle + \beta^b|-\boldsymbol{b}\rangle, \\ |-\boldsymbol{k}^a\rangle &= \gamma^a|+\boldsymbol{a}\rangle + \delta^a|-\boldsymbol{a}\rangle, \quad &|-\boldsymbol{k}^b\rangle = \gamma^b|+\boldsymbol{b}\rangle + \delta^b|-\boldsymbol{b}\rangle, \end{aligned} \tag{17.15}$$

where again, superscripts label particles and

$$\alpha^a \equiv \frac{\sin\theta^a}{\sqrt{2-2\cos\theta^a}}, \quad \alpha^b \equiv \frac{\sin\theta^b}{\sqrt{2-2\cos\theta^b}}, \tag{17.16}$$

and so on. Note that here $\{|+\boldsymbol{a}\rangle, |-\boldsymbol{a}\rangle\}$ is a preferred basis for $\mathcal{Q}^a$, the qubit associated with Alice's electron, and $\{|+\boldsymbol{b}\rangle, |-\boldsymbol{b}\rangle\}$ is a preferred basis for $\mathcal{Q}^b$, the qubit associated with Bob's electron.

Given (17.15), we readily find

$$|\Phi\rangle = \tfrac{1}{\sqrt{2}} \begin{array}{l} \{(\alpha^a\gamma^b - \gamma^a\alpha^b)|+\boldsymbol{a}\rangle \otimes |+\boldsymbol{b}\rangle + (\alpha^a\delta^b - \gamma^a\beta^b)|+\boldsymbol{a}\rangle \otimes |-\boldsymbol{b}\rangle \\ +(\beta^a\gamma^b - \delta^a\alpha^b)|-\boldsymbol{a}\rangle \otimes |+\boldsymbol{b}\rangle + (\beta^a\delta^b - \delta^a\beta^b)|-\boldsymbol{a}\rangle \otimes |-\boldsymbol{b}\rangle\} \end{array}. \tag{17.17}$$

The Bell Inequality

Now recall that the focus of attention here is the classical Bell inequality (17.9). Considering the CM single particle counterfactual probability $P_{CM}(\mathcal{A}^+, \mathcal{B}^-)$, this translates into the QM outcome probability $P_{QM}(+\boldsymbol{a}, +\boldsymbol{b}|\Phi)$ in the extended SG experiment (involving both Alice and Bob). This means that the classical Bell inequality (17.9) is replaced by the assertion that

$$P_{QM}(+\boldsymbol{a}, +\boldsymbol{b}|\Phi) + P_{QM}(+\boldsymbol{b}, +\boldsymbol{c}|\Phi) - P_{QM}(+\boldsymbol{a}, +\boldsymbol{c}|\Phi) \geqslant 0. \tag{17.18}$$

The first term in this expression is just the squared modulus of the coefficient of the tensor product term $|+\boldsymbol{a}\rangle \otimes |+\boldsymbol{b}\rangle$ in (17.17). Hence we deduce

$$P_{QM}(+\boldsymbol{a}, +\boldsymbol{b}|\Phi) = \tfrac{1}{2}|\alpha^a\gamma^b - \gamma^a\alpha^b|^2. \tag{17.19}$$

We can simplify this expression by using rotational symmetry, orienting our Cartesian coordinates along the direction of vector $\boldsymbol{a}$. Then we find

$$P_{QM}(+\boldsymbol{a}, +\boldsymbol{b}|\Phi) = \tfrac{1}{2}\sin^2(\tfrac{1}{2}\theta^{ab}), \tag{17.20}$$

where θ^{ab} is the angle between $\boldsymbol{a}$ and $\boldsymbol{b}$. A similar calculation for the other two terms in (17.15) gives

$$\sin^2(\tfrac{1}{2}\theta^{ab}) + \sin^2(\tfrac{1}{2}\theta^{bc}) - \sin^2(\tfrac{1}{2}\theta^{ac}) \geqslant 0. \tag{17.21}$$

[14] We can ignore the fact that electrons are identical and obey Fermi–Dirac statistics, because the spreading of the beam prior to the electrons entering either $SG(\boldsymbol{a})$ or else $SG(\boldsymbol{b})$ has introduced a form of classical labelling.

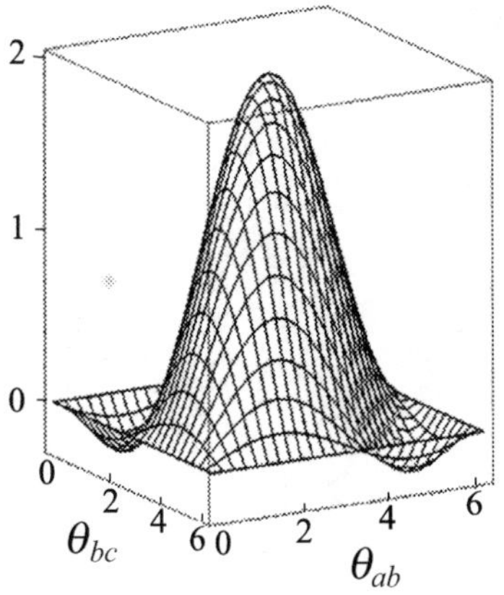

Figure 17.3. Plot of the function $f(\theta_{ab}, \theta_{bc})$ over a suitable domain.

The problem is that we can find vectors $\boldsymbol{a}$, $\boldsymbol{b}$, and $\boldsymbol{c}$ for which (17.21) is wrong. For example, take these vectors to lie in a plane, with $\theta^{ab} = \pi/3$, $\theta^{bc} = \pi/3$, $\theta^{ac} = \theta^{ab} + \theta^{bc} = 2\pi/3$. Then

$$\sin^2(\tfrac{1}{2}\tfrac{\pi}{3}) + \sin^2(\tfrac{1}{2}\tfrac{\pi}{3}) - \sin^2(\tfrac{1}{2}\tfrac{2\pi}{3}) = \tfrac{1}{4} + \tfrac{1}{4} - \tfrac{3}{4} = -\tfrac{1}{4}. \tag{17.22}$$

In Figure 17.3 we show a plot of the function $f(\theta^{ab}, \theta^{bc}) \equiv \sin^2(\frac{1}{2}\theta^{ab}) + \sin^2(\frac{1}{2}\theta^{bc}) - \sin^2(\frac{1}{2}\theta^{ab} + \frac{1}{2}\theta^{bc})$ over a range of possibilities. There are two distinct regions where the function value is negative, while the HV calculation predicts that there should be no such regions.

There have been many experiments related to the one discussed here that have shown violations of Bell's inequalities, a frequently quoted one being that of Aspect and others using photons (Aspect et al., 1982).

There is now not much doubt among the majority of physicists that classical counterfactuality has been shown empirically to be a false principle in physics. It remains an excellent principle as far as relative external context (the wider Universe) is concerned: we can usually go to work and remain confident that our house will still be there when we get back in the evening.

There remains a relatively small group of committed HV theorists who continue to probe this issue and have come up with classically based possible explanations for the observed empirical violations of Bell's inequalities. Experimentalists continue to test these loopholes, and have reached the point where the HV classically motivated "explanations" seem more unpalatable than the quantum theory they are trying to circumvent.

17.5 The QDN Calculation

In this section we apply QDN to the eSG scenario discussed above. The relevant stage diagram is Figure 17.4.

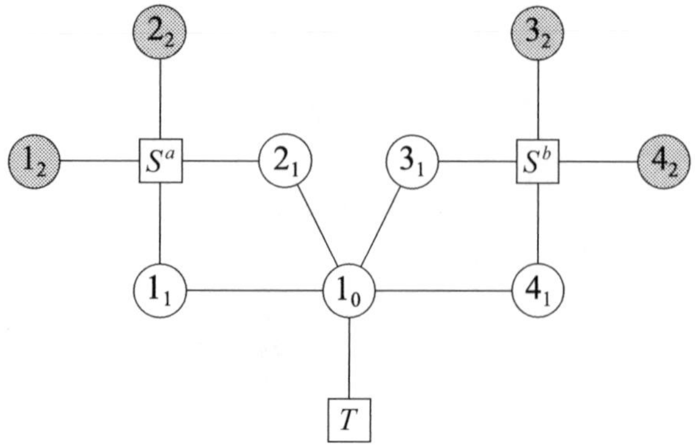

Figure 17.4. The QDN stage diagram of the enhanced SG experiment.

It was found that the total state at stage Σ_1, corresponding to the entangled state (17.14) needs to be described as a state in the tensor product of a four-dimensional internal spin space and a rank-four quantum register.

Stage $\mathbf{\Sigma_0}$

The preparation switch stage Σ_0 creates a beam of spin-zero, two electron states, represented by the total state $|\Psi_0\rangle \equiv |s_0^1\rangle \otimes \widehat{\mathbb{A}}_0^1 \mathbf{0}_0$. Here, $|s_0^1\rangle$ represents a normalized state in the one-dimensional Hilbert space of two-electron spin zero states.

Stage $\mathbf{\Sigma_0}$ to Stage $\mathbf{\Sigma_1}$

By the first stage, Σ_1, the beam has split into two entangled spin-half subbeams. This stage is *before* any of the subbeams enter their respective SG apparatus. Bitification requires each of these sub-beams to be associated with *two* signal qubits,[15] so we require a rank-four quantum register at that stage, as stated above.

The dynamics is given by the rule

$$U_{1,0}\left\{|s_0^1\rangle \otimes \widehat{\mathbb{A}}_0^1 \mathbf{0}_0\right\} = \frac{1}{\sqrt{2}}\left\{\begin{array}{l} |+\boldsymbol{k}_1^a\rangle \otimes |-\boldsymbol{k}_1^b\rangle \otimes \widehat{\mathbb{A}}_1^1 \widehat{\mathbb{A}}_1^3 - \\ |-\boldsymbol{k}_1^a\rangle \otimes |+\boldsymbol{k}_1^b\rangle \otimes \widehat{\mathbb{A}}_1^2 \widehat{\mathbb{A}}_1^4 \end{array}\right\} \mathbf{0}_1. \quad (17.23)$$

Stage Σ_1 to Stage Σ_2

Using (17.15) and from Figure 17.4 we have

$$\begin{aligned} U_{2,1}\left\{|+\boldsymbol{k}_1^a\rangle \otimes |-\boldsymbol{k}_1^b\rangle \otimes \widehat{\mathbb{A}}_1^1 \widehat{\mathbb{A}}_1^3 \mathbf{0}_1\right\} &= \left\{\alpha^a|+\boldsymbol{a}_2\rangle \otimes \widehat{\mathbb{A}}_2^1 + \beta^a|-\boldsymbol{a}_2\rangle \otimes \widehat{\mathbb{A}}_2^2\right\} \\ &\quad \times \left\{\gamma^b|+\boldsymbol{b}_2\rangle \otimes \widehat{\mathbb{A}}_2^3 + \delta^b|-\boldsymbol{b}_2\rangle \otimes \widehat{\mathbb{A}}_2^4\right\} \mathbf{0}_2, \\ U_{2,1}\left\{|-\boldsymbol{k}_1^a\rangle \otimes |+\boldsymbol{k}_1^b\rangle \otimes \widehat{\mathbb{A}}_1^2 \widehat{\mathbb{A}}_1^4 \mathbf{0}_1\right\} &= \left\{\gamma^a|+\boldsymbol{a}_2\rangle \otimes \widehat{\mathbb{A}}_2^1 + \delta^a|-\boldsymbol{a}_2\rangle \otimes \widehat{\mathbb{A}}_2^2\right\} \\ &\quad \times \left\{\alpha^b|+\boldsymbol{b}_2\rangle \otimes \widehat{\mathbb{A}}_2^3 + \beta^b|-\boldsymbol{b}_2\rangle \otimes \widehat{\mathbb{A}}_2^4\right\} \mathbf{0}_2. \end{aligned} \quad (17.24)$$

[15] This is on account of the fact that entanglement can be detected, given the right apparatus.

This is all that is needed to run our computer algebra program MAIN, which gives us much more information than we need here. Our interest originated in $P_{CM}(+\boldsymbol{a}, -\boldsymbol{b})$. Context tells us that we need to calculate the probability that detectors 1_2 and 3_2 are each in their respective signal state at stage Σ_2. Program MAIN gives us the answer

$$\Pr(\widehat{\mathbb{A}}_2^1\widehat{\mathbb{A}}_2^3\mathbf{0}_2|\Psi_0) = \tfrac{1}{2}|\alpha^a\gamma^b - \gamma^a\alpha^b|^2, \tag{17.25}$$

which is precisely (17.19), the result of the standard QM calculation. The same argument applies to the other terms in the Bell inequality (17.18).

Our conclusion is that QDN gives the same results as standard QM, and hence the same prediction of violations of Bell inequalities.

18

Change and Persistence

18.1 Introduction

Our interest in this chapter is in *change*, which is another way of discussing *persistence*, or the apparent endurance over intervals of time of spatially extended structures.

Persistence is the phenomenon that underpins all human activity. Given the rule that we may think of as the *first law of time*, the dictum of Heraclitus that *everything changes*, persistence is our name for those remarkable processes that appear to circumvent that law and give stability to our lives. We should be interested, for instance, in the fact that when we wake up each morning, we feel that we are the same individuals who went to sleep the previous night. Indeed, without persistence in one form or another, nothing would make sense, including logic, rational thought, and mathematics, for these depend on comparisons of standards and rules that persist in memory with information that we acquire in process time.

Persistence is necessary for physics to make sense. It is emphatically *not* a metaphysical topic but of the greatest relevance to science, including quantum mechanics (QM). Persistence is implicit in Wheeler's dictum that only acts of observation are meaningful in physics, because observers and their apparatus have to endure long enough to make observations. In quantized detector networks (QDN), the persistence of observers and their apparatus long enough to perform experiments should be regarded as axiomatic.

The subject of persistence is a deep and complex one, and will probably never be fully understood. Several reasons contribute to this.

Observers Themselves are Subject to Change

Contrary to the general implicit belief that the laws of physics transcend the subjective behavior of observers, it is our thesis that this is a vacuous proposition. It cannot be proved; it is just an assertion, albeit a very useful one. A more pragmatic view is the one held throughout this book, that the laws of physics are contextual to observers. Without observers, science means nothing in a literal sense. But according to the first law of time, observers themselves change over time. This includes their memories and the conditioning that they have; these are carried over from one stage to another. If the processes transporting those memories and belief structures change those memories and belief structures, then the laws of physics *as understood by the observers at the time* change as well. Five hundred years ago, science knew little of Newton's laws of motion; two hundred years ago, science did not know that there was such a thing as quantum mechanics.

We deal with the variability of observers by recognizing that truth values of propositions, including the laws of physics, are contextual to observers at the time of observation only. That the laws of physics themselves may change as the Universe expands is not a fanciful idea: the possibility that the so-called constants of nature have changed has been looked at by theorists such as Dirac (1938b) and Magueijo (2003). A complete theory of observation would take into account dynamical interaction between systems under observation, observers, and the Universe that contains those observers.

Time Scales Are Important

Give the contextuality of persistence, what are the factors that contribute to that contextuality? One of them is *time scale*, for the time scales over which objects are said to persist is a crucial factor in creating the illusion. Indeed, persistence may be thought of as a comparison of two time scales: the first is associated with the SUOs being investigated, and the second is the time scale that the observer regards as significant *for their purposes.*

Example 18.1 A scientist investigating the flight of insects needs to employ video equipment that can adequately capture the motion of insect wings. There will be no point in using a video frame speed of 25 frames per second if an insect under observation flaps its wings 100 times per second.

Sheer Complexity

Another crucial factor associated with persistence is that observers and SUOs are hideously complex phenomena. Indeed, the concepts of science are attempts to organize that complexity into sufficiently simple forms that the brain can interpret relatively quickly and easily.

Example 18.2 The human eye has a detecting screen known as the retina, over which are detectors known as rods and cones. There are perhaps 120 million rods and 6 million cones in the typical human eye. When light from suitable sources strikes the retina, many of these detectors will not register a signal. Of those that do, complex chemical processes in the rods and cones occur, resulting in signals being passed into specialized nerve layers below the retina. These raw signals from the rods and cones are then processed in those nerve layers before more complex signals are passed further on into the brain, where much more complex processes of comparison and pattern recognition must be occurring. The templates that the brain uses in these comparisons appear to be relatively stable themselves, so much so that the brain does its best to match the incoming signals to what it, the brain, has already prepared and expected for. When this process fails in some way, optical illusions can occur, such as a failure to recognize individuals whom we have met before. The simplicity that we think we see around us is a total illusion.

Quantum State Fragility

A critical factor contributing to persistence is atomic stability: atoms are generally stable, at least long enough to create the illusion of persistence. This brings us to perhaps *the* greatest enigma in physics, one that is necessary for persistence in the first place and one that could not be resolved by classical mechanics (CM). According to CM principles, an accelerating electric charge should dissipate energy by electromagnetic radiation. This led to the classical prediction that hydrogen atoms should be unstable, contrary to empirical evidence.

In order to bypass this prediction, Bohr constructed a model of hydrogen based on an unexplained veto of such radiation (Bohr, 1913). More refined QM, such as wave mechanics and quantum field theory, account for many structural aspects of atoms, but a completely satisfactory, reductionist explanation of atomic stability does not exist. For instance, the postulate of normalized wave functions in Schrödinger mechanics that leads to quantized atomic energy levels in hydrogen is manifestly an emergent one based on an implicit assumption of a persistent exophysical observer dealing only with normalized states. In interacting quantum field theories too, there is generally no proof that the vacuum, or state of lowest energy, exists; such a state is normally postulated.

The stability of atoms helps create the illusions of persistence. But below that classical layer of illusion is a seething mass of change. Of relevance here is the idea that the classical world view is good when quantum phases are so random and disorganized that a classical average approach can be relied on. The greatest developments in QM have occurred principally in the theoretical understanding and empirical control of quantum phase. For example, Feynman's path integral formulation of QM is based on a specific mathematical recipe for adding quantum phases associated with different dynamical paths. On the empirical side, the various quantum optics experiments discussed in the present book demonstrate the point excellently.

Our ambition to understand persistence will be limited. We shall discuss the persistence of labstates in a carefully controlled environment. Specifically, given an initial labstate $\boldsymbol{\Psi}_m$ at stage Σ_m, suppose it is allowed to evolve through a quantized detector network to labstate $\boldsymbol{\Psi}_n$ at a later[1] stage Σ_n, under the evolution given by $\boldsymbol{\Psi}_n = \mathbb{U}_{n,m}\boldsymbol{\Psi}_m$. A natural question to ask is: *how different are states $\boldsymbol{\Psi}_m$ and $\boldsymbol{\Psi}_n$?*

This question is not simple to answer, for several reasons.

The Parallel Transport Problem

An analogous problem arises in general relativity (GR), where there is a need to make a comparison of directions at different places. Specifically, given two points P and Q in a GR spacetime manifold, there is a tangent space associated with each point, denoted T_P for the tangent space at P, and T_Q for that at Q. Suppose we pick a vector $\boldsymbol{v}_P$ in T_P and vector $\boldsymbol{v}_Q$ in T_Q. The question is: how can we compare $\boldsymbol{v}_P$ and $\boldsymbol{v}_Q$? The problem is that these are vectors in two different vector spaces. Somehow, a method of "bringing them together" has to be found, so that a proper comparison can be made.

A solution to this problem was found by the mathematicians Christoffel, Levi-Civita, and others, who developed Riemannian (differential) geometry in the nineteenth and early twentieth centuries. Their approach was to introduce the notion of a *connection*, a rule for relating basis vectors in T_P to those in T_Q. Such a rule involves *parallel transport*, which means defining what it means to transport a vector from one tangent space to another in a manifold "without changing direction."

There is no unique connection over a manifold. Fortunately for Einstein, who from about 1911 onward was developing GR, there is a natural and unique connection for any manifold that has a metric, or distance, rule. The so-called *metric connection* can be derived from the metric tensor of the spacetime concerned, and GR deals specifically with manifolds with Lorentzian signature metrics. The metric connection in GR has been found over the last century to give excellent empirical predictions, such as the precession of the orbit of the planet Mercury, the gravitational deflection of light, and galactic lensing. Accordingly, it is the connection in general use to this day, although alternatives such as those with torsion are occasionally proposed.

In QDN, an important feature that assists us here is that empirical context gives a preferred basis B_m for the quantum register $\mathcal{Q}_m$ at stage Σ_m and likewise for stage Σ_n. Therefore, the parallel transport problem seems to have a natural resolution.

However, QDN does not insist on the constancy of quantum register dimensions from stage to stage. When discussing persistence, we need to ensure that the dimensions of the registers concerned are the same, that is, $\dim \mathcal{Q}_{n+1} = \dim \mathcal{Q}_n$. Otherwise, we would be trying to relate labstates in vector spaces that were

[1] In this context, stage Σ_n *later* than stage Σ_m always means $n > m$.

inherently dissimilar. This dimensionality issue does not arise in the case of Riemannian geometry because the dimension of each tangent space in a manifold is precisely the same as the dimension of the manifold over which they are defined.

We shall henceforth assume that when an observer is measuring the change in a state of an SUO, it will be under carefully controlled conditions such that $\dim \mathcal{Q}_n = \dim \mathcal{Q}_m \equiv r$. For instance, we might be interested in some branch of macroscopic quantum mechanics, such as superfluidity or superconductivity. Changes in quantum register dimension in that context would correspond to fluctuations in the numbers of electrons or other particles in the SUO, which would be a manifest breakdown of persistence.

Do We Know What We Are Doing?

One possible reason for the difficulty we have in understanding persistence is that perhaps we really don't understand the problem. It may be the case that QM is the wrong branch of physics needed to understand persistence. A comparison of $\boldsymbol{\Psi}_m$ and $\boldsymbol{\Psi}_n$ is a comparison of quantum states: these do not represent a reality that "exists" in a classical sense. As emphasized before in this book, quantum states represent the statistics of empirical context and are not snapshots of reality, while the persistence that we see around us is an illusion based on a near instantaneous comparison of signals received on our retinas and patterns in our memories. Whether QM is right way to discuss that process of comparison is debatable. The assumption that it is leads to vacuous concepts such as "wave functions for the Universe."

This issue has long been discussed in QM, examples being Ehrenfest's theorem (Ehrenfest, 1927) and decoherence theory (Zurek, 2002). These are viewed by us as attempts to bridge the gap known as the Heisenberg cut, from the relative internal contextual (RIC) side where QM is good and the relative external context (REC) where a classical description is good. We view these attempts as misguided, because REC is the domain proper of emergent physics, and we should not expect any reductionist approach to "explain" emergence. We note that despite initial hopes that decoherence would "explain" why the world around us appears classical, these expectations have not been met and have been criticized (Kastner, 2016).

As in other situations, the resolution of our problem comes from looking at what goes on in the laboratory, not in our theories. When all is said and done, the only things that matter in physics are the signals received in our detectors. That is where we can understand persistence and change. We shall continue our investigation into persistence, therefore, along the lines of comparing changes in labstate outcomes.

Correlation of Preferred Basis States

Given a computational basis representation (CBR) $B_m \equiv \{\boldsymbol{i}_m : 0 \leqslant i < 2^r\}$ for $\mathcal{Q}_m$ and a CBR $B_n \equiv \{\boldsymbol{i}_n : 0 \leqslant i < 2^r\}$ for $\mathcal{Q}_n$, we need to relate the individual elements of each preferred basis. Clearly, there will be no use attempting to relate these basis states if the quantum registers have unrelated contexts. For instance, if the stage Σ_m apparatus is observing photon signals and stage Σ_n is observing electron signals, then there may be no natural relationships between them. Therefore, the discussion requires us to correlate contexts.

Assuming this is the case, then since the labeling of CBR elements is arbitrary, we can assume that the ordering of elements in each preferred basis is correlated contextually. This means for instance that $\mathbf{0}_m$ is correlated to $\mathbf{0}_n$, that $\mathbf{1}_m$ is correlated to $\mathbf{1}_n$, and so on.

The Null Evolution Operator

Given the above provisos, an important operator will be the *null evolution operator* $\mathbb{N}_{n,m}$, defined by

$$\mathbb{N}_{n,m} \equiv \sum_{i=0}^{2^r-1} \boldsymbol{i}_n \overline{\boldsymbol{i}_m}. \tag{18.1}$$

This will have the critical role, rather like the metric connection in GR, of defining a concept of parallel transport in QDN, as we see from the following argument. Given an initial labstate

$$\boldsymbol{\Psi}_m = \sum_{i=0}^{2^r-1} \Psi_m^i \boldsymbol{i}_m, \tag{18.2}$$

we find

$$\tilde{\boldsymbol{\Psi}}_n \equiv \mathbb{N}_{n,m} \boldsymbol{\Psi}_m = \sum_{i=0}^{2^r-1} \Psi_m^i \boldsymbol{i}_n, \tag{18.3}$$

which is a carbon copy of $\boldsymbol{\Psi}_m$ but at stage $\boldsymbol{\Sigma}_n$.

18.2 Comparisons

The discussion now reduces to a comparison of the naturally evolved labstate $\boldsymbol{\Psi}_n \equiv \mathbb{U}_{n,m} \boldsymbol{\Psi}_m$ and the persistent image state $\tilde{\boldsymbol{\Psi}}_n \equiv \mathbb{N}_{n,m} \boldsymbol{\Psi}_m$. The problem is that there is no natural measure of difference between two quantum register states in the same register, no measure of *distance*, that survives all criticism. The following are some possibilities.

The Born Measure of Similarity

In standard QM Hilbert space theory, the inner product (Φ, Ψ) of two normalized state vectors in the same Hilbert space has an empirical significance: if $(\Phi, \Psi) = 0$, the two states are regarded as totally different. More generally, the square modulus $P(\Phi|\Psi) \equiv |(\Phi|\Psi)|^2$ of this amplitude has the Born interpretation as the conditional probability of a positive answer if tested for state Φ given prepared state Ψ.

We may, with some justification, refer to $P(\Phi|\Psi)$ as the *Born measure of similarity*.

The problem with the Born measure of similarity is that it can be far too crude for quantum register states, as the following example illustrates.

Example 18.3 Consider a rank-billion quantum register $\mathcal{Q}^{[10^9]}$. The states $\Psi \equiv \widehat{\mathbb{A}}^1\widehat{\mathbb{A}}^2\ldots\widehat{\mathbb{A}}^{999999999}\widehat{\mathbb{A}}^{10^9}\mathbf{0}$ and $\Phi \equiv \widehat{\mathbb{A}}^1\widehat{\mathbb{A}}^2\ldots\widehat{\mathbb{A}}^{999999999}\mathbf{0}$ have zero inner product, but differ in only one signal out of a billion. By any conventional, heuristic measure of similarity, these two states would be regarded as very similar though not identical, but the Born measure of similarity gives them as totally different.

We conclude that the Born measure of similarity is not good enough, in the context of this chapter and the next, as a measure of similarity.

The Hamming Measure of Dissimilarity

The problem of comparing two quantum register states in the same quantum register has an interesting parallel in the world of computing, cryptography, and information science. In those subjects, a frequent problem is to compare two strings of symbols, such as $S^1 \equiv \underline{\alpha Q55\{7}$ and $S^2 \equiv \underline{aP56[7}$. There are two possible cases: either the strings have the same number of elements (six in this case), or they have different numbers of elements. We shall discuss only the equal case.

Equal-Length Strings

In his study of the transmission of information over telephone networks, Hamming discussed the problem of identifying and then correcting errors in transmission (Hamming, 1950). This scenario is very much like a quantum experiment, with essentially the same architecture, including an information void. There are observers (the *speaker* and the *listener*) and their preparation devices and final state detectors (the telephones). The information void here consists of extensive transmission lines and modules, such as telephone exchanges, signal amplifiers, and so on.

Suppose $S \equiv \underline{s^1 s^2 \ldots s^r}$ is the message that is sent (the prepared state) and $T \equiv \underline{t^1 t^2 \ldots t^r}$ is the message that is actually received (the outcome state). A natural question is: *how different are these two strings?* Hamming devised a geometrical method of quantifying the difference between two character strings of equal length. He defined a distance, or metric, $d_H(S,T)$ between S and T according to the rule

$$d_H(S,T) \equiv \text{number of matched pairs } (s^i, t^i), i = 1, 2, \ldots, r, \text{ for which } s^i \neq t^i. \tag{18.4}$$

Example 18.4 For $S \equiv \underline{\alpha Q55\{7}$ and $T \equiv \underline{aP56[7}$ we find $d_H(S,T) = 4$, as these strings of length six differ everywhere except in their third and sixth elements.

The interpretation of the Hamming distance between two equal length strings is that it is the minimum number of single character replacements in one string that would convert it into the other. In Example 18.4, we can convert S into T by the four replacements $\alpha \to a$, $Q \to P$, $5 \to 6$, and $\{ \to [$.

Exercise 18.5 Prove that the Hamming metric is a true metric in the sense of a metric space.

Suppose we prepare a labstate $\boldsymbol{\Psi}_m$ in a rank-r quantum register $\mathcal{Q}_m$ at stage Σ_m, and pass it through an information void, until it is received as outcome $\boldsymbol{\Psi}_n$ in an identical rank quantum register $\mathcal{Q}_n$ at stage Σ_n, where $n > m$.[2] The two registers are copies of each other, having the same rank. Moreover, their preferred bases are contextually correlated: computational basis element $\boldsymbol{i}_m$ means the same to the speaker (the observer at stage $\boldsymbol{\Sigma}_m$) as $\boldsymbol{i}_n$ means to the listener (the observer at stage Σ_n). In other words, speaker and listener understand the same language.

Taking account of the parallel transport problem discussed above, we need to compare like with like. We choose to compare the persistent image $\tilde{\boldsymbol{\Psi}}_n \equiv \mathbb{N}_{n,m}\boldsymbol{\Psi}_m$ in $\mathcal{Q}_n$ of the original state with the transmitted state $\boldsymbol{\Psi}_n \equiv \mathbb{U}_{n,m}\boldsymbol{\Psi}_m$, also in $\mathcal{Q}_n$.

The problem is compounded here by *superposition*: neither $\boldsymbol{\Psi}_n$ nor $\tilde{\boldsymbol{\Psi}}_n$ are classical, but thay are complex superpositions of classical information (which is essentially what preferred basis elements are). This is a factor that Hamming did not face.

In the following, we drop the temporal index, as it does not play an essential role.

Hamming Distance between Preferred Basis Elements

In the simplest case, case, suppose $\boldsymbol{\Psi} = \boldsymbol{i}$ and $\tilde{\boldsymbol{\Psi}} = \boldsymbol{j}$, where $\boldsymbol{i}$ and $\boldsymbol{j}$ are elements of the computational basis representation (CBR) for $\mathcal{Q}$. Each basis element carries classical information, that is, corresponds to a signal state that could actually be observed in a run. We calculate the Hamming distance $d_H(\boldsymbol{i}, \boldsymbol{j})$ between these two elements by first converting each integer into its associated binary string and then working out the Hamming distance using the rule given in (18.4).

That means applying the process of binary decomposition. Specifically, we write

$$\begin{aligned} i &= i^{[1]} + i^{[2]}2 + i^{[3]}2^2 + \cdots + i^{[r]}2^{r-1}, \\ j &= j^{[1]} + j^{[2]}2 + j^{[3]}2^2 + \cdots + j^{[r]}2^{r-1}, \end{aligned} \tag{18.5}$$

[2] Note that the transmitted state is *not* an actual outcome (which is *classical information*), but a quantum state immediately before it is looked at for an outcome, and is therefore best thought of as a different form of information, referred to by us as *quantum information*.

Table 18.1 *The Hamming distance between elements of a rank-three quantum register basis*

	$\underline{000}$	$\underline{100}$	$\underline{010}$	$\underline{110}$	$\underline{001}$	$\underline{101}$	$\underline{011}$	$\underline{111}$
$\mathbf{0} \equiv \underline{000}$	0	1	1	2	1	2	2	3
$\mathbf{1} \equiv \underline{100}$	1	0	2	1	2	1	3	2
$\mathbf{2} \equiv \underline{010}$	1	2	0	1	2	3	1	2
$\mathbf{3} \equiv \underline{110}$	2	1	1	0	3	2	2	1
$\mathbf{4} \equiv \underline{001}$	1	2	2	3	0	1	1	2
$\mathbf{5} \equiv \underline{101}$	2	1	3	2	1	0	2	1
$\mathbf{6} \equiv \underline{011}$	2	3	1	2	1	2	0	1
$\mathbf{7} \equiv \underline{111}$	3	2	2	1	2	1	1	0

where the coefficients $i^{[k]}, j^{[k]}$, $k = 1, 2, \ldots, r$, are each either zero or one. Next, we form the strings $S^i \equiv \underline{i^{[1]}i^{[2]} \ldots i^{[r]}}$, $S^j \equiv \underline{j^{[1]}j^{[2]} \ldots j^{[r]}}$. Finally, we calculate the Hamming distance $d_H(S^i, S^j)$.

Table 18.1 gives the Hamming distance between pairs of elements in the preferred basis for a rank-three quantum register.

We note the following:

1. The maximally saturated state $\mathbf{7} \equiv \underline{111}$ is furthest away in Hamming distance terms from the ground state $\mathbf{0} \equiv \underline{111}$,
2. All elements of the same signality σ are a Hamming distance σ from the signal ground state.
3. For any two register basis elements, the Hamming distance between them defines what can be thought of as a *relative signality*: if any one of these elements were chosen to be the new ground state, then the other state would have signality equal to its Hamming distance from that new ground state. This underlines the fact that the ground state in a QDN register is not intrinsic to the apparatus but *defined* contextually by the observer. The choice of signal ground state is *not* dictated by lowest energy state.

The power of Hamming's approach is that his distance rule is a genuine metric, so that all the theorems of metric spaces can be applied. Lest this appear trivial, we should consider that quantum registers modeling real situations may have immense rank, such as that modeling a superconducting quantum interferometer. In such situations, the Hamming metric distance approach may well prove useful.

The above approach is classical, in that it covers classical register states completely. QM, however, involves superpositions of register basis states, so we are faced with the problem of defining the equivalent of a Hamming distance between arbitrary, normalized quantum register states in the same register.

The Hamming Operator

For a given rank-r quantum register $\mathcal{Q}$ with CBR $\{\boldsymbol{i} : i = 0, 1, 2, \ldots, 2^r - 1\}$ we define the *Hamming operator* $\mathbb{H}$ by

$$\overline{\boldsymbol{i}}\mathbb{H}\boldsymbol{j} \equiv d_H(i, j), \tag{18.6}$$

where $d_H(\boldsymbol{i}, \boldsymbol{j})$ is the Hamming distance between elements $\boldsymbol{i}$ and $\boldsymbol{j}$. Then using completeness, we have the representation

$$\mathbb{H} = \sum_{i,j=0}^{2^r-1} \boldsymbol{i} d_H(i,j) \overline{\boldsymbol{j}}. \tag{18.7}$$

We would in the first instance like to use this operator to attempt several definitions of "distance" between more general quantum register states. Examples are the following.

Quantum Hamming Measure of Dissimilarity

Given the CBR expansions $\boldsymbol{\Phi} \equiv \sum_{i=0}^{2^r-1} \Phi^i \boldsymbol{i}$ and $\boldsymbol{\Psi} \equiv \sum_{i=0}^{2^r-1} \Psi^i \boldsymbol{i}$ we define the quantum Hamming measure of dissimilarity $H_1(\boldsymbol{\Phi}, \boldsymbol{\Psi})$ as the magnitude of the matrix element $\overline{\boldsymbol{\Phi}} \mathbb{H} \boldsymbol{\Psi}$, that is,

$$H_1(\boldsymbol{\Phi}, \boldsymbol{\Psi}) \equiv \left| \sum_{i,j=0}^{2^r-1} \Phi^{i*} d_H(i,j) \Psi^j \right|. \tag{18.8}$$

A variant proposal is to define the quantum Hamming measure of dissimilarity H_2 as

$$H_2(\boldsymbol{\Phi}, \boldsymbol{\Psi}) \equiv \sum_{i,j=0}^{2^r-1} \Pr(\boldsymbol{i}|\boldsymbol{\Phi}) d_H(i,j) \Pr(\boldsymbol{j}|\boldsymbol{\Psi}) = \sum_{i,j=0}^{2^r-1} |\Phi^i|^2 d_H(i,j) |\Psi^j|^2. \tag{18.9}$$

Neither (18.8) nor (18.9) is a genuine metric, because they do not return zero in general when $\boldsymbol{\Phi} = \boldsymbol{\Psi}$. However, that is not necessarily a bad thing. After all, quantum states are not objective things. The fact that neither $H_1(\boldsymbol{\Psi}, \boldsymbol{\Psi})$ nor $H_2(\boldsymbol{\Psi}, \boldsymbol{\Psi})$ is not zero in general reflects the fact that labstates are not classical. If they were, they would be represented by a single element of the CBR, and in that case, both measures of dissimilarity vanish.

18.3 Signal Correlation Measure of Change

Suppose an observer is able to prepare and investigate, separately, two labstates $\boldsymbol{\Psi}$ and $\boldsymbol{\Phi}$ in a rank-r quantum register $\mathcal{Q}$. This means that the observer can examine each of these labstates separately and measure the answer to any maximal or partial question about either of these states. Another way of saying this is that the observer can determine, empirically, the frequencies of finding either labstate in a given preferred basis state. In reality, that is *all* that an observer can do.

With such information, the observer can meaningfully discuss differences in signal outcome probabilities, and it is on that basis that we now proceed. First we review briefly the concept of *correlation*.

Correlations in Statistics

The problem of comparing two data sets occurs frequently in statistics. Suppose we have two sets of data, X and Y, each consisting of n real numbers, such that the ith element of X is x^i and the ith element of Y is y^i. The question often arises: how "close" are the two sequences $\{x^1, x^2, \ldots, x^n\}$ and $\{y^1, y^2, \ldots, y^n\}$?

This question is a variant of the one originally addressed by Hamming. Statisticians have developed the concept of *correlation* in a variety of forms that attempt to give a number, usually between -1 and $+1$, that gives the degree of similarity, dependence, or correspondence between the two data sets (or sequences). A well-known correlation coefficient is the Pearson correlation coefficient, which in the case we are discussing is known as a *sample correlation coefficient* $C_{X,Y}$ defined by

$$C_{X,Y} \equiv \frac{\sum_{i=1}^{n}(x^i - \overline{x})(y^i - \overline{y})}{\sqrt{\sum_{j=1}^{n}(x^j - \overline{x})^2 \sum_{k=1}^{n}(y^k - \overline{y})^2}}, \tag{18.10}$$

where $\overline{x}$ and $\overline{y}$ are the respective sample averages. If a sample correlation coefficient has value $+1$, 0, or -1, the two samples are said to be perfectly correlated, uncorrelated, or perfectly anticorrelated, respectively.

Detector Correlations

In our case, our interest is directly related to detector physics. The "data" values in which we are interested, that is, the analogues of the x^i and y^i values in the above, are *truth* values, suitably modified to give correlations. What greatly helps here is that the values are binary. Suppose we look at two detectors and compare their signal status. If they both register a signal, or if they both register ground (no signal), then we can say they are perfectly correlated. Otherwise, they are perfectly anticorrelated.

More commonly, we might take many such joint readings. Suppose the probability of finding perfect correlation is p. Therefore, the probability of anticorrelation is $1 - p$. Assigning a value $+1$ for every case of perfect correlation and a value of -1 for every case of perfect anticorrelation, we deduce that the average correlation C will be given by $C = 2p - 1$.

Single Qubit Temporal Correlation

Quantum registers in practice may be vast, and the description of change and persistence of labstates will be accordingly complicated. It is reasonable to focus attention on the simplest SUO first, in order to appreciate what we are faced with.

Consider an experiment that is dealing with an SUO represented by a single qubit. Suppose that, by stage Σ_0, the observer has prepared a labstate $\boldsymbol{\Psi}_0$ in the rank-one quantum register $\mathcal{Q}_0$ given by the CBR expression

$$\boldsymbol{\Psi}_0 \equiv \alpha \mathbf{0}_0 + \beta \mathbf{1}_0, \tag{18.11}$$

where $|\alpha|^2 + |\beta|^2 = 1$.

Now suppose the labstate $\boldsymbol{\Psi}_0$ is allowed to evolve undisturbed to stage Σ_1 under semi-unitary evolution, into labstate $\boldsymbol{\Psi}_1$ in a rank-one quantum register

$\mathcal{Q}_1$. Then we write $\mathbf{\Psi}_1 \equiv \mathbb{U}_{1,0}\mathbf{\Psi}_0$, where the evolution operator $\mathbb{U}_{1,0}$ is semi-unitary. Taking the matrix representation of this operator in standard form, shown in Eq. (11.26), with the phase E set to zero, we have the rules

$$\begin{aligned} \mathbb{U}_{1,0}\mathbf{0}_0 &= a\mathbf{0}_1 - b^*\mathbf{1}_1, \\ \mathbb{U}_{1,0}\mathbf{1}_0 &= b\mathbf{0}_1 + a^*\mathbf{1}_1, \end{aligned} \tag{18.12}$$

where $|a|^2 + |b|^2 = 1$.

We now come to a critical point, one that impinges on the quantum Hamming measures of dissimilarity discussed above. Consider a persistent array of detectors from stage Σ_m to stage Σ_n, where $n > m$, such that detector i_m at stage Σ_m is regarded as the same as i_n at stage Σ_n. Given the labstates $\mathbf{\Psi}_m$ and $\mathbf{\Psi}_n$, there are two very different quantities that could be measured:

Correlation of Probabilities

Suppose the observer measured P_m^i, the probability of a positive signal in i_m and, separately, P_n^i, the probability of a signal in i_n. These two quantities give important information about what goes on at detector i at two separate stages. But there is no direct correlation between those two pieces of information. Every observation of i_m at stage Σ_m cannot possibly be influenced by what could happen at i_n at stage Σ_n, and vice-versa, simply because each observation in any given run necessarily precludes the other observation, in that run.

The argument goes as follows. If the observer looks at stage Σ_m to see what the signal status of detector i_m is, then that stops the run and so i_n cannot be investigated *during that run.* Likewise, if the observer decides to see what the signal status is of detector i_n at stage Σ_n, that means they cannot have looked at i_m at stage Σ_m, because that would have stopped that run immediately. In brief, a comparison of the signal status i_m and i_n in any given run is ruled out. The only information the observer has are the two probabilities P_m^i and P_n^i, and these have been measured separately over many different runs.

If $\widehat{P}_m^i \equiv 1 - P_m^i$, $\widehat{P}_n^i \equiv 1 - P_n^i$, we define the *correlation of probabilities* $P_{n,m}^i$ as follows:

$$\begin{aligned} P_{n,m}^i &\equiv P_n^i P_m^i + \widehat{P}_n^i \widehat{P}_m^i - P_n^i \widehat{P}_m^i - \widehat{P}_n^i P_m^i \\ &= (2P_n^i - 1)(2P_m^i - 1) = C_n^i C_m^i, \end{aligned} \tag{18.13}$$

where C_n^i and C_m^i are detector i correlations at stage Σ_n and Σ_m, respectively.

Probability of Correlation

What the observer is really after is the probability of finding a signal at i_n knowing for sure that there was a signal in i_m. This is then a genuine temporal correlation, denoted $C_{n,m}^i$. In the following, we simplify the discussion by taking the register to consist of just one detector.

Definition 18.6 Given initial state $\mathbf{\Psi}_m \equiv \alpha \mathbf{0}_m + \beta \mathbf{1}_m$, where $|\alpha|^2 + |\beta|^2 = 1$, and the evolution operator is $\mathbb{U}_{n,m}$, the *temporal correlation* $C_{n,m}$ is given by

$$\begin{aligned} C_{n,m} &\equiv \Pr(\mathbf{0}_m|\mathbf{\Psi}_m)\Pr(\mathbf{0}_n|\mathbf{0}_m) + \Pr(\mathbf{1}_m|\mathbf{\Psi}_m)\Pr(\mathbf{1}_n|\mathbf{1}_m) \\ &\quad - \Pr(\mathbf{1}_m|\mathbf{\Psi}_m)\Pr(\mathbf{0}_n|\mathbf{1}_m) - \Pr(\mathbf{0}_m|\mathbf{\Psi}_m)\Pr(\mathbf{1}_n|\mathbf{0}_m) \\ &= 2\Pr(\mathbf{0}_m|\mathbf{\Psi}_m)\Pr(\mathbf{0}_n|\mathbf{0}_m) + 2\Pr(\mathbf{1}_m|\mathbf{\Psi}_m)\Pr(\mathbf{1}_n|\mathbf{1}_0)_n, \\ &= 2|\alpha|^2|\overline{\mathbf{0}}_n \mathbb{U}_{n,m}\mathbf{0}_m|^2 + 2|\beta|^2|\overline{\mathbf{1}}_n \mathbb{U}_{n,m}\mathbf{1}_m|^2 - 1. \end{aligned} \tag{18.14}$$

We point out that the probabilities $\Pr(\mathbf{0}_m|\mathbf{\Psi}_m) \equiv |\alpha|^2$ and $\Pr(\mathbf{1}_m|\mathbf{\Psi}_m) \equiv |\beta|^2$ are found during the calibration process, prior to the experiment proper starting.

From (18.12), we readily find that

$$C_{n,m} = 2|a|^2 - 1, \tag{18.15}$$

which, significantly, is independent of the components α, β, of the initial labstate. This result makes sense: if the dynamics is trivial, in that the only effect is to multiply the initial labstate by some complex phase factor, then $|a| = 1$ and the correlation is perfect, that is, $C_{n,m} = 1$. If, conversely, the dynamics flips the initial labstate to an orthogonal labstate, then the correlation is -1.

We will use the above result (18.15) in the next chapter on the Leggett–Garg inequalities.

The generalization of these calculations to higher rank registers will undoubtedly be more complicated but, in principle, amenable to the same logic as above: in all cases, the observer should decide what it is that they can actually measure, and then the formalism will give unambiguous predictions.

A final point we need to make is that we expect our apparatus to be large rank and calibrated. Then the observables will be (say) the *up* and *down* components of spin, and interest will be in spin correlations. The QDN analysis in such a case assigns *two* qubits, one for spin *up* and one for spin *down*. This effectively doubles the size of the quantum register. If we labeled spin up by index $2i$, then spin down would be indexed by $2i + 1$, and our temporal correlations would involve signals in detectors $2i$ and $2i + 1$.

19
Temporal Correlations

19.1 Introduction

Our aim in this chapter is to extend the Bell inequality discussion in Chapter 17 to its temporal analogue, known as the *Leggett–Garg (LG) inequality*.

Bell inequalities involve spatial nonlocality, that is, signal observations distributed over space. It was shown by Leggett and Garg that analogous inequalities involving temporal nonlocality can be formulated (Leggett and Garg, 1985). We shall discuss one of these, known as the *LG inequality*.

We saw in Chapter 17 that Bell inequalities are based on certain classical mechanics (CM) assumptions about the nature of reality. Likewise, the LG inequality is based on two CM principles that are not incorporated into quantum mechanics (QM).

Definition 19.1 The principle of *macrorealism* asserts that if a macroscopic (that is, large-scale) system under observation (SUO) can be observed to be in one of two or more macroscopically distinct states, then it will always be in one or another of those states at any given time, not in a quantum superposition of those states, even when it is not being observed.

This principle is a foundational principle in CM but is incompatible with QM in at least two ways. First, it is vacuous, as it asserts the truth of a proposition in the absence of empirical validation: how can we define the concept of *always*, without introducing counterfactuality? Second, it is violated in quantum theory, for instance, in path integral calculations.

This principle in embedded in the nexus of issues explored in this book. The quantized detector network (QDN) version of it takes the form "If any SUO can be observed in any number of possible states, then it will be observed in one of them in any run." This is a near tautology. The classical version says the same thing but makes an additional assertion about something going on in the absence of observation, which is the vacuous element that QM cannot accept.

Definition 19.2 The principle of *noninvasive measurability* asserts that the actual state an SUO is in can always be determined cost-free, that is, without having any effect on that state or on its subsequent dynamical evolution.

This principle is subtle. We experience its apparent validity all the time in our ordinary lives: we look at objects and they appear not to change by those acts of observation. However, quantum physics tells us otherwise, for all information comes to us via quantum processes, and those involve quanta.[1] Newton's third law, *action and reaction are equal and opposite and act on different bodies*, really does have its quantum counterpart: *no observation leaves the observed completely unchanged.*

The LG inequality involves temporal correlations, which measure and compare changes in observables. We shall focus our attention on the temporal correlations of the signal states of dynamically evolving bits and qubits.

19.2 Classical Bit Temporal Correlations

Without loss of content or generality, we shall restrict the discussion in this section to an SUO consisting of N interacting classical bits. In such discussions, it is not necessary to say what these bits represent physically. What matters is that if the observer chose to look at any one of these bits, that bit would be observed to be in precisely one of two possible states. We will refer to these states as *up* and *down*. A useful mental image is that the *up* state represents a raised flag denoting a signal, while the *down* state represents a lowered flag denoting an absence of a signal.

Consider an experiment where at initial stage Σ_0, all N of these bits are definitely in the *up* state. We then allow the state of the SUO to evolve until stage Σ_1 and then we look at the state of the SUO at that stage.

Suppose that at stage Σ_1 we find a total of N^{up} bits in the *up* state and $N^{down} \equiv N - N^{up}$ bits in the *down* state. A natural question is: *how different is the state of the SUO at stage Σ_1 compared with its state at Σ_0?*

There are many ways to answer this question, that is, to define temporal change. Before we can do that, we need to clarify a metaphysical point, concerning the notions of *permanence* and *identity.*

Change Is Contextual

Up to now, our exposition of QDN has emphasized that *everything changes.* But on reflection, that is a vacuous inconsistency, for change can be measured

[1] We have not been much concerned with Planck's constant $\hbar$ so far in this book, but it comes in here. Changes occurring to states of SUO when they are observed are quantified by that unit of action, which is relatively negligible on our ordinary real-life (emergent) scales of measurement.

only by observers. But those observers themselves change, according to their own subjective experience. Therefore, any change in an SUO has to be carefully distinguished from the changes going on naturally in the laboratory.

Leggett and Garg chose *temporal correlations*. We discussed these for single qubits in the previous chapter.

Definition 19.3 If a classical bit is in the same state at stage Σ_n as it was at an earlier stage Σ_m, then the states are said to be *perfectly correlated*, with a correlation $C_{n,m} = +1$. If, conversely, a classical bit is in a different state at stage Σ_n compared with its state at stage Σ_m, then the two states are said to be *perfectly anticorrelated*, with a correlation $C_{n,m} = -1$.

Our interest will be in the average correlation $\overline{C}_{1,0}$. From the information given above, we find

$$\overline{C}_{1,0} = \frac{N^{up}}{N} - \frac{N^{down}}{N} = \frac{2N^{up}}{N} - 1. \tag{19.1}$$

By inspection, it is easy to see that the average correlation satisfies the constraints $-1 \leqslant \overline{C}_{1,0} \leqslant 1$.

19.3 The Classical Leggett–Garg Inequality

The LG inequality is based on a three-stage experiment, as follows. An SUO consisting of r classical bits, each in its *up* state at stage Σ_0, evolves to stage Σ_1, and the average correlation $\overline{C}_{1,0}$ is measured. The state is then allowed to evolve to stage Σ_2, where two new average correlations are measured. One of these is $\overline{C}_{2,1}$ and the other is $\overline{C}_{2,0}$.

We note that (1) each bit is followed from stage Σ_0 to stage Σ_1 and then finally to stage Σ_2 and its *up* or *down* state is observed noninvasively and recorded at each stage, and (2) according to the CM principles of macrorealism and noninvasive measurability, the observer will have all the information needed to calculate these three ensemble average quantities precisely and without error. Because the dynamics is assumed deterministic, only one run is required. If on the other hand, the dynamics is classical stochastic, or the initial state is random, then the argument can adjusted to take probabilities into account, with exactly the same result.

Given these three correlations, we define the LG correlation $\overline{K}$ by

$$\overline{K} \equiv \overline{C}_{1,0} + \overline{C}_{2,1} - \overline{C}_{2,0}. \tag{19.2}$$

We derive the LG inequality $-3 \leqslant \overline{K} \leqslant 1$ as follows.

Given an SUO of r bits, noninvasive measurability allows us to track any or all of the bits in the SUO over the three stages, without interfering in the dynamics. Suppose we tracked the ith bit in the SUO over the three stages Σ_0, Σ_1, and Σ_2.

Table 19.1 *Calculation of the possible values of the LG correlation K^i for the ith bit*

signal status at Σ_0	up	up	up	up
signal status at Σ_1	up	up	down	down
signal status at Σ_2	up	down	up	down
$C^i_{1,0}$	1	1	−1	−1
$C^i_{2,1}$	1	−1	−1	1
$C^i_{2,0}$	1	−1	1	−1
K^i	1	1	−3	1

Table 19.1 shows the possible *up* or down states of that bit at each stage, the correlations associated with those bit states, and the LG correlation for that bit.

By inspection of Table 19.1, we see that the LG correlation $K^i \equiv C^i_{1,0} + C^i_{2,1} - C^1_{2,0}$ for the ith bit lies in the interval $[-3, 1]$. This is perfectly general, being valid for any interaction whatsoever between the bits in that SUO. Since this is true for any bit in the SUO, we conclude that the average $\overline{K}$ satisfies the same condition. Hence we arrive at the LG condition

$$-3 \leqslant \overline{K} \leqslant 1. \tag{19.3}$$

There is no way that the LG inequality could ever be violated classically, given the principles of macrorealism and noninvasive measurability.

19.4 Qubit Temporal Correlations

Extending the above classical bit discussion to the quantum case involves significant differences. Specifically, each bit in the ensemble is replaced by a qubit, and then all of the qubits are tensored together to create a quantum register of rank r.

As will be appreciated by now, such registers are exceedingly complicated structures. We will assume in the first instance that the qubits in the quantum register do not interact with each other, but can interact with other elements of their environment. This assumption means that we can meaningfully discuss the evolution of a single qubit. We shall follow the evolution of a typical single qubit in a noninteracting ensemble from initial stage Σ_0 to intermediate stage Σ_1 and then to final stage Σ_2.

We shall discuss the correlation $C_{1,0}$ of two signal observations, at stage Σ_0 and stage Σ_1. Assuming that the semi-unitary operator acting on the chosen qubit is in the standard form

$$U_{1,0} = \begin{bmatrix} a_1 & -b_1^* \\ b_1 & a_1^* \end{bmatrix}, \tag{19.4}$$

where $|a_1|^2 + |b_1|^2 = 1$ and we ignore an overall phase, we use the results of the previous chapter to find

$$C_{0,1} = |a_1|^2 - |b_1|^2 = 2|a_1|^2 - 1. \tag{19.5}$$

As noted previously, this result is independent of the initial state of the qubit.

Now consider a further evolution from stage Σ_1 to stage Σ_2 with evolution operator

$$U_{2,1} = \begin{bmatrix} a_2 & -b_2^* \\ b_2 & a_2^* \end{bmatrix}, \tag{19.6}$$

where $|a_2|^2 + |b_2|^2 = 1$. Then $C_{2,1} = 2|a_2|^2 - 1$.

According to QM principles, evolution from stage Σ_0 to stage Σ_2 is given by the evolution operator $U_{2,0} = U_{2,1}U_{0,1}$. We find

$$U_{2,0} = \begin{bmatrix} a_2 & -b_2^* \\ b_2 & a_2^* \end{bmatrix} \begin{bmatrix} a_1 & -b_1^* \\ b_1 & a_1^* \end{bmatrix} = \begin{bmatrix} a_1a_2 - b_1b_2^* & -a_1^*b_2^* - b_1^*a_2 \\ a_1b_2 + b_1a_2^* & a_1^*a_2^* - b_1^*b_2 \end{bmatrix}, \tag{19.7}$$

from which we deduce

$$C_{2,0} = 2|a_1a_2 - b_1b_2^*|^2 - 1. \tag{19.8}$$

19.5 QDN Spin Correlation

In this section we show how QDN deals with spin correlation. The first thing to note in the above calculation is that according to QDN principles, a single persistent qubit representing a *calibrated* detector would show no changes in its signal status, by definition. Because we wish to discuss an actual spin, such as that of a proton, which would show changes in its signal status, bitification tells us that we need *two* detector qubits, one for proton spin *up* and the other for proton spin *down*. QDN qubits do not model angular momentum or spin *per se* but *yes/no* logic.

Suppose then that we want to investigate temporal spin correlation for a single proton. Following the discussion in previous sections, we will have a three-stage process, represented by Figure 19.1.

The initial labstate $\mathbf{\Psi}_0$ is taken to be given by

$$\mathbf{\Psi}_0 \equiv (\alpha\widehat{\mathbb{A}}_0^1 + \beta\widehat{\mathbb{A}}_0^2)\mathbf{0}_0, \tag{19.9}$$

where 1_0 detects spin *up* and 2_0 detects spin *down*.

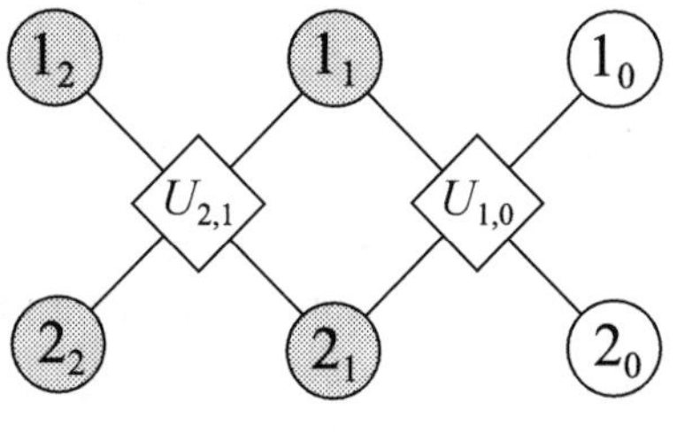

Figure 19.1. Stage diagram for a proton temporal spin correlation.

We encounter here an example of a contextual subspace. Our rank-two quantum register has four basis states, but only two of these are needed to describe the spin dynamics. For example, semi-unitary evolution from stage Σ_0 to stage Σ_1 is given by

$$\begin{aligned} \mathbb{U}_{1,0}\widehat{\mathbb{A}}_0^1\mathbf{0}_0 &= a_1\widehat{\mathbb{A}}_1^1\mathbf{0}_1 - b_1^*\widehat{\mathbb{A}}_1^2\mathbf{0}_1, \\ \mathbb{U}_{1,0}\widehat{\mathbb{A}}_0^2\mathbf{0}_0 &= b_1\widehat{\mathbb{A}}_1^1\mathbf{0}_1 + a_1^*\widehat{\mathbb{A}}_1^2\mathbf{0}_1, \end{aligned} \tag{19.10}$$

and similarly for evolution to stage Σ_2.

It is not hard to see that in this case, the required temporal correlation coefficient $C_{1,0}$ is given by $C_{1,0} = 2|a_1|^2 - 1$, exactly as above. Our conclusion is that, as elsewhere, QDN reproduces the standard QM results without any difficulty. Note that the bitification process appears to make QDN quantum registers unduly larger than the standard QM counterparts. That is not seen as anything other than an advantage, because it permits the modeling of more situations, such as labstates of signality greater than just one.

19.6 The Leggett–Garg Correlation

From the above correlations, using Definition 19.2 we find

$$K = 2|a_1|^2 + 2|a_2|^2 - 2|a_1 a_2 - b_1 b_2^*|^2 - 1. \tag{19.11}$$

Now the parameters of an experiment, such as $a_1, a_2, \ldots$ are under the control of the experimentalist. It is reasonable therefore to assume that the parameters a_1, b_1, a_2, and b_2 can be chosen to be whatever we wish, subject to the unitarity constraints $|a_1|^2 + |b_1|^2 = |a_2|^2 + |b_2|^2 = 1$.

Consider the reparametrization

$$\begin{aligned} a_1 &= \cos\theta_1, b_1 = \sin\theta_1 e^{i\phi_1}, \\ a_2 &= \cos\theta_2, b_2 = \sin\theta_2 e^{i\phi_2}, \end{aligned} \tag{19.12}$$

where θ_1, θ_2, ϕ_1, and ϕ_2 are real. Now take $\phi_1 = \phi_2$. Plotting K as θ_1 and θ_2 range from $-\pi$ to π each gives Figure 19.2.

It is clear that there are values of the parameters where K exceeds the classical limit $+1$. In fact, the maximum value of K over the region shown is 1.5, in clear violation of the Leggett–Garg upper bound of one. This is an entirely nonclassical result that is the temporal analogue of the violation of Bell inequalities in experiments such as those of Aspect (Aspect et al., 1982) and others.

19.7 Understanding the Leggett–Garg Prediction

The predicted QM violation of the Leggett–Garg inequality is another demonstration of the thesis running throughout this book: that empirical truth is contextual and that QM is best regarded as a statement of the laws of entitlement, of what we can legitimately say about SUOs, rather than an explicit description of those SUOs.

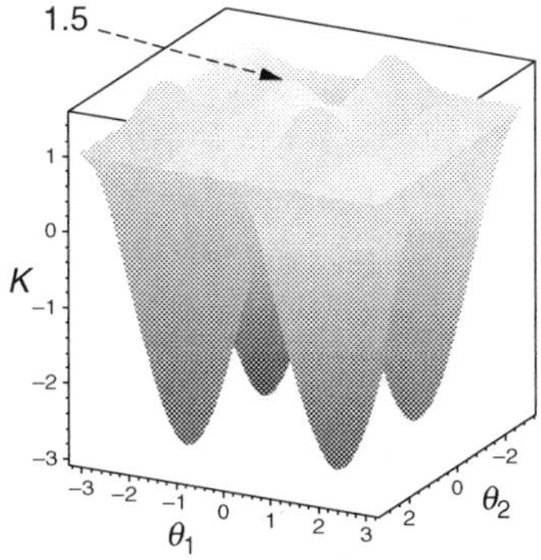

Figure 19.2. Values of K versus angle parameters θ_1, θ_2.

In view of the great predictive success of QM, it would be unwise to believe that we can replace it. At best, we can enhance it. But that does not answer the question, what is the origin of the discrepancy between the classical LG inequality and the predicted QM violation of it?

Our view is that there must be hidden assumptions somewhere that, if identified, would explain the discrepancy. Since QM has never yet been proved empirically incorrect while classical mechanics has, there must be something that we have missed in the way we think classically. It is not QM that is wrong, but our preconditioned classical way of thinking about physical reality that is not quite right.

The answer is given clearly by Wheeler's participatory principle that *if we have not actually done something, we should not make any undue assumption about it.*

With this in mind, consider the LG correlation K discussed above. It is calculated from three separate correlations that classically could be calculated from a single run. This is because of the assumed principle of noninvasive measurability that is one of the standard assumptions of classical mechanics.

But, according to the requirements of quantum experimentation, the correlation $C_{2,0}$ cannot involve any observation at stage Σ_1. The evolution operator $U_{2,0}$ has to be applied on the strict understanding that no attempt is made to extract information between initial stage Σ_0 and final stage Σ_2.

The essential fact is that the principle of noninvasive measurability cannot be assumed to hold in QM (although as discussed in Chapter 25, there are some experiments that can be described in such terms). Classically, our normal human conditioning is to think automatically that the stage-Σ_1 signals observed in the measurement of the correlation $C_{1,0}$ should play a role in $C_{2,0}$. But how can they? They are not observed when the SUO evolves undisturbed from stage Σ_0 to stage Σ_2.

In order to calculate the LG quantity K, we would have to perform three separate "subexperiments," one for each of the three correlations involved. These are separate experiments with separate contexts. Classically, we could believe that we could get away with just one experiment. Quantum mechanically, we have to recognize context is critical and perform three subexperiments.

The explanation then is that in the $C_{2,0}$ subexperiment, there is a lack of which-path information at stage time Σ_1, analogous to the architecture of the double-slit experiment, where an interference pattern on a screen is observed provided no attempt is made to determine through which slit the particle had gone.

20

The Franson Experiment

20.1 Introduction

Experimental violations of the Bell and Leggett–Garg inequalities studied in Chapters 17 and 19, respectively, show that quantum states of systems under observation (SUOs) cannot be interpreted classically. Interpretations that claim to do this require contextually incomplete modifications of classical principles, such as a breakdown of classical locality. Such modifications merely serve to make standard quantum mechanics (QM) more appealing.

However, that is only one side of the observer–SUO fence. In this chapter, we start to explore the other side of that fence, where the observer and their apparatus live. Our focus is a thought experiment proposed by Franson (Franson, 1989), which suggests that apparatus cannot always be treated classically. Three scenarios are discussed, each involving different time scales, with corresponding different outcomes.

20.2 The Franson Experiment

In the following, we refer to "photons" as if they were actual particles, but that is only a convenient and intuitive way of describing what at the end of the day are just clicks in detectors. There are experiments, for example, where it is not reasonable to think of the sources of photons (such as certain crystals) as point like (Paul, 2004).

Franson's proposed experiment FRANSON has parameters that can alter outcomes significantly. There are three significant choices of parameters, leading to three distinct scenarios labeled FRANSON-1, FRANSON-2, and FRANSON-3. The basic architecture consists of a coherent pair of photons, produced at a localized source S, sent in opposite directions toward a pair of separated Mach–Zehnder interferometers, as shown in Figure 20.1. Each photon passes through its own interferometer and, depending on path taken, can suffer a change in

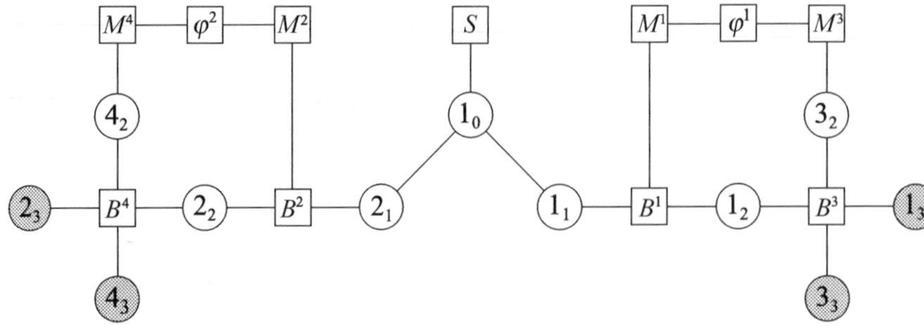

Figure 20.1. FRANSON-1: the Franson experiment for $\Delta T < \tau_2 \ll \tau_1$. S is the photon pair source, the B^i are beam splitters, the M^i are mirrors, and the ϕ^i are phase changers.

phase ϕ and a time delay ΔT. The time delay ΔT is assumed the same for each interferometer in any given run. The phase changes ϕ^1, ϕ^2 associated with the different interferometers can be altered by the observer, but are fixed before and during each run.

The experiment hinges on the relationship between three characteristic times (Franson, 1989).

Coherence Time τ_1

In quantum optics experiments, a finite electromagnetic wave train of length L moving at the speed of light c takes a time $\tau \equiv L/c$ to pass a given point. Such a time is known as a *coherence time*. In the Franson experiment, the coherence time τ_1 is that associated with the production of the photon pair by stage Σ_1. It is a characteristic of the photon pair source S and of the collimation procedures applied subsequently. While τ_1 cannot be altered, it can be determined empirically. We shall assume that τ_1 is the same for each photon.

Emission Time τ_2

The second characteristic time is τ_2, the effective time interval within which both photons in a pair can be said to have been emitted. This can be measured during calibration by coincidence observations of detectors 1_3 and 2_3 with all beam-splitters removed. It is assumed τ_2 can be determined empirically and that $\tau_2 \ll \tau_1$. This last inequality is crucial to FRANSON because when this inequality holds, the observer has no way of knowing when a photon pair was created during the relatively long time interval τ_1.[1] It is this lack of knowledge that leads to quantum interference in the FRANSON-3 scenario discussed below. The spectacular aspect of FRANSON is that unlike the double-slit experiment,

[1] Of course, "photon creation" is a vacuous metaphysical picture extrapolated from observations done *after* the source has been triggered. There is no evidence, based on the given apparatus, for the belief that anything has been "created" at S.

where the observer does not know from which point in space a photon came, here the observer does not know at which point in time the photon pair was produced.

Travel Time Difference ΔT

The third characteristic time is ΔT, the time difference between a photon traveling along the short arm of its interferometer and along its long arm. This time is adjustable but fixed during each run and is assumed the same for each of the two interferometers in that run.

There are three scenarios we shall discuss: FRANSON-1, for which $\Delta T \ll \tau_2$; FRANSON-2, for which $\tau_1 \ll \Delta T$; and FRANSON-3, for which $\tau_2 \ll \Delta T \ll \tau_1$. In Franson's original analysis (Franson, 1989), photon spin did not play a role. Therefore, in all scenarios considered here, photon spin is assumed fixed once a given photon pair has been created.

20.3 FRANSON-1: $\Delta T \ll \tau_2$

The relevant figure for this scenario is Figure 20.1. We discuss the experiment in terms of its stages, as follows.

Stage Σ_0

The initial total state is

$$|\mathbf{\Psi}_0) \equiv |s_0\rangle \otimes \widehat{\mathbb{A}}_0^1 \mathbf{0}_0, \tag{20.1}$$

where $|s_0\rangle$ represents the initial source spin state.

Stage Σ_0 to Stage Σ_1

By stage Σ_1, the initial state has split into a correlated pair of photons moving in opposite directions. The creation of this pair by this stage is represented by the action of the contextual evolution operator $U_{1,0}$:

$$|\mathbf{\Psi}_1) \equiv U_{1,0}|\mathbf{\Psi}_0) = |s_1\rangle \otimes \widehat{\mathbb{A}}_1^1 \widehat{\mathbb{A}}_1^2 \mathbf{0}_1, \tag{20.2}$$

where $|s_1\rangle$ represents the combined spin state of the photon pair at this stage.

Stage Σ_1 to Stage Σ_2

The stage-Σ_1 photons pass through beam splitters B^1 and B^2 as shown. One output channel from each beam splitter leads directly to a final beam splitter, while the other output channel is deflected by a mirror through a phase changer before being deflected onto that final beam splitter. The four beam splitters B^i, $i = 1, 2, 3, 4$, are parametrized by real transmission and reflection coefficients t^i, r^i respectively, according to the prescription given by Eq. (11.28).

The labstate evolution is given by

$$U_{2,1}\left\{|s_1\rangle \otimes \widehat{\mathbb{A}}_1^1\widehat{\mathbb{A}}_1^2\mathbf{0}_1\right\} = |s_2\rangle \otimes \left\{t^1\widehat{\mathbb{A}}_2^1 + ir^1e^{i\phi^1}\widehat{\mathbb{A}}_2^3\right\}\left\{t^2\widehat{\mathbb{A}}_2^2 + ir^2e^{i\phi^2}\widehat{\mathbb{A}}_2^4\right\}\mathbf{0}_2$$
$$= |s_2\rangle \otimes \left\{\begin{array}{l} t^1t^2\mathbf{3}_2 + ir^1t^2e^{i\phi^1}\mathbf{6}_2 + \\ ir^2t^1e^{i\phi^2}\mathbf{9}_2 - r^1r^2e^{i(\phi^1+\phi^2)}\underline{\mathbf{12}}_2 \end{array}\right\}, \tag{20.3}$$

where ϕ^1 and ϕ^2 are total phase change factors due to the increased path length of the long arms of the interferometers and phase-shift plates introduced in those long arms by the observer. In the last line in (20.3), we show the computation basis representation (CBR) of the expression in the previous line.

Stage Σ_2 to Stage Σ_3

There are four terms to consider in the transition from Σ_2 to Σ_3:

$$U_{3,2}\left\{|s_2\rangle \otimes \widehat{\mathbb{A}}_2^1\widehat{\mathbb{A}}_2^2\mathbf{0}_2\right\} = |s_3\rangle \otimes \left\{t^3\widehat{\mathbb{A}}_3^1 + ir^3\widehat{\mathbb{A}}_3^3\right\}\left\{t^4\widehat{\mathbb{A}}_3^2 + ir^4\widehat{\mathbb{A}}_3^4\right\}\mathbf{0}_3,$$
$$U_{3,2}\left\{|s_2\rangle \otimes \widehat{\mathbb{A}}_2^3\widehat{\mathbb{A}}_2^2\mathbf{0}_2\right\} = |s_3\rangle \otimes \left\{t^3\widehat{\mathbb{A}}_3^3 + ir^3\widehat{\mathbb{A}}_3^1\right\}\left\{t^4\widehat{\mathbb{A}}_3^2 + ir^4\widehat{\mathbb{A}}_3^4\right\}\mathbf{0}_3,$$
$$U_{3,2}\left\{|s_2\rangle \otimes \widehat{\mathbb{A}}_2^1\widehat{\mathbb{A}}_2^4\mathbf{0}_2\right\} = |s_3\rangle \otimes \left\{t^3\widehat{\mathbb{A}}_3^1 + ir^3\widehat{\mathbb{A}}_3^3\right\}\left\{t^4\widehat{\mathbb{A}}_3^4 + ir^4\widehat{\mathbb{A}}_3^2\right\}\mathbf{0}_3,$$
$$U_{3,2}\left\{|s_2\rangle \otimes \widehat{\mathbb{A}}_2^3\widehat{\mathbb{A}}_2^4\mathbf{0}_2\right\} = |s_3\rangle \otimes \left\{t^3\widehat{\mathbb{A}}_3^3 + ir^3\widehat{\mathbb{A}}_3^1\right\}\left\{t^4\widehat{\mathbb{A}}_3^4 + ir^4\widehat{\mathbb{A}}_3^2\right\}\mathbf{0}_3. \tag{20.4}$$

This is all the information needed for our computer algebra program MAIN to evaluate the answers to all maximal questions. These answers turn out to be complicated, long polynomials in the t^i and r^i parameters, so are not listed here. However, setting them to the empirically useful value $t^i = r^i = 1/\sqrt{2}$, $i = 1, 2, 3, 4$, as assumed by Franson (Franson, 1989), gives the relative coincidence rates

$$\Pr\left(\widehat{\mathbb{A}}_3^1\widehat{\mathbb{A}}_3^2\mathbf{0}_3|\Psi_0\right) = \sin^2\left(\tfrac{1}{2}\phi^1\right)\sin^2\left(\tfrac{1}{2}\phi^2\right),$$
$$\Pr\left(\widehat{\mathbb{A}}_3^3\widehat{\mathbb{A}}_3^2\mathbf{0}_3|\Psi_0\right) = \cos^2\left(\tfrac{1}{2}\phi^1\right)\sin^2\left(\tfrac{1}{2}\phi^2\right),$$
$$\Pr\left(\widehat{\mathbb{A}}_3^1\widehat{\mathbb{A}}_3^4\mathbf{0}_3|\Psi_0\right) = \sin^2\left(\tfrac{1}{2}\phi^1\right)\cos^2\left(\tfrac{1}{2}\phi^2\right),$$
$$\Pr\left(\widehat{\mathbb{A}}_3^3\widehat{\mathbb{A}}_3^4\mathbf{0}_3|\Psi_0\right) = \cos^2\left(\tfrac{1}{2}\phi^1\right)\cos^2\left(\tfrac{1}{2}\phi^2\right). \tag{20.5}$$

Each of these rates shows angular dependence due to independent "photon self-interference" within each separate interferometer. This form of interference will be referred to as *local*. There are no *global* interference effects involving both interferometers and no post selection of data is required.

20.4 FRANSON-2: $\tau_1 \ll \Delta T$

In this variant of the Franson experiment, the photon wave trains 3_2, 4_2 reflected at B^1 and B^2, respectively, travel along the long arms of their respective interferometers at the speed of light or less, depending on the medium through which

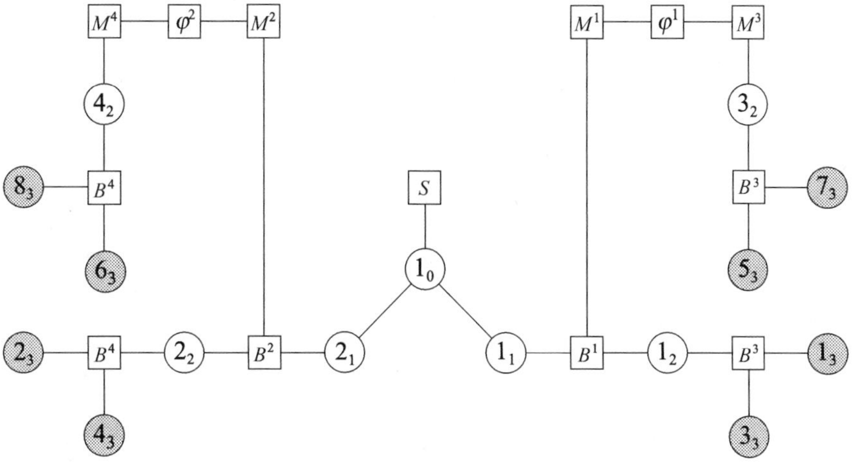

Figure 20.2. FRANSON-2: the Franson experiment for $\tau_2 \ll \tau_1 \ll \Delta T$.

they move. Because now the travel time difference ΔT is very much greater than the coherence time τ_1, these wave trains arrive at B^3 and B^4 long after the transmitted wave trains 1_2 and 2_2 have impinged on B^3 and B^4, respectively. In consequence, no local or global interference can take place. In fact, the observer can now obtain total information concerning the timing of each coincidence outcome in every run of the experiment and know precisely what path was taken by each photon.

Under this circumstance, the four original detectors $1_3, 2_3, 3_3$, and 4_3 used in FRANSON-1 now have to be regarded as *eight* separate detectors, i_3, $i = 1, 2, \ldots, 8$, as shown in Figure 20.2. The first four of these register photon clicks from short-path photons, while the last four signal clicks from those that have traveled the long paths. This information is specific to each photon and does not involve any photon pairs.

Significantly, the final-stage quantum register involved in this scenario and the next one, FRANSON-3, is 256-dimensional. However, our computer algebra program MAIN has no difficulty dealing with this because it has been encoded to process only the relevant contextual Hilbert spaces, and these are of greatly reduced dimensions.

This demonstrates a fundamental point about apparatus. In the conventional usage of apparatus, experimentalists tend to regard their equipment as having some sort of "transtemporal" identity, or persistence. In Figure 20.2, for example, beam splitters B^3 and B^4 would most likely persist in the laboratory as material objects, during the long interval ΔT between their interaction with wave-trains 1_2 and 2_2 and with the delayed wave-trains 3_2 and 4_2. Even classically, however, this need not be the case. It is conceivable that ΔT could be so long, such as several years, that the beam splitters could be destroyed and rebuilt at leisure between the observation of any short-arm photons and any long-arm photons.

Whatever the actuality in the laboratory, from a quantum point of view, the beam splitters receiving short- and long-arm photons should be considered as completely separate pieces of equipment in this scenario (but not in the next). In other words, apparatus and how it is used is time dependent. The analysis in the next section shows that the rules for doing this can be quite nonclassical and appear to violate the ordinary rules of causality.

For the FRANSON-2 scenario, $\tau_1 \ll \Delta T$, the dynamics follows the same rules as in the previous section up to the transition from stage Σ_2 to stage Σ_3. At this point, the transformation rules have to take into account the possibility that the observer could know the timings of all events. The rules for this transition are now

$$\begin{aligned}
U_{3,2}\left\{|s_2\rangle \otimes \widehat{\mathbb{A}}_2^1\widehat{\mathbb{A}}_2^2\mathbf{0}_2\right\} &= |s_3\rangle \otimes \left\{t^3\widehat{\mathbb{A}}_3^1 + ir^3\widehat{\mathbb{A}}_3^3\right\}\left\{t^4\widehat{\mathbb{A}}_3^2 + ir^4\widehat{\mathbb{A}}_3^4\right\}\mathbf{0}_3,\\
U_{3,2}\left\{|s_2\rangle \otimes \widehat{\mathbb{A}}_2^3\widehat{\mathbb{A}}_2^2\mathbf{0}_2\right\} &= |s_3\rangle \otimes \left\{t^3\widehat{\mathbb{A}}_3^5 + ir^3\widehat{\mathbb{A}}_3^7\right\}\left\{t^4\widehat{\mathbb{A}}_3^2 + ir^4\widehat{\mathbb{A}}_3^4\right\}\mathbf{0}_3,\\
U_{3,2}\left\{|s_2\rangle \otimes \widehat{\mathbb{A}}_2^1\widehat{\mathbb{A}}_2^4\mathbf{0}_2\right\} &= |s_3\rangle \otimes \left\{t^3\widehat{\mathbb{A}}_3^1 + ir^3\widehat{\mathbb{A}}_3^3\right\}\left\{t^4\widehat{\mathbb{A}}_3^6 + ir^4\widehat{\mathbb{A}}_3^8\right\}\mathbf{0}_3,\\
U_{3,2}\left\{|s_2\rangle \otimes \widehat{\mathbb{A}}_2^3\widehat{\mathbb{A}}_2^4\mathbf{0}_2\right\} &= |s_3\rangle \otimes \left\{t^3\widehat{\mathbb{A}}_3^5 + ir^3\widehat{\mathbb{A}}_3^7\right\}\left\{t^4\widehat{\mathbb{A}}_3^6 + ir^4\widehat{\mathbb{A}}_3^8\right\}\mathbf{0}_3.
\end{aligned} \tag{20.6}$$

which should be compared with (20.4).

In this scenario, we find sixteen nonzero coincidence rates, each of the form $\Pr(\widehat{\mathbb{A}}_3^i\widehat{\mathbb{A}}_3^j\mathbf{0}_3|\Psi_0)$, where $i = 1, 3, 5, 7$ and $j = 2, 4, 6, 8$. All of them are independent of ϕ^1 and of ϕ^2. For example, $\Pr(\widehat{\mathbb{A}}_3^1\widehat{\mathbb{A}}_3^6|\Psi_0) = (t^1r^2t^3t^4)^2$, and so on. In the case of symmetrical beam-splitters, where $t^i = r^i = 1/\sqrt{2}$, all 16 rates are equal to $1/16$.

For FRANSON-2, the detectors behave in a manner consistent with the notion that photons are classical-like particles propagating along definite paths.

We should comment on the choice of stage diagram for FRANSON-2, as this impacts significantly on the encoding of program MAIN. Our choice of introducing detectors $5_3, 6_3, 7_3$, and 8_3 seems inevitable but puzzling, because it could be claimed that in a real experiment, there would only be four final stage detectors, not eight. However, our observation above that the long-arm photon signals could occur perhaps many years after the short-arm signals addresses that point: the observer will have real evidence that one set of signals has arrived long before the other set. The fact that the same atoms could persist in the configuration of detectors $1_3, 2_3, 3_3$, and 4_3 until they did service as detectors $5_3, 6_3, 7_3$, and 8_3 is irrelevant.

A related point is that, for simplicity, we chose to model all of the eight final-stage detectors to be associated with stage Σ_3, but that is not necessary. It would perhaps have been more logical to assign the long-arm detectors to a separate, later stage, but that would have been an inessential complication in programming that would make no difference to the calculations or to the conclusions drawn from them.

FRANSON-2 is an important illustration that the observer–apparatus side of quantum physics can contain real surprises. The next variant, FRANSON-3, is even more surprising.

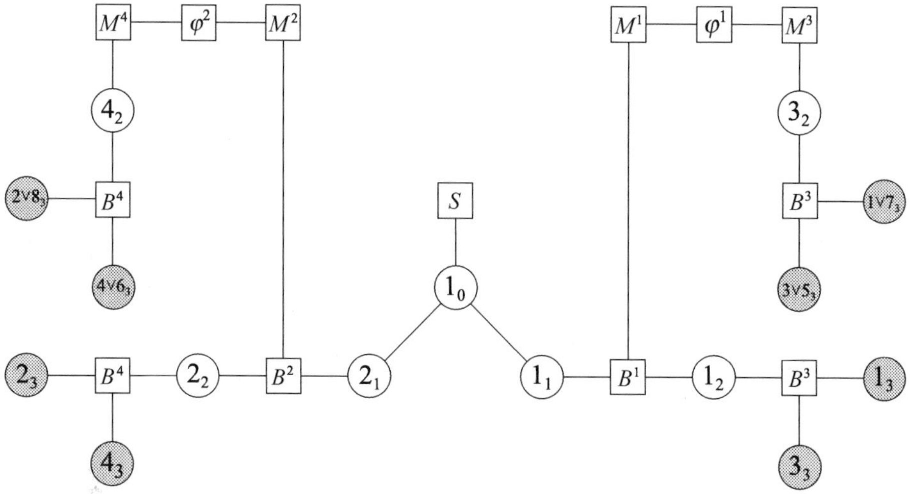

Figure 20.3. FRANSON-3: the Franson experiment for $\tau_2 \ll \Delta T \ll \tau_1$.

20.5 FRANSON-3: $\tau_2 \ll \Delta T \ll \tau_1$

This is the actual scenario discussed by Franson (Franson, 1989). In the following, S stands for "short path" and L for "long path." The fundamental change induced by the observer's setting of ΔT such that $\tau_2 \ll \Delta T \ll \tau_1$ is that, unlike the previous scenario, the observer cannot now use individual times of detector clicks to establish which of the coincidences S–S or L–L has occurred in a given run of the experiment. This presumes that a given run lasts at least as long as the coherence time, and that coincidence observations are being made at times greater than a critical time $\tau_c \equiv \tau_s + \Delta T$ into any given run. Here τ_s is the earliest time that a wave train following an S path could take to a stage Σ_3 detector. Coincidence observations of interference involving any L path at any time earlier than τ_c into a given run will not occur, simply because any wave train taking such a path would still be on its way to stage Σ_3 detectors.

FRANSON-3 is discussed and modeled assuming observations are made at times later than τ_c. In that regime, it will then not be possible for the observer to use detection times to distinguish between S–S correlations and L–L correlations. However, the observer will be able to filter out the two separate S–L correlations.

The relevant diagram is Figure 20.3, which is identical to Figure 20.2 except now 5_3 is replaced by $(3\vee5)_3$, 7_3 is replaced by $(1\vee7)_3$, 6_3 is replaced by $(4\vee6)_3$, and 8_3 is replaced by $(2\vee8)_3$, where for example $3\vee5$ means "3 or 5." Which alternative is taken depends on the contextual information available in principle to the observer.

The dynamics for this scenario is identical to that for the previous one, except for the last equation in (20.6), which is replaced by

$$U_{3,2}\left\{|s_2\rangle \otimes \widehat{\mathbb{A}}_2^3\widehat{\mathbb{A}}_2^4\mathbf{0}_2\right\} = |s_3\rangle \otimes \left\{t^3\widehat{\mathbb{A}}_3^3 + ir^3\widehat{\mathbb{A}}_3^1\right\}\left\{t^4\widehat{\mathbb{A}}_3^4 + ir^4\widehat{\mathbb{A}}_3^2\right\}\mathbf{0}_3. \quad (20.7)$$

With this amended information, program MAIN gives the following nonzero coincidence rates:

$$\begin{aligned} \Pr\left(\widehat{\mathbb{A}}_3^1\widehat{\mathbb{A}}_3^2\mathbf{0}_3|\Psi_0\right) &= \Pr\left(\widehat{\mathbb{A}}_3^3\widehat{\mathbb{A}}_3^4\mathbf{0}_3|\Psi_0\right) = \tfrac{1}{4}\cos^2\left(\tfrac{1}{2}\phi^1 + \tfrac{1}{2}\phi^2\right), \\ \Pr\left(\widehat{\mathbb{A}}_3^2\widehat{\mathbb{A}}_3^3\mathbf{0}_3|\Psi_0\right) &= \Pr\left(\widehat{\mathbb{A}}_3^1\widehat{\mathbb{A}}_3^4\mathbf{0}_3|\Psi_0\right) = \tfrac{1}{4}\sin^2\left(\tfrac{1}{2}\phi^1 + \tfrac{1}{2}\phi^2\right), \end{aligned} \tag{20.8}$$

and eight others that each have value 1/16, for the symmetric case $t^i = r^i = 1/\sqrt{2}$, $i = 1, 2, 3, 4$.

These results are a spectacular demonstration of temporal nonlocality as well as spatial non-locality: interference is occurring between S–S and L–L outcomes.

Note that in actual FRANSON-3 type experiments, the observer would have to measure the times at which coincidence clicks were obtained during each run and then post-select, that is, filter out, those coincidences corresponding to the $\{S\text{–}S,\ L\text{–}L\}$ processes and those corresponding to $\{S\text{–}L,\ L\text{–}S\}$.

20.6 Conclusions

Since the publication of Franson's original paper, there has been great interest in empirical confirmation of FRANSON-3 predictions. While there is still some room for debate concerning the interpretation of the experiment, the results of Kwiat et al. (1993) vindicate Franson's prediction, which corresponds to $\Pr(\widehat{\mathbb{A}}_3^1\widehat{\mathbb{A}}_3^2\mathbf{0}_3|\Psi_0) = \frac{1}{4}\cos^2(\frac{1}{2}\phi^1 + \frac{1}{2}\phi^2)$ in our approach.

Assuming the quantum theoretical interpretation of this experiment is correct, then there is an extraordinary lesson to be learned, not about systems under observation in particular, but about the rules concerning the use of apparatus and how these can differ spectacularly from those expected classically. The interference of the S–S and L–L amplitudes in FRANSON-3 cannot be envisaged in a classical way to occur locally in time. Any attempt to think about such interference in terms of photons as actual particles would lead to bizarre concepts that would never be acceptable conventionally. The conclusion is that a two-photon state is not equivalent under all circumstances to a state with two separate but entangled photons. A more recent quantum optics experiment with similar conclusions has been reported by Kim (Kim, 2003).

The weight of evidence points to the conclusion that quantum outcome amplitudes are dynamically correlated with *contextual information* held by the observer. When some information is absent, then quantum interference can occur. This supports the position of Heisenberg and Bohr concerning the fundamental principles and interpretation of quantum physics. Quantum optics experiments such as FRANSON are providing more and more evidence that QM is not just a theory of SUOs but also a fundamental perspective on the laws of observation in physics. It is our considered view that the surface of those laws has only been scratched to date.

21

Self-intervening Networks

21.1 Introduction

In this chapter, we discuss some of the differences between apparatus and systems under observation (SUOs). The discussion is phrased in terms of *self–intervening networks*, wherein partial quantum outcomes in an early stage of an apparatus can alter future configurations of that apparatus.

To explain what we mean, we introduce the following classification of experiments. Note that reference to *apparatus* here includes real and virtual detectors, and real and virtual modules. Virtual detectors and modules are mathematical fictions that are introduced into the formalism for convenience.

Type-Zero Experiments

Many important experiments involve signal propagation through "empty space." We classify these as type zero (T0). The feature of T0 experiments qualifying for this classification is that the observer has no control of the modules in the information void. Examples are experiments in astrophysics, where the observer can choose which source to observe and how they observe it, but has no influence on the physical properties of whatever lies between source and detector. If there are gas clouds between source and detector, the observer can only recognize that fact after the experimental data are analyzed. An important modern variant of T0 experiments involves map location via the Global Positioning System (GPS): signals sent from Earth to geostationary satellites and received by mobile devices back on Earth have traveled through Earth's atmosphere and through empty space, where special relativistic and general relativistic effects have to be taken into account (Ashby, 2002).

Before the rise of experimental science, most "experiments" were based on visual observations, which are principally of T0 classification. The advent of the telescope and the microscope did not change things in this respect. Indeed, it has been suggested that science did not progress in antiquity precisely because of the philosophical principle that the only valid experiments could be of T0, as these do not introduce the artificiality of constructed modules into observation, which (it was argued) would give a false perspective on "reality." The logic of that argument is superb, but it is a dead-end principle in science.

Several sciences such as geology, archaeology, and, indeed, cosmology started off with periods of basic observation rather than experimentation. During such initial stages of these sciences, the observations could be classified as T0 experiments. In mechanics, the apocryphal dropping by Galileo of two spheres of different mass from the Leaning Tower of Pisa can be classified as of type T0.

Type-One Experiments

Many quantum experiments involve time-independent modules, introduced by an observer into an information void, which appear to persist unchanged in the laboratory over the course of each run of an experiment. By this we mean that for each run of such an experiment, the apparatus that prepares the initial state, shields it from the environment during that run, and detects the outcome state is considered fixed during each stage of a run. In other words, such apparatus persists over time scales significantly greater than the time scale of each run. We classify such experiments as *type one* (T1). These differ from T0 experiments in that the observer explicitly introduces real modules, such as beam splitters and mirrors, into the information void.

The illusion of persistence is a convenient mental device, because all experiments are done in process time and everything changes. However, the concept of persistence is a potent and necessary one that is needed to make sense of what science is all about. It is formalized in quantized detector networks (QDN) by the use of null evolution, as discussed in Section 7.11 and Chapter 18. In standard quantum mechanics (QM), T1 experiments are described in terms of time-independent Hamiltonian operators in Schrödinger equations.

In practice, all experiments have some degree of T1 behavior. For instance, the very action of constructing apparatus before an experiment starts carries the hidden assumption that the constructed apparatus will persist long enough to be useful. That assumption has some amusing aspects. For example, no one would consider constructing parts of their apparatus out of ice, unless the environment made that feasible, as in the Antarctic.

In quantum optics, the double-slit (DS) experiment, Mach–Zehnder interferometer experiments, and quantum eraser experiments (Walborn et al., 2002) are all of this type. So too are high-energy particle-scattering experiments such as those conducted at the Large Hadron Collider at CERN.

After T0, T1 experiments are technically the easiest to construct, and many historical experiments were of this type. T1 experiments allow the focus of attention to be entirely on the dynamical evolution of the states of an SUO, because such states are conventionally regarded as the only objects of interest in physics.

Type-Two Experiments

An important class of experiment, referred to here as *type two* (T2), involves some time dependence in the apparatus that is controlled by factors external to the apparatus, either by the observer directly or by factors external to the laboratory. When controlled by the observer, such experiments require a more advanced technological level than in T1 experiments, such as advanced electronics and computerization. In standard quantum mechanics (QM), T2 experiments are described in terms of time-dependent Hamiltonian operators in Schrödinger equations.

Spin-echo magnetic resonance experiments are of this type, because the experimentalist arranges for certain magnetic fields to be rotated precisely, while additionally, the sample environment introduces random external influences related to local temperature. An example of random changes controlled by the experimentalist are the delayed-choice experiments such as that of Jacques et al. (2007), discussed in Section 14.5, where carefully arranged random changes are made during each run. QM typically describes such experiments via time-dependent Hamiltonian that may have stochastic elements.

T2 experiments are always going to be more interesting than T1 experiments for two reasons: T1 experiments can be regarded as limiting cases of T2 experiments where time dependence is turned off, and because T2 experiments have the potential to reveal more information about the dynamics of systems under observation than T1. Schwinger's source theory in relativistic quantum field theory shows that, in principle, T2 experiments allow for the extraction of all possible information about quantum systems (Schwinger, 1969, 1998a,b,c).

T1 and T2 experiments may be collectively labeled as *exophysical*, because they involve classical apparatus interventions that are external in origin to the SUO. In such experiments, the apparatus is classically well defined at each instant of time during each run, even in those situations where it changes randomly. Therefore, a classical Block Universe (Price, 1997) account of apparatus during each run of a T1 or T2 experiment may be possible.

Type-Three Experiments

In this chapter, we explore a third type of quantum experiment, referred to as *type three* (T3). In such an experiment, the apparatus is modified internally by the quantum dynamics of the SUO, rather than externally by the observer or the environment.

There is an interesting point to be made here about experiments conducted at high-energy particle-scattering laboratories such as the Large Hadron Collider (LHC). Such experiments have fixed sources and detectors and do not have real, physical modules in the interaction region. From that perspective they are T0 experiments. However, the observers can arrange for their particle beams to intersect and allow particle collisions, so from that perspective, such experiments are of type T1. However, the interactions that occur are entirely quantum processes, so they are of also type T3 in an essential way. Indeed, that is the whole point of such experiments.

In Chapter 25 we discuss such experiments as processes where virtual modules are created in the information void by the dynamics of the SUOs. For example, two-particle to two-particle scattering amplitudes discussed in high-energy particle physics have all the characteristics of beam splitters.

Type-Four Experiments

In recent years, there has been interest in experiments that contain T2 and T3 aspects. We have in mind experiments of the Eliztur–Vaidman bomb-tester variety, discussed in Chapter 25. Such experiments can be classified as *type-four* (T4) experiments

A question that we address toward the end of this chapter is whether T3 experiments can always be given a classical Block Universe account or whether something analogous to superpositions of different *apparatus* has to be envisaged. This is not to be confused with the superposition of labstates that we have discussed up to now.

This question is related to the rules of quantum information extraction as they are currently understood in QM. These rules state that quantum interference can occur in the absence of classical which-path information, the most well-known example of this being the DS experiment.

The question here is what precisely does a lack of which-path information mean? *If* such as thing as a photon actually passed through one of the slits, would it leave any trace in principle? Even if it did, it would have to be essentially unobservable, at least far below the scales of classical mechanical detection. In that case, an observer of the interference pattern would indeed be unaware of such an interaction. In essence, such a scenario would be welcomed by supporters of the Hidden Variables (HV) interpretation of empirical physics.

This supposition seems wrong to us for the following reason. In the DS experiment, there are three cases: (1) a photon actually goes through one of the slits but leaves no trace whatsoever; (2) a photon actually goes through one of the slits and leaves a trace that cannot be detected; or (3) we avoid the use of the photon concept.

Our inclination is to discount case 1 on the grounds that any assertion that something "exists" or has occurred, but no evidence for it can ever be found, is vacuous and can be dismissed by Hitchens' razor.[1]

As for case 2, the idea that it is the mere lack of suitable photon-slit interaction detection technology itself that produces interference patterns on a remote screen seems be wrong on two counts. First, there is now sufficient evidence against the

[1] "What can be asserted without evidence can be dismissed without evidence," a modern version of the Latin proverb "*Quod gratis asseritur, gratis negatur*" (what is freely asserted can be freely deserted).

notion that photons are particles in the conventional sense (Paul, 2004). Second, the idea that interference patterns could occur when one observer was unaware of actual which-path information while another observer was aware of it seems inconsistent.[2] It seems likely that there has to be something deeper in the origin of quantum interference. We do not believe it lies in the hidden variables direction because of its contextual incompleteness.

Neutron interference experiments (Greenberger and YaSin, 1989) explore this question in that they are moving toward larger scales of interaction between SUOs and apparatus. In their experiment, the movement of mirrors involved in their quantum erasure scheme involves macroscopic numbers of atoms and molecules (Becker, 1998). In this case, the dynamical effects of the impact of a particle on a mirror is reversed by a second impact. What is amazing is the idea that all possible traces of the first impact could be completely erased, even though there could (in principle) be time for information from the first impact to be dissipated into the environment, thereby rendering the process irreversible.

This raises the question of what irreversibility actually means. Is a process irreversible in some absolute sense, or is it contextual and dependent on the observer? We have previously suggested that there are no absolutes in physics. Therefore, we should examine the possibility that what is irreversible to one observer need not be irreversible to another. This does not seem inconsistent or unreasonable, given that irreversibility and observers are emergent processes, the laws of which have not yet been understood in any significant way.

We focus exclusively on linear quantum evolution, i.e., quantum state evolution conforming to the principles of QM as discussed for example in Peres (1995), rather than appeal to any form of nonlinear quantum mechanics to generate self-intervention effects. We explore a number of type-three thought experiments involving photons, which act as either quantum or classical objects at various times. As quantum objects they pass through beam-splitters and suffer random outcomes as a result. As classical objects they are used to trigger the switching on or off of macroscopic apparatus, a switching that determines the subsequent quantum evolution of other photons.

We shall not discuss the nature of photons per se, except to say that they are referred to as particles for convenience only: our ideology and formalism treats them as signals in detectors (Jaroszkiewicz, 2008a). Everything is idealized here, it being assumed that all detectors operate with 100 percent efficiency and that photon polarizations and wavelengths can be adjusted wherever necessary to make the scenarios discussed here physically realizable. The experiments we discuss are not necessarily based on photons: other particles such as electrons could be used in principle.

[2] There is a possible loophole here: our comment involves *two* observers with access to different contextual information. What that means physically is a nontrivial question for any theory of observation. In the absence of any proper theory of such a scenario, we can at best express our opinion on the matter.

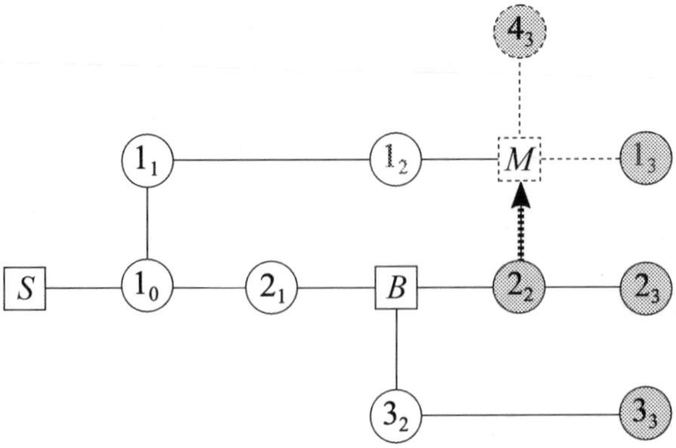

Figure 21.1. If detected, photon 2_2 triggers the switching-on of mirror M that deflects 1_2 into 4_3 rather than 1_3.

21.2 Experiment SI-1: Basic Self-intervention

To illustrate the sort of experiment we are interested in, we start with the experiment shown schematically in Figure 21.1. By stage Σ_1, a correlated, nonentangled two-photon state $\boldsymbol{\Psi}_1 \equiv \widehat{\mathbb{A}}_1^1 \widehat{\mathbb{A}}_1^2 \mathbf{0}_1$ is created by source S. Channel 2_1 is subsequently passed through beam splitter B into channels 2_2 or 3_2. If 2_2 is detected physically at stage Σ_2, then a macroscopic mechanism triggers mirror M to be swung into place so as to deflect 1_2 onto 4_3 rather than 1_3. The spin of the photons is neglected in this analysis but could easily be included in the discussion and encoded into our computer algebra program MAIN if it were required.

The labstate $\boldsymbol{\Psi}_2$ at stage Σ_2 is given by

$$\boldsymbol{\Psi}_2 = \widehat{\mathbb{A}}_2^1 (t \widehat{\mathbb{A}}_2^2 + ir \widehat{\mathbb{A}}_2^3) \mathbf{0}_2, \tag{21.1}$$

where t and r are beam splitter B parameters satisfying $t^2 + r^2 = 1$.

Note that in Figure 21.1, detector 2_2 is shown shaded, indicating that the observer looks at it and definitely ascertains its signal status, thereby extracting classical information. If there is no signal there, then the mirror M is not swung into place to intercept the channel from 1_2, and so a signal is certainly registered in 1_3. On the other hand, if there is a signal in 2_2, mirror M swings into place and deflects channel 1_2 into detector 4_3.

Semi-unitary evolution from stage Σ_2 to stage Σ_3 is given by

$$\mathbb{U}_{3,2} \widehat{\mathbb{A}}_2^1 \widehat{\mathbb{A}}_2^2 \mathbf{0}_2 = \widehat{\mathbb{A}}_3^4 \widehat{\mathbb{A}}_3^2 \mathbf{0}_3, \quad \mathbb{U}_{3,2} \widehat{\mathbb{A}}_2^1 \widehat{\mathbb{A}}_2^3 \mathbf{0}_2 = \widehat{\mathbb{A}}_3^1 \widehat{\mathbb{A}}_3^3 \mathbf{0}_3. \tag{21.2}$$

Note that in the program MAIN, null evolution carries 2_2 to 2_3 and 3_2 to 3_3, and the program reports detector status at stage Σ_3. The predicted nonzero outcome probabilities are as expected:

$$\Pr(\widehat{\mathbb{A}}_3^1 \widehat{\mathbb{A}}_3^3 \mathbf{0}_3 | \boldsymbol{\Psi}_0) = r^2, \ \Pr(\widehat{\mathbb{A}}_3^2 \widehat{\mathbb{A}}_3^4 \mathbf{0}_3 | \boldsymbol{\Psi}_0) = t^2. \tag{21.3}$$

There is nothing unexpected in these results, but there is an unusual confluence of classical and quantum interaction: on the quantum side of the experiment, mirror M is activated in a classically deterministic way only if and when quantum outcome 2_2 occurs, and that is unpredictable.

One interpretation of this experiment is that it is a form of delayed choice experiment: the mirror M is activated only if and when 2_2 has been detected, and this is *after* 1_2 is on its way. The difference between this experiment and previous analysis of delayed choice is that the "choice" is made via quantum processes occurring within the experiment itself, which motivates our chapter title "Self-Intervening Networks."

21.3 Experiment SI-2: Double Self-intervention

The next variant experiment, SI-2, is shown in Figure 21.2. Source S creates a correlated photon pair $1_1 2_1$ by stage $\mathbf{\Sigma}_1$. These are directed into beam splitters B^1 and B^2 as shown. B^2 is used as a trigger. If a signal is detected at 3_2, that triggers the classical switching-on of beam splitter B^3, which intercepts 1_2 and channels it into 5_3 or 6_3 instead of it passing on to 1_3. Conversely, if a signal is observed at 4_2, beam splitter B^4 is switched on, rather than B^3. Now 2_2 is channeled by B^4 into 7_3 or 8_3 rather than passing on to 2_3.

The dynamics is worked out by the methods outlined in previous experiments. The final stage rank is now eight, which presents no problem for program MAIN. Nonzero correlations are found to be

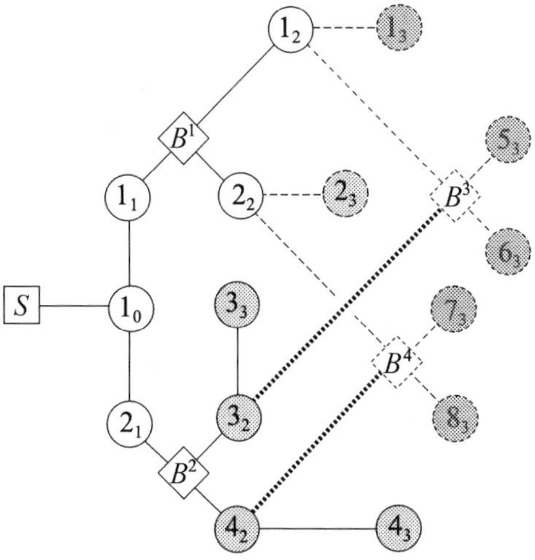

Figure 21.2. Experiment SI-2: in this scheme, no quantum interference occurs, because complete which-path information can be worked out from the signal pattern in each run.

$$\begin{aligned}
&\Pr(\widehat{\mathbb{A}}_3^2\widehat{\mathbb{A}}_3^3\mathbf{0}_3|\mathbf{\Psi}_0) = (r^1r^2)^2, && \Pr(\widehat{\mathbb{A}}_3^1\widehat{\mathbb{A}}_3^4\mathbf{0}_3|\mathbf{\Psi}_0) = (t^1t^2)^2,\\
&\Pr(\widehat{\mathbb{A}}_3^3\widehat{\mathbb{A}}_3^5\mathbf{0}_3|\mathbf{\Psi}_0) = (t^1r^2r^3)^2, && \Pr(\widehat{\mathbb{A}}_3^3\widehat{\mathbb{A}}_3^6\mathbf{0}_3|\mathbf{\Psi}_0) = (t^1r^2t^3)^2,\\
&\Pr(\widehat{\mathbb{A}}_3^4\widehat{\mathbb{A}}_3^7\mathbf{0}_3|\mathbf{\Psi}_0) = (r^1t^2r^4)^2, && \Pr(\widehat{\mathbb{A}}_3^4\widehat{\mathbb{A}}_3^8\mathbf{0}_3|\mathbf{\Psi}_0) = (r^1t^2t^4)^2.
\end{aligned} \tag{21.4}$$

Individual detector signal probabilities can be worked out from these correlations. For example, $\Pr(1_3)$, the probability that detector 1_3 will be in its signal state rather than its ground state at the final stage Σ_3, is given by

$$\Pr(1_3) \equiv \Pr(\widehat{\mathbb{A}}_3^2\widehat{\mathbb{A}}_3^3\mathbf{0}_3|\mathbf{\Psi}_0) + \Pr(\widehat{\mathbb{A}}_3^3\widehat{\mathbb{A}}_3^5\mathbf{0}_3|\mathbf{\Psi}_0) + \Pr(\widehat{\mathbb{A}}_3^3\widehat{\mathbb{A}}_3^6\mathbf{0}_3|\mathbf{\Psi}_0) = (r^2)^2, \tag{21.5}$$

and so on.

There are no interference effects predicted in this experiment.

21.4 Experiment SI-3: Interfering Single Self-intervention

The third scenario, SI-3, is shown in Figure 21.3. In this case, the initial photon pair is passed through a pair of beam splitters B^1 and B^2 exactly as in experiment SI-2. The difference lies in the next stage. Channels 1_2 and 2_2 are sent off over sufficiently long optical paths so as to allow interference between channels 3_2 and 4_2 in beam splitter B^3. If a signal is registered in 4_3, then beam splitter B^4 is swung into place to intercept 1_3 and 2_3, so that they can interfere and be observed at 5_4 and 6_4. If B^4 is not triggered, then 1_3 and 2_3 are channeled on to 1_4 and 2_4, respectively.

There is a point about beam splitter B^3 that needs attention. This experiment deals with signality-two amplitudes, so there is an issue at beam splitter B^3 in that the possibility exists that a photon comes in from 3_2 and another from 4_2. This is a case of *beam splitter saturation*, referred to in Chapter 11. Exercise 11.1 shows that if B^3 is *calibrated*, and if the evolution is semi-unitary such that

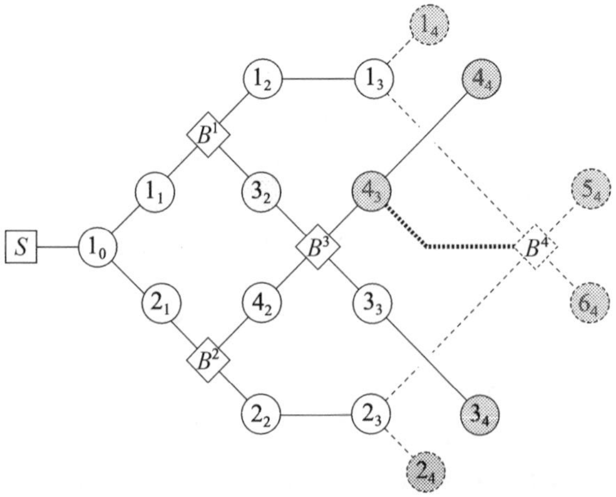

Figure 21.3. Interference at beam splitter B^3 can trigger beam splitter B^4.

signality one is conserved, then the only scenario permitted is that a signal is detected in 3_4 *and* in 4_4. In that case, even though beam splitter B^4 is triggered, no signals are detected at 5_4 or 6_4.

The beam splitter saturation scenario shows that there is a fundamental difference between *photon self-interference*, which is a signality-one process, and *two-photon scattering*, which is a signality-two process. The former is a manifestation of the quantum superposition principle, whereas the latter depends on the details of photon–matter coupling. For instance, the lowest order amplitude for photon–photon scattering in quantum electrodynamics is proportional to α^2, where α is the fine structure constant. We mention in passing that the correct calculation of the photon–photon scattering cross section requires delicate treatment of divergent integrals, in order to bring the predictions in line with cosmological observations (Liang and Czarnecki, 2011). This is a good example of the need to factor in detector physics properly in relativistic quantum field theory.

The dynamics for experiment SI-3 is handled in the usual way in program MAIN, with the following predictions:

$$\begin{aligned} &\Pr(\widehat{\mathbb{A}}_4^1\widehat{\mathbb{A}}_4^2\mathbf{0}_4|\mathbf{\Psi}_0) = (t^1t^2)^2, && \Pr(\widehat{\mathbb{A}}_4^1\widehat{\mathbb{A}}_4^3\mathbf{0}_4|\mathbf{\Psi}_0) = (t^1r^2r^3)^2, \\ &\Pr(\widehat{\mathbb{A}}_4^4\widehat{\mathbb{A}}_4^5\mathbf{0}_4|\mathbf{\Psi}_0) = (t^1r^2t^3r^4 + r^1t^2r^3t^4)^2, && \Pr(\widehat{\mathbb{A}}_4^2\widehat{\mathbb{A}}_4^3\mathbf{0}_4|\mathbf{\Psi}_0) = (r^1t^2t^3)^2, \\ &\Pr(\widehat{\mathbb{A}}_4^4\widehat{\mathbb{A}}_4^6\mathbf{0}_4|\mathbf{\Psi}_0) = (t^1r^2t^3r^4 - r^1t^2r^3t^4)^2, && \Pr(\widehat{\mathbb{A}}_4^3\widehat{\mathbb{A}}_4^4\mathbf{0}_4|\mathbf{\Psi}_0) = (r^1r^2)^2. \end{aligned} \tag{21.6}$$

In this variant experiment, two of the outcome correlations, $\Pr(\widehat{\mathbb{A}}_4^4\widehat{\mathbb{A}}_4^5\mathbf{0}_4|\mathbf{\Psi}_0)$ and $\Pr(\widehat{\mathbb{A}}_4^4\widehat{\mathbb{A}}_4^6\mathbf{0}_4|\mathbf{\Psi}_0)$, show interference that has essential contributions from beam splitters B^1 and B^2 in a manner that seems impossible to explain in terms of photons as classical particles.

In all experiments where quantum interference takes place, there inevitably has to be some which-path uncertainty somewhere. This does not occur in variant experiments SI-1 or SI-2 but does occur in SI-3. Program MAIN shows that the interference at 5_4 and 6_4 is photon self-interference and occurs only when a single photon enters B^3 and triggers B^4. Whenever that happens, the observer cannot say whether it was actually 3_2 or 4_2 that passed into B^3, and that is the origin of the missing which-path information that is responsible for the interference effect at 5_4 and 6_4. The signality-two beam splitter saturation possibility triggers B^4 but does not lead to any signals at 5_4 or 6_4.

Experiment SI-3 involves apparatus change that produces interference and can be identified as doing so: after each run, it will be clear whether B^4 was triggered, and the observer will then be sure that it was that beam splitter that created the interference.

21.5 Schrödinger's Cat

The Schrödinger's cat scenario is a well-known illustration of the dangers of taking a too-literal view of what a quantum wave function represents (Schrödinger,

1935). In brief, the architecture is this. A living cat is placed in a sealed, isolated box along with a flask of poison and a mechanism that contains a radioactive sample. The mechanism will release the poison if it is triggered by the decay of any one of the atoms in the sample. An observer external to the box closes the box and leaves it alone for 1 hour. At that time, the observer opens the box and ascertains whether or not the cat is still alive or now dead.

The issue here is that according to standard quantum wave mechanics, the wave function for the combined system of cat plus radioactive sample will develop from a separable state of a {live cat and undecayed sample} to a entangled superposition of a {live cat and undecayed sample} with a {dead cat and decayed sample}. This prompts questions, such as "How can a cat exist in a superposition of a live state and a dead state?" and so on, which have, over the decades since Schrödinger's article, generated a plethora of confusing views about the nature of reality.

This scenario is one we can label T3. It is a complex example of a self-intervening network, presented in simple terms.

For us, such questions are vacuous. All of them. If context is taken properly into account, then no such questions need be raised. It suffices to look at what the external observer can do, what they are actually doing, and when they actually do it. The QDN prescription is clear. The observer prepares the combined {cat/box/poison sample} SUO at stage Σ_0. On the basis of their contextual information, such as the decay probability associated with the radioactive sample, the observer can, just after closing the box, use standard quantum theory to estimate the probability of the cat being alive when the box is opened at stage Σ_1, that is, 1 hour later. What the observer eventually believes has actually happened inside the box depends on the classical information that is allowed to accumulate there during the hour. For instance, if a decay occurs, not only could it trigger the poison to be released to kill the cat, but the lab time at which this occurred could be registered. When the observer finally opened the box, if they found the cat dead, then the registered time of death would be available.

It is such examples that make it clear that "the" information void is not an absolute concept. It is contextual. In this case, the external observer has no information about what is going on inside the box during the given hour that the experiment runs. Therefore, a quantum description has to suffice. If after opening the box, sufficient classical information is recovered from it to enable the observer to reconstruct events inside that box during that hour, then a rather classical description can be given.

At no point is the observer empirically justified in asserting that the cat is in a quantum superposition of alive and dead. The observer may be justified in imagining a mathematical description of the state of the system in such terms. That is not the same as objective reality.

There are two further points to make here.

A Cat Is a Living Process

It is rather fanciful to think of a cat in terms of a pure quantum state, albeit with a vast number of degrees of freedom. A cat is a living process and therefore not amenable to a reductionist description. In particular, a cat needs an extensive environment to maintain itself: an atmosphere to breathe in, a source of warmth, and so on.

Suppose it was argued that such an environment could be provided inside the box. Yes, but then the observer could not allow that environment to be contaminated by any factors external to the box. What about gravity? What about temperature? What about the expansion of the Universe? Can we really arrange for total shielding of the inside of the box from the outside?

Reductionism in high-energy particle physics seems to work because elementary particle interactions appear to be amenable to shielding during scattering experiments, but even this has been questioned. For instance, it is often imagined that electrons are elementary point particles with a well-identified physical mass in vacuo, represented by a simple pole in their momentum space propagator. But radiative corrections due to "soft" photons (associated with the low-frequency end of the electromagnetic spectrum) appear to turn that pole into a branch cut (in the complex plane), with the consequence that the concept of "electron mass" is problematical. On that basis, an electron cannot be imagined as a single particle, but an extraordinarily complex system of a charge surrounded by an indeterminate cloud of photons.[3]

Our point is that a highly complex emergent concept such as a living cat might not be amenable to any form of quantum state description, because quantum processes manifest themselves only when phases are carefully controlled. The same comment applies to the so-called quantum mind program of investigation by Hameroff and others, in which brain function is modeled as a form of quantum computation (Hameroff, 1999).

One way of seeing why such a program would not work would be if we described the brain as affected by *contextual overcompleteness*, that is, subject to so many external, extraneous contextual factors that cannot be eliminated, such as temperature variations, pressure fluctuations, and so on, that phase control becomes meaningless and decoherence dominates.

Life and Death Are Not Classic Binary Alternatives

Zero and one (or their equivalents) are regarded as the only two possible alternative states of a classical bit. This degree of simplicity is so powerful and useful that we tend to apply it to many complex situations: we won the war or we

[3] It is such considerations that make it obvious to us that elementary particle physics is really a branch of emergent physics and should be approached as such, rather than as a reductionist theory from which emergence could be understood.

lost it; he is a good driver or he is a bad driver; the cat is alive or it is dead. In reality, wars may be lost by both sides, a man may drive well in his village but be reckless on the motorway; a cat may be in a vegetative state that is not dead but is not really worth calling life.

All of these concerns should be addressed before simplistic arguments based on contextually incomplete views of quantum mechanics are given.

22
Separability and Entanglement

22.1 Introduction

In this chapter, we discuss the structure of signal amplitudes in terms of the separability and entanglement of quantum register states. We review the concepts of *subregisters*, *splits*, *partitions*, factorizable Hilbert spaces, the separability of states, *separations*, and *entanglements*.

All Hilbert spaces in this chapter are taken to be complex and finite dimensional, and denoted by capital calligraphic Latin letters, such as $\mathcal{H}$. We do not restrict the discussion to quantum bits, but ultimately, it is those in which we are most interested.

Suppose $\mathcal{X}$ and $\mathcal{Y}$ are two finite-dimensional Hilbert spaces with a surjective linear map U from $\mathcal{X}$ to $\mathcal{Y}$.[1] Suppose further that U has the property that it preserves norm, that is, if for any element x in $\mathcal{X}$ we have $\|Ux\|_{\mathcal{Y}} = \|x\|_{\mathcal{X}}$, where $\|x\|_{\mathcal{X}} \equiv \sqrt{(x,x)_{\mathcal{X}}}$ is the Hilbert space "length" of x in $\mathcal{X}$ and $\|y\|_{\mathcal{Y}} \equiv \sqrt{(y,y)_{\mathcal{Y}}}$ is the Hilbert space "length" of y in $\mathcal{Y}$. Then U is called an *isometric isomorphism*. Under these conditions, it is necessarily the case that $\mathcal{X}$ and $\mathcal{Y}$ have the same dimension and, as far as basic Hilbert space properties are concerned, are identical. We denote this form of equality by $\mathcal{X} \simeq \mathcal{Y}$.

22.2 Quantum Registers

Hilbert spaces per se are central to quantized detector networks (QDN) but that is on the mathematical side. Physics brings in empirical context that adds an extra flavor to the discussion, requiring the mathematics of tensor products and quantum registers. We define a quantum register as the tensor product of two or more finite-dimensional Hilbert spaces, referred to as *subregisters*. In principle,

[1] Here *surjective* means that $U\mathcal{X} = \mathcal{Y}$, where $U\mathcal{X}$ is the image of $\mathcal{X}$ in $\mathcal{Y}$ under the linear map U. This condition is not imposed in our definition of semi-unitary operators (which are linear maps).

subregisters can have any dimension, including one. In the following, we shall rule out one-dimensional Hilbert spaces as being of no empirical interest to us.

Example 22.1 Let $\mathcal{U}$, $\mathcal{V}$, and $\mathcal{W}$ be three complex, finite-dimensional Hilbert spaces of dimensions $3, 2$, and 4, respectively. Consider the tensor product $\mathcal{X}$ defined as

$$\mathcal{X} \equiv \mathcal{U}^{(1)} \otimes \mathcal{W} \otimes \mathcal{U}^{(2)} \otimes \mathcal{V}, \tag{22.1}$$

where $\mathcal{U}^{(1)}$ and $\mathcal{U}^{(2)}$ are copies of $\mathcal{U}$. Then $\mathcal{X}$ is a complex Hilbert space of dimension $3 \times 4 \times 3 \times 2 = 72$.

The number of subregisters in a given quantum register $\mathcal{H}$ is called the *rank* of that register and denoted $rank\mathcal{H}$. In Example 22.1, $\mathcal{X}$ has rank four. A rank-one Hilbert space will be called an *atom*.

The dimension of a rank-r register must have at least r non-trivial factors. For example, a rank-two register of dimension 63 is the tensor product of a 7-dimensional atom with a nine-dimensional atom, or the tensor product of a 21-dimensional atom with a 3-dimensional atom.

The subregister concept, and by implication, that of rank, is contextual, because it is possible to encounter situations where a subregister could be thought of as a register itself, with its own subregisters. In other words, atoms can have constituents.

Example 22.2 *Positronium* is generally described as an unstable bound state of an electron and a positron. When viewed as a single particle, positronium comes in two forms: *para-positronium* and *ortho-positronium*. The former has a particle spin angular momentum classification $j = 0$ (spin zero), described by a singlet spin state in a one-dimensional Hilbert space $\mathcal{H}^{\text{para}}$, while the latter has spin $j = 1$ (spin one) described by spin states in a three-dimensional Hilbert space $\mathcal{H}^{\text{ortho}}$.

On the other hand, a nonrelativistic analysis of positronium in terms of its two constituents would lead to its spin being modeled in terms of a four-dimensional Hilbert space $\mathcal{H}^{\text{electron}} \otimes \mathcal{H}^{\text{positron}}$, a rank-two tensor product of an electron spin space $\mathcal{H}^{\text{electron}}$, and a positron spin space $\mathcal{H}^{\text{positron}}$.

Mathematically, we may write

$$\mathcal{H}^{\text{para}} \oplus \mathcal{H}^{\text{ortho}} \simeq \mathcal{H}^{\text{electron}} \otimes \mathcal{H}^{\text{positron}}, \tag{22.2}$$

where the symbol $\oplus$ denotes a *direct sum* of vector spaces. The direct sum of two Hilbert spaces is a Hilbert space with dimension equal to the sum of the two Hilbert spaces being summed. The group theory of such bound states and their decomposition leads in this case to the rule $\mathbf{1} \oplus \mathbf{3} \simeq \mathbf{2} \otimes \mathbf{2}$ (Lichtenberg, 1970).

Throughout this chapter, we adhere to our convention that subscripts label stages while superscripts label subregisters. Stages concern dynamics, which is not the focus in the present chapter, so we avoid subscripts in this chapter.

As we have seen before in this book, the ordering of subregisters in a tensor product will not usually be regarded as significant; i.e., $\mathcal{H}^a \otimes \mathcal{H}^b$ will mean the same mathematically as $\mathcal{H}^b \otimes \mathcal{H}^a$, as well as physically. What is important is the fact that each subregister can be identified, meaning that labels are physically significant.

The reason for this is based on the physics of observation. For instance, given two detectors D^a and D^b, we may represent them by two Hilbert spaces $\mathcal{H}^a$ and $\mathcal{H}^b$, respectively. There will in general be no natural way of ordering these detectors, and so there is no logical reason to order $\mathcal{H}^a$ and $\mathcal{H}^b$ in a tensor product. What matters is the labeling, which can be regarded as part of the empirical contextual information always available to observers.

With these comments in mind, we henceforth suppress the tensor product symbol $\otimes$, so that $\mathcal{H}^a\mathcal{H}^b$ and $\mathcal{H}^b\mathcal{H}^a$ both mean $\mathcal{H}^a \otimes \mathcal{H}^b$.

Given two Hilbert spaces $\mathcal{H}^a$ and $\mathcal{H}^b$ of dimension d^a and d^b, respectively, the tensor product $\mathcal{H}^a\mathcal{H}^b$ is a Hilbert space of rank two and dimension d^ad^b. We define $\mathcal{H}^{[ab]}$ as a rank-one Hilbert space (that is, an atom) of dimension d^ad^b that is isometrically isomorphic to $\mathcal{H}^a\mathcal{H}^b$, that is,

$$\mathcal{H}^{[ab]} \simeq \mathcal{H}^a\mathcal{H}^b. \tag{22.3}$$

Specifically, $\mathcal{H}^{[ab]}$ is the same mathematically as $\mathcal{H}^a\mathcal{H}^b$, but its physical context, that it is a tensor product, is now ignored.

Likewise, given three Hilbert spaces $\mathcal{H}^a$, $\mathcal{H}^b$, $\mathcal{H}^c$, we define $\mathcal{H}^{[abc]}$ to be an atom isometrically isomorphic to $\mathcal{H}^a\mathcal{H}^b\mathcal{H}^c$, and so on for higher rank registers. Note the trivial identity $\mathcal{H}^{[a]} \simeq \mathcal{H}^a$.

As we have indicated previously, the order of superscripts in the above is not significant, so $\mathcal{H}^{[ab]} = \mathcal{H}^{[ba]}$, and so on.

22.3 Splits

A *split* is any convenient way of grouping the subregisters in a quantum register into two or more factor registers, or *atoms*, each of which is regarded for the purposes of that split as a Hilbert space of rank one, that is, a Hilbert space not itself split into two or more factor registers. For large-rank quantum registers, very many different splits will be possible.

Example 22.3 The rank-three register $\mathcal{H}^a\mathcal{H}^b\mathcal{H}^c$ may be split in five distinct ways:

$$\mathcal{H}^a\mathcal{H}^b\mathcal{H}^c \simeq \mathcal{H}^a\mathcal{H}^{[bc]} \simeq \mathcal{H}^b\mathcal{H}^{[ac]} \simeq \mathcal{H}^c\mathcal{H}^{[ab]} \simeq \mathcal{H}^{[abc]}. \tag{22.4}$$

Note that one of these ways is the original register itself.

The number of ways of splitting a rank-n quantum register is the same as the number of ways B_n of partitioning a given set of cardinality n, a historically important topic in combinatorial mathematics. The B_n are called *Bell numbers*[2] and satisfy many curious and interesting relations in diverse fields, such as probability, game theory, and number theory. For instance,

$$B_{n+1} = \sum_{k=0}^{n} \frac{n!}{k!(n-k)!} B_k, \quad B_0 = 1, \tag{22.5}$$

from which we find the sequence $\{B_n, n = 1, 2, \ldots\} = \{1, 2, 5, 15, 52, \ldots\}$. An explicit formula for B_n is given by Dobinski's formula,

$$B_n = \frac{1}{e} \sum_{k=0}^{\infty} \frac{k^n}{k!}. \tag{22.6}$$

22.4 Partitions

Consider a set $\{\mathcal{H}^a : a = 1, 2, \ldots, r\}$ of Hilbert spaces, denoting the dimension of $\mathcal{H}^a$ by d^a. Then the rank-r quantum register $\mathcal{H}^{[r]} \equiv \mathcal{H}^1\mathcal{H}^2 \ldots \mathcal{H}^r$ is a vector space of dimension $d^{[r]} \equiv d^1 d^2 \ldots d^r$ that contains both entangled and separable states.

The classification of states in such a register into separable or entangled types is too limited for us, so we introduce the more useful concepts of *separations* and *entanglements*. We explain our terminology starting with the separable sets.

Separations

For any two subregisters $\mathcal{H}^a, \mathcal{H}^b$ of the quantum register $\mathcal{H}^{[r]}$, such that $a \neq b$, we define the *rank-two separation* $\mathcal{H}^{ab}$ to be the subset of the tensor product $\mathcal{H}^a\mathcal{H}^b$ consisting of all separable elements in that tensor product, that is,

$$\mathcal{H}^{ab} \equiv \{\phi^a \psi^b : \phi^a \in \mathcal{H}^a, \ \psi^b \in \mathcal{H}^b\}. \tag{22.7}$$

By definition, $\mathcal{H}^{ab}$ includes the zero vector 0^{ab} of the tensor product $\mathcal{H}^a\mathcal{H}^b$. Note that $\mathcal{H}^{ab} = \mathcal{H}^{ba}$.

The separation concept readily generalizes to higher rank separations as follows. Pick an integer k in the interval $[1, r]$ and then select k different elements $a^1, a^2, \ldots, a^k$ of this interval. Then the rank-k separation $\mathcal{H}^{a^1 a^2 \ldots a^k}$ is the subset of $\mathcal{H}^{a^1}\mathcal{H}^{a^2} \ldots \mathcal{H}^{a^k}$ given by

$$\mathcal{H}^{a^1 a^2 \ldots a^k} \equiv \left\{\psi^{a^1} \psi^{a^2} \ldots \psi^{a^k} : \psi^{a^i} \in \mathcal{H}^{a^i}, \quad 1 \leqslant i \leqslant k\right\}. \tag{22.8}$$

Every element of a rank-k separation has k factors. A rank-one separation of a subregister is by definition equal to that subregister, and is therefore a Hilbert space in its own right. Separations of rank greater than one, however, cannot

[2] After E. T. Bell (Bell, 1938).

be Hilbert spaces because they do not contain entangled states, which all tensor products of rank two or more necessarily do.

Rank-Two Entanglements

We can now construct the *entanglements*, which are defined in terms of complements. Starting with the lowest rank possible, we define the rank-two entanglement $\mathcal{H}^{\overline{ab}}$ to be the complement of the separation $\mathcal{H}^{ab}$ in the tensor product $\mathcal{H}^a\mathcal{H}^b$, that is,

$$\mathcal{H}^{\overline{ab}} \equiv \mathcal{H}^a\mathcal{H}^b - \mathcal{H}^{ab}. \tag{22.9}$$

Note that $\mathcal{H}^{\overline{ab}}$ cannot be a vector space because it does not contain the zero vector.

The original tensor product space considered as a set is therefore the union

$$\mathcal{H}^a\mathcal{H}^b = \mathcal{H}^{ab} \cup \mathcal{H}^{\overline{ab}} \tag{22.10}$$

of the set $\mathcal{H}^{ab}$ of all separable states and the set $\mathcal{H}^{\overline{ab}}$ of all entangled states. These two subsets are disjoint and neither is a vector space.

Separation Products

The generalization of the above to larger rank entanglements is straightforward, but first it will be useful to extend our notation to include the concept of *separation product*.

Suppose A and B are subsets of Hilbert spaces $\mathcal{H}^a$ and $\mathcal{H}^b$, respectively, where $a \neq b$. We define the *separation product* $A \bullet B$ to be the subset of $\mathcal{H}^a\mathcal{H}^b$ given by

$$A \bullet B \equiv \{\psi\phi : \psi \in A,\ \phi \in B\}. \tag{22.11}$$

Properties of the separation product are:

1. The separation product is symmetric, that is, $A \bullet B = B \bullet A$. This means that the separation product is not equivalent to the Cartesian product $A \times B$, which is the set of all ordered pairs of elements.
2. $\mathcal{H}^{ab} = \mathcal{H}^a \bullet \mathcal{H}^b$. Note that this is not the same thing as $\mathcal{H}^a\mathcal{H}^b$, which in our notation is the tensor product of $\mathcal{H}^a$ and $\mathcal{H}^b$.
3. The separation product is associative, commutative, and cumulative, i.e.

$$\begin{gathered}\left(\mathcal{H}^a \bullet \mathcal{H}^b\right) \bullet \mathcal{H}^c = \mathcal{H}^a \bullet \left(\mathcal{H}^b \bullet \mathcal{H}^c\right) \equiv \mathcal{H}^{abc} \\ \mathcal{H}^{ab} \bullet \mathcal{H}^c = \mathcal{H}^{abc}.\end{gathered} \tag{22.12}$$

The separation product can also be defined to include entanglements. For example,

$$\mathcal{H}^{\overline{ab}} \bullet \mathcal{H}^c = \left\{\phi\psi : \phi \in \mathcal{H}^{\overline{ab}}, \psi \in \mathcal{H}^c\right\}. \tag{22.13}$$

Significantly, while the separation product of two separations is a separation, the separation product of two entanglements is *not* an entanglement, that is,

$$\mathcal{H}^{\overline{ab}} \bullet \mathcal{H}^{\overline{cd}} \neq \mathcal{H}^{\overline{abcd}}. \tag{22.14}$$

A further notational simplification is to use a single $\mathcal{H}$ symbol, incorporating the separation product symbol $\bullet$ with indices directly, as the following example illustrates.

Example 22.4 Given Hilbert spaces $\mathcal{H}^a, \mathcal{H}^b, \ldots, \mathcal{H}^i$, we may write

$$\mathcal{H}^{\overline{cd}\bullet\overline{hi}\bullet abefg} \equiv \mathcal{H}^{ab} \bullet \mathcal{H}^{\overline{cd}} \bullet \mathcal{H}^{efg} \bullet \mathcal{H}^{\overline{hi}}. \tag{22.15}$$

Other expansions are possible, given that separations such as $\mathcal{H}^{efg}$ can be expanded further as separation products.

Associativity of the separation product applies to both separations and entanglements, as can be readily proved.

Higher Rank Entanglements

We can now define higher rank entanglements, such as $\mathcal{H}^{\overline{abc}}$, $\mathcal{H}^{\overline{abcd}}$, and so on. These are defined in terms of complements, in the same way that $\mathcal{H}^{\overline{ab}}$ was defined.

Example 22.5 Consider the rank-three tensor product $\mathcal{H}^a\mathcal{H}^b\mathcal{H}^c$ and the following disjoint subsets: $\mathcal{H}^{abc}$, $\mathcal{H}^{a\bullet\overline{bc}}, \mathcal{H}^{b\bullet\overline{ac}}, \mathcal{H}^{c\bullet\overline{ab}}$. These are all separable in one way or another. For instance, $\mathcal{H}^{abc}$ is completely separable, while the other three subsets referred to are partially separable. If we remove all those subsets from $\mathcal{H}^a\mathcal{H}^b\mathcal{H}^c$, then what is left will be completely entangled, which is what we want to define. Hence we define the rank-three entanglement $\mathcal{H}^{\overline{abc}}$ as the complement

$$\mathcal{H}^{\overline{abc}} \equiv \mathcal{H}^a\mathcal{H}^b\mathcal{H}^c - \mathcal{H}^{abc} \cup \mathcal{H}^{a\bullet\overline{bc}} \cup \mathcal{H}^{b\bullet\overline{ac}} \cup \mathcal{H}^{c\bullet\overline{ab}}. \tag{22.16}$$

We will refer to a set such as $\mathcal{H}^{\overline{abc}}$ as a rank-three entanglement, and so on. In general, higher rank entanglements such as $\mathcal{H}^{\overline{abcd}}$ in the above require a deal of filtering out of separations and cross-entanglements from the original tensor product Hilbert space for their definition to be possible, which accounts partly for the fact that entanglements are generally not as conceptually simple or intuitive as separations. From this we can appreciate just how complicated the entanglement structure of qubit registers can be.

Exercise 22.6 Show that the rank-four entanglement $\mathcal{H}^{\overline{abcd}}$ is given by

$$\begin{aligned}\mathcal{H}^{\overline{abcd}} \equiv \mathcal{H}^a\mathcal{H}^b\mathcal{H}^c\mathcal{H}^d \\ -\mathcal{H}^{abcd} \cup \mathcal{H}^{ab\bullet\overline{cd}} \cup \mathcal{H}^{ac\bullet\overline{bd}} \cup \mathcal{H}^{ad\bullet\overline{bc}} \cup \mathcal{H}^{bc\bullet\overline{ad}} \cup \mathcal{H}^{bd\bullet\overline{ac}} \cup \mathcal{H}^{cd\bullet ab} \\ \cup \mathcal{H}^{\overline{ab}\bullet\overline{cd}} \cup \mathcal{H}^{\overline{ac}\bullet\overline{bd}} \cup \mathcal{H}^{\overline{ad}\bullet\overline{bc}} \cup \mathcal{H}^{a\bullet\overline{bcd}} \cup \mathcal{H}^{b\bullet\overline{acd}} \cup \mathcal{H}^{c\bullet\overline{abd}} \cup \mathcal{H}^{d\bullet\overline{abc}}.\end{aligned} \tag{22.17}$$

Entanglement Partitions

The decomposition of a quantum register $\mathcal{H}$ into the union of disjoint separations and entanglements will be called an *entanglement partition* of $\mathcal{H}$, and each element of that partition will be referred to as a *partition element*. We can now list the various entanglement partitions that we have encountered, along with the number of partition elements in each:

rank	register	number of partition elements
1	$\mathcal{H}^a$	$1 = B_1$
2	$\mathcal{H}^a\mathcal{H}^b$	$2 = B_2$
3	$\mathcal{H}^a\mathcal{H}^b\mathcal{H}^c$	$5 = B_3$
4	$\mathcal{H}^a\mathcal{H}^b\mathcal{H}^c\mathcal{H}^d$	$15 = B_4$

(22.18)

A calculation gives 52 partition elements in the entanglement partition of a rank-five quantum register, which we recognize as B_5, the 5th Bell number. The conjecture, then, is that the entanglement partition of a rank-r quantum register has B_r partition elements. This was proved by Eakins (Eakins, 2004), and shows that the entanglement structure of large-rank quantum registers will be too complicated to deal with without computer assistance.

The relationships between separations and entanglements are subtle, as are the relationships between splits and entanglement partitions. Although the number of partitions in the entanglement partition of a rank-r quantum register is the same as the number of splits and given by the Bell numbers, splits and partitions cannot coincide for $r > 1$. Every factor register in a split is a vector space, whereas no partition element is a vector space. Both splits and partitions are essential and unavoidable features in quantum mechanics and hence in the QDN account of quantum measurement and causal set structure.

A final simplification in this line of investigation is to use the above superscript notation to label the various elements of entanglements and separations. So, for example, $\Psi^{abc\bullet\overline{de}\bullet\overline{fgh}}$ denotes a state in the partition element $\mathcal{H}^{abc\bullet\overline{de}\bullet\overline{fgh}}$ and so on. Translated into direct terms, this means the following. First, this state is an element of the rank-eight quantum register

$$\mathcal{H}^{[8]} \equiv \mathcal{H}^a\mathcal{H}^b\mathcal{H}^c\mathcal{H}^d\mathcal{H}^e\mathcal{H}^f\mathcal{H}^g\mathcal{H}^h. \tag{22.19}$$

Second, we may write this state in the factorized form

$$\Psi^{abc\bullet\overline{de}\bullet\overline{fgh}} = \psi^a\psi^b\psi^c\psi^{\overline{de}}\psi^{\overline{fgh}}, \tag{22.20}$$

where $\psi^a \in \mathcal{H}^a$, $\psi^b \in \mathcal{H}^b$, $\psi^c \in \mathcal{H}^c$, $\psi^{\overline{de}} \in \mathcal{H}^{\overline{de}}$, and $\psi^{\overline{fgh}} \in \mathcal{H}^{\overline{fgh}}$.

Exercise 22.7 Investigate the superposition (vector addition) properties of separations and entanglements.

22.5 Quantum Zipping

We may avoid the material presented in this chapter when dealing with most situations discussed so far in this book, because those tend to involve relatively low-rank quantum registers with subregisters of relatively low dimension. However, we envisage the application of QDN or its analogues to large-rank, large-dimensionality contexts. Then it will be necessary to have some organization principles available. The split, partition, separation, and entanglement concepts discussed above appear unavoidable in this respect.

Consider the inner product between two pure labstates in a quantum register $\mathcal{H}$. Each state will come from a specific partition. Depending on the details of the two partitions concerned, this inner product may or may not factorize. This is because factor states can only take inner products in combinations that lie in the same factor Hilbert space of some split of the $\mathcal{H}$, a process we refer to as *quantum zipping*. Using the notation for splits, partitions, separations, and entanglements given in the previous chapter, the following example illustrates the point.

Example 22.8 Consider the rank-eight quantum register $\mathcal{H}^{[1...8]}$. By inspection, the inner product of the states $\boldsymbol{\Psi}^{123\bullet\overline{456}\bullet\overline{78}}$ and $\boldsymbol{\Phi}^{145\bullet\overline{23}\bullet\overline{678}}$ takes the factorized form

$$\overline{\boldsymbol{\Phi}}^{145\bullet\overline{23}\bullet\overline{678}}\boldsymbol{\Psi}^{123\bullet\overline{456}\bullet\overline{78}} = (\overline{\phi}^1\psi^1)(\overline{\phi^{23}}\psi^2\psi^3)(\overline{\phi}^4\overline{\phi}^5\overline{\phi}^{\overline{678}}\psi^{\overline{456}\bullet\overline{78}}), \quad (22.21)$$

which cannot be simplified further. Figure 22.1 illustrates the factorization structure of this inner product. The numbers 1 to 8 represent the individual atoms of the quantum register, not necessarily detectors.

This example is reasonably simple, as the two labstates involved can be ordered as shown. In general, more complex patters will occur, and then the corresponding zip diagrams will be more complicated.

The separation and entanglement structure of labstates should be of importance in any experiment. A significant application of these concepts is to *causal sets*, studied in the next chapter.

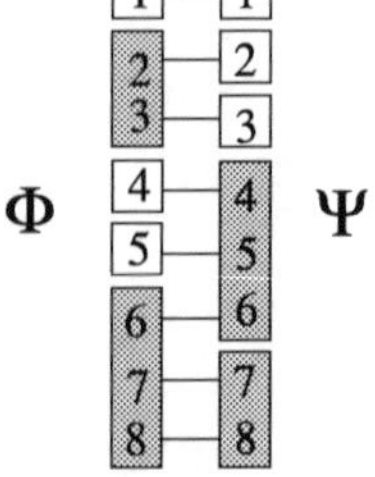

Figure 22.1. Quantum zipping.

23
Causal Sets

23.1 Introduction

In this chapter we apply the concepts presented in Chapter 22 to a topic that has been of much theoretical interest in recent decades: *causal set theory* (CST).

It is interesting and useful to review the status of CST relative to other foundational topics in mathematical physics that have been and remain of prime significance to empirical physics, namely, special relativity (SR), general relativity (GR), and quantum mechanics (QM). Those great frameworks are built on foundations that presuppose the continuity of space and time.

Despite this, it is an empirical fact that we are surrounded by discreteness in the form of atoms, so it is natural to see how that discreteness fits in with the continuity of space and time. We have attempted in this book to show that the application of discrete principles to QM can go far beyond the basic Planck quantum of action. We have gone as far as discretizing empirical time, in the form of the *stage* concept. However, as far as relativity in either of its manifestations is concerned, the case for any form of discreteness has not yet been established empirically, although great efforts are being made in that direction in the program known as *quantum gravity*. We have particularly in mind the discretization approaches to GR of Regge (Regge, 1961) and Penrose (Penrose, 1971).

Regge's approach is a classically based discretization of the Riemannian manifold concept, so has at best a generalized propositional classification (GPC) of two.[1] Penrose's spin network paradigm started as a quantum description of spacetime and has led to much work in the area of spin networks in the hope of unifying GR and QM. However, workers in spin networks and related branches of quantum gravity have focued their attention on the mathematical

[1] When Regge's approach is used to regularize Feynman path integrals in quantum gravity, its GPC drops to one, as it then becomes an exercise in mathematical physics.

complexities and hardly any on the role of observers and their apparatus, leading to a GPC classification of quantum gravity as one, that is, as mathematical physics and not physics. Claims made that such programs represent empirical physics have not been empirically substantiated to date; quantum gravity at this time remains empirically vacuous, despite the best efforts of many theorists over many decades.[2]

In various attempts to account for the existence of space, time, and matter in the Universe, physicists often adopt one of two opposing viewpoints. These may be labeled *bottom-up* and *top-down*, reflecting the basic difference between reductionist physics and emergent physics. QM makes an appearance in both approaches because while it really models the emergent side of observation, its mathematical structure has a strong reductionist flavor.

A number of bottom-up approaches to cosmology proceed from the assertion that at its most basic level, the Universe can be represented by a vast collection of discrete events embedded in some sort of mathematical space. For example, the *pregeometric* approach asserts that conventional classical time and space and the classical reality that we appear to experience all emerge on macroscopic scales due to the complex connections between certain fundamental microscopic, unobserved and unobservable, pregeometric entities. This approach was pioneered by Wheeler (Wheeler, 1980) and has received attention more recently by Stuckey (Stuckey, 1999).

A bottom-up approach to cosmology that we can relate to in quantized detector networks (QDN) is the *causal set hypothesis*, which asserts that spacetime itself is discrete at the most fundamental level. In CST models it is postulated that classical, discrete events are generated either randomly or through some agency, though neither the nature of these events nor the mechanism generating them is explained or discussed in detail. QDN comes in cost-free at this point. In this chapter we show how the particular mathematical properties and dynamical principles of QDN generates causal set structures naturally, without any extra assumptions.

Although it seems natural to generate causal sets by the discretization of a pseudo-Riemannian spacetime manifold of fixed dimension, as is done in lattice gauge theories and Regge's approach to GR (Regge, 1961), causal set events need not in principle be regarded as embedded in some background spacetime of fixed dimension d. One view taken by some theorists is that conventional (i.e., physical) spacetime emerges in some appropriate limit as a consequence of the causal set relations between discrete events. If so, then it is reasonable to expect that in the correct continuum limit, metric structure should emerge naturally (Brightwell and Gregory, 1991). An even more intriguing hypothesis

[2] This is not the same thing as QM over classical curved GR background spacetimes, for which there is some empirical evidence, such gravitational lensing and the need to correct GPS signals because of gravitational curvature and time dilation (Ashby, 2002).

is the suggestion that the dimension of this emergent spacetime might be scale dependent (Bombelli et al., 1987), making the model potentially compatible with GR in four spacetime dimensions, string theory and p-brane cosmology, and higher dimensional Kaluza–Klein theories. Such a hypothesis fits in with the general philosophy taken in this book that physics is contextual: different effective theories may explain different experiments.

The Origin Problem

One of the problems we have in accepting any pregeometric approach is what we may call the *origin problem*. This is the problem of explaining where any foundational concept comes from. It seems scientifically inconsistent to try to explain empirical physics on the basis of unverifiable hypotheses.

Mathematics came to terms with the origin problem in set theory by admitting that there are concepts, such as those of set, integer, infinity, and infinitesimal, that cannot be explained further and have to be taken as primitive concepts that cannot be derived from deeper principles and have to be treated as axiomatic. We have taken the same approach in our concept of *primary observer*. It is not inconsistent to base our foundations on such a concept, because we ourselves are primary observers: we can point to real, physical examples of the concept.

Our approach to causal set structure then is predicated on the concept of a primary observer. Any causal set structure being discussed will be contextual to the primary observer involved and their apparatus, so we do not conjecture any absolute pregeometric structure.

It turns out that the separation and entanglement properties of quantum register labstates discussed in the previous chapter have enough structure in them to provide the necessary causal set attributes of interest here. A novel feature of our approach is the occurrence of *two* distinct but interleaved causal set structures, in contrast to the one normally postulated in CST. One of these causal sets arises from the entanglement and separation properties of labstates themselves, while the other arises from the split and partition properties of the quantum registers and operators representing modules. In our view, these two distinct causal sets correspond to the two distinct classes of information transmission observed in physics. One of these involves nonlocal quantum correlations in signal labstates and does not respect Einstein locality. The other involves classical information transmission equivalent to local operations carried out by the observer on apparatus, and that does respect Einstein locality.

In the next section we review classical CST. Then we show how the separation and entanglement concepts discussed in the previous chapter lead to a natural definition of the concepts of families, parents, and siblings used in CST. Then we show how two sorts of causal set structure arise in QDN, one of which is associated with labstates and the other with modules.

23.2 Causal Sets

A number of authors (Bombelli et al., 1987; Brightwell and Gregory, 1991; Markopoulou, 2000; Requardt, 1999; Ridout and Sorkin, 2000) have discussed the idea that spacetime could be discussed in terms of causal sets. In the causal set paradigm, the universe is envisaged as a set $\mathcal{C} \equiv \{x, y, \ldots\}$ of objects (or *events*) that may have a particular binary relationship among themselves denoted by the symbol $\prec$, which may be taken to be a mathematical representation of a temporal ordering. For any two different elements x, y, if neither of the relations $x \prec y$ nor $y \prec x$ holds, then x and y are said to be *relatively spacelike*, *causally independent*, or *incomparable* (Howson, 1972). The objects in $\mathcal{C}$ are usually assumed to be the ultimate description of spacetime, which in the causal set hypothesis is often postulated to be discrete (Ridout and Sorkin, 2000). Minkowski spacetime is an example of a causal set with a continuum of elements (Brightwell and Gregory, 1991), with the possibility of extending the relationship $\prec$ to include the concept of *null* or *lightlike* relationships.

The causal set paradigm supposes that for given elements x, y, z of the causal set $\mathcal{C}$, the following relations hold:

$$\begin{array}{rll} \forall\, x, y, z \in \mathcal{C}, & x \prec y \text{ and } y \prec z \Rightarrow x \prec z & \text{(transitivity)} \\ \forall x, y \in \mathcal{C}, & x \prec y \Rightarrow y \nprec x & \text{(asymmetry)} \\ \forall x \in \mathcal{C}, & x \nprec x. & \text{(irreflexivity)} \end{array} \tag{23.1}$$

A causal set may be represented by a Hasse diagram such as shown in Figure 23.1 (Howson, 1972). In a typical Hasse diagram, the events are shown as labeled circles or spots and the relation $\prec$ as a solid line or link between the events, with the "temporal" ordering running from bottom to top. Stage diagrams in QDN are a form of Hasse diagram, with the added complication of including modules.

Hasse diagrams can be discussed in familial terms. For example, in Figure 23.1, event 1_0 is the *ancestor* of all subsequent events; 2_2 and 3_2 are the *parents* of their *child* 2_3; 3_2 and 4_2 are *siblings*; and so on.

One method of generating a causal set is via a process of "sequential growth" (Ridout and Sorkin, 2000). At each step of the growth process a new element is created at random, and the causal set is developed by considering the relations

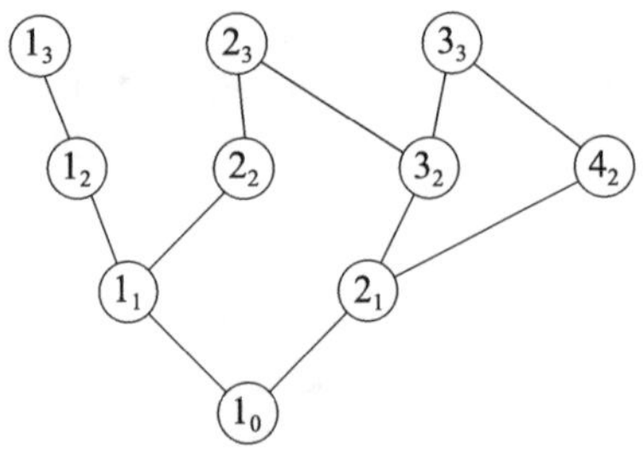

Figure 23.1. A typical Hasse diagram.

between this new event and those already in existence. Specifically, either the new event y may be related to another event x as $x \prec y$, or x and y are said to be unrelated. Thus the ordering of the events in the causal set is as defined by the symbol $\prec$, and it is by a succession of these orderings, i.e., the growth of the causet, that constitutes the passage of time. The relation $x \prec y$ is hence interpreted as the statement: "y is to the future of x." Further, the set of causal sets that may be constructed from a given number of events can be represented by a Hasse diagram of Hasse diagrams (Ridout and Sorkin, 2000).

The importance of CST is that in the large-scale limit of very many events, causal sets may yield all the properties of continuous spacetimes, such as metrics, manifold structure, and even dimensionality, all of which should be determined by the dynamics (Bombelli et al., 1987). For example, it should be possible to use the causal order of the set to determine the topology of the manifold into which the causet is embedded (Brightwell and Gregory, 1991). This is the converse of the usual procedure of using the properties of the manifold and metric to determine the lightcones of the spacetime, from which the causal order may in turn be inferred.

Distance may be introduced into the analysis of causal sets by considering the length of paths between events (Bombelli et al., 1987; Brightwell and Gregory, 1991). A *maximal chain* is a set $\{a_1, a_2, ..., a_n\}$ of elements in a causal set $\mathcal{C}$ such that, for $1 \leq i \leq n$, we have $a_i \prec a_{i+1}$ and there is no other element b in $\mathcal{C}$ such that $a_i \prec b \prec a_{i+1}$. We may define the path length of such a chain as $n - 1$. The distance $d(x, y)$ between comparable (Howson, 1972) elements x, y in $\mathcal{C}$ may then be defined as the maximum length of path between them, i.e., the "longest route" allowed by the topology of the causet to get from x to y. This implements Riemann's notion that ultimately, distance is a counting process (Bombelli et al., 1987). For incomparable elements, it should be possible to use the binary relation $\prec$ to provide an analogous definition of distance, in much the same way that light signals may be used in special relativity to determine distances between space-like separated events.

In a similar way, "volume" and "area" in the spacetime may be defined in terms of numbers of events within a specified distance. Likewise, it should be possible to give estimates of dimension in terms of average lengths of path in a given volume. An attractive feature of causal sets is the possibility that different spatial dimensions might emerge on different physical scales (Bombelli et al., 1987), whereas in conventional theory, higher dimensions generally have to be put in by hand.

23.3 QDN and Causal Sets

Our approach to causal sets in QDN differs from the above in that we do not assume any structure to the information void, such as spacetime per se.

Our discrete set structure arises from the separation and entanglement properties of labstates and modules.

Furthermore, in the QDN approach, various relations assumed in the "sequential growth" mentioned above must be interpreted carefully. In quantum physics, past, present, and future can never have the status they have in Block Universe models. At best we can only talk about conditional probabilities, such as asking for the probability of a possible future stage *if* we assumed we were in a given present stage. This corresponds directly to the meaning of the Born interpretation of probability in QM, where all probabilities are conditional.

Another point is that the causal set relations discussed by Sorkin and others imply that the various elements $a, b, \ldots, z$ have an independent existence outside the relations themselves and that these relations merely reflect some existing attributes. This is a vacuous Block Universe perspective that QDN does not accept.

A further criticism of classical CST comes from dynamics. In some diagrams, relatively space-like events with no previous causal connection are permitted to be the parents of the same event. The question arises then, given two such unrelated events at a given "time" n, how any information from either of them could ever coincide, that is, be brought together to be used to create any mutual descendants. The only sensible answer is that there must be some external agency organizing the flow of that information. But the whole point of standard CST, however, is that there is no external space in which these events are embedded, or any external "memory," observer, or information store correlating such information. It is not clear then how such processes could be encoded into the dynamics. In our QDN approach, however, we have no such issue: the primary observer is such an agency, manipulating apparatus in order to produce such processes.

23.4 Quantum Causal Set Rules

Stage diagrams are greatly useful in understanding the architecture of apparatus. Similarly, causal set diagrams are greatly useful in understanding the pattern of causality as labstates "flow" through that apparatus. With reference to Figure 23.2, the conventions in causal set diagrams are as follows.

Time

Stages go up the page, starting from stage Σ_0.

Separations

These are represented by squares surrounding single integers identifying separations. As with detectors on a given stage in a stage diagram, labeling is arbitrary. Our convention then is to label atoms left to right.

Entanglements

A rank-r entanglement is represented by a left to right rectangle equivalent to r adjacent squares, labeled by the atoms involved in the entanglement.

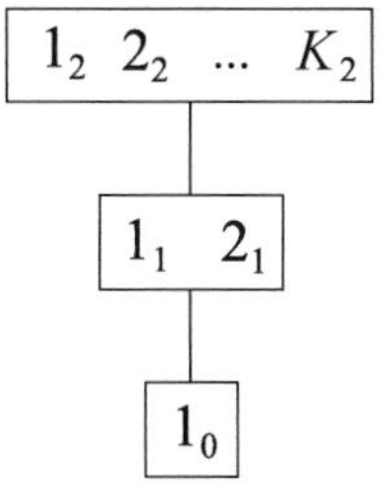

Figure 23.2. The causal set diagram for the DS experiment.

Lines
These connect elements of the causal set as required from the modules involved.

23.5 Case Study 1: The Double-Slit Experiment

We are now in a position to reexamine the QDN analysis of the double-slit (DS) experiment given in Chapter 10 and analyze its causal set structure.

Looking at Figure 10.2, we would get the impression that it is a Hasse diagram as it stands and so the causal set structure of the experiment would appear to be obvious. This is not what we are after, however, because the causal set structure we are interested in involves labstates and not detectors. These are two separate concepts.

Our point is this. The stage Σ_1 labstate given in Eq. (10.13) is a signality-one state with two identifiable components $\alpha^1\widehat{\mathbb{A}}_1^1\mathbf{0}_1$ and $\alpha^2\widehat{\mathbb{A}}_1^2\mathbf{0}_1$ added together, giving the impression that this is not an entangled state. If we return to a more explicit signal basis representation, the entanglement at stage Σ_1 becomes clear. If instead of (10.13) we write

$$\boldsymbol{\Psi}_1 = \alpha^1\mathbf{1}_1^1\mathbf{0}_1^2 + \alpha^2\mathbf{0}_1^1\mathbf{1}_1^2, \tag{23.2}$$

then it is clear that $\boldsymbol{\Psi}_1$ is actually an entangled state, an element of the entanglement $\mathcal{Q}_1^{\overline{12}}$, one of the partition elements of the quantum register $\mathcal{Q}_1^{[12]} \equiv Q_1^1 Q_1^2$.

Note that in the above, the superscripts refer to the two slits, labeled 1 and 2.

Remark 23.1 In standard QM, the state at stage Σ_1 in the DS experiment would also usually be written in nonentangled form $|\Psi_1\rangle = \alpha^1|\text{slit } 1\rangle + \alpha^2|\text{slit } 2\rangle$, masking the fact that it is actually an entangled state (contextual on the observer's ability to look at both slits simultaneously). This is a far more subtle point than it appears and goes to the heart of the difference between classically conditioned thinking and the sort required to make sense of quantum processes.

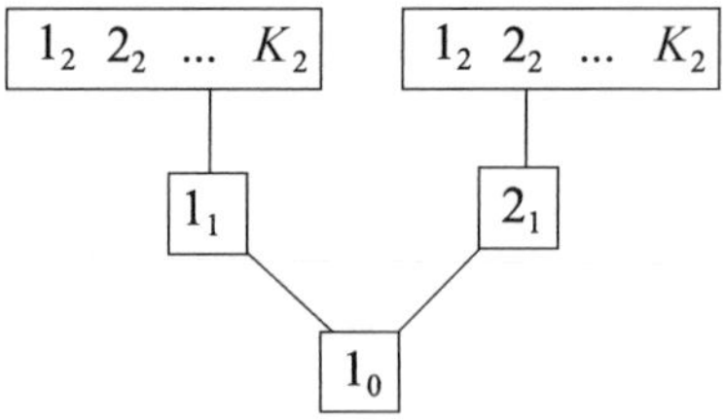

Figure 23.3. The causal set diagram for the monitored DS experiment.

The stage Σ_2 labstate requires some commentary also, because Figure 10.2 is misleading in the same respect as at stage Σ_1. The final stage labstate $\mathbf{\Psi}_2$ is an element of the entanglement $\mathcal{Q}_2^{\overline{123\ldots K}}$, where K is the number of detectors in the detecting screen S. The final labstate may be written as a sum of computational basis representation states:

$$\mathbf{\Psi}_2 = \sum_{j=1}^{K} \left\{ \sum_{a=1}^{2} \alpha^a U_{2,1}^{j,a} \right\} \underline{\mathbf{2}^{j-1}_2}. \tag{23.3}$$

With this analysis of the entanglement structure of the DS experiment, the appropriate QDN causal set diagram is given by Figure 23.2. Temporally successive stages run bottom to top. Separations are in square boxes, and entanglements are in rectangular boxes, labeled by the atoms involved.

23.6 Case Study 2: The Monitored Double-Slit Experiment

The interference pattern observed in the DS experiment disappears when which-path information is available to the observer, that is, when it is possible to identify which path a photon had taken from source to detector. In this situation, known as the monitored double-slit experiment, the labstate at stage Σ_1 can no longer be considered entangled. In consequence, there is doubling of final screen elements in the relevant causal set diagram, as shown in Figure 23.3. While the two slit elements at stage Σ_1 are not considered entangled once which-path information is obtained post each run, the labstate for stage Σ_1 may be considered entangled from the perspective of the observer at stage Σ_0, that is, *before* any which-path information is obtained at stage Σ_2. This underlines the fact that such a causal set diagram changes with context.

It should be noted that the two stage-Σ_1 separations 1_1 and 2_1 in the monitored DS experiment are physically disjoint, in that they start outcome branches that do not occur in the same run. They are classically distinct. The stage-Σ_1 labstate is not the tensor product of elements from these two separations; neither is it a superposition of those two separations.

A significant point to make here is that Figure 23.3 is not a Block Universe diagram.

23.7 Case Study 3: Module Causal Set Structure

We mentioned above that QDN leads naturally to another causal set structure, one associated with the modules in a given experiment. For a given experiment, this causal set structure is associated with the links between real and virtual detector nodes, and on that account will be referred to as the *dual causal set.* It is at this point that our stage diagram convention of representing modules by boxes, rather than the circles associated with the detectors, becomes useful.

Given a conventional stage diagram, such as Figure 13.3, the two-photon interferometer, the first step is to fill in the missing module boxes. These occur when signals propagate through the information void directly. In Figure 23.4 we show how Figure 13.3 is amended: the new modules V^1 and V^2 represent information void propagation.

The next step is to remove all clear circles (that is, except final stage detectors), join up the module boxes according to how they interact with the real and virtual detectors, and then readjust the resulting diagram into a vertical Hasse form. Figure 23.4 gives Figure 23.5, the module Hasse diagram for the two

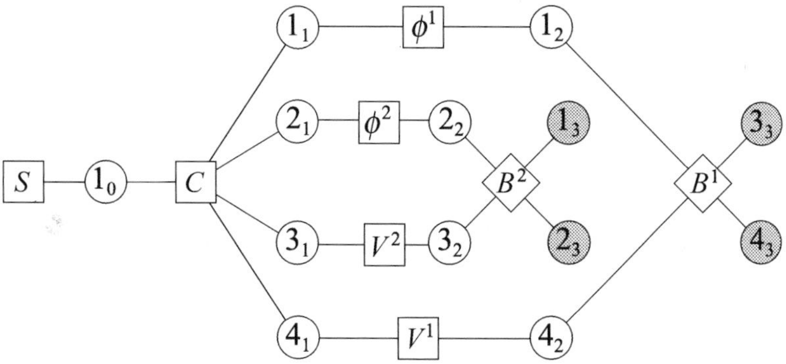

Figure 23.4. The amended stage diagram for the two photon interferometer.

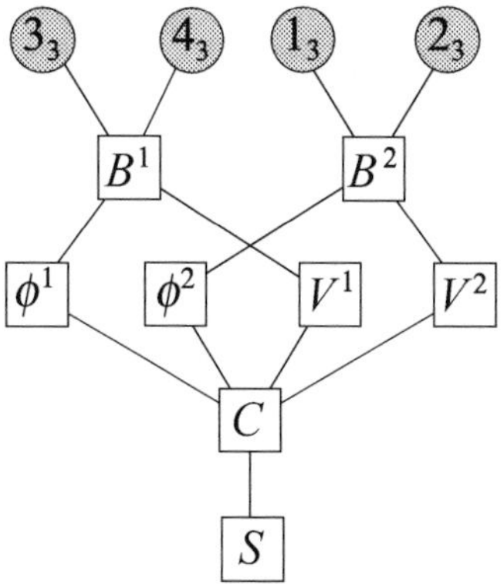

Figure 23.5. Two-photon interferometer module causal set diagram.

photon interferometer. This casual set diagram is entirely classical: there are no entanglements.

The reader is invited to analyze the other experiments discussed in this book in terms of their labstate and apparatus causal set structures. It will become apparent that there are significant complexities in even relatively simple experiments.

24
Oscillators

24.1 Introduction

Our aim in this chapter is to show how quantized detector networks (QDN) describe the quantized one-dimensional bosonic (harmonic) oscillator. The adjective *bosonic* here refers to the fact that in the standard quantum theory of this system under observation (SUO), the phase space operators of position and momentum satisfy commutation properties, in contrast to the *fermionic* oscillator, whose corresponding variables satisfy anticommutation properties and which is studied at the end of this chapter.

The one-dimensionality referred to here is somewhat misleading: it refers to the classical mechanics (CM) theory of a single particle moving in one spatial dimension under the influence of an attractive quadratic force potential. In the quantized version of the same SUO, there are infinitely many degrees of freedom, requiring the use of an infinite-rank quantum register. This is one of the few occasions in QDN in which we refer to the concept of infinity.

Because the rank of the oscillator's quantum register is infinite, we find that the QDN representation comes with a tremendous amount of mathematical overkill and redundancy. Specifically, the Hilbert space dimension of such a quantum register is far bigger than that actually needed to describe a quantized bosonic oscillator, the vast bulk of states in the register being what we shall call *transbosonic*.

As a Hilbert space, the infinite-dimension quantum register is nonseparable, which means that it has no *denumerable* (that is, countable) basis, unlike the Hilbert space $\mathcal{H}^{HO}$ of energy eigenstates used to describe the standard QM bosonic oscillator. While the improper position coordinate basis $\{|x\rangle : x \text{ real}\}$ is nondenumerable, the normalized energy eigenstate basis for $\mathcal{H}^{HO}$ is countable. We will show how the QDN labstates that are the analogues of standard QM oscillator states are restricted to a subspace of the register of measure zero and remain there from stage to stage.

The degree of "overkill" in the QDN description of the oscillator is no different in essence to that encountered in the quantum field theory (QFT) description of the oscillator. QFT is a *many particle theory* that allows for the possibility of more than one oscillator to be excited at the same time. This is in contrast to Heisenberg's matrix mechanics (Heisenberg, 1925) and Schrödinger's wave mechanics (Schrödinger, 1926), which are one-particle descriptions of the oscillator. Being based on signal detector principles, QDN can readily describe arbitrary numbers of oscillator-type signals, over and above quantum superpositions of one-particle signal states. This underlines the point that QDN looks much more like a halfway house to QFT rather than another representation of standard QM. The transbosonic states referred to above are really a reflection of that fact.

24.2 The Classical Oscillator Register

The first step in the QDN description of the harmonic oscillator is to define an infinite-rank classical bit register $\mathcal{R}^{[\infty]}$. This is an infinite, countable collection of classical bits, each bit being labeled by a distinct nonnegative integer n running from zero to infinity. We shall call $\mathcal{R}^{[\infty]}$ the *classical oscillator register* and write $\mathcal{R}^{[\infty]} \equiv B^0 B^1 B^2 \ldots$, where B^i is the ith classical bit.

Here we adopt the convention that $\mathcal{R}^{[\infty]}$ is the *Cartesian product* of the binary sets B^0, B^1, and so on.

Our QDN approach to the harmonic oscillator raises a significant issue. Up to this point, we have argued that no experiment deals with actual infinities, so QDN has exploited that fact. Indeed, finiteness seems to be a strength of QDN rather than a weakness. We should explain why we now introduce an infinite-rank quantum register to describe one of the simplest of SUOs, the harmonic oscillator.

We have found that whenever such issues arise in QDN, the answers are to be found by looking at what happens in the laboratory and what the observer actually does. QDN discusses detectors, not the supposed objects being detected. There are two contrasting facts about apparatus, however. The first fact is that relative to any real observer, all apparatus comes in discrete packages, at any empirical level.[1] At the highest empirical level, which we may call the emergent level, a detector has a single objectivized identity, which is as a detector.[2] In QDN we model this by a single binary detector. At the lowest empirical level, which we may call the reductionist level, a real detector is described as a vast but countable number of discrete components called molecules and atoms. At that level, it need not be recognizable as a detector at all.

The second fact is that despite the countability associated with the emergent and reductionist levels, there are three scenarios where infinity makes an appearance.

[1] By this we mean what is observable, not what is theorized.

[2] If we view reductionism as the counterbalance to emergence, then the highest empirical level corresponds to the lowest reductionist level, where minimal mathematical details are given.

Modeling Space Is Not Easy

The great mathematical physicist Schwinger gave the following succinct statement on how he viewed QM, apparatus, space, and time:

> The mathematical machinery of quantum mechanics is a symbolic expression of the laws of atomic measurement, abstracted from the specific properties of individual techniques of measurement. In particular, the space-time manifold that is the background of any quantum-mechanical description is an idealization of the function of a measurement apparatus to define a macroscopic frame of reference. (Schwinger, 1958)

This is in accord with the principles of QDN. If we wish to model space-time itself, rather than just apparatus that is *in* space-time (a rather different proposition altogether), then we are faced with the concept of *indefiniteness* rather than infinity. By this we mean that there no natural, obvious, or observable limits to space and time as far as any observer is concerned. Even if we discretized space and time coordinates, thereby eliminating continuity, how many points would we include in our modeling?

Apparatus Depends on Continuous Parameters

Continuity cannot be eliminated even when we have very simple apparatus. For example, in the Stern–Gerlach experiment, the orientation of the main magnetic field is parametrized by three angles in a continuum of angles, and there is no natural way of discretizing any of those.

Quantum Process in the Information Void

QDN uses, but does not derive from reductionist principles, any transition amplitudes from stage to stage: it is designed to work with the architecture of processes and how those amplitudes are related to observable signals. The calculation of such amplitudes will usually (but not invariably) involve working models of empty space (the vacuum) that depend on continuity.

Every classical bit B^i has two elements, $\mathbf{0}^i$ and $\mathbf{1}^i$, representing "*no signal*" and "*signal*" respectively. A rank-r classical register has 2^r elements, so $\mathcal{R}^{[\infty]}$ contains an infinite number of classical states. Representing these requires dealing with infinite sets.

The mathematical difficulty here is that these sets are nondenumerable, that is, are not countable. The paradox is that while the bits making up the register are each of finite cardinality, and the bits themselves can be counted (our labeling proves this), the number of states is not countable. Mathematically, this is the same phenomenon that allows mathematicians to express every rational real number in terms of a recurring decimal expansion, but none of the irrationals.

The Signal Basis Representation

The most natural representation of a classical state in $\mathcal{R}^{[\infty]}$ is the signal basis representation (SBR), which we have met before. An arbitrary state $\boldsymbol{\Psi}$ in $\mathcal{R}^{[\infty]}$ can be defined in the form

$$\boldsymbol{\Psi} \equiv \boldsymbol{i}^0\boldsymbol{i}^1\boldsymbol{i}^2\ldots = \prod_{a=0}^{\infty} \boldsymbol{i}^a, \qquad \boldsymbol{i}^a = \mathbf{0}^a \text{ or } \mathbf{1}^a \text{ for } a = 0, 1, 2, \ldots \tag{24.1}$$

Every such state therefore corresponds to a unique *binary sequence* $S_{\boldsymbol{\Psi}} \equiv \{i^0, i^1, i^2, \ldots\}$, consisting of an infinite string of ones and zeros.

Example 24.1 The register state

$$\boldsymbol{\Psi} \equiv \mathbf{0}^0\mathbf{0}^1\mathbf{1}^2\mathbf{1}^3\mathbf{0}^4\mathbf{1}^5\mathbf{0}^6\mathbf{1}^7\mathbf{1}^8\mathbf{0}^9\mathbf{1}^{10}\mathbf{1}^{11}\mathbf{0}^{12}\mathbf{1}^{13}\mathbf{1}^{14}\mathbf{0}^{15}\mathbf{1}^{16}\ldots \tag{24.2}$$

corresponds to the binary sequence

$$S_{\boldsymbol{\Psi}} \equiv \{0, 0, 1, 1, 0, 1, 0, 1, 0, 1, 1, 0, 1, 1, 0, 1, \ldots\}. \tag{24.3}$$

The nondenumerability of $\mathcal{R}^{[\infty]}$ creates a potential problem when we come to quantization, because the Hilbert space $\mathcal{Q}^{[\infty]}$ corresponding to $\mathcal{R}^{[\infty]}$ is an infinite tensor product. Such Hilbert spaces are always nonseparable (Streater and Wightman, 1964), which means that they have no countable basis. This contrasts with the fact that the Hilbert space of the standard QM quantized bosonic oscillator has a complete, countable basis set, consisting of energy eigenstates.

Fortunately, in the QDN analysis for the harmonic oscillator, we can restrict our attention to a set of special operators, referred to here as *bosonic operators* over the infinite-rank quantum register $\mathcal{Q}^{[\infty]}$, which have the merit that, in physical applications dealing with a single oscillator, the problems of nonseparability can be avoided. A similar phenomenon occurs in relativistic quantum field theory (Streater and Wightman, 1964; Klauder and Sudarshan, 1968).

To understand how this comes about, we first classify each state in $\mathcal{R}^{[\infty]}$ as one of three possible types. Two of these types form countable subsets of the register, while the third type forms a nondenumerable subset. These types correspond, roughly speaking, to the integers, the rationals, and the irrationals, respectively, in the real number system. This can be seen by the following heuristic arguments.

Finite Countable States

The first type, the set of all *finite countable* states in $\mathcal{R}^{[\infty]}$, consists of states associated with binary sequences that consist of zeros after some given finite element J, which depends on the sequence. For example, the state $\mathbf{1}^0\mathbf{1}^1\mathbf{0}^2\mathbf{1}^3\mathbf{0}^4\mathbf{0}^5\mathbf{0}^6\mathbf{0}^7\mathbf{0}^8\mathbf{0}^9\mathbf{0}^{10}\ldots$ is finite countable ($J = 4$), whereas the state corresponding to the infinitely recurring sequence

$$\mathbf{1}^0\mathbf{0}^1\mathbf{1}^2\mathbf{0}^3\mathbf{1}^4\mathbf{0}^5\mathbf{1}^6\mathbf{0}^7\mathbf{1}^8\mathbf{0}^9\mathbf{1}^{10}\ldots \tag{24.4}$$

is not finite countable. For a finite countable state $\boldsymbol{i}^0\boldsymbol{i}^1\boldsymbol{i}^2\ldots\boldsymbol{i}^J\mathbf{0}^{J+1}\mathbf{0}^{J+2}\ldots$ a modification of the computational basis map (5.14) maps this state to the integer $i^0 + 2i^1 + 2^2i^2 + \cdots + 2^Ji^j$, which is finite.

Recurring Sequence States

The second type, the *recurring sequence* states in $\mathcal{R}^{[\infty]}$, consists of those sequences that would be finite countable sequences but for the fact that the infinite string of zeros after J is replaced by some nontrivial recurring binary sequence. Recurring sequence states cannot be classified by finite integers using the computational map (5.14). However, we can use another map, defined by

$$\boldsymbol{i}^0\boldsymbol{i}^1\boldsymbol{i}^2\ldots \to i^0 + \frac{i^1}{2^1} + \frac{i^2}{2^2} + \cdots \tag{24.5}$$

to map such states into the interval $[0, 2]$. We shall call this the *continuum map*. It is easy to see that, in fact, *all* states in $\mathcal{R}^{[\infty]}$ can be mapped into the interval $[0, 2]$ via the continuum map. The signal ground state $\mathbf{0}^0\mathbf{0}^1\mathbf{0}^2\ldots$ maps into 0, while the *fully occupied state* $\mathbf{1}^0\mathbf{1}^1\mathbf{1}^2\ldots$ maps into 2. All other states necessarily map into the open interval $(0, 2)$.

It is not hard to see that finite countable states and recurring sequence states map into the rationals via the continuum map, but not in a one-to-one way. For example, the finite countable state $\mathbf{1}^0\mathbf{0}^1\mathbf{0}^2\mathbf{0}^3\ldots$ maps into the number 1 by the continuum map, which is also the value mapped from the recurring sequence state $\mathbf{0}^0\mathbf{1}^1\mathbf{1}^2\mathbf{1}^3\ldots$

Remark 24.2 In standard decimal-based arithmetic, it is generally asserted that the infinitely recurring decimal $0.\dot{9} \equiv 0.9999\ldots$ is "equal" to the number 1. While the register states $\mathbf{1}^0\mathbf{0}^1\mathbf{0}^2\mathbf{0}^3\ldots$ and $\mathbf{0}^0\mathbf{1}^1\mathbf{1}^2\mathbf{1}^3\ldots$ map to the same value 1 via the continuum map, physically, these are two very different states. This underlines the difference between pure mathematics ($0.\dot{9} = 1$) and physics ($\mathbf{1}^0\mathbf{0}^1\mathbf{0}^2\mathbf{0}^3\ldots \neq \mathbf{0}^0\mathbf{1}^1\mathbf{1}^2\mathbf{1}^3\ldots$).

Not all pure mathematicians would agree that $0.\dot{9} = 1$ is an absolute equality; some would argue that the difference $1 - 0.\dot{9}$ is an infinitesimal. But not every mathematician accepts the concept of infinitesimals as sound.

Irrational Sequence States

The problem with nondenumerability arises because of the existence of the third type of infinite binary sequence. This consists of all those binary sequences that are not recurring, such as

$$\{1, 1, 1, 0, 1, 1, 1, 1, 0, 1, 1, 1, 1, 1, 0, 0, 0, 0, 0, 0, 0, 0, 0, 1, 1, \ldots\}, \tag{24.6}$$

an example based on the successive digits in the decimal representation of π. There are infinitely many such sequences and they cannot be counted, as they correspond to the irrationals, which is easy to prove.

Our conclusion is, therefore, that states in $\mathcal{R}^{[\infty]}$ cannot be classified by the integers alone. Instead, we may use the sequence corresponding to each state as an index. Specifically, if S is the binary sequence

$$S \equiv \{s^a : s^a = 0 \text{ or } 1 \text{ for } a = 0, 1, 2, \ldots\}, \tag{24.7}$$

then the corresponding state $\boldsymbol{S}$ in $\mathcal{R}^{[\infty]}$ is given uniquely by the expression

$$\boldsymbol{S} \equiv \boldsymbol{s}^0 \boldsymbol{s}^1 \boldsymbol{s}^2 \ldots = \prod_{a=0}^{\infty} \boldsymbol{s}^a. \tag{24.8}$$

Remark 24.3 The *zero sequence* $Z \equiv \{0, 0, 0, \ldots\}$ corresponds to the **signal ground state**, denoted $\mathbf{0}^0\mathbf{0}^1\mathbf{0}^2\ldots$ in the occupation representation and $\mathbf{0}$ in the computation representation. This state is **not** the QDN analogue of the conventional oscillator ground state $|0\rangle$.

Given two binary sequences S, T corresponding to register states $\boldsymbol{S}$ and $\boldsymbol{T}$, respectively, their "inner product" $\overline{\boldsymbol{S}}\boldsymbol{T}$ is defined in the obvious way, viz.

$$\overline{\boldsymbol{S}}\boldsymbol{T} = \left\{\prod_{a=0}^{\infty} \overline{\boldsymbol{s}}^a\right\}\left\{\prod_{b=0}^{\infty} \boldsymbol{t}^b\right\} \equiv \prod_{a=0}^{\infty} \overline{\boldsymbol{s}}^a \boldsymbol{t}^a \equiv \delta^{ST}, \tag{24.9}$$

where the generalized Kronecker symbol δ^{ST} takes the value unity if and only if the binary sequences S and T are identical; otherwise, it is zero.

Definition 24.4 Two binary sequences $S \equiv \{s^a : a = 0, 1, 2, \ldots\}, T \equiv \{t^a : a = 0, 1, 2, \ldots\}$ are identical if and only if $s^a = t^a$ for all nonnegative integers, that is, for $a = 0, 1, 2, \ldots$

We take it as an axiom that, given two binary sequences S and T, the product $\prod_{a=0}^{\infty} \overline{\boldsymbol{s}}^a \boldsymbol{t}^a$ is 1 if the sequences are identical and 0 otherwise.

24.3 Quantization

Quantization amounts to associating the set of states in $\mathcal{R}^{[\infty]}$ as a (preferred) basis $B^{[\infty]}$ for a nonseparable Hilbert space $\mathcal{Q}^{[\infty]}$. States in $\mathcal{Q}^{[\infty]}$ will be called *quantum register states* and are of the form

$$\boldsymbol{\Psi} = \sum_S \Psi(S)\, \boldsymbol{S}, \tag{24.10}$$

where the summation is over all possible infinite binary sequences S and the coefficients $\Psi(S)$ are complex. The Hilbert space inner product is defined in the obvious way, that is,

$$\begin{aligned} \overline{\boldsymbol{\Psi}}\boldsymbol{\Phi} &= \left(\sum_S \overline{\Psi}(S)\overline{\boldsymbol{S}}\right)\left(\sum_T \Phi(T)\boldsymbol{T}\right) \\ &= \sum_S \sum_T \overline{\Psi}(S)\Phi(T)\underbrace{\overline{\boldsymbol{S}}\boldsymbol{T}}_{\delta^{ST}} = \sum_S \Psi^*(S)\,\Phi(S). \end{aligned} \tag{24.11}$$

As discussed above, finite countable sequences can be mapped into the integers via the computational map (5.14). For a sequence S that maps into integer n, we can use the notation $\boldsymbol{n}$ rather than $\boldsymbol{S}$ to denote the corresponding quantum register state.

24.4 Bosonic Operators

The quantum register states in QDN required to represent the standard discrete set of quantized bosonic oscillator energy eigenstates $\{|n\rangle : n = 0, 1, 2, 3, \ldots\}$ form a subset of the finite countable states. We shall call any normalized element of this subset a *bosonic state*. To identify this subset, we need to filter out of the set of all finite countable states those states that are redundant. To do this, we first define the *bosonic projection operators* as follows.

Each component B^a of the classical oscillator register $\mathcal{R}^{[\infty]}$ is a classical bit with two elements. Quantization starts with the interpretation of the two bit states $\mathbf{0}^a$, $\mathbf{1}^a$ in B^a as the two preferred basis quantum outcomes states of a quantum bit, Q^a, representing the two states, *ground* and *signal*, of a detector. Each quantum bit Q^a is a two-dimensional Hilbert space with its own set of projection and signal operators $\left\{\boldsymbol{P}^a, \widehat{\boldsymbol{P}}^a, \boldsymbol{A}^a, \widehat{\boldsymbol{A}}^a\right\}$ satisfying Table 24.1, a generalization of Table 4.1.

The *quantized bosonic register* $\mathcal{Q}^{[\infty]}$ is the infinite-dimensional Hilbert space given by the commuting tensor product of all the individual qubits, i.e.,

$$\mathcal{Q}^{[\infty]} \equiv Q^0 \otimes Q^1 \otimes Q^2 \otimes \cdots = Q^0 Q^1 Q^2 \ldots \qquad (24.12)$$

Here and below we shall drop the tensor product symbol $\otimes$, it being understood we are dealing with commuting tensor products.

Definition 24.5 The set of *bosonic filter operators* $\{\widehat{\mathbb{P}}^a_B : a = 0, 1, 2, \ldots\}$ is defined by the commuting tensor product

$$\widehat{\mathbb{P}}^a_B \equiv \left\{\prod_{b \neq a}^{\infty} \boldsymbol{P}^b\right\} \widehat{\boldsymbol{P}}^a = \boldsymbol{P}^0 \boldsymbol{P}^1 \ldots \boldsymbol{P}^{a-1} \widehat{\boldsymbol{P}}^a \boldsymbol{P}^{a+1} \ldots, \quad a = 0, 1, 2, \ldots \qquad (24.13)$$

Each bosonic projection operator $\widehat{\mathbb{P}}^a_B$ defines a one-dimensional Hilbert subspace of $\mathcal{Q}^{[\infty]}$ with basis $\{\mathbf{2}^a\}$ in the computational basis representation. Eigenstates of these operators with eigenvalue $+1$ will be called *bosonic eigenstates*.

Table 24.1 *Products of signal bit operators*

	$\boldsymbol{P}^a$	$\widehat{\boldsymbol{P}}^a$	$\boldsymbol{A}^a$	$\widehat{\boldsymbol{A}}^a$
$\boldsymbol{P}^a$	$\boldsymbol{P}^a$	0	$\boldsymbol{A}^a$	0
$\widehat{\boldsymbol{P}}^a$	0	$\widehat{\boldsymbol{P}}^a$	0	$\widehat{\boldsymbol{A}}^a$
$\boldsymbol{A}^a$	0	$\boldsymbol{A}^a$	0	$\boldsymbol{P}^a$
$\widehat{\boldsymbol{A}}^a$	$\widehat{\boldsymbol{A}}^a$	0	$\widehat{\boldsymbol{P}}^a$	0

Theorem 24.6 *The bosonic filter operators satisfy the multiplicative rule*

$$\widehat{\mathbb{P}}^a_B \widehat{\mathbb{P}}^b_B = \delta^{ab} \widehat{\mathbb{P}}^a_B, \qquad a, b \in \mathbb{N} \equiv \{0, 1, 2, \ldots\}. \tag{24.14}$$

The element $\mathbf{2}^n$ in $\mathcal{Q}^{[\infty]}$ corresponds to the energy eigenstate $|n\rangle$ in the standard quantum theory of the oscillator.

The bosonic filter operator $\widehat{\mathbb{P}}^a_B$ should not be confused with the signal projection operators $\widehat{\mathbb{P}}^a \equiv \left\{\prod_{b \neq a}^{\infty} \boldsymbol{I}^b\right\} \widehat{\boldsymbol{P}}^a$, where $\boldsymbol{I}^b$ is the identity operator for qubit Q^b. The difference is based on logic: an eigenstate of $\widehat{\mathbb{P}}^a$ will return a positive signal if detector a is examined, regardless of whatever signal state any other detector is in, whereas an eigenstate of $\widehat{\mathbb{P}}^a_B$ will return a positive signal in detector a only if all the other detectors are each in their signal ground state.

Given the bosonic projection operators $\widehat{\mathbb{P}}^a$, the next step is to define the *bosonic identity operator*

$$\mathbb{I}_B \equiv \sum_{a=0}^{\infty} \widehat{\mathbb{P}}^a_B. \tag{24.15}$$

This operator satisfies the idempotency condition required of any projection operator, that is,

$$\mathbb{I}_B \mathbb{I}_B = \mathbb{I}_B \tag{24.16}$$

and plays the role of a "bosonic filter," passing only those states and operators associated with the quantum register that have the desired properties associated with the harmonic oscillator.

Definition 24.7 A quantum register operator $\mathbb{O}$ is *bosonic* if and only if it commutes with the bosonic identity, i.e.,

$$\mathbb{O} \text{ bosonic } \Leftrightarrow [\mathbb{O}, \mathbb{I}_B] = 0. \tag{24.17}$$

Definition 24.8 A *bosonic state* is defined to be any vector in $\mathcal{Q}^{[\infty]}$ that is an eigenstate of $\mathbb{I}_B$ with eigenvalue $+1$. All other states in $\mathcal{Q}^{[\infty]}$ will be referred to as *transbosonic*.

The set of all bosonic states is denoted $\mathcal{Q}_B$ and is the QDN analogue of the Hilbert space of quantum oscillator states in standard QM.

Examples of transbosonic states are the signal ground state $\mathbf{0}$ and all those finite countable elements $\boldsymbol{p}$ of the computational basis B where p is not some power of two. In fact, almost all elements in the quantum register are transbosonic. It can be readily verified that linear superpositions of bosonic states are always bosonic states, while the linear superposition of any transbosonic state with any other state in the register is always transbosonic.

The importance of the bosonic operators is that when they are applied to bosonic states, the result is always a bosonic state, which can be easily proved.

24.5 Quantum Register Oscillator Operators

We now in a position to discuss how we map the standard quantum oscillator into the quantum register.

In the standard QM of the bosonic oscillator, the most important operators are the ladder operators a and $a^\dagger$. These satisfy the commutation relation

$$\left[a, a^\dagger\right] = 2I, \tag{24.18}$$

where I is the identity operator in the standard oscillator Hilbert space $\mathcal{H}$ and we take Planck's constant $\hbar$ and the oscillator constant ω to be unity. These ladder operators have the representations

$$\begin{aligned} a &= \sum_{n=0}^{\infty} |n\rangle \sqrt{(n+1)\,2} \langle n+1|, \\ a^\dagger &= \sum_{n=0}^{\infty} |n+1\rangle \sqrt{(n+1)\,2} \langle n|, \end{aligned} \tag{24.19}$$

where the states $|n\rangle$, $n = 1, 2, 3, \ldots$ are the usual orthonormal excited states of the oscillator ground state $|0\rangle$. The key to the QDN description is the observation that these states are identified one-to-one with the bosonic states $\mathbf{2}^n$ discussed above, namely,

$$|n\rangle \leftrightarrow \mathbf{2}^n, \quad n = 0, 1, 2, \ldots \tag{24.20}$$

We note that

$$|0\rangle \leftrightarrow \mathbf{1}^0\mathbf{0}^1\mathbf{0}^2\mathbf{0}^3 \ldots = \mathbf{1}, \quad |1\rangle \leftrightarrow \mathbf{0}^0\mathbf{1}^1\mathbf{0}^2\mathbf{0}^3 \ldots = \mathbf{2}, \tag{24.21}$$

and so on. A particularly significant observation is that the quantum register signal ground state $\mathbf{0}$ is *not* the oscillator ground state $|0\rangle$. This is one of the reasons we felt it necessary to introduce nonstandard notation for labstates: keeping a clear distinction between those and conventional QM states is more than a matter of notion but reflects deeper interpretational issues.

To find a quantum register representation of the ladder operators, we first introduce some auxiliary notation. We define the register operators

$$\mathbb{P} \equiv \left\{\prod_{a=0}^{\infty} \boldsymbol{P}^a\right\}, \quad \mathbb{A}^a \equiv \left\{\prod_{b \neq a}^{\infty} \boldsymbol{I}^b\right\} \boldsymbol{A}^a, \quad \widehat{\mathbb{A}}^a \equiv \left\{\prod_{b \neq a}^{\infty} \boldsymbol{I}^b\right\} \widehat{\boldsymbol{A}}^a. \tag{24.22}$$

None of these operators is bosonic.

Exercise 24.9 Prove that the operators $\mathbb{P}$, $\mathbb{A}^a$, and $\widehat{\mathbb{A}}^a$ do not commute with the bosonic identity $\mathbb{I}_B$ defined by Eq. (24.15).

The operators $\widehat{\mathbb{A}}^a$ can be used to generate bosonic states from the signal ground state $\mathbf{0}$, which is transbosonic.[3] Specifically, we have

$$\mathbf{2}^a = \widehat{\mathbb{A}}^a \mathbf{0}, \quad a = 0, 1, 2, \ldots \tag{24.23}$$

With these definitions we construct the operators

$$\widehat{\mathbb{B}}^a \equiv \widehat{\mathbb{A}}^a \mathbb{P} \mathbb{A}^{a+1}, \quad \mathbb{B}^a \equiv \widehat{\mathbb{A}}^{a+1} \mathbb{P} \mathbb{A}^a. \tag{24.24}$$

Remarkably, these operators are bosonic, as can be readily proved from the fact that

$$\mathbb{A}^a \mathbb{I}_B = \mathbb{A}^a. \tag{24.25}$$

We find for $a = 0, 1, 2, \ldots$

$$\begin{aligned}
\mathbb{B}^a &\equiv \left\{ \bigotimes_{b \neq a, a+1}^{\infty} \boldsymbol{P}^b \right\} \boldsymbol{A}^a \widehat{\boldsymbol{A}}^{a+1} = \boldsymbol{P}^0 \boldsymbol{P}^1 \ldots \boldsymbol{P}^{a-1} \boldsymbol{A}^a \widehat{\boldsymbol{A}}^{a+1} \boldsymbol{P}^{a+2} \boldsymbol{P}^{a+3} \ldots \\
\widehat{\mathbb{B}}^a &\equiv \left\{ \bigotimes_{b \neq a, a+1}^{\infty} \boldsymbol{P}^b \right\} \widehat{\boldsymbol{A}}^a \boldsymbol{A}^{a+1} = \boldsymbol{P}^0 \boldsymbol{P}^1 \ldots \boldsymbol{P}^{a-1} \widehat{\boldsymbol{A}}^a \boldsymbol{A}^{a+1} \boldsymbol{P}^{a+1} \boldsymbol{P}^{a+3} \ldots
\end{aligned} \tag{24.26}$$

These operators satisfy the relations

$$\widehat{\mathbb{B}}^a \mathbb{B}^b = \delta^{ab} \mathbb{P}^{a+1}, \quad \mathbb{B}^a \widehat{\mathbb{B}}^b = \delta^{ab} \mathbb{P}^a. \tag{24.27}$$

Then the standard ladder operators $a, a^\dagger$ take the quantum register representation

$$\begin{aligned}
a \leftrightarrow a_B &\equiv \sum_{n=0}^{\infty} \sqrt{2(n+1)} \mathbb{B}^n, \\
a^\dagger \leftrightarrow a_B^\dagger &\equiv \sum_{n=0}^{\infty} \sqrt{2(n+1)} \widehat{\mathbb{B}}^n.
\end{aligned} \tag{24.28}$$

These operators are bosonic and have the commutation relation

$$\left[a_B, a_B^\dagger \right] = 2 \mathbb{I}_B. \tag{24.29}$$

Because these register ladder operators commute with the bosonic identity $\mathbb{I}_B$, any states that they create from bosonic states are also bosonic. We find for example

$$|n\rangle \equiv \frac{(a^\dagger)^n}{\sqrt{2^n n!}} |0\rangle \leftrightarrow \frac{\left(a_B^\dagger\right)^n}{\sqrt{2^n n!}} \mathbf{1} = \mathbf{2}^n, \quad n = 0, 1, 2, \ldots \tag{24.30}$$

Note that a_B annihilates the bosonic ground state $\mathbf{1}$, giving the zero vector 0 in the quantum register $\mathcal{Q}^{[\infty]}$, not the signal ground state $\mathbf{0}$.

[3] The signal ground state $\mathbf{0}$ is not the same as the oscillator ground state $|0\rangle$, which is identified with the quantum register state $\mathbf{2}^0 = \mathbf{1}$ in the computational basis representation and $\mathbf{1}^0 \mathbf{0}^1 \mathbf{0}^2 \mathbf{0}^3 \ldots$ in the occupation basis representation.

The standard QM bosonic oscillator Hamiltonian operator $\widehat{H} \equiv \frac{1}{2}a^\dagger a + \frac{1}{2}$ has quantum register representation

$$\widehat{H} \leftrightarrow \mathbb{H}_B \equiv \frac{1}{2}a_B^\dagger a_B + \frac{1}{2}\mathbb{I}_B = \sum_{n=0}^{\infty}\left(n + \frac{1}{2}\right)\mathbb{P}_B^n. \tag{24.31}$$

This operator commutes with the bosonic identity $\mathbb{I}_B$, and therefore, any states in the quantum register that start off bosonic at initial time remain bosonic, if they evolve under this Hamiltonian.

24.6 Comparison with Quantum Field Theory

Several factors indicate that the QDN formalism is a halfway house between fixed-particle number Schrödinger–Dirac quantum mechanics and the multiparticle formalism of quantum field theory (QFT). First, the QDN strategy focuses not on particles but on signals. There is no intrinsic requirement to conserve signality except in those experiments where there is a physical reason, such as the conservation of some quantum number, such as electric charge. Second, the existence of transbosonic states in the QDN register $\mathcal{Q}^{[\infty]}$ is a clear marker that the QDN formalism can accommodate the equivalent of QFT multiparticle states.

An important similarity between QDN and QFT is the existence of the Fock vacuum state $|0\rangle$ in QFT and the signal ground state $\mathbf{0}$ in QDN. We recall that in QFT, Fock space $\mathcal{F}$ is defined by expansions of the form

$$\mathcal{F} \equiv \{|0\rangle\} \oplus \mathcal{H} \oplus \mathfrak{S}(\mathcal{H} \otimes \mathcal{H}) \oplus \cdots, \tag{24.32}$$

that is, as the infinite direct sum of appropriately symmetrized tensor products of copies of a single-particle Hilbert space $\mathcal{H}$. Looking at the above presented QDN version of the bosonic oscillator, we see that the QDN signal ground state $\mathbf{0}$ corresponds to the QFT vacuum $|0\rangle$; the QDN bosonic state space $\mathcal{Q}_B$ corresponds to $\mathcal{H}$; and the transbosonic states in QDN correspond to the multiparticle states in QFT.

The QDN formalism may be interpreted as an attempt to encode Fock's vision of QFT into a mathematical formalism that is based on detectors rather than SUOs. In the next chapter, we go further in this respect by extending the QDN formalism to allow for the possibility of constructing the signal ground state $\mathbf{0}$ itself through the creation of apparatus in the laboratory from a state of nonexistence, denoted $\boldsymbol{\emptyset}$, that corresponds to the information void that we have focused on in other chapters. The information void represents the state of a laboratory in which there are no detectors whatsoever. However, this certainly does not mean that there is no observer or laboratory.

24.7 Fermionic Oscillators

In this section we extend the above discussion of the bosonic oscillator to the fermionic oscillator, that is, a system under observation (SUO) that requires anticommuting degrees of freedom in its classical (Martin, 1959b; Casalbuoni, 1976a) and quantum formulations (Candlin, 1956; Martin, 1959a; Casalbuoni, 1976b).

Not long after the discovery of quantum mechanics by Heisenberg and Schrödinger, Jordan and Wigner showed how to describe fermions in quantum register terms (Jordan and Wigner, 1928; Bjorken and Drell, 1965). Their construction of local fermionic quantum field operators requires tensor product contributions from all of the qubits in a quantum register. In a QDN approach to fermionic quantum fields (Eakins and Jaroszkiewicz, 2005), their techniques were used to describe fermionic quantum fields using an infinite-rank quantum register associated with a net of detectors distributed throughout all of physical space. Because the Jordan–Wigner construction requires nontrivial contributions from all qubits in the register, fermionic fields are manifestly and inherently nonlocal in QDN.

In contrast with the bosonic oscillator studied above, we may restrict attention to a finite-rank quantum register $\mathcal{Q}^{[N]} \equiv Q^1 Q^2 \dots Q^N$.

The most convenient basis here is the signal basis representation. We follow the approach outlined in Bjorken and Drell (1965) for the construction of fermionic operators.

We use all the bit operators discussed previously and introduce a new one, denoted $\boldsymbol{\sigma}^a$, for each qubit $a = 1, 2, \dots, N$ in the register, defined by

$$\boldsymbol{\sigma}^a \equiv \boldsymbol{P}^a - \widehat{\boldsymbol{P}}^a, \quad a = 1, 2, \dots, N. \tag{24.33}$$

Next, we define a set of nonlocal operators, α^a, $\widehat{\alpha}^a$, $a = 1, 2, \dots, N$:

$$\begin{aligned}
\alpha^1 &\equiv \boldsymbol{A}^1 \boldsymbol{I}^2 \boldsymbol{I}^3 \dots \boldsymbol{I}^N, \\
\alpha^a &\equiv \boldsymbol{\sigma}^1 \boldsymbol{\sigma}^2 \dots \boldsymbol{\sigma}^{a-1} \boldsymbol{A}^a \boldsymbol{I}^{a+1} \dots \boldsymbol{I}^N, \\
\widehat{\alpha}^1 &\equiv \widehat{\boldsymbol{A}}^1 \boldsymbol{I}^2 \boldsymbol{I}^3 \dots \boldsymbol{I}^N, \\
\widehat{\alpha}^a &\equiv \boldsymbol{\sigma}^1 \boldsymbol{\sigma}^2 \dots \boldsymbol{\sigma}^{a-1} \widehat{\boldsymbol{A}}^a \boldsymbol{I}^{a+1} \dots \boldsymbol{I}^N, \quad a = 1, 2, \dots, N.
\end{aligned} \tag{24.34}$$

These are the required "fermionic field operators." They satisfy the anticommutation relations

$$\begin{aligned}
\alpha^a \alpha^b + \alpha^b \alpha^a &= 0, \qquad \widehat{\alpha}^a \widehat{\alpha}^b + \widehat{\alpha}^b \widehat{\alpha}^a = 0, \\
\alpha^a \widehat{\alpha}^b + \widehat{\alpha}^b \alpha^a &= \delta^{ab} \mathbb{I}, \qquad 1 \leqslant a, b, \leqslant N.
\end{aligned} \tag{24.35}$$

where $\mathbb{I} \equiv \boldsymbol{I}^1 \boldsymbol{I}^2 \dots \boldsymbol{I}^N$ is the register identity operator.

Exercise 24.10 Use the signal bit algebra listed in Table 4.1 to prove (24.35).

The α^a, $\widehat{\alpha}^b$ operators have the following properties, which are easy to prove:

$$\begin{aligned} \alpha^a \mathbf{0} &= 0, \\ \widehat{\alpha}^a \mathbf{0} &\neq 0, \\ \widehat{\alpha}^a \widehat{\alpha}^a \mathbf{0} &= 0, \quad a = 1, 2, \ldots, N. \end{aligned} \tag{24.36}$$

This means that we can reconstruct all the relevant structure of fermionic field theory, the operators $\widehat{\alpha}^a$ and α^a playing the role of fermionic creation and annihilation operators.

A significant feature of the above anticommutation relations is that they are achieved via the use of the nonlocal operators and do not invoke Grassmannian (anticommuting) numbers to do so.

25

Dynamical Theory of Observation

25.1 Introduction

In this chapter we discuss how quantized detector network (QDN) theory can be extended to cover the creation and decommissioning of apparatus, from the perspective of observers and their laboratories. This extension will be referred to as *extended* QDN. It allows us to discuss bit power sets, laboratories, the universal register, contextual vacua, and the creation of quantized detector apparatus. This extended formalism is used to describe the Elitzur–Vaidman bomb-tester experiment and the Hardy paradox experiment.

A central concept running through this book is that of the *observer*, the enigmatic "*I*" of *I think therefore I am*. The problem is that, despite the many triumphs of quantum mechanics (QM), the physics of the observer and observation is still not well understood.

Regardless of how observers are defined and whether classical or quantum principles are involved, physicists generally believe that classical information in some form is extracted from systems under observation (SUOs) in actual physics experiments. In all branches of science, their language reflects this belief: experimentalists talk of measuring an electron's spin or the mass of a new particle, and so on.

The conceptual issues in quantum mechanics such as wave–particle duality, quantum interference, and nonlocality gave the first indication that all might not be well with this perspective. We need only to look at the photon concept to appreciate some of the problems with the idea that photons are particles (Paul, 2004). There are experiments where a photon (that is, a signal) is detected from a crystal, but the atom "from whence it came" cannot be identified, suggesting that a cooperative process is involved, rather like the phonon concept in the physics of crystals.

The problem we have with the photon-as-particle concept is the gap in logic. If the particle interpretation is taken literally, then obvious questions about the physical structure or its equivalent of such particles spring up. Does such a particle have a "surface"? If it has a surface, what is that surface made of? These and other simple (minded) questions quickly lead to the conclusion that the particle concept is a convenient and economical objectification of context. In other words, a useful illusion.

Some particle concepts appear better in this respect than others. For instance, fermion number is conserved in astrophysical processes, but photon number is not. Photons from the sun cannot be said to have traveled from the deep interior, whereas we could say so about neutrinos.

So much for the signals that observers detect. What about the observers themselves? It is surely too great an assumption to think of them as a mere auxiliary phenomenon in physics. Without observers, physics is a vacuous subject.

What seems to be missing is a comprehensive dynamical theory of observation that would treat observers and SUOs more on the same footing. In such a theory, observers would be subject to the same laws of physics as the SUOs that they were observing. Such a theory would be capable of accounting for the creation and annihilation of observers and their apparatus, as well as states of SUOs, because in the real world, that is what goes on and nothing lasts forever.

We do not have such a theory. That may come only once the physics of emergence has been much better understood. However, we can make a start with what we have at present in what we think may be a useful direction (Jaroszkiewicz, 2010). In this chapter, we develop an extension of the binary truth value concepts explored in earlier chapters, aimed at describing the possible states of apparatus in a more general form than having just two states, *yes* or *no*.

25.2 Power Bits

We have up to this point identified the two possible normal signal states, *ground* and *signal*, of a functioning binary detector as the two elements of a bit, denoted **0** and **1**, respectively. As we have mentioned, however, bits are not vector spaces and there seems to be no meaning to the addition of bit state **0** to bit state **1**, or even of the multiplication of a bit state by a real or complex number.

There is in fact a way of defining bit state addition, of a kind, in terms of set theory. We recall that the *power set* $\mathcal{P}(S)$ of a set is the set of all possible subsets of S including the empty set $\emptyset$ and S itself. Recall also that the number of elements in the power set of a set of cardinality c is 2^c. Therefore, we expect the power set $\mathcal{P}(B)$ of a bit B to have four distinct, nonempty elements. These are

$$\mathcal{P}(B) = \{\underset{\sim}{\mathbf{0}}, \underset{\sim}{\mathbf{1}}, \underset{\sim}{\mathbf{2}}, \underset{\sim}{\mathbf{3}}\}, \tag{25.1}$$

where we define

$$\underset{\sim}{\mathbf{0}} \equiv \{\mathbf{0}\}, \quad \underset{\sim}{\mathbf{1}} \equiv \{\mathbf{1}\}, \quad \underset{\sim}{\mathbf{2}} \equiv \{\mathbf{0}, \mathbf{1}\}, \quad \underset{\sim}{\mathbf{3}} \equiv \{\emptyset\}. \tag{25.2}$$

In this scheme, the set $\{\emptyset\}$ is a nontrivial element of $\mathcal{P}(B)$ and counts as one element of the power set. We shall refer to the power set of a bit as a *power bit*.

In this chapter, we shall work in terms of the elements of $\mathcal{P}(B)$ rather than with the elements of B itself, identifying elements $\underset{\sim}{\mathbf{0}}$ and $\underset{\sim}{\mathbf{1}}$ of $\mathcal{P}(B)$ as synonymous with bit states $\mathbf{0}$ and $\mathbf{1}$ of B, respectively. The value of using $\mathcal{P}(B)$ rather than B itself is that the elements of the former are sets, so we can use the set properties of *union* and *intersection* to make propositions about power bits equivalent to those in the algebra of sets (Howson, 1972).

Interpretation

Before we proceed further, however, we need to resolve the following problem: the power set $\mathcal{P}(B)$ of a bit B appears to have too many elements for a physical interpretation. Logic suggests that only the elements $\underset{\sim}{\mathbf{0}}$ and $\underset{\sim}{\mathbf{1}}$ of the power set are actually needed in an experiment. What could the elements $\underset{\sim}{\mathbf{2}}$ and $\underset{\sim}{\mathbf{3}}$ represent?

It has been our experience that many such questions in QDN are answered by looking at the situation in question and identifying any hidden assumptions. Consider an experimentalist who is going into a laboratory in order to determine the signal status of a certain detector. Even before they could get an answer, one glaring fact has to be true: that the detector actually *exists*.

Now whether or not a detector actually exists in a laboratory ("existence" being defined as a detector actually being found and recognized as such by an observer coming into that laboratory) is surely an empirical question, the answer to which the observer cannot assume is "*yes, it exists,*" *before* they enter the laboratory. This then is where one of our two extra elements of the power set $\mathcal{P}(B)$ finds a natural interpretation. We interpret the element $\underset{\sim}{\mathbf{3}} \equiv \{\emptyset\}$ as a state of physical nonexistence, in the laboratory, of the detector in question. We shall call it the *void state* of a detector.

It will undoubtedly seem strange to assign a state in our formalism to a condition of nonexistence of a detector. But if we want to construct a theory where apparatus can be created or destroyed, then that is precisely what we have to do.

As for the remaining element, $\underset{\sim}{\mathbf{2}} \equiv \{\mathbf{0}, \mathbf{1}\}$, of the power set $\mathcal{P}(\mathcal{B})$, the ambiguity in the elements it has ($\mathbf{0}$ and $\mathbf{1}$) fits in nicely with an obvious physical condition that a detector could be in: a state where the apparatus exists physically but is not operating normally, to the extent that it is not registering either a ground state or a signal state. For instance, such a detector could be faulty. Or it could have been deliberately taken off-line (or decommissioned). We shall call such a condition of a detector the *faulty state*, the *off-line state*, or the *decommissioned state*.

Given the above four states of a detector, we can find four natural power bit questions, denoted $\tilde{\boldsymbol{i}}$, $i = 0, 1, 2, 3$. These are explained in Table 25.1, along with the answers they induce when asked of each of the four power bit states.

Table 25.1 *The extended QDN questions and answers*

Question in words	**Symbol**	$\underset{\sim}{\mathbf{0}}$	$\underset{\sim}{\mathbf{1}}$	$\underset{\sim}{\mathbf{2}}$	$\underset{\sim}{\mathbf{3}}$
Is this detector in its ground state?	$\tilde{\mathbf{0}}$	1	0	0	0
Is this detector in its signal state?	$\tilde{\mathbf{1}}$	0	1	0	0
Is this detector off-line?	$\tilde{\mathbf{2}}$	0	0	1	0
Is this detector in its void state?	$\tilde{\mathbf{3}}$	0	0	0	1

We may summarize these answers by the rule

$$\tilde{\boldsymbol{i}}\underset{\sim}{\boldsymbol{j}} = \delta^{ij}, \qquad 0 \leqslant i, j \leqslant 3. \tag{25.3}$$

25.3 Power Bit Operators

A *power bit operator* is any mapping from the power set $\mathcal{P}(B)$ back into the power set. Given an element $\underset{\sim}{\boldsymbol{i}}$ of $\mathcal{P}(B)$ and a power bit operator $\underset{\cdots}{\boldsymbol{O}}$, then we denote the value of the operator's action on $\underset{\sim}{\boldsymbol{i}}$ by $\underset{\cdots}{\boldsymbol{O}}\underset{\sim}{\boldsymbol{i}}$. There is a total of $4^4 = 256$ different bit operators and only a few will be of use to us.

25.4 Matrix Representation

A useful way of representing power bit operators is via matrices. The elements $\underset{\sim}{\boldsymbol{i}}$ of $\mathcal{P}(B)$ may be represented by column matrices $[\underset{\sim}{\boldsymbol{i}}]$ given by

$$\underset{\sim}{\mathbf{0}} \rightleftharpoons [\underset{\sim}{\mathbf{0}}] \equiv \begin{bmatrix} 1 \\ 0 \\ 0 \\ 0 \end{bmatrix}, \quad \underset{\sim}{\mathbf{1}} \rightleftharpoons [\underset{\sim}{\mathbf{1}}] \equiv \begin{bmatrix} 0 \\ 1 \\ 0 \\ 0 \end{bmatrix}, \quad \text{and so on.} \tag{25.4}$$

We represent the action of bit operator $\underset{\cdots}{\boldsymbol{O}}$ on $\underset{\sim}{\boldsymbol{i}}$ by the action of a *power bit matrix* $[\underset{\cdots}{\boldsymbol{O}}]$ on a column matrix $[\underset{\sim}{\boldsymbol{i}}]$, such that

$$\underset{\cdots}{\boldsymbol{O}}\underset{\sim}{\boldsymbol{i}} \rightleftharpoons [\underset{\cdots}{\boldsymbol{O}}][\underset{\sim}{\boldsymbol{i}}] \equiv [\underset{\cdots}{\boldsymbol{O}}\underset{\sim}{\boldsymbol{i}}]. \tag{25.5}$$

In this matrix representation the dual elements $\tilde{\boldsymbol{i}}$ are represented by the row matrices

$$\begin{aligned} \tilde{\mathbf{0}} &\rightleftharpoons \begin{bmatrix} 1 & 0 & 0 & 0 \end{bmatrix}, \quad \tilde{\mathbf{1}} \rightleftharpoons \begin{bmatrix} 0 & 1 & 0 & 0 \end{bmatrix}, \\ \tilde{\mathbf{2}} &\rightleftharpoons \begin{bmatrix} 0 & 0 & 1 & 0 \end{bmatrix}, \quad \tilde{\mathbf{3}} \rightleftharpoons \begin{bmatrix} 0 & 0 & 0 & 1 \end{bmatrix}. \end{aligned} \tag{25.6}$$

This is consistent with the question and answer relations (25.3).

We can use the power bit matrix representation to define the operational meaning of the dyadics $\underset{\sim}{\boldsymbol{i}}\tilde{\boldsymbol{j}}$. These can then serve as formal basis elements in the expansion of power bit operators. Given a power bit operator defined by (25.5), we can write it as the formal (dyadic) expression $\underset{\dots}{\boldsymbol{O}} = \sum_{i=0}^{3} \underset{\dots}{\boldsymbol{O}}\, \underset{\sim}{\boldsymbol{i}}\, \widetilde{\boldsymbol{i}}$.

Products of power bit operators are defined in the natural way: given power bit operators $\underset{\dots}{\boldsymbol{M}}$, $\underset{\dots}{\boldsymbol{N}}$, we define their "product" $\underset{\dots}{\boldsymbol{M}}\underset{\dots}{\boldsymbol{N}}$ by its action on any element $\underset{\sim}{\boldsymbol{i}}$ of the power set $\mathcal{P}(B)$ according to the rule $(\underset{\dots}{\boldsymbol{M}}\underset{\dots}{\boldsymbol{N}})\underset{\sim}{\boldsymbol{i}} \equiv \underset{\dots}{\boldsymbol{M}}\{\underset{\dots}{\boldsymbol{N}}\underset{\sim}{\boldsymbol{i}}\}$. This product rule is associative but not commutative, which can be seen from the matrix representation.

25.5 Special Operators

In the following, we have to take into account that a bit power set does not contain a *zero* element, because such a set is not a vector space. The void element $\underset{\sim}{\mathbf{3}}$ is not physically meaningless quantity, either: it represents a definite state in a laboratory. Therefore, the bit operators we discuss in this section cannot map elements of a bit power set to any state other than $\underset{\sim}{\mathbf{0}}$, $\underset{\sim}{\mathbf{1}}$, $\underset{\sim}{\mathbf{2}}$, or $\underset{\sim}{\mathbf{3}}$.

The operators we define below have been constructed on the basis of what seems reasonable to us. The following bit operators seem physically reasonable, practically useful, and unavoidable in the kind of formalism we are aiming for.

Identity $\underset{\dots}{\boldsymbol{I}}$

This operator maps every element back into itself, i.e., $\underset{\dots}{\boldsymbol{I}}\,\underset{\sim}{\boldsymbol{i}} = \underset{\sim}{\boldsymbol{i}}$. Its matrix elements are given by the Kronecker delta, that is, $[\underset{\dots}{\boldsymbol{I}}]_{ij} = \delta_{ij}$.

Annihilator $\underset{\dots}{\boldsymbol{Z}}$

This operator maps any element $\underset{\sim}{\boldsymbol{i}}$ of the power set $\mathcal{P}(B)$ into the void state $\underset{\sim}{\mathbf{3}} \equiv \{\emptyset\}$, that is, $\underset{\dots}{\boldsymbol{Z}}\underset{\sim}{\boldsymbol{i}} = \underset{\sim}{\mathbf{3}}$, $i = 0, 1, 2, 3$. Its matrix representation is

$$[\underset{\dots}{\boldsymbol{Z}}] = \begin{bmatrix} 0 & 0 & 0 & 0 \\ 0 & 0 & 0 & 0 \\ 0 & 0 & 0 & 0 \\ 1 & 1 & 1 & 1 \end{bmatrix}. \tag{25.7}$$

The annihilator has a fundamental role in our theory: it represents the process of destroying and removing from the laboratory all traces of an already physically existing detector.

The Power Bit Signal Projection Operators $\underset{\dots}{\boldsymbol{P}}$ and $\underset{\dots}{\widehat{\boldsymbol{P}}}$

These operators have the action

$$\begin{array}{llll} \underset{\dots}{\boldsymbol{P}}\underset{\sim}{\mathbf{0}} = \underset{\sim}{\mathbf{0}}, & \underset{\dots}{\boldsymbol{P}}\underset{\sim}{\mathbf{1}} = \underset{\sim}{\mathbf{3}}, & \underset{\dots}{\boldsymbol{P}}\underset{\sim}{\mathbf{2}} = \underset{\sim}{\mathbf{3}}, & \underset{\dots}{\boldsymbol{P}}\underset{\sim}{\mathbf{3}} = \underset{\sim}{\mathbf{3}}, \\ \underset{\dots}{\widehat{\boldsymbol{P}}}\underset{\sim}{\mathbf{0}} = \underset{\sim}{\mathbf{3}}, & \underset{\dots}{\widehat{\boldsymbol{P}}}\underset{\sim}{\mathbf{1}} = \underset{\sim}{\mathbf{1}}, & \underset{\dots}{\widehat{\boldsymbol{P}}}\underset{\sim}{\mathbf{2}} = \underset{\sim}{\mathbf{3}}, & \underset{\dots}{\widehat{\boldsymbol{P}}}\underset{\sim}{\mathbf{3}} = \underset{\sim}{\mathbf{3}}, \end{array} \tag{25.8}$$

so their matrix representations are

$$[\underset{\cdots}{\boldsymbol{P}}] = \begin{bmatrix} 1 & 0 & 0 & 0 \\ 0 & 0 & 0 & 0 \\ 0 & 0 & 0 & 0 \\ 0 & 1 & 1 & 1 \end{bmatrix}, \quad [\underset{\cdots}{\widehat{\boldsymbol{P}}}] = \begin{bmatrix} 0 & 0 & 0 & 0 \\ 0 & 1 & 0 & 0 \\ 0 & 0 & 0 & 0 \\ 1 & 0 & 1 & 1 \end{bmatrix}. \tag{25.9}$$

These operators are idempotent, namely, $\underset{\cdots}{\boldsymbol{P}}\underset{\cdots}{\boldsymbol{P}} = \underset{\cdots}{\boldsymbol{P}}$, $\underset{\cdots}{\widehat{\boldsymbol{P}}}\underset{\cdots}{\widehat{\boldsymbol{P}}} = \underset{\cdots}{\widehat{\boldsymbol{P}}}$ and *power bit orthogonal*, namely, $\underset{\cdots}{\boldsymbol{P}}\underset{\cdots}{\widehat{\boldsymbol{P}}} = \underset{\cdots}{\widehat{\boldsymbol{P}}}\underset{\cdots}{\boldsymbol{P}} = \underset{\cdots}{\boldsymbol{Z}}$. In this context, the annihilator $\underset{\cdots}{\boldsymbol{Z}}$ plays the role of a zero element.

The Power Bit Signal Creation and Annihilation Operators $\underset{\cdots}{\boldsymbol{A}}$, $\underset{\cdots}{\widehat{\boldsymbol{A}}}$

These operators are defined principally by their action on the normal signal states $\underset{\sim}{\mathbf{0}}$ and $\underset{\sim}{\mathbf{1}}$:

$$\begin{aligned} &\underset{\cdots}{\boldsymbol{A}}\underset{\sim}{\mathbf{0}} = \underset{\sim}{\mathbf{3}}, \quad \underset{\cdots}{\boldsymbol{A}}\underset{\sim}{\mathbf{1}} = \underset{\sim}{\mathbf{0}}, \quad \underset{\cdots}{\boldsymbol{A}}\underset{\sim}{\mathbf{2}} = \underset{\sim}{\mathbf{3}}, \quad \underset{\cdots}{\boldsymbol{A}}\underset{\sim}{\mathbf{3}} = \underset{\sim}{\mathbf{3}}, \\ &\underset{\cdots}{\widehat{\boldsymbol{A}}}\underset{\sim}{\mathbf{0}} = \underset{\sim}{\mathbf{1}}, \quad \underset{\cdots}{\widehat{\boldsymbol{A}}}\underset{\sim}{\mathbf{1}} = \underset{\sim}{\mathbf{3}}, \quad \underset{\cdots}{\widehat{\boldsymbol{A}}}\underset{\sim}{\mathbf{2}} = \underset{\sim}{\mathbf{3}}, \quad \underset{\cdots}{\widehat{\boldsymbol{A}}}\underset{\sim}{\mathbf{3}} = \underset{\sim}{\mathbf{3}}, \end{aligned} \tag{25.10}$$

which gives the matrix representations

$$[\underset{\cdots}{\boldsymbol{A}}] = \begin{bmatrix} 0 & 1 & 0 & 0 \\ 0 & 0 & 0 & 0 \\ 0 & 0 & 0 & 0 \\ 1 & 0 & 1 & 1 \end{bmatrix}, \quad [\underset{\cdots}{\widehat{\boldsymbol{A}}}] = \begin{bmatrix} 0 & 0 & 0 & 0 \\ 1 & 0 & 0 & 0 \\ 0 & 0 & 0 & 0 \\ 0 & 1 & 1 & 1 \end{bmatrix}. \tag{25.11}$$

These operators are *power bit nilpotent*, that is, $\underset{\cdots}{\boldsymbol{A}}\underset{\cdots}{\boldsymbol{A}} = \underset{\cdots}{\widehat{\boldsymbol{A}}}\underset{\cdots}{\widehat{\boldsymbol{A}}} = \underset{\cdots}{\boldsymbol{Z}}$, with $\underset{\sim}{\boldsymbol{Z}}$ once again playing the role of a zero element.

The product rules for the operators $\underset{\cdots}{\boldsymbol{P}}, \underset{\cdots}{\widehat{\boldsymbol{P}}}, \underset{\cdots}{\boldsymbol{A}}$, and $\underset{\cdots}{\widehat{\boldsymbol{A}}}$ are given in Table 25.2. Comparison with Table 4.1 shows that these tables are isomorphic, provided the zero operator $\boldsymbol{Z}$ in Table 4.1 is identified with the annihilator $\underset{\cdots}{\boldsymbol{Z}}$ in Table 25.2.

Construction Operator $\underset{\cdots}{\boldsymbol{C}}$

This operator acts on every element $\underset{\sim}{\boldsymbol{i}}$ of power bit set and sets it to the signal ground state in readiness for observation, i.e., $\underset{\cdots}{\boldsymbol{C}}\underset{\sim}{\boldsymbol{i}} = \underset{\sim}{\mathbf{0}}$, $i = 0, 1, 2, 3$. There are

Table 25.2 *The product table for the special operators*

	$\underset{\cdots}{\boldsymbol{P}}$	$\underset{\cdots}{\widehat{\boldsymbol{P}}}$	$\underset{\cdots}{\boldsymbol{A}}$	$\underset{\cdots}{\widehat{\boldsymbol{A}}}$
$\underset{\cdots}{\boldsymbol{P}}$	$\underset{\cdots}{\boldsymbol{P}}$	$\underset{\cdots}{\boldsymbol{Z}}$	$\underset{\cdots}{\boldsymbol{A}}$	$\underset{\cdots}{\boldsymbol{Z}}$
$\underset{\cdots}{\widehat{\boldsymbol{P}}}$	$\underset{\cdots}{\boldsymbol{Z}}$	$\underset{\cdots}{\widehat{\boldsymbol{P}}}$	$\underset{\cdots}{\boldsymbol{Z}}$	$\underset{\cdots}{\widehat{\boldsymbol{A}}}$
$\underset{\cdots}{\boldsymbol{A}}$	$\underset{\cdots}{\boldsymbol{Z}}$	$\underset{\cdots}{\boldsymbol{A}}$	$\underset{\cdots}{\boldsymbol{Z}}$	$\underset{\cdots}{\boldsymbol{P}}$
$\underset{\cdots}{\widehat{\boldsymbol{A}}}$	$\underset{\cdots}{\widehat{\boldsymbol{A}}}$	$\underset{\cdots}{\boldsymbol{Z}}$	$\underset{\cdots}{\widehat{\boldsymbol{P}}}$	$\underset{\cdots}{\boldsymbol{Z}}$

two scenarios. If a detector is in its void state, then it does not exist, so the action of the construction operator represents the physical construction of that detector, set up in its ground state in the laboratory, prior to any experiment. It is assumed that facilities exist in the laboratory for this. Alternatively, if the detector already exists, then the construction operator resets it to its ground state if it is normal or repairs it and sets it to its ground state if it is faulty.

The construction operator is represented by the matrix

$$[\underset{\cdots}{\boldsymbol{C}}] = \begin{bmatrix} 1 & 1 & 1 & 1 \\ 0 & 0 & 0 & 0 \\ 0 & 0 & 0 & 0 \\ 0 & 0 & 0 & 0 \end{bmatrix}. \tag{25.12}$$

Decommissioning Operator $\underset{\cdots}{\boldsymbol{D}}$

This operator represents the action of decommissioning an already existing detector, setting it into its decommissioned state $\underset{\sim}{\mathbf{2}}$. This operator does *not* reset states $\underset{\sim}{\mathbf{0}}$, $\underset{\sim}{\mathbf{1}}$, or $\underset{\sim}{\mathbf{2}}$ to the void state $\underset{\sim}{\mathbf{3}}$ because in the real world, there will invariably be some remaining information in the form of *debris* that will inform the observer that apparatus has been decommissioned. This is an important feature of our discussion toward the end of this chapter of the Elitzur–Vaidman bomb-tester experiment and Hardy's paradox experiment.

The decommissioning operator is represented by the matrix

$$[\underset{\cdots}{\boldsymbol{D}}] = \begin{bmatrix} 0 & 0 & 0 & 0 \\ 0 & 0 & 0 & 0 \\ 1 & 1 & 1 & 0 \\ 0 & 0 & 0 & 1 \end{bmatrix}. \tag{25.13}$$

25.6 The Laboratory

In extended QDN it is assumed that an observer exists in a physical environment referred to as the *laboratory*, Λ. This will have the facilities for the construction or introduction of apparatus consisting of a number of detectors. At any given stage Σ_n, the observer will associate a state known as a *generalized labstate* to the collection of detectors at that time. This state could be a pure state or a mixed state. We shall restrict our attention in this chapter to pure labstates for reasons of space.

Extended QDN is based on what really happens in laboratories. Suppose an observer walks into a laboratory, intending to look at a given detector. There are the following four classical scenarios.

No Detector

The detector in question does not exist. The observer will see this; it is an empirical fact that has physical significance. This absence is represented by extended labstate $\underset{\sim}{\mathbf{3}}$.

Faulty Detector

The detector in question exists physically, but is faulty, off-line, or decommissioned. This situation is represented by extended labstate $\underset{\sim}{\mathbf{2}}$.

Ground State

The detector in question exists physically, is in perfect working order, and is in its ground state. This is represented by extended labstate $\underset{\sim}{\mathbf{0}}$.

Signal State

The detector in question exists physically, is in perfect working order, and is in its signal state. This is represented by extended labstate $\underset{\sim}{\mathbf{1}}$.

25.7 The Universal Register

The power set approach to detectors allows us to think of an absence of a detector in Λ as an observable fact that can be representable mathematically. The state corresponding to an absent detector is represented by the element $\underset{\sim}{\mathbf{3}}$ of its associated power set. We can do this for any number of absent detectors. Therefore, we can represent a complete absence of any detectors whatsoever by an infinite collection of such elements.

Remark 25.1 This is one of the very few places in QDN that we permit an infinity to enter the discussion; it costs us nothing in material terms and is a purely theoretical construct that nevertheless represents a very real state of existence of a laboratory, *before* any detectors have been constructed in it.

Such a condition corresponds to an observer without any apparatus, i.e., an empty laboratory. We denote this labstate by the symbol $\underset{\sim}{\boldsymbol{\emptyset}}$ and call it the *information void*, or just the *void*. It represents a potential for the existence of any detector or detectors, relative to a given observer.

If the observer's laboratory Λ is in its void state $\underset{\sim}{\boldsymbol{\emptyset}}$, that does not mean that the laboratory Λ or the observer do not exist, or that there are no systems under observation in Λ. It means simply that the observer has no current means of acquiring any information. An empty laboratory devoid of any detectors is a physically meaningful concept, but one with no interesting empirical content.

The information void can be thought of as one element in an infinite set called the *universal register* $\boldsymbol{\Omega}$, the Cartesian product of an infinite number of bit power sets. We write

$$\underset{\sim}{\boldsymbol{\emptyset}} \equiv \prod_{\alpha}^{\infty} \underset{\sim}{\mathbf{3}}^{\alpha}, \in \boldsymbol{\Omega} \equiv \prod_{\alpha}^{\infty} \mathcal{P}(B^{\alpha}), \tag{25.14}$$

where the index α could in principle be discrete, continuous, or a combination of both.

The cardinality of the universal register is a measure of how many power sets could belong to it. That depends on the imagination, and may be assumed to be infinity. Precisely what sort of Cantorian cardinality it should be is not clear, but that is a vacuous concept anyway. If we thought in terms of Λ sitting in continuous space, then we expect at least the cardinality, $\mathfrak{c}$, of the continuum. However, that is a metaphysical statement, because there would not be enough energy in the universe to create a continuum of detectors.[1] What helps us immeasurably here is that real observers can only ever deal with finite numbers of detectors. We can generally ignore all nonexistent potential detectors, but keep in mind that an absence of a detector can be significant in some circumstances, such as in the Renniger thought experiment (Renniger, 1953).

The product notation in (25.14) reflects the relationship between collections of power sets and the tensor products of qubit spaces that we encounter in the quantum version of this approach, discussed later. In these products, ordering is not significant, since labels keep track of the various terms. An arbitrary classical labstate $\underset{\sim}{\boldsymbol{\Psi}}_n$ in the universal register $\boldsymbol{\Omega}_n$ at stage Σ_n will be of the form $\prod_\alpha^\infty \underset{\sim}{\psi}_n^\alpha$, where $\underset{\sim}{\psi}_n^\alpha$ is one of the four elements of $\mathcal{P}(B_n^\alpha) \equiv \{\underset{\sim}{\mathbf{0}}_n^\alpha, \underset{\sim}{\mathbf{1}}_n^\alpha, \underset{\sim}{\mathbf{2}}_n^\alpha, \underset{\sim}{\mathbf{3}}_n^\alpha\}$.

Operators acting on universal register states will be denoted in blackboard bold font with three dots below, and act as follows. If $\underset{\dots}{\boldsymbol{O}}_n^\alpha$ is a power bit operator acting on elements of $\mathcal{P}(B_n^\alpha)$, then $\underset{\dots}{\mathbb{O}}_n \equiv \prod_\alpha^\infty \underset{\dots}{\boldsymbol{O}}_n^\alpha$ acts on an arbitrary classical state $\underset{\sim}{\boldsymbol{\Psi}}_n \equiv \prod_\alpha^\infty \underset{\sim}{\psi}_n^\alpha$ in $\boldsymbol{\Omega}_n$ according to the rule

$$\underset{\dots}{\mathbb{O}}_n \underset{\sim}{\boldsymbol{\Psi}}_n = \left\{\prod_\alpha^\infty \underset{\dots}{\boldsymbol{O}}_n^\alpha\right\}\left\{\prod_\beta^\infty \underset{\sim}{\psi}_n^\beta\right\} \equiv \prod_\alpha^\infty \underset{\dots}{\boldsymbol{O}}_n^\alpha \underset{\sim}{\psi}_n^\alpha. \tag{25.15}$$

For every classical register state $\underset{\sim}{\boldsymbol{\Psi}} \equiv \prod_\alpha^\infty \underset{\sim}{\psi}^\alpha$ there will be a corresponding dual register state $\widetilde{\boldsymbol{\Psi}} \equiv \prod_\alpha^\infty \widetilde{\psi}^\alpha$, where $\widetilde{\psi}^\alpha$ is dual to $\underset{\sim}{\psi}^\alpha$. Classical register states including the void satisfy the orthonormality condition

$$\widetilde{\boldsymbol{\Phi}}\underset{\sim}{\boldsymbol{\Psi}} \equiv \left\{\prod_\alpha^\infty \widetilde{\phi}^\alpha\right\}\prod_\beta^\infty \underset{\sim}{\psi}^\beta = \prod_\alpha^\infty \widetilde{\phi}^\alpha \underset{\sim}{\psi}^\alpha = \prod_\alpha^\infty \delta^{\phi^\alpha \psi^\alpha}. \tag{25.16}$$

Classical register states $\underset{\sim}{\boldsymbol{\Phi}}$, $\underset{\sim}{\boldsymbol{\Psi}}$ that differ in at least one bit power set element therefore satisfy the rule $\widetilde{\boldsymbol{\Phi}}\underset{\sim}{\boldsymbol{\Psi}} = 0$.

25.8 Contextual Vacua

In conventional classical mechanics or Schrödinger–Dirac quantum mechanics, empty space is generally not represented by any specific mathematical object.

[1] In other words, the Universe could not observe itself completely.

In relativistic quantum field theory (RQFT), however, empty space is represented by the *vacuum*, a normalized vector in an infinite-dimensional Hilbert space. It has physical properties such as zero total momentum, zero total electric charge, and so on, which although bland are physically significant attributes nevertheless. In RQFT, particle states are represented by the application of particle creation operators to the vacuum.

In our approach we encounter an analogous concept. Starting with an information void $\underset{\sim}{\emptyset}_n$, we represent the construction of a collection of detectors in the laboratory Λ_n at stage Σ_n by the application of a corresponding number of construction operators $\underset{\cdots}{\mathbb{C}}{}^i_n$ acting on that information void.

Example 25.2 The construction of detector i_n in its signal ground state in a previously empty laboratory is represented by the element $\underset{\cdots}{\mathbb{C}}{}^i_n \underset{\sim}{\emptyset}_n$ of the universal register $\boldsymbol{\Omega}_n$ at stage Σ_n, where

$$\underset{\cdots}{\mathbb{C}}{}^i_n \equiv \left\{ \prod_{j \neq i}^{\infty} \underset{\cdots}{\boldsymbol{I}}{}^j_n \right\} \underset{\cdots}{\boldsymbol{C}}{}^i_n. \tag{25.17}$$

More generally, an extended labstate $\underset{\sim}{\mathbf{0}}{}^{[r]}_n$ consisting of a number r of detectors each in its signal ground state at stage Σ_n is given by

$$\underset{\sim}{\mathbf{0}}{}^{[r]}_n \equiv \underset{\cdots}{\mathbb{C}}{}^1_n \underset{\cdots}{\mathbb{C}}{}^2_n \cdots \underset{\cdots}{\mathbb{C}}{}^r_n \underset{\sim}{\emptyset}_n, \tag{25.18}$$

where without loss of generality we label the detectors involved from 1 to r. Such a state will be said to be a *rank-r ground state*, or *contextual vacuum state*.

We can now draw an analogy between the vacuum of RQFT and the rank-r ground states in our formalism. The physical three-dimensional space of conventional physics would correspond to a contextual vacuum of extremely large rank, *if* physical space were relevant to the experiment. This would be the case for discussions involving particle scattering or gravitation, for example. For many experiments, however, such as the Stern–Gerlach experiment and quantum optics networks, physical space would be considered part of the relative external context and therefore could be ignored for the purposes of those experiments. It all depends on what the observer is trying to do.

In the real world there is more than one observer, so a theory of observation should take account of that fact. That is readily done in extended QDN. For example, the mutual contextual vacuum for two distinct observers A and B for which some commonality of time had been established would be represented by an element in $\boldsymbol{\Omega}_n$ of the form

$$\underset{\sim}{\mathbf{0}}{}^{A,B}_n \equiv \underset{\cdots}{\mathbb{C}}{}^{A,1}_n \underset{\cdots}{\mathbb{C}}{}^{A,2}_n \cdots \underset{\cdots}{\mathbb{C}}{}^{A,r_A}_n \underset{\cdots}{\mathbb{C}}{}^{B,1}_n \underset{\cdots}{\mathbb{C}}{}^{B,2}_n \cdots \underset{\cdots}{\mathbb{C}}{}^{B,r_B}_n \underset{\sim}{\emptyset}_n, \tag{25.19}$$

and so on. If subsequent dynamics was such that the detectors of observer A never sent signals to those of observer B and vice versa, then to all intents and purposes we could discuss each observer as if we were primary and they were separate secondary observers. If on the other hand some signals did pass between them, then that would be equivalent to having only one secondary observer.

If no commonality of time or other context has been established between primary observers, then there can be no physical meaning to (25.19). This is an important point in cosmology, where there are frequent discussions about multiple universe "bubbles" beyond the limits of observation. The mere fact that astronomers have received light from extremely distant galaxies establishes a context between the signal sources in those galaxies and the detectors used by the astronomers and validates the use of general relativity (GR) all the way to those regions of spacetime. If, on the other hand, no such signals have been received, then there is no such context. Therefore, relative to astronomers today, the universe beyond the horizon of observation can be meaningfully represented only by an information void, not a spatial vacuum. Something may be there, but we should not discuss it as if we had access to any form of information about it, such as its spacetime structure.

Much the same concern must be raised about black hole physics. That requires careful analysis of the contextual relation between observers outside the critical (Schwarzschild) radius and those who were assumed to be inside it, if any such context can be found. An event horizon is a boundary between observers on one side and observers on the other. For all intents and purposes, observers on one side of such a horizon have to regard those on the other side as in an information void.

This raises deep questions and concerns about the use of coordinate patches in general relativity and black hole physics that may be related to the problems of "quantizing gravity." We have asserted throughout this book that quantum mechanics is much more like a theory of entitlement than about the structure of SUOs. With this in mind, we should always ask the question: how do we *know* this or that? Consider GR. It is standard to postulate a line element without asking what could have been the source of that spacetime structure, or whether there are physically enigmatic possibilities, such as closed time-like curves, as in Gödel's spacetime (Gödel, 1949). There may be event horizons in our solutions to Einstein's equations. Our point is this: what empirical entitlement do we have to discuss two regions of spacetime separated by an event horizon, such as happens in the case of the Schwarzschild metric? Contextual completeness should tell us that such a scenario is vacuous. Trying to quantize any theory under such circumstances seems to be asking for trouble.

25.9 Experiments

Long before any experiment can begin, an observer starts off with a laboratory Λ in its void state $\underset{\sim}{\emptyset}$. Then at some stage long before any runs can be taken, specific apparatus consisting of a finite number r of detectors has to be constructed in Λ. We will assume without loss of generality that these are all functioning normally and in their ground state, so the labstate at that point is given by the equivalent of the right-hand side of (25.18). All of this is necessary before state preparation.

Modules

We have so far discussed only the detectors. Similar comments can be made about the modules that sit in the information void. The difference is that labstates are directly associated with detectors, both real or virtual, whereas modules are almost invariably classical in their representation, at least in the formulation of QDN that we have discussed in this book. Therefore, we have not invested much effort in constructing a dynamical theory of modules, although that is expected in a more complete theory.

There is an intriguing relationship here with Feynman diagrams in RQFT. Consider quantum electrodynamics (QED) and the simplest Feynman diagram for electron–proton scattering, Figure 25.1. In QED, the particle source and final state detectors are taken to be at remote past and future times, respectively, and propagation is in the information void (the vacuum). However, the quantum fields are interacting in QED. This particular Feynman diagram can be interpreted as a process where a virtual module (the photon, represented by the wavy line in Figure 25.1) is created not by the observer, but by the quantum process itself. This module has many of the characteristics of a beam splitter, discussed in Chapter 11. Because this module is created via the quantum dynamics in the information void, it is reasonable to classify the experiment as type T3, as defined in Chapter 21.

Care has to be taken not to take Feynman diagrams too literally. They represent individual terms in infinite perturbation expansions or in asymptotic expansions of complete amplitudes. Those complete amplitudes will satisfy all the requirements of scattering matrix theory (S-matrix) (Eden et al., 1966), and they will behave as virtual quantum modules.

The Contextual Register

According to what we said earlier, external context involving detectors in their void state $\underset{\sim}{\mathbf{3}}$ can be ignored; they do not exist in any physical sense. Therefore, we need only discuss those detectors that are in states $\underset{\sim}{\mathbf{0}}$, $\underset{\sim}{\mathbf{1}}$, or $\underset{\sim}{\mathbf{2}}$. A further simplification is that in real experiments, observers generally filter out observations

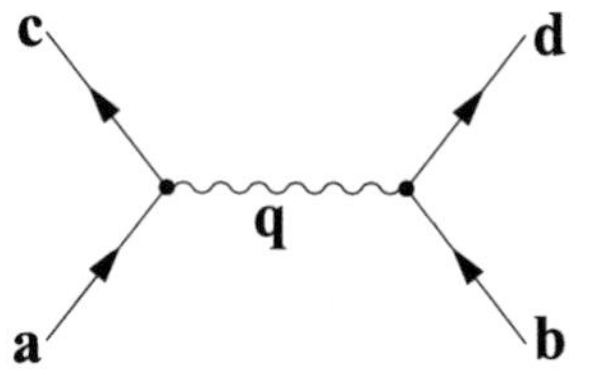

Figure 25.1. Lowest order Feynman diagram for electron–proton scattering.

from faulty detectors (assuming these have been identified) by postselecting only those labstates that contain the *normal bit states* $\underset{\sim}{\mathbf{0}}$ or $\underset{\sim}{\mathbf{1}}$. We shall confine our attention to such normal labstates until we deal with applications to quantum mechanics.

Given this condition, we can restrict our discussion at any given stage Σ_n to the *physical register* $\underset{\sim}{\mathcal{R}}_n$, a subset of the universal register $\boldsymbol{\Omega}_n$. If there are r_n working detectors in the laboratory at stage Σ_n, then $\underset{\sim}{\mathcal{R}}_n$ consists of 2^{r_n} normal states. This is the extended QDN version of the classical registers $\mathcal{R}_n$ discussed in earlier chapters. We may represent these normal states with the extended version of the computational basis and signal basis representations discussed previously, For example, a typical normal extended basis labstate is of the form

$$\underset{\sim}{\boldsymbol{i}}_n \equiv \underset{\sim}{\boldsymbol{i}}_n^1 \underset{\sim}{\boldsymbol{i}}_n^2 \cdots \underset{\sim}{\boldsymbol{i}}_n^{r_n}, \tag{25.20}$$

where $i^j = 0$ or 1 for $j = 1, 2, \ldots, r_n$, and

$$i_n \equiv i_n^1 + 2i_n^2 + \cdots + 2^{r_n - 1} i_n^{r_n}. \tag{25.21}$$

The physical register $\underset{\sim}{\mathcal{R}}_n$ represents all those detectors in the laboratory Λ at stage Σ_n that exist relative to the observer and are not faulty.

Virtually all of the concepts, such as signality, encountered with binary registers are encountered with physical registers, with some enhancements that are particular to enhanced QDN. One such is that the power bit operators $\underset{\cdots}{\boldsymbol{A}}$ and $\underset{\cdots}{\widehat{\boldsymbol{A}}}$ defined in (25.10) are not adjoints of each other, whereas the bit operators $\boldsymbol{A}$ and $\widehat{\boldsymbol{A}}$ defined in (4.13) are mutual adjoints. It was in anticipation of that fact that we chose the hat (circumflex) notation for $\widehat{\boldsymbol{A}}$ and not $\boldsymbol{A}^\dagger$.

Physical Register Signal Operators

Given a rank-r physical register $\underset{\sim}{\mathcal{R}}_n$ we define the physical register signal operators

$$\underset{\cdots}{\mathbb{A}}_n^i \equiv \left\{ \prod_{j \neq i}^{\infty} \underset{\cdots}{\boldsymbol{I}}_n^j \right\} \underset{\cdots}{\boldsymbol{A}}_n^i, \quad \underset{\cdots}{\widehat{\mathbb{A}}}_n^i \equiv \left\{ \prod_{j \neq i}^{\infty} \underset{\cdots}{\boldsymbol{I}}_n^j \right\} \underset{\cdots}{\widehat{\boldsymbol{A}}}_n^i, \qquad 1 \leqslant i \leqslant r. \tag{25.22}$$

There are several interesting processes that can be described by these extended register operators. For example, an application of the operator $\underset{\cdots}{\mathbb{A}}_n^i$ to the rank-r contextual vacuum $\underset{\sim}{\mathbf{0}}_n^{[r]} \equiv \underset{\cdots}{\mathbb{C}}_n^1 \underset{\cdots}{\mathbb{C}}_n^2 \cdots \underset{\cdots}{\mathbb{C}}_n^r \underset{\sim}{\emptyset}_n$ gives a rank-$(r-1)$ contextual vacuum:

$$\underset{\cdots}{\mathbb{A}}_n^i \underset{\cdots}{\mathbb{C}}_n^1 \underset{\cdots}{\mathbb{C}}_n^2 \cdots \underset{\cdots}{\mathbb{C}}_n^r \underset{\sim}{\emptyset}_n = \underset{\cdots}{\mathbb{C}}_n^1 \underset{\cdots}{\mathbb{C}}_n^2 \cdots \underset{\cdots}{\mathbb{C}}_n^{i-1} \underset{\cdots}{\mathbb{C}}_n^{i+1} \cdots \underset{\cdots}{\mathbb{C}}_n^r \underset{\sim}{\emptyset}_n. \tag{25.23}$$

This is a nonzero labstate in $\boldsymbol{\Omega}_n$ but is not an element of the original physical register $\underset{\sim}{\mathcal{R}}_n^{[r]}$. What has happened is analogous to the convention qubit register result $\mathbb{A}_n^i \mathbf{0}_n = 0$. In the extended QDN case, we do not get zero but the equivalent of it: the action of $\underset{\dots}{\mathbb{A}}{}^{i}_{n}$ on the signal ground state (contextual vacuum) $\underset{\sim}{\mathbf{0}}{}^{[r]}_{n}$ of $\underset{\sim}{\mathcal{R}}{}^{[r]}$ maps it into the ground state of a different physical register, one of rank $r-1$, a state orthogonal to every state in $\underset{\sim}{\mathcal{R}}{}^{[r]}$.

25.10 Quantization

The signal operators and the corresponding projection operators $\underset{\dots}{\mathbb{P}}{}^{i}_{n}, \underset{\dots}{\widehat{\mathbb{P}}}{}^{i}_{n}$ can be used to discuss all the quantum processes covered in this book, and will be used in the discussions on the bomb-tester and Hardy paradox below. An additional feature we make use of is that we can now factor into the mathematical representation the construction operators $\underset{\dots}{\mathbb{C}}{}^{i}_{n}$ and the decommissioning operators $\underset{\dots}{\mathbb{D}}{}^{i}_{n}$ explicitly.

In the next two sections we show how the extended QDN formalism describes experiments where the apparatus changes in one way or another during the experiment. In particular, the faulty/off-line state plays an important role in these experiments.

25.11 The Elitzur–Vaidman Bomb-Tester Experiment

In this experiment, a stockpile of active (A) and dud (D) bombs is analyzed, one by one, in order to find as many unexploded active bombs as possible. The approach follows that discussed in Elitzur and Vaidman (1993). The stage diagram shown in Figure 25.2 is a modified Mach–Zehnder interferometer. B^1 and B^2 are beam splitters, M is a mirror, and T is the actual bomb-testing triggering device that explodes the bomb if it is an active bomb *and* a signal could have been detected at 2_1.[2]

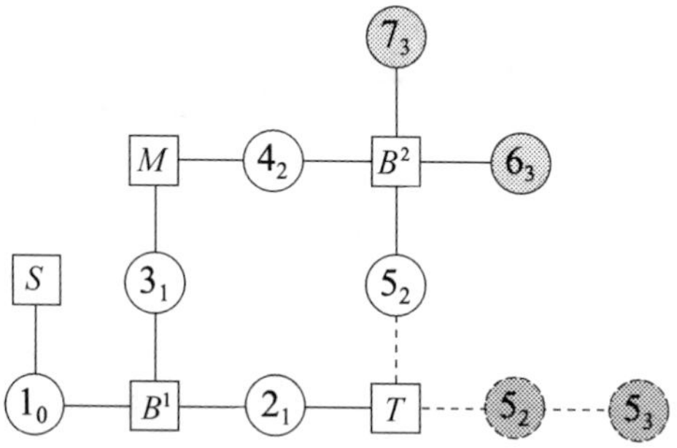

Figure 25.2. The bomb-tester network.

[2] One of the reasons this experiment is significant and to some extent baffling is the strange mix of classical and quantum counterfactuality. Analyzing it in terms of *which-path* information seems the best approach, because such information is not controversial and its extent can be established directly from a stage diagram in general.

The process goes as follows. Source S prepares a single photon state at 1_0, which is then passed into beam splitter B^1. Transmitted channel 2_1 feeds onto the bomb tester module T while reflected channel 3_1 is deflected by mirror M toward beam splitter B^2.

The action at module T is critical and is based on a specific assumption that sits on the knife edge between the classical and quantum worlds. For each run of the experiment, a bomb is taken from the untested stockpile of bombs (which includes active and dud bombs) and is placed in contact with module T. If the bomb is a dud (D), then 2_1 is not prevented from being deflected onto 5_2, and thence onto beam splitter B^2, where it can interfere with 4_2.

If on the other hand the bomb is active (A), then either channel 3_1 is realized *and* the bomb does not explode, or channel 3_1 is not realized, and then certainly the bomb explodes. In this latter eventuality, channel 5_2 is blocked.

This experiment involves a randomly changing apparatus network, because the bomb being tested in a given run is replaced by a new, randomly active or dud bomb from the unused stockpile. In addition, the network is self-intervening in the case of an active bomb coinciding with realized channel 2_1. This experiment is a combination of T2 and T3 experiments (these are discussed in Chapter 21). The experiment looks like T2 in that the probability of a bomb being active or dud in each run introduces an explicit time dependence in the experiment as a whole, but that probability is determined by factors external to the apparatus. It has also the characteristics of a T3 experiment, in that beam splitter B^1 has random outcomes, and these determine whether module T is triggered. Therefore, we may classify this experiment as T4.

There are two distinct networks to consider separately, because whether the bomb is active or dud is part of relative external context and not a quantum process requiring extended QDN. Therefore, we deal with each of the networks separately, using a pure labstate description for each initially. Then we use a density matrix approach to consider the combined experiment.

Our formulation of this scenario differs in two respects from those considered previously in this book. First, we shall use extended QDN. Second, we shall introduce *persistence*, in that all real and virtual detectors will be created before stage Σ_0 and persist to the final stage.

Creation of the Apparatus

Prior to the creation of the apparatus, well before stage Σ_0, the signal state of the laboratory Λ is its void state $\underset{\sim}{\emptyset}$ in the universal register $\boldsymbol{\Omega}$.[3] We may take the apparatus to be created in its ground state (equivalent to a contextual vacuum)

[3] The choice of the words *its* and *the* in this sentence is deliberate. A void state is contextual, whereas we can assume that there is just one enormous universal register that describes the totality of possible contextual existence and nonexistence that could ever be imagined, under all possible circumstances. This latter assumption costs us nothing.

before stage Σ_0, that is, before the start of each run, so we would normally expect the apparatus to exist and be in its quiescent, no-signal state,

$$\underset{\sim}{\mathbf{0}}_0 \equiv \underset{\dots}{\mathbb{C}}_0^1 \underset{\dots}{\mathbb{C}}_0^2 \underset{\dots}{\mathbb{C}}_0^3 \underset{\dots}{\mathbb{C}}_0^4 \underset{\dots}{\mathbb{C}}_0^5 \underset{\dots}{\mathbb{C}}_0^6 \underset{\dots}{\mathbb{C}}_0^7 \underset{\sim}{\boldsymbol{\emptyset}}_0. \tag{25.24}$$

This deserves some explanation, because sites $1_0, 2_1, 3_1, 4_2$, and 5_2 are not real detectors. So the question arises, what does it mean to "create" them?

The answer is given by Schwinger, quoted in Chapter 24. The modules S, B^1, B^2, M, and T really do "exist" relative to the observer, and it is that very existence that creates, as it were, the context for virtual detector sites $1_0, 2_1, 3_1, 4_2$, and 5_2 to be meaningful. We may refer to this as "*Schwinger counterfactual reality.*" It is not metaphysics but actually part and parcel of the process of running any experiment, as we have commented before. The observer would have had to place real detectors in those sites in order to calibrate the given modules, *before* the actual runs had started (and then taken those calibration detectors away). Otherwise, there could be no degree of confidence in what the observer believed about the labstates being studied.

In addition to the creation of real and virtual detector sites, the preparation device S prepares a photon signal in 1_0, so the initial extended labstate $\underset{\sim}{\boldsymbol{\Psi}}_0$ is given by $\underset{\sim}{\boldsymbol{\Psi}}_0 \equiv \underset{\dots}{\widehat{\mathbb{A}}}{}_0^1 \underset{\sim}{\mathbf{0}}_0$. This is the initial extended labstate for each of the two networks we now consider.

In the following, it is useful to label the respective transmission coefficient and reflection coefficient associated with each beam splitter separately, in order to see how information is flowing through the networks. We work in extended QDN throughout.

Runs with a Dud Bomb

Stage Σ_0 to Stage Σ_1

By inspection of Figure 25.2 we have

$$\underset{\sim}{\boldsymbol{\Psi}}_1 \equiv \underset{\dots}{\mathbb{U}}_{1,0} \underset{\sim}{\boldsymbol{\Psi}}_0 = (t^1 \underset{\dots}{\widehat{\mathbb{A}}}{}_1^2 + ir^1 \underset{\dots}{\widehat{\mathbb{A}}}{}_1^3) \underset{\sim}{\mathbf{0}}_1. \tag{25.25}$$

Stage Σ_1 to Stage Σ_2

In this scenario, the bomb is a dud, so any signal entering module T from port 2_1 is transmitted directly to 5_2. Taking the mirror M not to change any phase, we then have

$$\underset{\sim}{\boldsymbol{\Psi}}_2 \equiv \underset{\dots}{\mathbb{U}}_{2,1} \underset{\sim}{\boldsymbol{\Psi}}_1 = (t^1 \underset{\dots}{\widehat{\mathbb{A}}}{}_2^5 + ir^1 \underset{\dots}{\widehat{\mathbb{A}}}{}_2^4) \underset{\sim}{\mathbf{0}}_2. \tag{25.26}$$

Stage Σ_2 to Stage Σ_3

Using

$$\underset{\dots}{\mathbb{U}}_{3,2} \underset{\dots}{\widehat{\mathbb{A}}}{}_2^5 \underset{\sim}{\mathbf{0}}_2 = (t^2 \underset{\dots}{\widehat{\mathbb{A}}}{}_3^7 + ir^2 \underset{\dots}{\widehat{\mathbb{A}}}{}_3^6) \underset{\sim}{\mathbf{0}}_3, \tag{25.27}$$

$$\underset{\dots}{\mathbb{U}}_{3,2} \underset{\dots}{\widehat{\mathbb{A}}}{}_2^4 \underset{\sim}{\mathbf{0}}_2 = (t^2 \underset{\dots}{\widehat{\mathbb{A}}}{}_3^6 + ir^2 \underset{\dots}{\widehat{\mathbb{A}}}{}_3^7) \underset{\sim}{\mathbf{0}}_3, \tag{25.28}$$

we end up with the final state outcome probabilities

$$\Pr(6_3|\underset{\sim}{\mathbf{\Psi}}_0) = |t^1 r^2 + r^1 t^2|^2, \quad \Pr(7_3|\underset{\sim}{\mathbf{\Psi}}_0) = |t^1 t^2 - r^1 r^2|^2. \tag{25.29}$$

The significance of this result is that for matched symmetric beam splitters, for which $t^1 = r^1 = t^2 = r^2 = 1/\sqrt{2}$, the prediction is that $\Pr(7_3|\underset{\sim}{\mathbf{\Psi}}_0) = 0$. The conclusion, that for symmetrical beam splitters a dud bomb never coincides with a positive signal in detector 7_3, is critical to the point of this experiment.

Runs with an Active Bomb

In this scenario, the bomb is active and could explode if triggered at module T. We shall take the decommissioning of 5_2 as a marker of that explosion: the observer will be able to see real debris in the laboratory if the bomb explodes.

Everything is the same as in the previous scenario with a dud bomb, up to stage Σ_1.

Stage Σ_1 to Stage Σ_2

In this scenario, the bomb explodes and decommissions the device T if a signal enters from 2_1. Hence we may write

$$\underset{\sim}{\mathbf{\Psi}}_2 \equiv \underset{\cdots}{\mathbb{U}}_{2,1}\underset{\sim}{\mathbf{\Psi}}_1 = (t^1 \underset{\cdots}{\mathbb{D}}{}_2^5 + ir^1 \underset{\cdots}{\widehat{\mathbb{A}}}{}_2^4)\underset{\sim}{\mathbf{0}}_2. \tag{25.30}$$

Stage Σ_2 to Stage Σ_3

The debris created by any bomb exploding at stage Σ_2 will certainly be transmitted on to stage Σ_3, so we have

$$\begin{aligned} \underset{\cdots}{\mathbb{U}}_{3,2}\underset{\cdots}{\mathbb{D}}{}_2^5\underset{\sim}{\mathbf{0}}_2 &= \underset{\cdots}{\mathbb{D}}{}_3^5\underset{\sim}{\mathbf{0}}_3, \\ \underset{\cdots}{\mathbb{U}}_{3,2}\underset{\cdots}{\widehat{\mathbb{A}}}{}_2^4\underset{\sim}{\mathbf{0}}_2 &= (t^2\underset{\cdots}{\widehat{\mathbb{A}}}{}_3^6 + ir^2\underset{\cdots}{\widehat{\mathbb{A}}}{}_3^7)\underset{\sim}{\mathbf{0}}_3, \end{aligned} \tag{25.31}$$

giving the final state

$$\underset{\sim}{\mathbf{\Psi}}_3 = t^1\underset{\cdots}{\mathbb{D}}{}_3^5\underset{\sim}{\mathbf{0}}_3 + ir^1(t^2\underset{\cdots}{\widehat{\mathbb{A}}}{}_3^6 + ir^2\underset{\cdots}{\widehat{\mathbb{A}}}{}_3^7)\underset{\sim}{\mathbf{0}}_3. \tag{25.32}$$

We conclude that in this scenario, the following probabilities hold:

$$\begin{aligned} &\text{probability of an explosion :} && (t^1)^2, \\ &\text{probability of no explosion and signal in } 6_3 : && (r^1t^2)^2, \\ &\text{probability of no explosion and signal in } 7_3 : && (r^1r^2)^2. \end{aligned} \tag{25.33}$$

In the symmetric case, we conclude that the probability of no explosion of an active bomb *and* a signal in detector 7_3 is $\frac{1}{4}$.

It is this outcome that allows the observer to find an unexploded, active bomb, because detector 7_3 does not trigger in the symmetric beam splitter scenario when the bomb in question is a dud.

Random Testing

Unfortunately, the observer does not know before each run whether a particular bomb is active or a dud. Consider a sequence of runs such that there is a (classical) probability ω^A of encountering an active bomb and a probability $\omega^D \equiv 1 - \omega^A$ of a dud. In this case we take a density matrix approach. At the nth stage, for $n = 0, 1, 2, 3$, we define the extended density matrix

$$\underset{\dots}{\rho}_n \equiv \omega^A \underset{\sim}{\Psi}_n^A \widetilde{\Psi}_n^A + \omega^D \underset{\sim}{\Psi}_n^D \widetilde{\Psi}_n^D. \tag{25.34}$$

Here $\underset{\sim}{\Psi}_n^A$ and $\underset{\sim}{\Psi}_n^D$ are the extended labstates for the active and dud scenarios, respectively, at stage Σ_n. It could be thought that this must be incorrect at stage Σ_0, because the observer has no information as to which scenario is in place. But in fact, that information could be established in principle *before* stage Σ_1 by, for example, triggering the bomb directly. It would explode in the active case and that would inform the observer that the initial state should have been $\underset{\sim}{\Psi}_n^A$ in the now aborted run. A similar remark holds for the dud case.

The overall probability of triggering detector 7_3 is given by

$$\Pr(7_3) = Tr\left\{ \underset{\dots}{\rho}_3 \underset{\dots}{\widehat{\mathbb{P}}}_3^7 \right\}, \tag{25.35}$$

where $\underset{\dots}{\widehat{\mathbb{P}}}_3^7$ is the signal projection operator for detector 7_3. We find

$$\Pr(7_3) = \omega^D (t^1 t^2 - r^1 r^2)^2 + \omega^A (r^1 r^2)^2. \tag{25.36}$$

In the symmetric case, we conclude that whenever detector 7_3 registers a signal, the bomb is active and can be stockpiled accordingly. A single sweep of the original stockpile of active and dud bombs will identify one-quarter of the active bombs without exploding them (the rest go up in smoke). Moreover, whenever detector 6_3 has triggered, there has been no explosion, but the observer cannot be sure which scenario has occurred. Therefore, that bomb can be retested.

In this way, the observer should be able to find up to one-third of the active bombs, the rest having exploded.

This thought experiment has been the motivation for real experiments that simulate in some way the action of module T in the above experiment, a class of experiment known as *interaction-free measurement* (Kwiat et al., 1994).

25.12 The Hardy Paradox Experiment

The Elitzur–Vaidman bomb-tester experiment discussed above may be interpreted as a simplified form of double-slit experiment, where the screen has only two sites and one of the slits can be blocked off or not, depending on whether a bomb is active or dud. This blocking off occurs in a classical way, because the uncertainty as to whether the bomb is active or dud is not intrinsic to the nature of the bomb but reflects the observer's ignorance of the nature of the bomb.

A variant of the bomb-tester experiment is known as the Hardy paradox experiment (Hardy, 1992). In this variant, the blocking-off of a slit occurs in an intrinsically random way because it is governed by quantum processes, contrasted

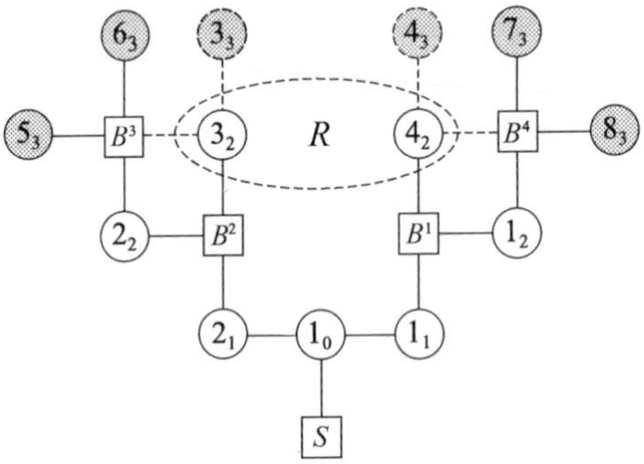

Figure 25.3. The Hardy paradox experiment.

to the bomb-tester experiment analyzed above, where the observer's ignorance about the nature of the bomb plays a role.

The Hardy paradox experiment shown in Figure 25.3 consists of an electron–positron pair 1_1, 2_1 produced by a source S. These particles are then deflected by magnets (not shown) to pass through the equivalent of two coupled Mach–Zehnder-type networks, as shown in Figure 25.3. Output ports 3_2 and 4_2 of beam splitters B^2 and B^1, respectively, are arranged to intersect in a confined region, R, where electron–positron annihilation may occur.

There are two possibilities if both particles are in region R at the same stage: either (1) they annihilate into a photon pair, or (2) they pass onto their respective beam splitters B^3 and B^4 as shown.

Case (1)

In case (1) and in conventional terminology, the annihilation effectively blocks off one of the entry ports in each of those beam splitters. This is equivalent to a virtual module in interaction region R, which would otherwise be a true information void. In this case, the photons produced by the annihilation are detected in 3_3 and 4_3.

Case (2)

In case (2), the virtual detectors 3_2 and 4_2 transmit their amplitudes onto beam splitters B^3 and B^4, respectively, as shown in Figure 25.3. Interference occurs in each of these beam splitters. In the symmetric case, a signal will be detected at 6_3 and not in 5_4, and a signal will be detected in 7_3 and not in 8_3.

The Hardy paradox experiment is intrinsically a pure quantum experiment and we can discuss it via pure labstates alone. An important point is that electron–positron annihilation is a well-known quantum process that occurs in nature, whereas the detonation mechanism of the Elitzur–Vaidman bomb-tester is left unspecified.

As with the bomb-tester experiment above, we start our analysis of the Hardy paradox experiment by first defining the contextual vacuum $\underset{\sim}{\mathbf{0}}_0 \equiv \underset{\cdots}{\mathbb{C}}{}^1_n \underset{\cdots}{\mathbb{C}}{}^2_n \cdots \underset{\cdots}{\mathbb{C}}{}^8_n \underset{\sim}{\emptyset}_0$ and then the initial state is given by $\underset{\sim}{\boldsymbol{\Psi}}_0 \equiv \underset{\cdots}{\widehat{\mathbb{A}}}{}^1_0 \underset{\sim}{\mathbf{0}}_0$.

The dynamics then goes as follows.

Stage $\boldsymbol{\Sigma_0}$ to Stage $\boldsymbol{\Sigma_1}$

The electron–positron pair produced by source S is split by magnetic fields ready to be deflected into beam splitters B^1 and B^2 as shown, giving extended labstate $\underset{\sim}{\boldsymbol{\Psi}}_1 \equiv \underset{\cdots}{\widehat{\mathbb{A}}}{}^1_1 \underset{\cdots}{\widehat{\mathbb{A}}}{}^2_1 \underset{\sim}{\mathbf{0}}_1$.

Stage $\boldsymbol{\Sigma_1}$ to Stage $\boldsymbol{\Sigma_2}$

$$\underset{\sim}{\boldsymbol{\Psi}}_2 \equiv \underset{\cdots}{\mathbb{U}}_{2,1} \underset{\sim}{\boldsymbol{\Psi}}_1 = (t^1 \underset{\cdots}{\widehat{\mathbb{A}}}{}^1_2 + i r^1 \underset{\cdots}{\widehat{\mathbb{A}}}{}^4_2)(t^2 \underset{\cdots}{\widehat{\mathbb{A}}}{}^2_2 + i r^2 \underset{\cdots}{\widehat{\mathbb{A}}}{}^3_2) \underset{\sim}{\mathbf{0}}_2. \qquad (25.37)$$

Stage $\boldsymbol{\Sigma_2}$ to Stage $\boldsymbol{\Sigma_3}$

The only novel factor concerns potential electron–positron annihilation, involving 3_2 and 4_2. We will explore the possibility of partial annihilation, by taking the evolution for this term to be given by

$$\underset{\cdots}{\mathbb{U}}_{3,2} \underset{\cdots}{\widehat{\mathbb{A}}}{}^4_2 \underset{\cdots}{\widehat{\mathbb{A}}}{}^3_2 \underset{\sim}{\mathbf{0}}_2 = \alpha \underset{\cdots}{\mathbb{D}}{}^4_3 \underset{\cdots}{\mathbb{D}}{}^3_3 \underset{\sim}{\mathbf{0}}_3 + \beta(t^4 \underset{\cdots}{\widehat{\mathbb{A}}}{}^8_3 + i r^4 \underset{\cdots}{\widehat{\mathbb{A}}}{}^7_3)(t^3 \underset{\cdots}{\widehat{\mathbb{A}}}{}^5_3 + i r^3 \underset{\cdots}{\widehat{\mathbb{A}}}{}^6_3) \underset{\sim}{\mathbf{0}}_3, \quad (25.38)$$

where $|\alpha|^2 + |\beta|^2 = 1$. These coefficients reflect the fact that electron–positron annihilation is a quantum process and does not always occur. Note the introduction of the destruction operators here, signifying that photons have been produced (these can be regarded as the debris from the annihilation of the virtual electron–positron detectors at 3_3 and 4_3).

The relevant information when fed into program MAIN gives the following correlations for the symmetric beam splitter case:

$$\begin{aligned} &\Pr(3_3, 4_3|1_0) = \tfrac{1}{4}|\alpha|^2, \\ &\Pr(5_3, 7_3|1_0) = \Pr(5_3, 8_3|1_0) = \Pr(6_3, 8_3|1_0) = \tfrac{1}{16}|1-\beta|^2, \\ &\Pr(6_3, 7_3|1_0) = \tfrac{1}{16}|3+\beta|^2. \end{aligned} \qquad (25.39)$$

When β is set to one, corresponding to no annihilation in region R, the prediction is that the only outcome is that both 6_3 and 7_3 show a signal. However, when annihilation is permitted, $\alpha \neq 0$ and the outcomes show characteristics of interaction-free measurement. This is deduced from conservation of electric charge. The argument goes as follows. If a signal is detected at, say, 5_3, then annihilation cannot have occurred, because signality is conserved in this experiment; annihilation would give a signal at 3_3 and at 4_3, and so a signal elsewhere would be ruled out. But a signal in 5_3 arises only if something is affecting the signal from 3_2, and the only factor that could affect that could be 4_2. Therefore, any signal in 5_3 is an "interaction-free" detection of the particle at 4_2. A similar remark holds for any signal observed at 8_3.

25.13 Implications and Comments

The application of our extended QDN formalism to the Elitzur–Vaidman bomb-tester and Hardy paradox experiments demonstrates that the concept of faulty or decommissioned states has physical significance.

An important point about this chapter is that the phenomena discussed show that specific reductionist details of interactions seem not to be relevant to the overall picture. For instance, in the bomb-tester experiment we have not said anything at all about what sort of module T really is. Likewise, we have not done any specific quantum electrodynamical calculation in the Hardy paradox. What seems important is the architecture of the apparatus involved, and that is an emergent aspect of these experiments.

Such considerations are what have led us to the conclusion that quantum mechanics is really a theory of contextuality. That the rules are unfamiliar and counterintuitive in many cases is a commentary not on reality, but on the inadequacy and limitations of our classical conditioning.

26
Conclusions

In the Preface to this book, we started with Alice and Bob having very different views of their observations. Alice used an optical telescope and reported nothing unusual about a distant galaxy that she could see. Bob, on the other hand, detected intense radio activity in that galaxy. Our question was: *who has the true view of that galaxy?*

We advised the reader that the answer is not *Alice*. Neither is it *Bob*. The answer is not *both of them*, nor is it *neither of them*. It is not a trick question either. So what is the answer that this book would supply?

Our answer will come presently.

While this book has set forth a definite mathematical perspective on empirical physics, it has contained quite a lot of commentary that might be disparaged as metaphysics or philosophy. Such commentary is generally frowned on in science, because it has no empirical content. It is vacuous.

In our defense, we point out the obvious: science is not a robotic activity; it is carried out by humans and these *are* driven by their metaphysical, philosophical, and emotional imperatives. For instance, the hard-core scientific view that quantum theory needs no interpretation but only application is itself a philosophy. It is no more than a conditioned response, based on opinion and subscription to current scientific norms. It is, indeed, a philosophy of how to do quantum mechanics. So we all do it, in one way or another. Our concern in this book is that the way that we do it should be based soundly on scientific principles. The quantized detector approach that we describe and use in this book is our attempt to do just that.

There is an excellent paper on all of this that has the provocative title "Quantum Theory Needs No Interpretation," by Fuchs and Peres (2000). It seems on the face of it to dismiss any sort of "interpretation" of quantum mechanics. In fact, close reading of it confirms (to us at least) the agenda that we have set out in this book. Yes, quantum mechanics needs no interpretation, *when it is being applied to processes that take place in the information void.*

But in our view, that is only half the story. We cannot exclude the observer and what they are doing. The information void is contextual: it is defined by whatever observer is involved, and by the apparatus that they are using. Schwinger said just that (see his quote on this in Chapter 24). It has long seemed obvious to us that the idea expressed by Schwinger really points to an extension of what physics means. Physics cannot be just the study of systems under observation (SUOs). The relationship of observers to those SUOs and to the Universe in which both observers and SUOs are embedded should surely be just as important an issue in physics as the perceived properties of SUOs.

That naturally leads to the realization that the physics of *emergence* should be at least as important in physics as the reductionist agenda is currently. Our explanation as to why it is not so regarded currently is that it is *hard*, much harder than standard reductionist approaches to physics, which absolutely rely on and utilize emergent concepts, but place them firmly outside the scope of investigation.

Fortunately, human opinions on what is significant carry no weight in real science: nature has a tendency to tell us when we are being complacent. Markers of this are evident throughout quantum mechanics: the randomness of quantum outcomes, the nonlocality of quantum correlations, the violations of Bell and Leggett–Garg inequalities, quantum interference, state reduction, the Kochen–Specker theorem (Kochen and Specker, 1967), entanglement, and more.

That brings us back to Alice and Bob. It was not a trick question: it was not a proper question in the first place, because it is contextually incomplete.

As given, there was no indication for whom this question could have a truth value. If you say that it is you, the reader, then our response is: what sort of "truth" is it that deals with an invented scenario? Alice and Bob don't exist. They were invented in the Preface for the purposes of demonstrating a fundamental point.

It may seem a glib, cheap point, but it is not. It points to the fact that "truth" does not exist in a vacuum. It requires an observer for whom that "truth" is empirically meaningful.

Contextual incompleteness is invariably the main ingredient in a number of theoretical disciplines that purport to "explain" reality, but in fact are as fictional in their scientific content as our image of Alice and Bob is. That is the essential point behind the article of Fuchs and Peres, mentioned above. If we could, we would retitle their paper: *quantum theory needs no interpretation, it needs to be extended to cover the observer.* That is what this book is all about.

A final comment about Alice and Bob. Our answer is that each of them is stating the truth, but only relative to their individual contexts. In that sense they are both making contextually true statements. But that is not the same as saying they both have the "true" view of the galaxy they are observing: there is no "true" view of that galaxy in an absolute sense. Reality is far too deep for that.

Appendix

A.1 QDN Notation

Symbol	Meaning
$\times$	Cartesian product, direct product
$\otimes$	tensor product
Σ_n	stage n
$\mathcal{A}_n$	apparatus at stage Σ_n
r_n	number of real or virtual detectors in $\mathcal{A}_n$ = the rank of $\mathcal{A}_n$
i_n	the ith real or virtual detector in $\mathcal{A}_n$
Q_n^i	the qubit representing i_n, $i = 1, 2, \ldots, r_n$
$\mathcal{Q}_n \equiv Q_n^1 Q_n^2 \ldots Q_n^{r_n}$	the quantum register at stage Σ_n : a rank-r_n tensor product
I_n	the observer's information about $\mathcal{Q}_n$
$H_n \equiv (\mathcal{Q}_n, I_n)$	Heisenberg net at stage Σ_n
2^{r_n}	dimension of $\mathcal{Q}_n \equiv \dim \mathcal{Q}_n$
B_n^i	the preferred basis for $\mathcal{Q}_n^i, i = 1, 2, \ldots, r$
$\mathcal{R}_n^{r_n} \equiv B_n^1 B_n^2 \ldots B_n^{r_n}$	the preferred basis for $\mathcal{Q}_n$: a Cartesian product
$\boldsymbol{k}_n$	kth element of computational basis B_n, $k = 0, 1, 2, \ldots, 2^{r_n} - 1$
$\overline{\boldsymbol{k}_n}$	dual of $\boldsymbol{k}_n$
$\boldsymbol{\Psi}_n$	pure labstate at stage Σ_n: an element of $\mathcal{Q}_n$
$\mathbb{A}_n^i$	ith signal destruction operator at stage Σ_n
$\widehat{\mathbb{A}}_n^i$	ith signal creation operator at stage Σ_n
$\mathbb{P}_n^i \equiv \mathbb{A}_n^i \widehat{\mathbb{A}}_n^i$,	ith no-signal projection operator
$\widehat{\mathbb{P}}_n^i \equiv \widehat{\mathbb{A}}_n^i \mathbb{A}_n^i$	ith signal projection operator
$S_n^i \equiv \{\mathbb{P}_n^i, \widehat{\mathbb{P}}_n^i, \mathbb{A}_n^i, \widehat{\mathbb{A}}_n^i\}$	ith signal set
$\mathbb{T}_{mn}^{ij} \equiv \boldsymbol{i}_m \overline{\boldsymbol{j}}_n$	transition operator

A.2 Lab Time and Frame Fields

In general relativity (GR), spacetime is modeled as a four-dimensional manifold with a Lorentz signature metric. In general, GR spacetimes cannot be covered by a single coordinate patch, particularly if there are closed time-like curves (CTCs) as in the case of the Gödel metric (Gödel, 1949). In spacetimes with CTCs, a global temporal foliation cannot be constructed, which means that a global laboratory perspective cannot be contemplated in such cases.

The principles of quantized detector networks (QDN) are well suited to deal with such issues. First, QDN is an endophysical approach to empirical physics, meaning that it does not attempt a global (exophysical) description of the Universe. In addition, finiteness is the order of the day, which means that actual physics laboratories are regarded as of finite extent and duration, and that no infinities are measurable. QDN does not normally attempt to discuss systems under observation (SUOs) in terms of an infinite number of real or virtual detectors. The QDN discussion of the bosonic and fermionic oscillators in Chapter 24 are given to illustrate the remarkable theoretical properties of infinite-rank quantum registers. In applications to real SUOs, QDN invariably involves finite-rank quantum registers.

In GR, a relatively localized laboratory description typically involves a complex of four *frame fields*, $\{e_\mu : \mu = 0, 1, 2, 3\}$ constituting the laboratory frame, or coordinate patch adapted to cover a real physical laboratory. These basis vectors are usually chosen to satisfy the orthonormality relations

$$g(e_\mu, e_\nu) = \eta_{\mu\nu}, \tag{A.1}$$

where g is the metric tensor and the $\eta_{\mu\nu}$ are the components of a 4×4 matrix displaying the Lorentz signature of a Minkowski metric:

$$[\eta_{\mu\nu}] \equiv \begin{bmatrix} 1 & 0 & 0 & 0 \\ 0 & -1 & 0 & 0 \\ 0 & 0 & -1 & 0 \\ 0 & 0 & 0 & -1 \end{bmatrix}. \tag{A.2}$$

The significant frame field is e_0, which is time-like and indicates the temporal foliation that dictates clock time over the laboratory. The three other frame fields are space-like and lie in the space-like hypersurfaces of relative simultaneity that the observer has set up in their laboratory using some chosen synchronization protocol.

QDN requires such a framework in order for quantum principles to be applicable: space and time have different roles in QM as it is encountered in the laboratory. The concept of *stage* is intimately linked to the existence of such a framework.

A.3 Lab Time and Stages

Consider an experiment of the Stern–Gerlach type, wherein a beam of particles is passed through a beam splitter feeding onto two detectors denoted A and B.

Suppose each run consists of a beam prepared at labtime reset to $t = 0$ with the observer looking at each detector separately, at labtime T_A in the case of detector A, and labtime T_B in the case of detector B. We are interested here in the possibility that T_B is very much greater than T_A. For instance, suppose T_A is of the order of a millionth of a second and T_B is a million years. Then it is an empirical fact that the two times are not simultaneous relative to the laboratory concerned, and then the question arises as to whether the observation at A could causally impact on the observation at B.

In the framework of special relativity, the answer is determined by the lightcone structure of Minkowski spacetime. If B is outside the forward lightcone centered on A, then Einstein causality tells us that A could not influence B. In that situation, an inertial frame could always be found in which these two events were simultaneous, relative to that frame. Since probabilities are related to outcome frequencies, which is a counting process of signals, we would expect the same outcomes in such a frame as in the original frame where T_A and T_B were vastly different.

We expect the same results would hold if B was inside the forward lightcone of A but adequate shielding was in place. In modern electronics, the problem of undesirable signal interference due to one detector affecting another is known as *crosstalk*. *Shielding* is our term for the elimination of crosstalk. It is a problem but one that in principle can be overcome. Indeed, it could be argued that the very existence of physically distinct persistent SUOs co-existing at the same time is evidence for shielding. The fact that atoms are generally stable and can be regarded as distinct is direct evidence for that.

This line of reasoning leads to the interesting idea that *space* itself is a manifestation of shielding. Consider two hydrogen atoms. According to standard nonrelativistic QM, the combined SUO consists of two identical protons and two identical electrons. If the atoms were separated by, say, 2 angstroms, we might be tempted to describe their combined wave function in terms of a two-proton, two-electron wave function, properly antisymmetrized on account of the indistinguishability and fermionic nature of the constituents. On the other hand, if the two atoms were, say, a light year apart, no one would ever think of these atoms as anything other than two separate SUOs, each described by a one-proton, one-electron wave function (in standard nonrelativistic QM).

This raises the interesting question: which comes first, space or shielding? According to Schwinger, quoted in Chapter 24, the space-time is an idealization of empirical context, so the implication is that the emergent concept of shielding comes before we can define the reductionist concept of space.

The above arguments tells us that labtime simultaneity or strict lightcone causality is not essential in signal detection; the important criterion is that different signal detectors should not interfere causally with, or have the possibility of interfering causally with, other signal detectors in the experiment. It is this condition, referred to as shielding, that defines our concept of *stage*. Signal detectors that are looked at in such a way constitute a single *stage* in an experiment. A stage is an collection of detectors that are looked at by an observer in such a way that no subset of those detectors can causally influence any other subset.

A stage is therefore a contextual classification of detectors that collectively plays the same role in QDN as events on a space-like hyperplane of "simultaneity" in some time-like foliation in general relativity.

The concept of stage undermines the contextually incomplete Block Universe concept: stages require the existence of observers, and the experimental protocol (context) to be given explicitly, something that the Block Universe has no place for. The "Consistent Histories" approach to QM is an interesting development of QM in that it deals with empirical propositions at multiple different observation times (Griffiths, 1984), and these could be regarded as stages.

A.4 Ensembles

The ensemble concept arises in physics for two reasons. First, no experiment normally validates a generalized proposition in a single run, even in classical physics, which deals with generalized propositions with a generalized proposition classification (GPC) of 2. Such propositions are based on the classical assertion that SUOs have precise qualities that can be quantified by exact measurement. Regardless of their view of that assertion, experimentalists will know that, in reality, experimental errors and inaccuracies are always present. To overcome this, multiple runs of the basic experimental protocol are generally performed and then averages and other statistical quantities established from the accumulated data. These multiple runs constitute ensembles of one kind or another.

The second reason for the use of ensembles is that quantum physics asserts that the outcome of any given run in an experiment is a random variable, so that statistical analysis based on the Born rule is required as a matter of principle and policy.

Ensembles come in several varieties, each with its particular spatiotemporal architecture, and each characterized by generally unstated, implicit context. Whichever kind of ensemble is chosen depends on several factors usually outside the experimentalist's control, such as limited resources and time. The following are two important kinds of ensemble.

Spatial Ensembles

A spatial ensemble is a physical collection in a given laboratory of multiple, mutually isolated (from each other) copies of a given SUO, such as atoms in a crystal, such that a basic run is performed on each copy once. The statistics for the experiment is then established by collecting the outcome data from each copy and assuming that outcome frequencies can be attributed to probabilities associated with the original SUO.

In experiments based on relatively localized laboratories, wherein physical conditions are relatively homogeneous, this ensemble concept is usually reasonable. However, three problems may arise. First, different copies may actually interact with each other. This is the case in magnetic resonance experiments, for instance, where an interaction between nuclear spins and their neighboring spin environment is all the point. The second problem is that inhomogeneities in the laboratory environment may invalidate the above logic. For instance, a spatial ensemble carrying out an Unruh-type experiment in an accelerating laboratory (in contrast to a freely falling one) will almost certainly display laboratory inhomogeneities in what looks like a local gravitational field. The third problem is one of economics. Some experiments cannot be based on the idealized spatial ensemble concept simply because each individual run may be too costly or too big in spatial terms to duplicate in any laboratory. An example is the Large Hadron Collider: there is only one particle accelerator and it is very big and very expensive.

In the case of the Large Hadron Collider, there is a modification of the spatial ensemble concept that works excellently: multiple copies of the same SUO (protons) are contained in a single circulating beam. Assumptions are then made that during the very brief time of interaction involved, each proton behaves as if it was isolated from the other protons in that beam and would interact with only one other proton in the opposing beam. The beam statistics of the Large Hadron Collider, are impressive: in a given run, each beam consists of 2,808 bunches of protons, and each bunch contains about 10^{11} protons.

Temporal Ensembles

Some experiments are too costly to perform via spatial ensembles, so the standard alternative is to use a temporal ensemble. In such an architecture, multiple runs of the same basic protocol are implemented in temporal succession using the same apparatus each time. In principle, an ideal temporal ensemble should be equivalent to an ideal spatial ensemble, but that is an assertion that can be challenged on the basis of cosmological evidence. It is now believed that the universe is expanding in an irreversible way, relative to all endophysical observers. Therefore, the environment around any laboratory is not quite the same during any given run of an experiment as any other run. Of course, such discrepancies are minute and could be laughed at as generating a pointless debate, were it not for one glaring fact: the expansion *does* have observable effects, namely, the red shift of light from distant galaxies.

The issue here is the relative scale of times: a comparison of the typical time τ to complete a given run, the interval T to the next run, and the age A of the Universe estimated from the observed Hubble constant. When τ and T are both negligible compared with A, as is almost always the case, then temporal ensembles should be as good as spatial ensembles.

The impact of cosmological expansion should not be too lightly dismissed. It has been speculated by scientists such as Dirac that the gravitational constant G and perhaps even the speed of light c may change over cosmological time scales (Dirac, 1938a). Therefore, any discussion of observed physical properties such as electron mass and other properties should take such issues into account if the context merits it.

We note that Peres did not consider temporal ensembles to be proper ensembles in QM (Peres, 1995), but the fact is that many experiments are indeed carried out via such ensembles.

A.5 Vector Spaces

A **vector space** $(V, \mathbb{F})$ over a field $\mathbb{F}$ such as the real numbers $\mathbb{R}$ or complex numbers $\mathbb{C}$ is a set V of elements $V \equiv \{\boldsymbol{a}, \boldsymbol{b}, \ldots\}$, known as vectors, with the following properties:

(i) There is a binary map $\dotplus : V \times V \to V$ such that, for any elements $\boldsymbol{a}, \boldsymbol{b} \in V$, the object $\boldsymbol{a} \dotplus \boldsymbol{b} \in V$. This is called *addition of vectors*, or just *vector addition*. The elements of V are called *vectors*.

Vector addition is commutative, i.e.,

$$\boldsymbol{a} \dotplus \boldsymbol{b} = \boldsymbol{b} \dotplus \boldsymbol{a}, \qquad \forall \boldsymbol{a}, \boldsymbol{b} \in V. \tag{A.3}$$

Vector addition is associative, i.e.,

$$\boldsymbol{a} \dotplus (\boldsymbol{b} \dotplus \boldsymbol{c}) = (\boldsymbol{a} \dotplus \boldsymbol{b}) \dotplus \boldsymbol{c}, \qquad \forall \boldsymbol{a}, \boldsymbol{b}, \boldsymbol{c} \in V. \tag{A.4}$$

(ii) There is a unique element in V, known as the *zero vector*, denoted by $\boldsymbol{0}_V$, such that

$$\boldsymbol{a} \dotplus \boldsymbol{0}_V = \boldsymbol{a}, \qquad \forall a \in V. \tag{A.5}$$

(iii) For every vector $\boldsymbol{a}$, there exists an *additive inverse*, denoted by $-\boldsymbol{a}$, such that

$$\boldsymbol{a} \dotplus (-\boldsymbol{a}) = \boldsymbol{0}. \tag{A.6}$$

These properties mean that V is an abelian group under vector addition.

(iv) For any $\boldsymbol{a} \in V$, $\lambda \in \mathbb{F}$, then the object $\lambda \boldsymbol{a} \in V$. This is known as *multiplication by a scalar*, or just *scalar multiplication*. In this context, the elements of $\mathbb{F}$ are called *scalars*.

Scalar multiplication satisfies the property

$$\lambda (\mu \boldsymbol{a}) = (\lambda \mu) \boldsymbol{a}, \qquad \lambda, \mu \in \mathbb{F}, \boldsymbol{a} \in V. \tag{A.7}$$

(v) Scalar multiplication is *distributive*, i.e.,

$$\begin{aligned} \lambda (\boldsymbol{a} \dotplus \boldsymbol{b}) &= (\lambda \boldsymbol{a}) \dotplus (\lambda \boldsymbol{b}), \\ (\lambda + \mu) \boldsymbol{a} &= (\lambda \boldsymbol{a}) \dotplus (\mu \boldsymbol{a}) \end{aligned} \tag{A.8}$$

It is standard practice to use V to mean $(V, \mathbb{F})$. The *ground field* $\mathbb{F}$ is generally understood, but it is important to know whether it is $\mathbb{R}$ or $\mathbb{C}$. In the former case we say V is a *real vector space*, while in the latter case we say V is a *complex vector space*. The space of *three-vectors* used to represent position in physical space is a real vector space, while the Hilbert space of quantum state vectors is a complex vector space.

It is easy to show that, $\forall\lambda \in \mathbb{F}$, $\lambda \mathbf{0}_V = \mathbf{0}_V$. Likewise, if $0_{\mathbb{F}}$ is the zero element in $\mathbb{F}$, then $0_{\mathbb{F}}\boldsymbol{a} = \mathbf{0}_V$, $\forall \boldsymbol{a} \in V$. It is important not to confuse the scalar zero $0_{\mathbb{F}}$ and the vector zero $\mathbf{0}_V$.

In practice, we do not bother to use a different symbol for vector addition, $\dot{+}$, in order to distinguish addition in the field, $+$. Henceforth, the same symbol, $+$, will be used for both.

We often modify the notation involving the additive inverse $-\boldsymbol{b}$, writing

$$\boldsymbol{a} + (-\boldsymbol{b}) = \boldsymbol{a} - \boldsymbol{b}, \tag{A.9}$$

thereby suggesting a new binary process known as *subtraction*. This is not necessary, but it is useful and should always be interpreted in terms of the addition of vectors.

What is immensely astounding is that the above theory, which may appear no more than a mathematician's game, seems necessary to describe empirically validated quantum physics, with the additional surprise that the field $\mathbb{F}$ is required to be $\mathbb{C}$ and not $\mathbb{R}$.

Subspaces of a Vector Space

Suppose U is a subset of a vector space V over some field $\mathbb{F}$. If U is a vector space over $\mathbb{F}$, using the same rules for vector addition and scalar multiplication as for V, then we say U is a *subspace* of V. Every subspace of V necessarily contains the zero vector $\mathbf{0}_V$ of V.

Given two subspaces U_1, U_2 of V, then the *intersection* $U_1 \cap U_2$ is the set of elements common to U_1 and U_2, and is also a subspace of V.

Spanning Sets

Suppose $S \equiv \{\boldsymbol{v}_1, \boldsymbol{v}_2, \ldots, \boldsymbol{v}_k\}$ is a set of vectors in some vector space V with ground field $\mathbb{F}$. Let M be the set of all vectors of the form $x^1\boldsymbol{v}_1 + x^2\boldsymbol{v}_2 + \cdots + x^k\boldsymbol{v}_k$, where $x^1, x^2, \ldots, x^k \in \mathbb{F}$. Then M is a subspace of V, *spanned* by S. We write

$$M \equiv [\boldsymbol{v}_1, \boldsymbol{v}_2, \ldots, \boldsymbol{v}_k]. \tag{A.10}$$

S is called a *spanning set* for M.

Linear Independence

An expression of the form $x^1\boldsymbol{v}_1 + x^2\boldsymbol{v}_2 + \cdots + x^k\boldsymbol{v}_k$, where $\boldsymbol{v}_1, \boldsymbol{v}_2, \ldots, \boldsymbol{v}_k \in V$ and x^1, x^2, $\ldots$, $x^k \in \mathbb{F}$ is called a *linear combination* of the vectors $\boldsymbol{v}_1, \boldsymbol{v}_2, \ldots, \boldsymbol{v}_k$.

If the x^i are not all zero, then it is called a *nontrivial* linear combination. Otherwise it is called *trivial.*

A set of vectors $\{\boldsymbol{v}_1, \boldsymbol{v}_2, \ldots, \boldsymbol{v}_k\}$ is *linearly dependent* if there exists a nontrivial linear combination equal to the zero vector $\mathbf{0}_V$. In other words, $\{\boldsymbol{v}_1, \boldsymbol{v}_2, \ldots, \boldsymbol{v}_k\}$ is linearly dependent if the equation

$$x^1\boldsymbol{v}_1 + x^2\boldsymbol{v}_2 + \cdots + x^k\boldsymbol{v}_k = \mathbf{0}_V \tag{A.11}$$

has a solution for which at least one of the x^i is nonzero.

A set of vectors $\{\boldsymbol{v}_1, \boldsymbol{v}_2, \ldots, \boldsymbol{v}_k\}$ is *linearly independent* if the only solution to Eq. (A.11) is $x^1 = x^2 = \cdots = x^k = 0_{\mathbb{F}}$.

Theorem A.1 *The nonzero vectors $\boldsymbol{v}_1, \boldsymbol{v}_2, \ldots, \boldsymbol{v}_n \in V$ are linearly dependent if and only if one of the vectors $\boldsymbol{v}_k$ is a linear combination of the preceding ones $\boldsymbol{v}_1, \boldsymbol{v}_2, \ldots, \boldsymbol{v}_{k-1}$.*

A single nonzero vector $\boldsymbol{v}$ is necessarily independent, since the $x\boldsymbol{v} = \mathbf{0}_V$ if and only if $x = 0_{\mathbb{F}}$.

$$\boxed{\text{A linearly independent spanning set is called a } \textit{basis.}} \tag{A.12}$$

Theorem A.2 *Any vector space that has a finite spanning set contains a basis.*

A vector space is *finite dimensional* if it has a finite basis (i.e., one consisting of a finite number of vectors). Hence, every vector space spanned by a finite spanning set is finite dimensional.

Theorem A.3 *If V is a finite-dimensional vector space with basis $\boldsymbol{e}_1, \boldsymbol{e}_2, \ldots, \boldsymbol{e}_n$ then every vector $\boldsymbol{v}$ in V can be expressed in one and only one way as a linear combination*

$$\boldsymbol{v} = x^1\boldsymbol{e}_1 + x^2\boldsymbol{e}_2 + \cdots + x^n\boldsymbol{e}_n = x^i\boldsymbol{e}_i \tag{A.13}$$

using the summation convention.

All bases of a finite-dimensional vector space have the same number of elements. The *dimension* $\dim V$ of a finite-dimensional vector space V is the number of elements of a basis.

Linear Transformations

Let U and V be two vector spaces, not necessarily of the same dimension, over the same field $\mathbb{F}$. A *linear transformation* (or *linear mapping*) T of U into V is a mapping that assigns to every $\boldsymbol{u} \in U$ a unique vector $T(\boldsymbol{u}) \in V$, such that

$$\begin{aligned} T(\boldsymbol{u}_1 + \boldsymbol{u}_2) &= T(\boldsymbol{u}_1) + T(\boldsymbol{u}_2), \qquad \forall \boldsymbol{u}_1, \boldsymbol{u}_2 \in U, \\ T(\lambda\boldsymbol{u}) &= \lambda T(\boldsymbol{u}), \qquad \forall \boldsymbol{u} \in U, \quad \forall \lambda \in \mathbb{F}. \end{aligned} \tag{A.14}$$

$T(\boldsymbol{u})$ is the *image* of $\boldsymbol{u}$ under T.

The set of all linear mappings of U into V is denoted $L(U, V)$.

Given a linear transformation $T(U,V)$, the set of all vectors $\boldsymbol{u} \in U$ such that

$$T\boldsymbol{u} = \mathbf{0}_V \tag{A.15}$$

is called the *kernel of* T and written $\ker T$.

The set of all vectors $T(\boldsymbol{u})$, $\boldsymbol{u} \in U$ is called the *image* of U under T, and is denoted by $T(U)$.

The following is a critical theorem in QDN.

Theorem A.4 *(Tropper, 1969) If $T \in L(U,V)$, then $\ker T$ is a subspace of U and $T(U)$ is a subspace of V, such that*

$$\dim \ker T + \dim T(U) = \dim U. \tag{A.16}$$

If $\dim \ker T$ is called the *nullity* of T and $\dim T(U)$ is called the *rank* of T, then the above theorem can be stated as

$$\textit{nullity} \text{ of } T + \textit{rank} \text{ of } T = \dim U. \tag{A.17}$$

Linear Functionals

A *linear functional* $\tilde{f}$ is a linear mapping of a vector space V into its ground field $\mathbb{F}$, such that

$$\tilde{f}(\alpha\boldsymbol{u} + \beta\boldsymbol{v}) = \alpha\tilde{f}(\boldsymbol{u}) + \beta\tilde{f}(\boldsymbol{v}) \in \mathbb{F}, \quad \forall\alpha,\beta \in \mathbb{F}, \quad \forall\boldsymbol{u},\boldsymbol{v} \in V. \tag{A.18}$$

Note that summation on the left-hand side is in V, while summation on the right-hand side is in $\mathbb{F}$. Denote the set of all linear functionals over V by $L(V,\mathbb{F})$.

When $\mathbb{F} = \mathbb{R}$, then $\tilde{f}$ is a *real-valued linear functional,* whereas if $\mathbb{F} = \mathbb{C}$ then $\tilde{f}$ is a *complex-valued linear functional.*

One-Forms

Given any two linear functionals $\tilde{f}$, $\tilde{g} \in L(V,\mathbb{F})$, define the linear combination $\alpha\tilde{f} + \beta\tilde{g}$, where $\alpha,\beta \in \mathbb{F}$ by the rule

$$\left(\alpha\tilde{f} + \beta\tilde{g}\right)(\boldsymbol{v}) \equiv \alpha\tilde{f}(\boldsymbol{v}) + \beta\tilde{g}(\boldsymbol{v}), \quad \alpha,\beta \in \mathbb{F}, \quad \boldsymbol{v} \in V. \tag{A.19}$$

With this rule, $L(V,\mathbb{F})$ is itself a vector space, known as the *dual vector space* and denoted by V^*. Elements of this vector space will be called *one-forms.*

The convention we shall follow as much as possible is that vectors will be represented by symbols in bold, such as $\boldsymbol{v}$, while one-forms will be denoted by symbols with a tilde, or a bar, such as $\tilde{\omega}$ or $\overline{\omega}$.

The one-form/vector relation is employed in QDN in our representation of questions and answers, as discussed in Chapter 2.

An important fact is that when V is finite dimensional, then V^* has the same dimension, namely, $\dim V^* = \dim V$.

Dual Basis

Suppose V is an n-dimensional vector space with basis $B(V) \equiv \{\boldsymbol{e}^a\colon a = 1, 2, \ldots, n\}$. Then an arbitrary vector $\boldsymbol{v} \in V$ can be written in the form

$$\boldsymbol{v} = \sum_{a=1}^{n} v^a \boldsymbol{e}^a, \tag{A.20}$$

where the $\{v^i\}$ are known as the *components* of $\boldsymbol{v}$ relative to the basis $B(V)$.

Given $B(V)$, we can always find a basis $B(V^*) \equiv \{\tilde{e}^a\colon a = 1, 2, \ldots, n\}$ for the dual space V^* such that $\tilde{e}^a(\boldsymbol{e}^b) = \delta^{ab}$. We call $B^*(V)$ the *conjugate basis.* This greatly simplifies calculations.

Bracket Notation

Given a vector $\boldsymbol{v} \in V$ and one-form $\tilde{\omega} \in V^*$, the bracket notation $\langle \tilde{\omega}, \boldsymbol{v} \rangle \equiv \tilde{\omega}(\boldsymbol{v})$ is often used. In quantum mechanics, vectors are often written as *kets*, such as $|\psi\rangle$, and dual vectors as *bra-vectors*, such as $\langle \phi|$, a notation used extensively by Dirac (Dirac, 1958). Then the "inner product" $\langle \phi|\psi\rangle$ is called the *bracket* of $|\psi\rangle$ and $\langle \phi|$.

Quantum mechanics vector spaces use a complex-valued field and the following rule is imposed: $\langle \phi|\psi\rangle^* = \langle \psi|\phi\rangle$.

Tensor Product Spaces

Suppose U and V are vector spaces over the same field $\mathbb{F}$. It is possible for U and V to be copies of the same vector space, but not necessarily so. If they were, we would simple label them U^1 and U^2, respectively. Significantly, U and V need not have the same dimension.

If $\boldsymbol{u} \in U$ and $\boldsymbol{v} \in V$, then the *direct product* $\boldsymbol{u} \otimes \boldsymbol{v}$ is identified with $(\boldsymbol{u}, \boldsymbol{v})$, an element of the Cartesian product space $U \times V$.

Example A.5 Consider the vector space of all real 2×2 matrices $M(2, \mathbb{R})$ and the vector space of all real 3×3 matrices $M(3, \mathbb{R})$. Then elements of the Cartesian product $M(2, \mathbb{R}) \times M(3, \mathbb{R})$ are given by *Kronecker products* of matrices. For example, if $A \equiv [A_{ab}] \in M(2, \mathbb{R})$ and $B \equiv [B_{ij}] \in M(3, \mathbb{R})$, then the Kronecker product $A \otimes B \equiv [C_{ai,bj}]$ is an array with double matrix indices, such that

$$C_{ai,bj} \equiv A_{ab} B_{ij}, \quad a, b = 1, 2, \quad i, j = 1, 2, 3. \tag{A.21}$$

Unfortunately, direct product vector spaces are **not** vector spaces themselves, which can be readily seen by considering linear combinations of arbitrary elements.

Because we need vector addition to represent superposition in QM, and we find ourselves dealing with tensor products, we overcome this problem by extending $U \times V$ to a larger space, the *tensor product* of U and V, denoted by $U \otimes V$. This tensor product space is defined to contain **all** linear combinations of elements of the Cartesian product space $U \times V$ and satisfies all the axioms of a vector space over the common ground field $\mathbb{F}$. Elements of $U \otimes V$ are either of the form $\boldsymbol{u} \otimes \boldsymbol{v}$ (which *is* in $U \times V$) or linear combinations of such direct products, which may or may not be in $U \times V$.

Elements of $U \otimes V$ that are of the form $\boldsymbol{u} \otimes \boldsymbol{v}$ for $\boldsymbol{u} \in U$, $\boldsymbol{v} \in V$ are called *separable*. Elements of $U \otimes V$ that are not separable are called *entangled.*

Entanglement is of great significance in quantum mechanics. The physically observable properties of entangled quantum states lie at the heart of the problems with the interpretation of quantum mechanics.

As with ordinary arithmetic, the tensor product operation $\otimes$ takes precedence over the vector summation operation $+$ in $U \otimes V$, so we may leave out the brackets and just write

$$(\boldsymbol{u}_1 \otimes \boldsymbol{v}_1) + (\boldsymbol{u}_2 \otimes \boldsymbol{v}_1) \equiv \boldsymbol{u}_1 \otimes \boldsymbol{v}_1 + \boldsymbol{u}_2 \otimes \boldsymbol{v}_2 \in U \otimes V. \tag{A.22}$$

Note that $U \otimes V$ is a vector space over the field $\mathbb{F}$ common to U and V. Multiplication by a scalar can be considered in several ways:

$$\lambda \{\boldsymbol{u} \otimes \boldsymbol{v}\} = (\lambda \boldsymbol{u}) \otimes \boldsymbol{v} = \boldsymbol{u} \otimes (\lambda \boldsymbol{v}) = \lambda \boldsymbol{u} \otimes \boldsymbol{v}. \tag{A.23}$$

Similarly,

$$\lambda \{\boldsymbol{u}_1 \otimes \boldsymbol{v}_1 + \boldsymbol{u}_2 \otimes \boldsymbol{v}_2\} = \lambda \boldsymbol{u}_1 \otimes \boldsymbol{v}_1 + \lambda \boldsymbol{u}_2 \otimes \boldsymbol{v}_2. \tag{A.24}$$

Denote the zero vectors in U, V, and $U \otimes V$ by $\mathbf{0}_U$, $\mathbf{0}_V$, and $\mathbf{0}_{U\otimes V}$, respectively. Then for any $\boldsymbol{u} \in U$, $\boldsymbol{v} \in V$, we have

$$\mathbf{0}_U \otimes \boldsymbol{v} = \boldsymbol{u} \otimes \mathbf{0}_V = \mathbf{0}_{U\otimes V}. \tag{A.25}$$

Likewise, if $0_{\mathbb{F}}$ is the zero element of the field $\mathbb{F}$, then

$$0_{\mathbb{F}}(\boldsymbol{u} \otimes \boldsymbol{v}) = \mathbf{0}_{U\otimes V}. \tag{A.26}$$

Rank

The *rank* of a tensor product space is the number of vector spaces in the product. For example, $U \otimes V$ is a *rank-two* tensor product space; $U^1 \otimes U^2 \otimes \cdots \otimes U^n$ is of rank n.

Elements of a given tensor product have the rank of that tensor product space. Hence scalars have rank zero, while vectors and one-forms have rank one.

Given a number of vectors spaces $V^1, V^2, \ldots, V^r$, then the rank-r tensor product space $V^1 \otimes V^2 \otimes \cdots \otimes V^r$ has dimension equal to the product of all the individual dimensions, that is,

$$\dim \left\{V^1 \otimes V^2 \otimes \cdots \otimes V^r\right\} = \left\{\dim V^1\right\} \left\{\dim V^2\right\} \ldots \left\{\dim V^r\right\}. \tag{A.27}$$

Likewise, if $V^{1*}, V^{2*}, \ldots, V^{s*}$ are dual vector spaces, we define the rank-s tensor product space $V^{1*} \otimes V^{2*} \otimes \cdots \otimes V^{s*}$ in an analogous way, and similarly for mixed tensor product spaces such as $V^1 \otimes V^{2*} \otimes \cdots$.

Separable Bases

If $\{\boldsymbol{u}^a : a = 1, 2, \ldots, \dim U\}$ is a basis for U and $\{\boldsymbol{v}^b : b = 1, 2, \ldots, \dim V\}$ is a basis for V, then a frequently useful basis for $U \otimes V$ is given by $\{\boldsymbol{u}^a \otimes \boldsymbol{v}^b : a = 1, 2, \ldots, \dim U,\ b = 1, 2, \ldots, \dim V\}$. Every element of this basis is a separable element of the tensor product space $U \otimes V$, so we call this a *separable basis*. From this we immediately conclude that

$$\dim\{U \otimes V\} = \{\dim U\} \cdot \{\dim V\}. \tag{A.28}$$

This result generalizes to higher rank tensor product spaces.

Hilbert spaces

Hilbert spaces are finite or infinite dimensional vector spaces with a complete inner product (Streater and Wightman, 1964).

References

Adler, S. L. 1995. *Quaternionic Quantum Mechanics and Quantum Fields.* International Series of Monographs on Physics, 88. Oxford University Press.

Adler, S. L. 2016. Does the Peres Experiment Using Photons Test for Hyper-complex (Quaternionic) Quantum Theories? *arXiv:quant-ph/1604.04950*, 1–5.

Afshar, S. S. 2005. Violation of the Principle of Complementarity, and Its Implications. Pages 229–244 of: *The Nature of Light: What Is a Photon?* Proceedings of SPIE, no. 5866.

Apollo Program Office. 1969. Apollo 11 (AS-506) Mission. *Mission Operation Report*, 1–109.

Ashby, Neil. 2002. Relativity and the Global Positioning System. *Physics Today*, May, 41–47.

Aspect, A., Grangier, P., and Roger, G. 1982. Experimental Realization of Einstein–Podolsky–Rosen–Bohm Gedankenexperiment: A New Violation of Bell's Inequalities. *Phys. Rev. Lett.*, **49**, 91–94.

Becker, L. 1998. A New Form of Quantum Interference Restoring Experiment. *Phys. Lett. A*, **249**, 19–24.

Bell, E. T. 1938. The Iterated Exponential Integers. *Ann. Math.*, **39**(3), 539–557.

Bell, J. S. 1964. On the Einstein–Podolsky–Rosen paradox. *Physics*, **1**, 195–200.

Bell, J. S. 1988. *Speakable and Unspeakable in Quantum Mechanics.* Cambridge University Press.

Berkeley, G. 1721. De Motu or Sive de Motus Principio & Natura et de Causa Communicationis Motuum [The Principle and Nature of Motion and the Cause of the Communication of Motions]. Translated by A. A. Luce Pages 253–276 of: Michael R. Ayers, *George Berkeley: Philosophical Works.* London: Everyman, 1993.

Bernstein, J. 2010. The Stern–Gerlach Experiment. *asXiv.org[physics.hist.ph]*, arXiv:1007.2435, 1–16.

Birrell, N., and Davies, P. 1982. *Quantum Fields in Curved Space.* Cambridge University Press.

Bjorken, J. D., and Drell, S. D. 1965. *Relativistic Quantum Fields.* McGraw-Hill.

Bohm, D. 1952. A Suggested Interpretation of the Quantum Theory in Terms of "Hidden Variables," I and II. *Phys. Rev.*, **85**, 166–193.

Bohr, N. 1913. On the Constitution of Atoms and Molecules. *Philos. Mag.*, **26**(1), 1–24.

Bombelli, L., Lee, J., Meyer, D., and Sorkin, R. 1987. Space-Time as a Causal Set. *Phys. Rev. Lett.*, **59**(5), 521–524.

Bondi, H., and Gold, T. 1948. The Steady-State Theory of the Expanding Universe. *MNRAS*, **108**, 252–270.

Born, M. 1926. Zur Quantenmechanik der Stossvorgänge [The Quantum Mechanics of the Impact Process (Collision Processes)]. *Zeitschrift fur Physik*, **37**, 863–867.

The Born–Einstein Letters. 1971. Trans. Irene Born. Macmillan.

Brandt, H. E. 1999. Positive Operator Valued Measure in Quantum Information Processing. *Am. J. Phys.*, **67**(5), 434–439.

Brightwell, G., and Gregory, R. 1991. Structure of Random Discrete Spacetime. *Phys. Rev. Lett.*, **66**(3), 260–263.

Burnham, D. C., and Weinberg, D. L. 1970. Observation of Simultaneity in Parametric Production of Optical Photon Pairs. *Phys. Rev. Lett.*, **25**, 84–86.

Candlin, D. J. 1956. On Sums over Trajectories for Systems with Fermi Statistics. *Nuovo Cimento*, **4**(2), 231–239.

Casalbuoni, R. 1976a. The Classical Mechanics for Bose–Fermi Systems. *Nuovo Cimento A Series 11*, **33**(3), 389–431.

Casalbuoni, R. 1976b. On the Quantization of Systems with Anticommuting Variables. *Nuovo Cimento A Series 11*, **33**(1), 115–125.

Cowan, C. L. Jr., Reines, F., Harrison, F. B., et al. 1956. Detection of the Free Neutrino: A Confirmation. *Science*, **124**(3212), 103–104.

Cramer, J. G. 1986. The Transactional Interpretation of Quantum Mechanics. *Rev. Mod. Phys.*, **58**(3), 647–688.

D-Wave Systems. 2016. *The D-Wave 2000Q Quantum Computer*. D-Wave Systems, 1–12.

de Broglie, L. 1924. Recherches sur la Théorie des Quanta [Researches on the Quantum Theory]. Ph.D. thesis, Faculty of Sciences at Paris University.

Deutsch, D. 1997. *The Fabric of Reality*. Penguin Press.

Deutsch, D. 1999. Quantum Theory of Probability and Decisions. *Proc. R. Soc*, **455**, 3129–3137.

DeWitt, B. S. 1975. Quantum Field Theory in Curved Spacetime. *Physics Reports C*, **19**(6), 295–357.

Dingle, H. 1967. The Case against Special Relativity. *Nature*, **216**, 119–122.

Dirac, P. A. M. 1938a. Classical Theory of Radiating Electrons. *Proc. Roy. Soc. A*, **167**, 148–169.

Dirac, P. A. M. 1938b. A New Basis for Cosmology. *Proc. Roy. Soc (London) A*, **165**(921), 199–208.

Dirac, P. A. M. 1958. *The Principles of Quantum Mechanics*. Clarendon Press.

Eakins, J. 2004. Classical and Quantum Causality in Quantum Field Theory, or, "The Quantum Universe." Ph.D. thesis, University of Nottingham.

Eakins, J., and Jaroszkiewicz, G. 2005. A Quantum Computational Approach to the Quantum Universe. Pages 1–51 of: Albert Reimer (ed.), *New Developments in Quantum Cosmology Research*. Horizons in World Physics, vol. 247. New York: Nova Science.

Eden, R. J., Landshoff, P. V., Olive, D. I., and Polkinghorne, J. C. 1966. *The Analytic S-Matrix*. Cambridge University Press.

Ehrenfest, P. 1927. Bemerkung über die angebote Gültigkeit der klassischen Mechanik Innerhalb der Quantenmechanik. *Zeitschrift für Physik*, **45**(7–8), 455–457.

Einstein, A. 1905a. Über einen die Erzeugung und Verwandlung des Lichtes betreffenden heuristischen Gesichtspunkt [Concerning an Heuristic Point of View Toward the Emission and Transformation of Light]. *Annalen der Physik*, **17**, 132–148. Translation into English *American Journal of Physics*, **33**(5), May 1965.

Einstein, A. 1905b. Zur Electrodynamik Bewgter Körper. *Annalen der Physik*, **17**, 891–921. *On the Electrodynamics of Moving Bodies*, translation in *The Principle of Relativity*, Dover Publications.

Einstein, A., Podolsky, B., and Rosen, N. 1935. Can Quantum-Mechanical Description of Physical Reality Be Considered Complete? *Phys. Rev.*, **47**, 777–780.

Elitzur, A. C., and Vaidman, L. 1993. Quantum-Mechanical Interaction-Free Measurements. *Found. Phys.*, **23**, 987–997.

Encyclopaedia Britannica. 2000. *Time*. CD Rom. Britannica.co.uk.

Everett, H. 1957. "Relative State" Formulation of Quantum Mechanics. *Rev. Mod. Phys.*, **29**(3), 454–462.

Feynman, R. P., and Hibbs, A. R. 1965. *Quantum Mechanics and Path Integrals*. New York: McGraw-Hill.

Feynman, R. P., Leighton, R. B., and Sands, M. 1966. *The Feynman Lectures on Physics: Quantum Mechanics*. Vol. III. Addison-Wesley.

FitzGerald, G. F. 1889. The Ether and the Earth's Atmosphere. *Science*, **13**, 390.

Franson, J. D. 1989. Bell Inequality for Position and Time. *Phys. Rev. Lett.*, **62**(19), 2205–2208.

Fuchs, C. A., and Peres, A. 2000. Quantum Theory Needs No Interpretation. *Physics Today*, March, 70–71.

Gell-Mann, M., and Pais, A. 1955. Behavior of Neutral Particles under Charge Conjugation. *Phys. Rev.*, **97**, 1387–1389.

Gerlach, W., and Stern, O. 1922a. Das magnetische Moment des Silberatoms. *Zeits. Phys.*, **9**, 353–355.

Gerlach, W., and Stern, O. 1922b. Der experimentelle Nachweis der Richtungsquantelung im Magnetfeld. *Zeits. Phys.*, **9**, 349–352.

Glauber, R. J. 1963a. Coherent and Incoherent States of the Radiation Field. *Phys. Rev.*, **131**(6), 2766 –2788.

Glauber, R. J. 1963b. Photon Correlations. *Phys. Rev. Lett.*, **10**(3), 84–86.

Glauber, R. J. 1963c. The Quantum Theory of Optical Coherence. *Physical Review*, **130**(6), 2529–2539.

Gödel, K. 1949. An Example of a New Type of Cosmological Solution of Einstein's Field Equations of Gravity. *Rev. Mod. Phys.*, **21**(3), 447–450.

Goldstein, H. 1964. *Classical Mechanics*. Addison-Wesley.

Greenberger, D., and YaSin, A. 1989. "Haunted" Measurements in Quantum Theory. *Found. Phys.*, **19**(6), 679–704.

Griffiths, R. B. 1984. Consistent Histories and the Interpretation of Quantum Mechanics. *J. Stat. Phys.*, **36**(12), 219–272.

Halligan, Peter, and Oakley, David. 2000. Greatest Myth of All. *New Scientist, 18 November*, 34–39.

Hameroff, S. 1999. Quantum Computation in Brain Microtubules? The Penrose–Hameroff "Orch-OR" Model of Consciousness. www.u.arizona.edu/~hameroff/royal.html, 1–30.

Hamming, R. W. 1950. Error Detecting and Error Correcting Codes. *Bell Syst. Tech. J.*, **29**(2), 147–160.

Hardy, L. 1992. Quantum Mechanics, Local Realistic Theories and Lorentz-Invariant Realistic Theories. *Phys. Rev. Lett.*, **68**(20), 2981–2984.

Heisenberg, W. 1925. Über Quantentheoretische Umdeutung Kinematischer und Mechanischer Beziehungen [Quantum-Theoretical Reinterpretation of Kinematic and Mechanical Relations]. *Zeits. Physik A Hadrons and Nuclei*, **33**(1), 879 –893.

Heisenberg, W. 1930. *The Physical Principles of the Quantum Theory*. Dover Edition, 1949 ed. University of Chicago Press.

Heisenberg, W. 1952. Questions of Principle in Modern Physics. Pages 41–52 of: *Philosophic Problems in Nuclear Science*. London: Faber and Faber.

Horne, M. A., Shimony, A., and Zeilinger, A. 1989. Two-Particle Interferometry. *Phys. Rev. Lett.*, **62**(19), 2209–2212.

Howson, A. G. 1972. *A Handbook of Terms Used in Algebra and Analysis*. Cambridge University Press.

Hoyle, F. 1948. A New Model for the Expanding Universe. *MNRAS*, **108**, 372–382.

Itano, W. M., Heinzen, D. J., Bollinger, J. J., and Wineland, D. J. 1990. Quantum Zeno Effect. *Phys. Rev. A*, **41**(5), 2295–2300.

Jacques, V., Wu, E., Grosshans, F., Treussart, F., Grangier, P., Aspect, A., and Roch, J.-F. 2007. Experimental Realization of Wheeler's Delayed-Choice Gedankenexperiment. *Science*, **315**(5814), 966–968. www.arxiv.org/abs/quant-ph/0610241.

Jacques, V., Lai, N. D., Zheng, D., Chauvat, F., Treussart, F., Grangier, P., and Roch, J.-F. 2008. Illustration of Quantum Complementarity Using Single Photons Interfering on a Grating. *New. J. Phys.*, **10**, 123009.

Jaroszkiewicz, G. 2004. Quantum Register Physics. *arXiv:quant-ph/0409094*.

Jaroszkiewicz, G. 2008a. Quantized Detector Networks: A Review of Recent Developments. *Int. J. Mod. Phys. B*, **22**(3), 123–188.

Jaroszkiewicz, G. 2008b. Quantized Detector Networks, Particle Decays and the Quantum Zeno Effect. *J. Phys. A: Math. Theor.*, **41**(9), 095301.

Jaroszkiewicz, G. 2010. Towards a Dynamical Theory of Observation. *Proc. Roy. Soc. A*, **466**(2124), 3715–3739.

Jaroszkiewicz, G. 2016. Principles of Empiricism and the Interpretation of Quantum Mechanics. Pages 139–173 of: M. Dugic, R. Kastner, and G. Jaroszkiewicz (eds.), *Quantum Structural Studies*. World Scientific.

Jaynes, E. T. 2003. *Probability Theory: The Logic of Science*. Cambridge University Press.

Joos, E. 2012. Decoherence. www.decoherence.de.

Jordan, P., and Wigner, E. P. 1928. Über Das Paulische Äquivalenzverbot. *Zeitschrift für Physik*, **47**, 631–651.

Kahneman, D. 2011. *Thinking, Fast and Slow*. Allen Lane.

Karyotakis, Y., and de Monchenault, G. H. 2002. A Violation of CP Symmetry in B Meson Decays. *Europhysics News*, May/June, 89–93.

Kastner, R. E. 2005. Why the Afshar Experiment Does Not Refute Complementarity. *Stud. Hist. Philos. M. P.*, **36**(4), 649–658.

Kastner, R. E. 2016. The Illusory Appeal of Decoherence in the Everettian Picture: Affirming the Consequent. *arXiv:1603.04845 [quant-ph]*, 1–7.

Kim, Y., Yu, R., Kulik, S., Shih, Y., and Scully, M. 2000. A Delayed Choice Quantum Eraser. *Phys. Rev. Lett.*, **84**, 1–5. *arXiv:quant-ph/9903047*.

Kim, Yoon-Ho. 2003. Two-Photon Interference Without Bunching Two Photons. *Phys. Lett. A*, **315**, 352–355.

Klauder, J. R., and Sudarshan, E. C. G. 1968. *Fundamentals of Quantum Optics*. W.A. Benjamin.

Klyshko, D. N., Penin, A. N., and Polkovnikov, B. F. 1970. Parametric Luminescence and Light Scattering by Polaritons. *JETP Lett.*, **11**(1), 5–8.

Kochen, S., and Specker, E. 1967. The Problem of Hidden Variables in Quantum Mechanics. *J. Math. Mechanics*, **17**, 59–87.

Koke, S., Grebing, C., Frei, H., Anderson, A., Assion, A., and Steinmeyer, G. 2010. Direct Frequency Comb Synthesis with Arbitrary Offset and Shot-Noise-Limited Phase Noise. *Nature Photonics*, **4**, 463–465.

Kracklauer, A. F. 2002. Time Contortions in Modern Physics. *arXiv:quant-ph/0206164*, 1–7. Draft for: Proceedings, The Nature of Time: Geometry, Physics and Perception; May 21–24, 2002, Tatranska Lomnica, Slovak Republic.

Kraus, K. 1974. *Operations and Effects in the Hilbert Space Formulation of Quantum Theory.* Berlin: Springer, pages 206–229.

Kraus, K. 1983. *States, Effects, and Operations: Fundamental Notions of Quantum Theory.* Lecture Notes in Physics (190). Berlin: Springer-Verlag.

Kwiat, P. G., Steinberg, A. M., and Chiao, R. Y. 1993. High-Visibility Interference in a Bell-Inequality Experiment for Energy and Time. *Physical Review A*, **47**(4), R2472–R2475.

Kwiat, P., Weinfurter, H., Herzog, T., Zeilinger, A., and Kasevich, M. 1994. Experimental Realization of "Interaction-Free" Measurements. In: K. V. Laurikainen, C. Montonen, and Sunnarborg (eds.), *Symposium on the Foundations of Modern Physics 1994.* Editions Frontières.

Larmor, J. J. 1897. On a Dynamical Theory of the Electric and Luminiferous Medium, Part 3, Relations with Material Media. *Phil. Trans. Roy. Soc.*, **190**, 205–300.

Laven, P. A., Taplin, D. W., and Bell, C. P. 1970. Television Reception over Sea Paths: The Effect of the Tide. *BBC Research Department Report*, 1–25.

Leech, J. W. 1965. *Classical Mechanics.* Methuen and Co.

Leggett, A. J., and Garg, A. 1985. Quantum Mechanics versus Macroscopic Realism: Is the Flux There When Nobody Looks? *Phys. Rev. Lett.*, **54**(9), 857–860.

Lehmann, H., Symanzik, K., and Zimmermann, W. 1955. Zur Formulierung Quantisierter Feldtheorien. *Il Nuovo Cimento*, **1**(1), 205–225.

Lewis, G. N. 1926. The Conservation of Photons. *Nature*, **118**, 874–875.

Liang, Y., and Czarnecki, A. 2011. Photon–Photon Scattering: A Tutorial. *arXiv:1111.6126v2 [hep-ph]*, 1–9.

Lichtenberg, D. B. 1970. *Unitary Symmetry and Elementary Particles.* Academic Press.

Lorentz, H. A. 1899. Simplified Theory of Electrical and Optical Phenomena in Moving Systems. *Proc. Acad. Science Amsterdam*, **I**, 427–443.

Ludwig, G. 1983a. *Foundations of Quantum Mechanics I.* New York: Springer.

Ludwig, G. 1983b. *Foundations of Quantum Mechanics II.* New York: Springer.

Magueijo, J. 2003. New Varying Speed of Light Theories. *Rept. Prog. Phys.*, **66**(11), 2025–2068.

Mardari, G. N. 2005. What Is a Quantum Really Like? In: G. Adenier, A. Y. Khrennikov, and T. M. Nieuwenhuizen (eds.), *Quantum Theory: Reconsideration of Foundations-3.* AIP Conference Proceedings, vol. 81.

Markopoulou, F. 2000. Quantum Causal Histories. *Class. Quant. Grav.*, **17**, 2059–2072.

Mars Climate Orbiter Mishap Investigation Board. 1999. Phase I Report. 1–44.

Martin, J. L. 1959a. The Feynman Principle for a Fermi System. *Proc. Roy. Soc.*, **A251**, 543–549.

Martin, J. L. 1959b. Generalized Classical Dynamics and the "Classical Analogue" of a Fermi Oscillator. *Proc. Roy. Soc.*, **A251**, 536–542.

Merli, P. G., Missiroli, G. F., and Pozzi, G. 1976. On the Statistical Aspect of Electron Interference Phenomena. *Am. J. Phys.*, **44**(3), 306–307.

Meschini, D. 2007. Planck-Scale Physics: Facts and Beliefs. *Found. Science*, **12**(4), 277–294.

Minkowski, H. 1908. Space and Time. A translation of an Address delivered at the 80th Assembly of German Natural Scientists and Physicians, at Cologne, 21 September, 75–91. Reprinted in H. A. Lorentz, A. Einstein, H. Minkowski, and H. Weyl, *The Principle of Relativity.* Dover.

Misra, B., and Sudarshan, E. C. G. 1977. The Zeno's Paradox in Quantum Theory. *J. Math. Phys.*, **18**(4), 756–763.

Newton, I. 1687. *The Principia (Philosophiae Naturalis Principia Mathematica).* University of California Press, 1999. New translation by I. B. Cohen and Anne Whitman. University of California Press, 1999.

Newton, I. 1704. *Opticks.* 1952 ed. Dover Publications.

Newton, I. 2006. Original letter from Isaac Newton to Richard Bentley. *The Newton Project* (online).

Nielsen, M. A., and Chuang, I. L. 2000. *Quantum Computation and Quantum Information.* Cambridge University Press.

Paris, M. G. A. 2012. The Modern Tools of Quantum Mechanics (A Tutorial on Quantum States, Measurement, and Operations). *Eur. Phys. J. Special Topics*, **203**, 61–86.

Paul, H. 2004. *Introduction to Quantum Optics.* Cambridge University Press. Translated from the German *Photonen. Eine Einführung in die Quantenoptik, 2. Auflage* (1999).

Penrose, R. 1971. Angular Momentum: An Approach to Combinatorial Spacetime. In: T. Bastin (ed.), *Quantum Theory and Beyond.* Cambridge University Press.

Peres, A. 1995. *Quantum Theory: Concepts and Methods.* Kluwer Academic.

Petkov, V. 2012. *Space and Time: Minkowski's Papers on Relativity.* Minkowski Institute Press, pages 1–37.

Planck, M. 1900a. On the Theory of the Energy Distribution Law of the Normal Spectrum. *Verhandl. Dtsch. Phys. Ges.*, **2**(17), 237–245.

Planck, M. 1900b. Ueber eine Verbesserung der Wien'schen Spectralgleichung [On an Improvement of Wein's Equation for the Spectrum]. *Verhandl. Dtsch. Phys. Ges.*, **2**(13), 202–204.

Poincaré, H. 1890. Sur Le Problème Des Trois Corps et Les Équations de la Dynamique. *Acta. Math.*, **13**, 1–270.

Price, H. 1997. *Time's Arrow.* Oxford University Press.

Procopio, L. M., et al. 2016. Experimental Test of Hyper-Complex Quantum Theories. *arXiv:1602.01624v2 [quant-ph].*

Regge, T. 1961. General Relativity Without Coordinates. *Il Nuovo Cimento*, **19**(3), 558–571.

Renniger, M. 1953. Zum Wellen-Korpuskel-Dualismus. *Zeits. für Physik*, **136**, 251–261.

Requardt, M. 1999. Space-Time as an Orderparameter Manifold in Random Networks and the Emergence of Physical Points. *gr-qc/99023031*, 1–40.

Ridout, D. P., and Sorkin, R. D. 2000. A Classical Sequential Growth Dynamics for Causal Sets. *Phys. Rev.*, **D61**, 024002. *arXiv: gr-qc/9904062.*

Rosa, R. 2012. The Merli–Missiroli–Pozzi Two-Slit Electron-Interference Experiment. *Phys. Perspect.*, **14**, 178–195.

Scarani, V., Tittel, W., Zbinden, H., and Gisin, N. 2000. The Speed of Quantum Information and the Preferred Frame: Analysis of Experimental Data. *Phys. Lett.*, **A276**, 1–7.

Schrödinger, E. 1926. Quantisierung als Eigenwertproblem (Erste Mitteilung). *Ann. Phys.*, **384**(4), 361–376.

Schrödinger, E. 1935. The Present Situation in Quantum Mechanics. *Naturwissenschaften*, **23**(48), 807–812. Original title: Die gegenw artige Situation in der

Quantenmechanik, reprinted in *Quantum Theory and Measurement*, edited by J. A. Wheeler and W. H. Zurek, Princeton University Press, 1983.

Schutz, B. 1980. *Geometrical Methods of Mathematical Physics.* Cambridge University Press.

Schwinger, J. 1958. Spin, Statistics and the TCP theorem. *Proc. N. A. S.*, **44**, 223–228.

Schwinger, J. 1969. *Particles and Sources.* Gordon and Breach.

Schwinger, J. 1998a. *Particles, Sources, and Fields.* Advanced Books Classics. Reading, MA: Perseus.

Schwinger, J. 1998b. *Particles, Sources, and Fields.* Advanced Books Classics, vol. 2. Reading, MA: Perseus.

Schwinger, J. 1998c. *Particles, Sources, and Fields.* Advanced Books Classics, vol. III. Reading, MA: Perseus.

Sen, R. N. 2010. *Causality, Measurement Theory and the Differentiable Structure of Space-Time.* Cambridge Monographs on Mathematical Physics. Cambridge University Press.

Shannon, C. E. 1948. A Mathematical Theory of Communication. *Bell Syst. Tech. J.*, **27**(July, October), 379–423, 623–656.

Sillitto, R. M., and Wykes, C. 1972. An Interference Experiment with Light Beams Modulated in Anti-Phase by an Electro-Optic Shutter. *Phys. Lett. A*, **39**(4), 333–334.

Sinha, U., et al. 2010. Ruling Out Multi-Order Interference in Quantum Mechanics. *Science*, **329**, 418 –421.

Snyder, H. S. 1947a. The Electromagnetic Field in Quantized Space-Time. *Phys. Rev.*, **72**(1), 68–71.

Snyder, H. S. 1947b. Quantized Space-Time. *Phys. Rev.*, **71**(1), 38–41.

Sorkin, R. D. 1994. Quantum Mechanics as Quantum Measure Theory. *Mod. Phys. Lett. A*, **9**, 3119–3128.

Streater, R. F., and Wightman, A. S. 1964. *PCT, Spin and Statistics, and All That.* W.A. Benjamin.

Stuckey, M. 1999. Pregeometry and the Trans-Temporal Object. In: R. Buccheri, V. Di Gesu, and M. Saniga (eds.), *Studies of the Structure of Time: from Physics to Psycho(Patho)Logy*, 121–128. Dordrecht: Kluwer.

Taylor, G. I. 1909. Interference Fringes with Feeble Light. *Proc. Camb. Philos. Soc.*, **15**, 114–115.

Tropper, A. M. 1969. *Linear Algebra.* Thomas Nelson and Sons.

Unruh, W. G. 1976. Notes on Black-Hole Evaporation. *Phys. Rev. D*, **14**(4), 870–892.

von Neumann, J. 1955. *The Mathematical Foundations of Quantum Mechanics.* Princeton University Press. Originally published as *Mathematische Grundlagen der Quantenmechanik*, Berlin: Springer, 1932.

Walborn, S. P., Cunha, M. O. Terra, Pádua, S., and Monken, C. H. 2002. Double–Slit Quantum Eraser. *Phys. Rev.*, **65**, 033818 1–6.

Wheeler, J. A. 1979. From the Big Bang to the Big Crunch. *Cosmic Search Magazine*, **1**(4). Interview with J. A. Wheeler.

Wheeler, J. A. 1980. Pregeometry: Motivations and Prospects. Pages 1–11 of: A. R. Marlow (ed.), *Quantum Theory and Gravitation.* New York: Academic Press.

Wheeler, J. A. 1983. *Quantum Theory and Measurement.* Princeton Series in Physics. Princeton University Press, pages 182–213.

Woit, P. 2006. *Not Even Wrong: The Failure of String Theory and the Search for Unity in Physical Law.* Basic Books.

Wu, C. S., Ambler, E., Hayward, R. W., Hoppes, D. D., and Hudson, R. P. 1957. Experimental Test of Parity Conservation in Beta Decay. *Phys. Rev.*, **105**(4), 1413–1415.

Zurek, W. 2002. Decoherence and the Transition from Quantum to Classical–Revisited. Los Alamos Science, (27), 2–24.

Index